MW00845089

Algebraic Number Theory and Fermat's Last Theorem

Updated to reflect current research and extended to cover more advanced topics as well as the basics, **Algebraic Number Theory and Fermat's Last Theorem, Fifth Edition** introduces fundamental ideas of algebraic numbers and explores one of the most intriguing stories in the history of mathematics—the quest for a proof of Fermat's Last Theorem. The authors use this celebrated theorem to motivate a general study of the theory of algebraic numbers, initially from a relatively concrete point of view. Students will see how Wiles's proof of Fermat's Last Theorem opened many new areas for future work.

New to the Fifth Edition

- Pell's Equation x^2-dy^2 = 1: all solutions can be obtained from a single 'fundamental' solution, which can be found using continued fractions.
- Galois theory of number field extensions, relating the field structure to that of the group of automorphisms.
- More material on cyclotomic fields, and some results on cubic fields.
- Advanced properties of prime ideals, including the valuation of a fractional ideal relative to a prime ideal, localisation at a prime ideal, and discrete valuation rings.
- Ramification theory, which discusses how a prime ideal factorises when the number field is extended to a larger one.
- A short proof of the Quadratic Reciprocity Law based on properties of cyclotomic fields.
- Valuations and p-adic numbers. Topology of the p-adic integers.

Written by preeminent mathematicians Ian Stewart and David Tall, this text continues to teach students how to extend properties of natural numbers to more general number structures, including algebraic number fields and their rings of algebraic integers. It also explains how basic notions from the theory of algebraic numbers can be used to solve problems in number theory.

Ian Stewart is an emeritus professor of mathematics at the University of Warwick and a fellow of the Royal Society. Dr. Stewart has been a recipient of many honours, including the Royal Society's Faraday Medal, the IMA Gold Medal, the AAAS Public Understanding of Science and Technology Award, the LMS/IMA Zeeman Medal, and the University of Warwick Chancellor's Medal. He has published more than 220 scientific papers and numerous books, including several bestsellers co-authored with Terry Pratchett and Jack Cohen that combine fantasy with nonfiction.

David Tall was an emeritus professor of mathematical thinking at the University of Warwick. He died in 2024. Dr. Tall has published numerous mathematics textbooks and more than 200 papers on mathematics and mathematics education. His research interests include cognitive theory, algebra, visualization, mathematical thinking, and mathematics education.

Algebraic Number Theory and Fermat's Last Theorem

Fifth Edition

Ian Stewart and David Tall
University of Warwick' United Kingdom

CRC Press is an imprint of the
Taylor & Francis Group, an **informa** business
A CHAPMAN & HALL BOOK

Designed cover image: Ian Stewart and David Tall

Fifth edition published 2025
by CRC Press
2385 NW Executive Center Drive, Suite 320, Boca Raton FL 33431

and by CRC Press
4 Park Square, Milton Park, Abingdon, Oxon, OX14 4RN

CRC Press is an imprint of Taylor & Francis Group, LLC

First edition published by Taylor & Francis Group, LLC 1979
Second edition published by Taylor & Francis Group, LLC 1987
Third edition published by Taylor & Francis Group, LLC 2002
Fourth edition published by Taylor & Francis Group, LLC 2015

Library of Congress Cataloging-in-Publication Data
Names: Stewart, Ian, 1945- author. | Tall, David Orme, author.
Title: Algebraic number theory and Fermat's last theorem / Ian Stewart, University of Warwick, United Kingdom, David Tall, University of Warwick, United Kingdom.
Description: Fifth edition. | Boca Raton, FL : C&H, CRC Press, 2025. | Includes bibliographical references and index.
Identifiers: LCCN 2024028878 (print) | LCCN 2024028879 (ebook) | ISBN 9781032602257 (hardback) | ISBN 9781032610931 (paperback) | ISBN 9781003462002 (ebook)
Subjects: LCSH: Algebraic number theory. | Fermat's last theorem.
Classification: LCC QA247 .S76 2025 (print) | LCC QA247 (ebook) | DDC 512.7/4--dc23/eng/20240705
LC record available at https://lccn.loc.gov/2024028878
LC ebook record available at https://lccn.loc.gov/2024028879

ISBN: 978-1-032-60225-7 (hbk)
ISBN: 978-1-032-61093-1 (pbk)
ISBN: 978-1-003-46200-2 (ebk)

DOI: 10.1201/978100346200

Typeset in CMR10 font
by KnowledgeWorks Global Ltd.

Access the Instructor and Student Resources: www.Routledge.com/ 9781032602257

Contents

VI Galois Theory and Other Topics 343

Preface to the Fifth Edition

The title of this book indicates a dual purpose. Our first aim is to introduce fundamental ideas about algebraic numbers. The second is to tell one of the most intriguing stories in the history of mathematics—the quest for a proof of Fermat's Last Theorem. We use this celebrated theorem to motivate a general study of the theory of algebraic numbers, from a reasonably concrete point of view. The range of topics that we cover is selected to allow students to make early progress in understanding the necessary concepts.

An algebraic number is a complex number that satisfies a polynomial equation with rational coefficients. Thus $\sqrt{2}$, the imaginary number i, more complicated expressions such as $\sqrt[3]{5} - 42\sqrt[5]{99}$ and even the humble $\frac{22}{7}$, are algebraic numbers. On the other hand, e and π are not algebraic numbers, though this was not proved until Charles Hermite dealt with e in 1873 and Ferdinand Lindemann dealt with π in 1882.

Algebraic number theory can be read in two distinct ways. One is the theory of numbers viewed algebraically, the other is the study of algebraic numbers. Both apply here. We illustrate how basic notions from the theory of algebraic numbers can be used to solve problems in number theory. However, our main focus is to extend properties of the natural numbers to more general number structures: algebraic number fields and their rings of algebraic integers. These structures have most of the standard properties that we associate with ordinary whole numbers, but subtler properties concerning primes and factorisation often fail to generalise.

•

Historical Motivation

A Diophantine equation is a polynomial equation, or a system of polynomial equations, that is to be solved in integers or rational numbers. The name comes from Diophantus of Alexandria, who—it is thought—lived around 250 and whose book *Arithmetica* systematised such concepts. The central problem of the present book concerns solutions of a very special Diophantine equation:

$$x^n + y^n = z^n$$

where the exponent n and the numbers x, y, z are positive integers. For $n = 2$ there are many integer solutions—in fact, infinitely many—which neatly relate to the theorem of Pythagoras. For $n \geq 3$, however, there appear to be no positive integer solutions. It is this assertion that became known as Fermat's Last Theorem.

One method of attack might be to situate the equation $x^n + y^n = z^n$ in the complex numbers, and to use the complex nth root of unity $\zeta = e^{2\pi i/n}$ to obtain the factorisation (valid for odd n)

$$x^n + y^n = (x + y)(x + \zeta y) \dots (x + \zeta^{n-1} y)$$

This approach entails introducing algebraic ideas, including the notion of factorisation in the ring $\mathbb{Z}[\zeta]$ of polynomials in ζ. This promising line of attack was pursued for a time in the 19th century, until it was discovered that this particular ring of algebraic numbers does not possess all of the properties that it 'ought to'. In particular, factorisation into 'primes' is not unique in this ring. (It fails, for instance, when $n = 23$, although this is far from obvious.) It took a while for this obstacle to be understood, and for its consequences to sink in, but as it did so, the theory of algebraic numbers was developed and refined, leading to substantial improvements in our knowledge of Diophantine equations. In particular, it became possible to prove Fermat's Last Theorem for a whole range of special cases. Subsequently, geometric methods and other approaches were introduced to make further gains, until, at the end of the 20th century, Andrew Wiles finally set the last links in place to establish a complete proof after a three hundred-year search.

To gain insight into this extended story we must assume a certain level of algebraic background. Our choice is to start with fundamental ideas that are usually introduced into algebra courses, such as commutative rings, groups and modules. These concepts smooth the way for the modern reader, but they were not explicitly available to the pioneers of the theory; indeed, they arose from their work when later mathematicians sought more general principles. The leading mathematicians in the 19th and early 20th centuries developed and used most of the basic results and techniques of linear algebra—for perhaps a hundred years—without ever defining an abstract vector space. There is no evidence that they suffered as a consequence of this lack of an explicit theory. This historical fact indicates that abstraction can be built only on an existing body of specific concepts and relationships. This suggests that students might profit from direct contact with concrete examples of number-theoretic concepts, so the text is interspersed with such examples.

The algebra that we introduce in the opening chapters—which is what we consider necessary for grasping the essentials of the struggle to prove Fermat's Last Theorem—is therefore not always as 'abstract' as it might be. We believe that in mathematics it is important to 'get your hands dirty'. This requires struggling with calculations in specific contexts, where the elegance of polished theory may disguise the essential nature of the mathematics. For instance, factorisation into primes in specific number fields displays the tendency of mathematical objects to take on a life of their own. In some situations something works, in others it does not, and the reasons why are often far from obvious. Without experiencing the struggle in person, it is hard to understand why the pioneers in algebraic number theory had such difficulties and how they overcame them. Of such frustrating yet stimulating stuff is the mathematical fabric woven.

Summary of Contents

We therefore do not begin with later theories that have proved to be of value in a wider range of problems, such as Galois theory, valuation rings, Dedekind domains, and the like. We deal with these more advanced and abstract topics later in the book. Our initial purpose is to involve students in performing calculations that help them build a platform for understanding the theory. We have therefore emphasised concrete examples and subfields of the complex numbers. More advanced concepts are developed as and when needed.

For organisational reasons rather than mathematical necessity, the book is divided into six parts.

Part I develops the basic theory from an algebraic standpoint. We assume a working knowledge of a variety of topics from algebra, reviewed in detail in Chapter 1. These include commutative rings and fields, ideals and quotient rings, factorisation of polynomials with real coefficients, field extensions, symmetric polynomials, modules, free abelian groups, and finite fields. Apart from these concepts we assume only some elementary results from the theory of numbers and, later, a superficial comprehension of multiple integrals.

The story really starts in Chapter 2, where we introduce the ring of integers of a number field and explore prime factorisation within it. Quadratic and cyclotomic fields are investigated in more detail, and the Euclidean imaginary fields are classified. To obtain some number-theoretic payoff early on, we discuss Pell's Equation $x^2 - dy^2 = 1$ where d is a squarefree integer and x, y are integers. (This chapter can be skipped if the main aim is Fermat's Last Theorem.)

We then consider factorisation of algebraic integers and see how the notion of a 'prime' p can be pulled apart into two distinct ideas. The first is the concept of being 'irreducible' in the sense that p has no factors other than 1 and p. The second is what we now call 'prime': that if p is a factor of the product ab, then it must be a factor of either a or b. In this sense, a prime must be irreducible, but an irreducible need not be prime. It turns out that factorisation into irreducibles is not always unique in a number field, but useful sufficient conditions for uniqueness may be found. The factorisation theory of ideals in a ring of algebraic integers is more satisfactory, in that every ideal is a unique product of prime ideals. The extent to which factorisation is not unique can be 'measured' by the 'class group' of fractional ideals modulo principal ones.

Part II emphasises the power of geometric methods arising from Minkowski's theorem on convex sets relative to a lattice. We prove this key result geometrically by looking at the torus that appears as a quotient of Euclidean space by the lattice concerned. As illustrations of these ideas we prove the two- and four-squares theorems of classical number theory; as the main applications we prove Dirichlet's Units Theorem and the finiteness of the class group.

Part III concentrates on applications of the theory thus far developed, beginning with some slightly *ad hoc* computational techniques for class numbers, and leading up to a special case of Fermat's Last Theorem that exemplifies the development of the theory by Kummer, prior to the final push by Wiles.

Part IV Provides preparation for an outline of some of the ideas involved in Wiles's proof of Fermat's Last Theorem; specifically, elliptic curves and elliptic functions. We introduce these ideas in terms of a simpler analogy: conic sections (with quadratic equations in $\mathbb{R}^2$ or $\mathbb{C}^2$) and trigonometric functions. We also review some basic projective geometry. We relate Pythagorean triangles to these geometric and analytic ideas. Then we introduce elliptic curves, which have a particular kind of cubic equation in $\mathbb{R}^2$ or $\mathbb{C}^2$ and are analogous to conics; associated elliptic functions, which are analogous to trigonometric functions; and the Diophantine equation associated with an elliptic curve, which is analogous to the equation for Pythagorean triangles. A key point is the natural, geometrically inspired, group structure on the rational points of an elliptic curve.

Part V describes the final breakthrough, when—after a long period of solitary thinking—Wiles finally put together his proof of Fermat's Last Theorem. Even this tale is not without incident. His first proof, announced in a lecture series in Cambridge, turned out to contain a subtle unproved assumption, and it took another year to rectify the error. However, the proof is finally in a form that has been widely accepted by the mathematical community. In this text we cannot give the full proof in all its glory: it is too long and too technical. Instead we discuss the new ingredients that make the proof possible: the ideas of elliptic curves and elliptic integrals, and the link that shows that the existence of a counterexample to Fermat's Last Theorem would lead to a mathematical construction involving elliptic integrals. The proof of the theorem rests upon showing that such a construction cannot exist. We end with a brief survey of some of the more important later developments, new conjectures, and open problems.

Part VI is devoted to new topics, not included in previous editions, described in more detail below. We place the new material here for two reasons:

- It is more advanced and depends on the older material (though not on Parts IV and V).
- Doing so allows anyone who has been teaching or learning from the previous edition to cover the same ground as before, without skipping chapters (although Chapter 4 is also new: it can safely be omitted, as already explained).

Changes for the Fifth Edition

The main changes to this edition were suggested by a panel of reviewers, and we have generally tried to follow their advice. We have retained all of the material in the fourth edition, except that the appendix on quadratic reciprocity has been replaced by a different proof using cyclotomic fields and the Ramanujan-Nagell Theorem is omitted. We have added many new exercises and deleted a few. They range in difficulty from very easy to distinctly hard. We do not annotate the exercises to distinguish hard ones from easier ones, because part of a mathematician's training should be to figure this out for themselves, but instructors may wish to warn students about the more challenging ones.

The main new topics, which occupy seven new chapters, are:

- An early application to number theory in the form of Pell's Equation $x^2 - dy^2 = 1$. We show that all integer solutions can be obtained from a single 'fundamental' solution using a process that goes back to

Brahmagupta in the year 628. We interpret this process geometrically in terms of 2×2 matrices that preserve the hyperbola $x^2 - dy^2 = 1$ in the plane $\mathbb{R}^2$ and give the classical proof of the existence of a fundamental solution using continued fractions.

This chapter appears in Part I. Part VI contains six topics:

- Galois theory of number field extensions. This relates the field structure to that of the group of automorphisms and is essential for more advanced work.
- More material on cyclotomic fields, and some results on cubic fields.
- Further properties of prime ideals, which in particular set up machinery for the last two chapters: in particular, the valuation of a fractional ideal relative to a prime ideal, the technique of localisation at a prime ideal, local rings, and properties of discrete valuation rings.
- Ramification theory, which discusses how a prime ideal factorises when the number field is extended to a larger one. There are three main possibilities: it remains prime (*inert*); it *splits* completely into a product of distinct prime ideals; or it *ramifies*, meaning that some prime factors occur to higher powers than 1.
- A short proof of the Quadratic Reciprocity Law based on properties of cyclotomic fields. This theorem is then used to illustrate key concepts in ramification theory.
- Valuations and p-adic numbers. Here the rational numbers $\mathbb{Q}$ are embedded in various complete metric spaces. The obvious ones are $\mathbb{R}$ and $\mathbb{C}$; the less familiar ones are the p-adic numbers $\mathbf{Q}_p$ and the associated ring of p-adic integers $\mathbf{Z}_p$. The latter are ultrametric spaces, and their topology is strikingly different from that of the real line and complex plane.

We acknowledge a considerable debt to numerous authors, whose previous contributions to the literature have been a valuable source of inspiration and information: definitions, theorems, proofs, exercises, and much more. The Reference section includes all of these sources.

Our notation for p-adic numbers differs slightly from the standard one. We use $\mathbb{Z}_n$ for the integers modulo n. Algebraic number theorists prefer the notation $\mathbb{Z}/n\mathbb{Z}$ or $\mathbb{Z}/n$. When $n = p$ is prime, this frees up $\mathbb{Z}_p$ for the ring of p-adic integers, with $\mathbb{Q}_p$ for the field of p-adic numbers. Since these notational inconsistencies affect only the final chapter, we have retained

$\mathbb{Z}_n$ for the integers modulo n, and use $\mathbf{Q}_p$ and $\mathbf{Z}_p$ for p-adic numbers and integers.

The appendix on quadratic reciprocity in the fourth edition has been replaced by a proof in Chapter 22. The appendix on Dirichlet's Units Theorem in the fourth edition is now included in the main text as Chapter 10. Short extra topics have been included in a few places, and the main text has if necessary been amended to keep the story consistent.

Known typographical errors have been corrected. No doubt new ones have been created...

Notation has been updated. The imaginary number $\sqrt{-1}$ is now indicated by Roman lowercase i, not italic i, to avoid confusion with other uses of i, especially as an index. A field extension is now denoted L/K rather than $L : K$. (However, its degree remains $[L : K]$ since this seems to be standard.) The maps involved in the Galois correspondence are denoted by Gal and Fix, which are easy to remember.

The summaries of more recent developments in Chapter 17 have also been updated. We have made minor tweaks to the exposition throughout. Some material has been moved but is otherwise unchanged. We have adopted UK English spelling, not American. Some terminology has been modernised ('residue field' for 'residue class field', for example), but tradition is strong in this area of mathematics. We have added more exercises, including some using computer algebra, an increasingly important skill. We do not specify any particular software package.

We use the standard 'end of proof' symbol □. We have retained the use of 'Gothic' or 'Fraktur' letters such as $\mathfrak{O}$ for rings of algebraic integers and $\mathfrak{p}$ for prime ideals, because this is traditional in the subject. We explain some of the more obscure ones like $\mathfrak{y}$, which is the lowercase 'y'. We have tried to avoid using two symbols that are easily confused; for example $\mathfrak{O}$ (Fraktur 'O') is standard for a ring of integers, but $\mathfrak{Q}$ (Fraktur 'Q') looks very similar, so we avoid it.

We continue to break the usual rule that displayed formulas should include appropriate punctuation, believing clean typography and avoidance of confusion in formulas to be more important. In our experience hardly anyone notices, but this decision is controversial, so be warned (or warn your students) if it bothers you.

We thank the reviewers for their excellent advice, but we emphasise that they bear no responsibility for how we have attempted to follow it.

Coventry and Kenilworth, May 2024 Ian Stewart and David Tall

Origins of Algebraic Number Theory

Numbers have fascinated the human race for millennia. The Pythagoreans studied many properties of the positive integers $1, 2, 3, \ldots$, and the famous theorem of Pythagoras, though geometrical, has a pronounced number-theoretic content. Earlier Babylonian civilisations had noted empirically many so-called Pythagorean triples, such as 3, 4, 5 and 5, 12, 13. These are natural numbers a, b, c such that

$$a^2 + b^2 = c^2 \tag{1}$$

A clay tablet from about 1500 BC includes the triple 4961, 6480, 8161, demonstrating the sophisticated techniques of the Babylonians.

The Ancient Greeks, though concentrating on geometry, continued to take an interest in numbers. Around 250 AD, Diophantus of Alexandria wrote a highly influential treatise on polynomial equations which studied solutions in fractions. Particular cases of these equations with natural number solutions have been called *Diophantine* equations to this day.

The study of algebra developed over the centuries, too. Indian and Chinese mathematicians dealt with increasing confidence with negative numbers and zero. In about 628 AD the Indian mathematician Brahmagupta studied the Diophantine equation that we now write as

$$x^2 - dy^2 = \pm 1 \tag{2}$$

where d is a squarefree integer, and he showed how to combine solutions to obtain new ones. Later this became known as Pell's Equation.

Meanwhile the Rashidun Caliphate conquered Alexandria in the 7th century, sweeping across North Africa and Spain. The ensuing civilisation brought an enrichment of mathematics with Muslim ingenuity grafted onto Greek and Hindu influence. The word 'algebra' itself derives from the Arabic title 'al jabr w'al muqābalah' (literally 'completing and balancing') of a book written by the Persian Al-Khwarizmi in about 825. In 1150 the Indian mathematician Bhāskara II described a method to solve equation 2 in integers and applied it to some difficult values of d. By the 13th century, peaceful coexistence of Islam and Christianity led to most Greek and Arabic classics being available in Latin translations.

In the 16th century, Girolamo Cardano used negative and imaginary solutions in his famous book *Ars Magna* (The Great Art), and in succeeding centuries complex numbers were used with greater understanding and flexibility.

Meanwhile the theory of natural numbers was not neglected. One of the greatest number theorists of the 17th century was Pierre de Fermat (1601–1665). His fame rests on his correspondence with other mathematicians, for he published very little. He would set challenges in number theory based on his own calculations; and at his death he left a number of theorems whose proofs were known, if at all, only to himself. The most notorious of these is a marginal note in his personal copy of Diophantus, written in Latin, which translates:

> To resolve a cube into two cubes, a fourth power into fourth powers, or in general any power higher than the second into two of the same kind, is impossible; of which fact I have found a remarkable proof. The margin is too small to contain it.

More precisely, Fermat asserted that, in contrast to Pythagorean triples, Equation (1), the equation

$$x^n + y^n = z^n \tag{3}$$

has no integer solutions x, y, z (other than the trivial ones with one or more of x, y, z equal to zero).

In the years following Fermat's death, almost all of his stated results were furnished with a proof. An exception was his claim that $F_n = 2^{2^n} + 1$ is prime for all positive integers n. In a letter to Pierre de Carcavi in 1659 he claimed a proof of this conjecture, but Leonhard Euler subsequently showed that this is false, observing that F_5 is divisible by 641. Even the great Fermat could make mistakes. But one by one, his other assertions were furnished with proofs, until by the mid-19th century only one elusive jewel remained. A proof of his statement about the non-existence of solutions of (3) for $n \geq 3$ exceeded the powers of all 19th century mathematicians. This beguiling and infuriating assertion, so simple to state, yet so subtle in its labyrinthine complexity, became known as 'Fermat's Last Theorem'. This romantic epithet is in fact doubly inappropriate for, without a proof, it was not a 'theorem', neither was it the last result that Fermat studied—only the last to remain unproved by other mathematicians.

Given that a proof is so elusive, is it really credible that Fermat could have possessed a genuine proof—a clever way of looking at the problem, which eluded later generations? Or had he made a subtle error, which passed unnoticed, so that his 'theorem' had no proof at all? No one knows for sure, but there is a strong consensus that if he did have what he thought was a proof, it would not survive modern scrutiny. Consensus and certainty are not the same thing, however.

Nevertheless, during the late 19th and early 20th centuries the name stuck, with its glow of romanticism—somehow lacking in the more appropriate title 'Fermat Conjecture'. It has the two classic ingredients of a problem that can capture the imagination of a wider public—a simple statement that can be widely understood, but whose proof or disproof defeats the greatest intellects.

By the 19th century the developing theory of algebra had matured to a state where it could usefully be applied in number theory. As it happened, Fermat's Last Theorem was not the main problem being attacked by number theorists at the time; for example, when Ernst Eduard Kummer made the all-important breakthrough that we are to describe in this text, he was working on a different problem: 'higher reciprocity laws'. At this stage it is worth making a minor diversion to look at this subject.

Euler had used algebraic numbers in passing to prove a few results in number theory, without noticing the issue of unique prime factorisation. Algebraic numbers entered number theory in a significant way when Carl Friedrich Gauss took an interest in higher reciprocity laws. This area of number theory had originated in 1783, when Euler noticed a remarkable numerical pattern. He investigated when an integer q is congruent to a perfect square modulo a prime p,

$$x^2 \equiv q \pmod{p}$$

If so, q is a *quadratic residue* of p. Concentrating on the case when p, q are distinct odd primes, he noticed that if at least one of p or q is of the form $4k + 1$, then q is a quadratic residue of p if and only if p is a quadratic residue of q. On the other hand, if both p, q are of the form $4k + 3$, then precisely one is a quadratic residue of the other. Extensive calculations supported this pattern, but Euler could not find a proof.

Because of the reciprocal nature of the relationship between p and q, this result became known as the *quadratic reciprocity law*. Adrien-Marie

Legendre attempted a proof in 1785 but assumed that certain arithmetic series contain infinitely many primes—a theorem whose proof turned out to be far deeper than the quadratic reciprocity law itself. Legendre also introduced the symbol

$$\left(\frac{q}{p}\right) = \begin{cases} 1 & \text{if } q \text{ is a quadratic residue of } p \\ -1 & \text{if not} \end{cases}$$

We now call this the Legendre symbol.

Gauss gave the first proof of the law of quadratic reciprocity in 1796, but he was dissatisfied because his method did not seem a natural way to attack so seemingly simple a theorem. He went on to give several more proofs, two of which appeared in his book *Disquisitiones Arithmeticae* (1801), a definitive text on number theory which still remains in print [49]. Between 1808 and 1832 Gauss continued to look for similar laws for powers higher than squares. When considering fourth powers, he discovered that his calculations were simplified by working over what we now call the Gaussian integers $a + b\mathrm{i}$ ($a, b \in \mathbb{Z}$, $\mathrm{i} = \sqrt{-1}$). This led him to develop a theory of prime factorisation for Gaussian integers. He proved that decomposition into primes is unique in that context, and from that he developed a law of biquadratic reciprocity. In the same way, he considered cubic reciprocity by using numbers of the form $a + b\omega$ where $\omega = e^{2\pi\mathrm{i}/3}$. These new types of number are of fundamental importance for Fermat's Last Theorem, and the study of their factorisation properties has proved a deep and fruitful source of methods and problems.

The numbers concerned are all examples of a particular type of complex number, namely one that is a solution of a polynomial equation

$$a_n x^n + \ldots + a_1 x + a_0 = 0$$

where all the coefficients a_j are integers. Such a complex number is said to be *algebraic*; if further $a_n = 1$ it is called an *algebraic integer*. Examples of algebraic integers include i (which satisfies $x^2 + 1 = 0$), $\sqrt{2}$ ($x^2 - 2 = 0$) and more complicated examples, such as the roots of $x^7 - 265x^3 + 7x^2 - 2x + 329 = 0$. The number $\frac{1}{2}\mathrm{i}$ (satisfying $4x^2 + 1 = 0$) is algebraic but not an integer. On the other hand, $\frac{1}{2} + \frac{1}{2}\sqrt{5}$ is an algebraic integer, since it satisfies the equation $x^2 - x - 1 = 0$. Moreover, some complex numbers, such as e and π, are not algebraic, although proofs of those statements are difficult. Indeed, in 1874 Georg Cantor used his (then controversial) notion that infinite sets can have different sizes, to prove that 'almost all' complex numbers are not algebraic—without specifying any particular one of these.

In the wider setting of algebraic integers, if n is odd then we can factorise a solution of Fermat's equation $x^n + y^n = z^n$ (if one exists) by introducing a complex nth root of unity $\zeta = e^{2\pi i/n}$ and writing (3) as

$$(x + y)(x + \zeta y) \dots (x + \zeta^{n-1} y) = z^n \tag{4}$$

If $\mathbb{Z}[\zeta]$ denotes the set of algebraic integers of the form $a_0 + a_1\zeta + \dots + a_r\zeta^r$ where each a_r is an ordinary integer, then this factorisation takes place in the ring $\mathbb{Z}[\zeta]$.

In 1847 the French mathematician Gabriel Lamé announced a proof of Fermat's Last Theorem. In outline, his proposal was to start from (4) and show that only the case where x, y have no common factors need be considered, and then to deduce that in this case $x+y, x+\zeta y, \dots, x+\zeta^{n-1}y$ have no common factors, that is, they are relatively prime. He then argued that a product of relatively prime numbers in (4) can equal an nth power only if each factor is an nth power. On this basis, Lamé claimed to have derived a contradiction.

Announcing a proof does not imply that it is one. Joseph Liouville immediately pointed out that Lamé had assumed unique factorisation. Later, Liouville read to the French Academy a paper by Kummer showing that in general this is false; in particular, it fails for $n = 23$. Over the summer of 1847 Kummer went on to devise his own proof of Fermat's Last Theorem for certain exponents n, surmounting the difficulties of non-uniqueness of factorisation by appealing to his theory of 'ideal' complex numbers, which originated in his work on higher reciprocity laws. In retrospect this theory can be viewed as introducing numbers from outside $\mathbb{Z}[\zeta]$ to use as factors when factorising elements within $\mathbb{Z}[\zeta]$. These 'ideal factors' restore a version of unique factorisation.

Using his theory of ideal numbers, Kummer proved Fermat's Last Theorem for a wide range of prime powers—the so-called 'regular' primes. Subsequently the theory began to take on a different form from that in which Kummer had left it. The key concept of an 'ideal'—a reformulation by Richard Dedekind of Kummer's 'ideal number'—gave the theory a major boost. These ideas evolved into a powerful machine with applications to many other problems in mathematics. In fact a large part of classical number theory can be expressed in the framework of algebraic numbers. This point of view was urged most strongly by David Hilbert in his *Zahlbericht* (Number Report) of 1897, which had an enormous influence on the development of number theory, see Reid [104]. Gray [54] provides a comprehensive history of the area, with a focus on the key developments in the 19th and early 20th centuries.

As a result, algebraic number theory today is a flourishing and important branch of mathematics, with deep methods and insights, and—most significantly—applications not only to number theory, but also to group theory, algebraic geometry, topology, and analysis. It was these wider links that eventually led to the final proof of Fermat's Last Theorem by Andrew Wiles. This long and difficult proof built on the work of many other mathematicians, using advanced techniques from elliptic functions, modular forms, and Galois representations.

As a teenager, fascinated by the simplicity of the statement of the theorem, Wiles had begun a long and mostly solitary journey in search of a proof. The event that triggered his final push was a conjecture put forward by two Japanese mathematicians, Yutaka Taniyama and Goro Shimura, who hypothesised a link between elliptic curves and modular forms. Their ideas were later refined by André Weil. This proposal became known as the Taniyama–Shimura–Weil Conjecture, and it was discovered that if this conjecture could be proved, then Fermat's Last Theorem could be deduced from it. At this point, Wiles leaped into action. He worked in solitude for seven years before he convinced himself that he had proved a special case of the Taniyama–Shimura–Weil Conjecture that was strong enough to imply Fermat's Last Theorem. He announced his result in a lecture in Cambridge on 23 June 1993.

When his proof was being checked, a query from a colleague revealed a gap, and Wiles accepted that some details required attention. It took him so long to do this that some questioned whether he had ever been close to the proof at all. However, in the autumn of 1994, working with his former student Richard Taylor, he finally realised that he could complete the proof satisfactorily. He released the proof for scrutiny in October 1994 and it was published in May 1995. Since then, the ideas used in the proof have been extended considerably, leading to other significant advances.

Fermat's Last Theorem probably has the distinction of being the theorem with the greatest number of false 'proofs', so the proof was scrutinised very carefully. This time, the ideas fitted together so tightly that experts in the mathematical community agreed that all was well. However, the proof uses techniques far beyond what would have been available to Fermat. So when he stated that he had found a proof that could not be fitted into the margin of his book, had he truly found a perceptive insight that has been missed by mathematicians for over three hundred fifty years? Or was it, as observed by the historian Dirk Struik [132], that 'even the great Fermat slept sometimes'?

I

Algebraic Methods

1

Algebraic Background

Fermat's Last Theorem is a special problem in the general theory of Diophantine equations—integer solutions of polynomial equations. To place the problem in context, we move to the wider realm of algebraic numbers, which arise as the real or complex solutions of polynomials with integer coefficients; we focus particularly on algebraic integers, which are solutions of polynomials with integer coefficients where the leading coefficient is 1. For example, the equation $x^2 - 2 = 0$ has no integer solutions, but it has two real solutions $x = \pm\sqrt{2}$. The leading coefficient of the polynomial $x^2 - 2$ is 1, so $\pm\sqrt{2}$ are algebraic integers. Similarly, the equation $x^2 + 1 = 0$ has no integer solutions, but it has two complex solutions $x = \pm\mathrm{i}$, and these are algebraic integers.

To operate with such numbers, it is useful to work in subsystems of the complex numbers that are closed under the usual operations of arithmetic. Such subsystems include subrings (which are closed under addition, subtraction and multiplication) and subfields (closed under all four arithmetic operations including division). Thus along with $\pm\sqrt{2}$ we consider the ring of all numbers $a + b\sqrt{2}$ for $a, b \in \mathbb{Z}$ and the field of all numbers $p + q\sqrt{2}$ for $p, q \in \mathbb{Q}$; for $x = \pm\mathrm{i}$ we consider the ring of all numbers $a + b\mathrm{i}$ for $a, b \in \mathbb{Z}$ and the field of all numbers $p + q\mathrm{i}$ for $p, q \in \mathbb{Q}$.

In this chapter we lay the foundations for algebraic number theory by considering some fundamental facts about rings, fields, and other algebraic structures, including abelian groups and modules, which are relevant to our theoretical development. We assume an acquaintance with elementary properties of groups, rings and fields, and a basic knowledge of linear algebra over an arbitrary field, up to simple properties of determinants. Familiar results at this level are stated without proof; results that we think might be less familiar are proved in full or in outline as appropriate. References are usually given for results that are not proved. Useful general references on abstract algebra, with emphasis on rings and fields, are Dummit and Foote [36], Fraleigh [41], Jacobson [68, 69], Lee [78], Sharpe [121], and Stewart and Tall [129]. For group theory, see Burn [15], Humphreys [65], Macdonald [83], Neumann *et al.* [99], Robinson [110], and Rotman [112], and Smith and Tabachnikova [124].

First we set up the ring-theoretic language, in particular the key notion of an ideal. Then we consider factorisation of polynomials over a ring, which in this book is usually—but not always—a subring of the complex numbers. Topics of central importance at this stage are factorisation of a polynomial over an extension field, and the theory of elementary symmetric polynomials. We review the concept of a module over a ring, which is analogous to—indeed, generalises—that of a vector space over a field, but is not quite as well behaved. Module-theoretic language greatly simplifies the more advanced areas of the theory. We prove some basic results about finitely generated abelian groups because they are vital when describing the additive group structure of the subrings of the complex numbers that occur. We end with basic properties of finite fields.

1.1 Rings and Fields

This section is largely about standardising notation and terminology for concepts we assume are familiar. We mostly avoid numbered definitions at this stage since there would be little else, but a few crucial concepts are given numbered definitions. From Chapter 2 onwards, important new concepts are generally given numbered definitions.

As standard notation, we denote the cardinality of a set S by $|S|$. For example, if G is a group then $|G|$ is the *order* of G. The same symbol also denotes the absolute value $|x|$ of a real number x and $|z|$ of a complex number z. Which meaning is intended should be clear from the context.

Here and throughout the book, composition of functions is defined so that $fg(x) = f(g(x))$, that is, we apply g first, *then* f. Often fg is written $f \circ g$, but we avoid this notation because it soon becomes cumbersome.

An especially important convention about terminology, which we use throughout without further mention, is:

Remark. Unless explicitly stated to the contrary, the term *ring* in this book always means a commutative ring R with identity element 1 (or 1_R).

Let R be a ring with no zero-divisors, so that in R, $a \neq 0$, $b \neq 0$ implies $ab \neq 0$. Assume also that $1 \neq 0$ in R. Then R is a *domain.* (Another common term is *integral domain*, but we omit 'integral' throughout.) An element a in a ring R is a *unit* if there exists $b \in R$ such that $ab = 1$. Suppose that $ab = ac = 1$. Then $c = 1c = abc = acb = 1b = b$. The unique b such that $ab = 1$ is denoted by a^{-1}, and ca^{-1} is also denoted by c/a. If $1 \neq 0$ in R and every nonzero element in R is a unit, then R is a *field.*

We use standard notation $\mathbb{N}$ for the set of natural numbers $0, 1, 2, \ldots$, $\mathbb{Z}$ for the integers, $\mathbb{Q}$ for the rationals, $\mathbb{R}$ for the reals and $\mathbb{C}$ for the complex numbers. Under the usual operations $\mathbb{Q}, \mathbb{R}, \mathbb{C}$ are fields, $\mathbb{Z}$ is a domain, and the set of natural numbers $\mathbb{N}$ is not even a ring. For $n \in \mathbb{N}$, $n > 0$, we denote the ring of integers modulo n by $\mathbb{Z}_n$. If n is composite, then $\mathbb{Z}_n$ has zero divisors, but for n prime, then $\mathbb{Z}_n$, is a field; see Fraleigh [41] p. 217.

Remark. We follow Bourbaki and include 0 as a natural number, so $\mathbb{N} = \{0, 1, 2, 3, \ldots\}$. Some sources now exclude 0 from $\mathbb{N}$.

At a more advanced level the symbol $\mathbb{Z}_p$ is standard for the p-adic integers when p is prime, and the integers modulo n are denoted by $\mathbb{Z}/n\mathbb{Z}$ or just $\mathbb{Z}/n$. To avoid confusion we use $\mathbf{Z}_p$ for the p-adic integers and $\mathbf{Q}_p$ for the p-adic numbers. These occur only in Chapter 23.

Our convention is that a *subring* S of a ring R is required to contain 1_R. We can check that S is a subring by demonstrating that $1_R \in S$, and if $s, t \in S$ then $s + t, -s, st \in S$. The subset S then forms a ring in its own right under the operations restricted from R, and $1_S = 1_R$. In the same way, if K is a field, then a subfield F of K is a subset that is a field under the operations restricted from K. We can check that F is a subfield of K by demonstrating that $1_K \in F$, and if $s,\ t \in F\ \ (s \neq 0)$ then $s + t,\ -s,\ st,\ s^{-1} \in F$.

The concept of an *ideal* is of central importance in this text. Recall that an ideal is a non-empty subset I of a ring R such that if $r, s \in I$, then $r - s \in I$, and if $r \in R$, $s \in I$ then $rs \in I$. We also require the concept of the *quotient ring* R/I of R by an ideal I. The elements of R/I are cosets $I + r$ of the additive group of I in R, with addition and multiplication defined by

$$\begin{aligned}(I + r) + (I + s) &= I + (r + s) \\ (I + r)(I + s) &= I + rs\end{aligned}$$

for all $r, s \in R$. For example, if $n\mathbb{Z}$ is the set of integer multiples of $n \in \mathbb{Z}$, then $\mathbb{Z}/n\mathbb{Z}$ is isomorphic to $\mathbb{Z}_n$.

A *homomorphism* $f : R_1 \to R_2$, where R_1 and R_2 are rings, is a function such that

$$\begin{aligned} f(1_{R_1}) &= 1_{R_2} \\ f(r+s) &= f(r)+f(s) \\ f(rs) &= f(r)f(s) \end{aligned}$$

for all $r, s \in R_1$. A *monomorphism* is an injective (1–1) homomorphism and an *isomorphism* is a bijective (1–1 and onto) homomorphism. We write $R \cong S$ to indicate that rings R, S (or other algebraic structures) are isomorphic.

The *kernel* and *image* of a homomorphism $f : R_1 \to R_2$ are defined in the usual way:

$$\begin{aligned} \ker f &= \{r \in R_1 \mid f(r) = 0\} \\ \operatorname{im} f &= \{f(r) \in R_2 \mid r \in R_1\} \end{aligned}$$

The kernel is an ideal of R_1; the image is a subring of R_2; and the *isomorphism theorem* states that there is a natural isomorphism from $R_1/\ker f$ to $\operatorname{im} f$. For details, see Fraleigh [41], Jacobson [68], or Sharpe [121].

If X and Y are subsets of a ring R we write $X + Y$ for the set of all elements $x + y\,(x \in X, y \in Y)$, and XY for the set of all finite sums $\Sigma x_i y_i\,(x_i \in X, y_i \in Y)$. When X and Y are both ideals, so are $X + Y$ and XY.

The sum $X+Y$ of two subsets can be generalised to an arbitrary collection $\{X_i\}_{i \in I}$ by defining $\Sigma_{i \in I} X_i$ to be the set of all finite sums $x_{i_1} + \cdots + x_{i_n}$ of elements $x_{i_j} \in X_i$.

We make the customary compression of notation with regard to $\{x\}$ and x, writing for example xY for $\{x\}Y$, $x+Y$ for $\{x\}+Y$, and 0 for $\{0\}$.

If R_1, R_2 are rings, the *direct product* $\mathbb{R}_1 \times R_2$ is a ring with operations $(r_1, r_2) + (s_1, s_2) = (r_1 + r_2, s_1 + s_2)$ and $(r_1, r_2)(s_1, s_2) = (r_1 r_2, s_1 s_2)$. For commutative rings, it is also called the *direct sum* and written $R_1 \oplus R_2$.

If I, J are ideals of R then $IJ, I \cap J$, and $I + J$ are also ideals. Clearly $IJ \subseteq I \cap J$. The next result (often used to define an 'internal' direct product) is trivial but very useful:

Proposition 1.1. *Let R be a ring with ideals I, J such that $I \cap J = 0$. Then $I + J \cong I \times J$.*

Proof: If $r \in I + J$ then $r = r_I + r_J$ where $r_I \in I$, $r_J \in J$. Moreover, this decomposition is unique. For suppose that $r_I + r_J = s_I + s_J$ with $s_I \in I$, $s_J \in J$. Then

$$r_I - s_I = s_J - r_J \in I \cap J = 0$$

so $r_I = s_I, r_J = s_J$.

Define $\phi : R \to I \times J$ by $\phi(r_I + r_J) = (r_I, r_J)$. Then

$$\begin{aligned}
\phi((r_I + r_J) + (s_I + s_J)) &= \phi((r_I + s_I) + (r_J + s_J)) \\
&= ((r_I + s_I), (r_J + s_J)) = (r_I, r_J) + (s_I, s_J) \\
&= \phi(r_I + r_J) + \phi(s_I + s_J), \\
\phi((r_I + r_J)(s_I + s_J)) &= \phi(r_I s_I + r_I s_J + r_J s_I + r_J s_J) \\
&= \phi((r_I r_J + s_I s_J) \quad \text{since } IJ \subseteq I \cap J = 0 \\
&= (r_I r_J, s_I s_J) = (r_I, s_I)(r_J, s_J) = \phi(r_I + r_J)\phi(s_I + s_J)
\end{aligned}$$

Therefore ϕ is a homomorphism. It is clearly a bijection, so it is an isomorphism. □

Corollary 1.2. *Let I, J be ideals of a ring S. Then*

$$\frac{I+J}{I \cap J} \cong \frac{I}{I \cap J} \times \frac{J}{I \cap J}$$

Proof: Let $R = I + J$ in Proposition 1.1 and work modulo $I \cap J$. □

The ideal *generated* by a subset X of R is the smallest ideal of R containing X; we denote this by $\langle X \rangle$. If $X = \{x_1, \ldots, x_n\}$, we write $\langle X \rangle$ as $\langle x_1, \ldots, x_n \rangle$. (Some writers use (X) where we have written $\langle X \rangle$, but then the last-mentioned simplification of notation would reduce to the notation for an n-tuple $(x_1, \ldots, x_n)$, so $\langle X \rangle$ is to be preferred.)

A simple calculation shows that

$$\langle X \rangle = XR = \sum_{x \in X} xR$$

The identity element 1_R is crucial in this equation, to ensure that $X \subseteq \langle X \rangle$.

If there exists a finite subset $X = \{x_1, \ldots, x_n\}$ of R such that $I = \langle X \rangle$, then we say that I is *finitely generated* as an ideal of R. If $I = \langle x \rangle$ for an element $x \in R$ we say that I is the *principal ideal* generated by x.

Example 1.3. (a) Let $R = \mathbb{Z}$, $X = \{4, 6\}$. Then $\langle 4, 6 \rangle$ is finitely generated. In fact $\langle 4, 6 \rangle$ contains $2 \cdot 4 - 6 = 2$, which easily implies that $\langle 4, 6 \rangle = \langle 2 \rangle$, so this ideal is principal. More generally, every ideal of $\mathbb{Z}$ is of the form $\langle n \rangle$ for some $n \in N$, hence principal.

(b) Let $\mathbb{Q}[x, y]$ be the ring of 2-variable polynomials over $\mathbb{Q}$, and let $I = \langle x, y \rangle$. Then I consists of all polynomials $p(x, y)$ such that $p(0, 0) = 0$. It is not principal since x and y have no common factor.

An ideal I of a ring R is *maximal* if the only ideals of R that contain I are R and I. It is important to understand that in general this does not imply that I contains every ideal of R except R. Similarly a *maximal principal ideal* is a principal ideal that is not contained in any principal ideal except for itself and the whole ring.

Although a maximal ideal need not contain every proper ideal, it can be proved that every proper ideal is contained in *some* maximal ideal. The proof uses Zorn's Lemma and is postponed until we need it; see Theorem 20.18.

If K is a field and R is a subring of K then R is a domain. Conversely, every domain D can be embedded in a field L; and there exists such an L consisting only of elements d/e where $d, e \in D$ and $e \neq 0$. Such an L, which is unique up to isomorphism, is called the *field of fractions* or *field of quotients* of D. See Fraleigh [41] theorem 26.1 p. 239.

Theorem 1.4. *Every finite domain is a field.*

Proof: Let D be a finite domain. Since $1 \neq 0$, then D has at least two elements. For $0 \neq x \in D$ the elements xy, as y runs through D, are distinct; for if $xy = xz$ then $x(y - z) = 0$ and so $y = z$ since D has no zero-divisors. Hence, by counting, the set of all elements xy is D. Thus $1 = xy$ for some $y \in D$, so D is a field. □

Every field has a unique minimal subfield, the *prime subfield* generated by $\{1\}$. This is isomorphic either to $\mathbb{Q}$ or to $\mathbb{Z}_p$ where p is a prime number. Correspondingly, we say that the *characteristic* of the field is 0 or p. In a field of characteristic p we have $px = 0$ for every element x, where as usual we write

$$px = (1 + 1 + \cdots + 1)x$$

where there are p summands 1; moreover, p is the smallest positive integer with this property. In a field of characteristic zero, if $nx = 0$ for some nonzero element x and integer n, then $n = 0$. We focus mainly on subfields of $\mathbb{C}$, which of course have characteristic zero, but fields of prime characteristic arise naturally from time to time.

For future use, we introduce:

Definition 1.5. The *multiplicative group* of a field K is the set of nonzero elements under the operation of multiplication in K. It is an abelian group, and we denote it by K^*.

We use without further comment an important property of $\mathbb{C}$: it is *algebraically closed*: given any polynomial p over $\mathbb{C}$ there exists $x \in \mathbb{C}$ such that $p(x) = 0$. For a proof see Stewart [128], chapter 2. Different proofs can be found in Hardy [57] p. 492 and Titchmarsh [137] p. 118.

Another concept that will prove useful is that of a *K-algebra*, for a field K. For our purposes this is a vector space A over K that is also a ring, whose operations are linked together. Addition in the ring is the same as in the vector space, with the same zero element, and the following conditions hold:

$$\begin{aligned} 1.x &= x \quad \text{for all } x \in A \\ 0.x &= 0 \quad \text{for all } x \in A \\ (\alpha x)y &= \alpha(xy) \quad \text{for all } x, y \in A, \alpha \in K \\ x(y+z) &= xy + xz \quad \text{for all } x, y, x \in A \end{aligned}$$

More generally, the condition that the ring structure of a K-algebra is commutative with 1 is often dropped, but we do not need this generalisation.

1.2 Factorisation of Polynomials

Later we consider factorisation in a more general context. Here we concentrate on factorising polynomials. First, we will make a few general remarks.

In a ring S, if we can write $a = bc$ for $a, b, c \in S$, then we say that b, c are *factors* of a. We also say that 'b divides a', and write

$$b \mid a$$

For any unit $e \in S$ we can always write

$$a = e(e^{-1}a)$$

so, trivially, a unit is a factor of all elements in S. If $a = bc$ where neither b nor c is a unit, then b and c are called *proper factors* and a is said to be *reducible*. In particular $0 = 0 \cdot 0$ is reducible.

If a is a unit and $a = bc$, then

$$1 = aa^{-1} = bca^{-1}$$

so b and c are both units, so a unit cannot have a proper factorisation. We therefore concentrate on factorisation of non-units. A non-unit $a \in S$ is said to be *irreducible* if it has no proper factors.

Now we turn our attention to the case $S = R[t]$, the ring of polynomials in an indeterminate t with coefficients in a ring R. The elements of $R[t]$ are expressions

$$r_n t^n + r_{n-1} t^{n-1} + \cdots + r_1 t + r_0$$

where $r_0, r_1, \ldots, r_n \in R$ and addition and multiplication are defined in the obvious way. (For a formal treatment of polynomials, and why not to use it, see Fraleigh [41] pp. 263–265.) We often write the terms in reverse order when this is more convenient. Since addition is commutative, this is permissible.

A *monic* polynomial is one with highest coefficient 1; that is, to be pedantic, some $r_m = 1$ and $r_k = 0$ for all $k > m$. (We usually drop all leading terms with zero coefficients, except for the zero polynomial 0. For example, $0.t^3 + t + 2$ is an entirely respectable polynomial, but it is the same as $t + 2$, which is monic.)

Given a nonzero polynomial

$$p = r_n t^n + \cdots + r_0$$

define the *degree* of p to be the largest value of m for which $r_m \neq 0$, and write it ∂p. Polynomials of degree $0, 1, 2, 3, 4, 5, \ldots$, are often referred to as *constant, linear, quadratic, cubic, quartic, quintic*, ... polynomials, respectively. In particular a constant polynomial is just a (nonzero) element of R.

If R is a domain, then

$$\partial pq = \partial p + \partial q$$

for nonzero p, q so $R[t]$ is also a domain. If $p = aq$ in $R[t]$, then $\partial p = \partial a + \partial q$ implies that

$$\partial q \leq \partial p$$

When R is not a field, it is possible to have a nontrivial factorisation in which $\partial p = \partial q$. For example

$$3t^2 + 6 = 3(t^2 + 2)$$

in $\mathbb{Z}[t]$, where neither 3 nor $t^2 + 2$ is a unit. This is because of the existence of non-units in R. However, if R is a field, then all (nonzero) constants in $R[t]$ are units and so if q is a proper factor of p for polynomials over a field, then $\partial q < \partial p$.

Division Algorithm and Euclidean Algorithm

Concentrate first on polynomials over a field K. The *division algorithm* for polynomials states that if $p, q \neq 0$ then

$$p = qs + r$$

where either $r = 0$ or $\partial r < \partial q$. The proof is by induction on ∂p and in practice is no more than long division of p by q leaving remainder r, which is either zero (in which case $q \mid p$) or of degree lower than q.

The division algorithm is used repeatedly in the *Euclidean algorithm*, which is a particularly efficient method for finding the *highest common factor* d of nonzero polynomials p, q. This is defined by the properties:

(a) $d \mid p, d \mid q$

(b) If $d' \mid p$ and $d' \mid q$ then $d' \mid d$.

These define d uniquely up to nonzero constant multiples.

To calculate d we first suppose that p, q are named so that $\partial p \geq \partial q$; then divide q into p to get

$$p = qs_1 + r_1 \qquad \partial r_1 < \partial q \leq \partial p$$

and continue in the following way:

$$\begin{aligned} q &= r_1 s_2 + r_2 & \partial r_2 &< \partial r_1 \\ r_1 &= r_2 s_3 + r_3 & \partial r_3 &< \partial r_2 \\ &\vdots \\ r_{n-2} &= r_{n-1} s_n + r_n & \partial r_n &< \partial r_{n-1} \end{aligned}$$

until reaching a zero remainder:

$$r_{n-1} = r_n s_{n+1}$$

The last nonzero remainder r_n is the highest common factor. (From the last equation $r_n \mid r_{n-1}$, and working back successively, r_n is a factor of $r_{n-2}, \ldots, r_1, p, q$, verifying (a). If $d' \mid p, d' \mid q$, then from the first equation, d' is a factor of $r_1 = p - qs_1$, and successively working down the equations, d' is a factor of $r_2, r_3, \ldots, r_n$, so $d' \mid r_n$, verifying (b).)

Beginning with the first equation, and substituting in those which follow, we find that $r_i = a_i p + b_i q$ for suitable $a_i, b_i \in K[t]$, and in particular the highest common factor $d = r_n$ is of the form

$$d = ap + bq \quad \text{for suitable } \ a, b \in K[t] \tag{1.1}$$

A useful special case is when $d = 1$, when p, q are called *coprime* and (1.1) gives

$$ap + bq = 1 \quad \text{for suitable } \ a, b \in K[t]$$

This technique for calculating the highest common factor can also be used to find suitable polynomials a, b.

Factorising a single polynomial p is by no means as straightforward as finding the highest common factor of two. It is known that every nonzero polynomial over a field K is a product of finitely many irreducible factors, and these are unique up to the order in which they are multiplied and up to constant factors: see Fraleigh [41] theorem 31.8 p.284; Stewart [128] theorem 3.14 p. 36 and theorem 3.18 p. 38. Finding these factors by hand is very much an *ad hoc* matter, although there are efficient computer algorithms to factorise polynomials in $\mathbb{Q}[t]$, and in $R[t]$ for some fields R. Linear factors are easiest, since $(x - \alpha) \mid p$ if and only if $p(\alpha) = 0$.

Repeated Factors

If $p(\alpha) = 0$, then α is called a *zero* of p (some sources use the term *root*). If $(t - \alpha)^m \mid p$ where $m \geq 2$, then α is a *repeated zero* and the largest such m is the *multiplicity* of α. Similarly a factor q of a polynomial p is *repeated* if $q^r \mid p$ for some $r \geq 2$. In particular q is repeated if its square divides p. Alternative terms are *multiple zero* and *multiple factor*.

To detect repeated zeros or repeated factors we use a method which, like much in this chapter, was more familiar around 1900 than it is now. Given a polynomial

$$f = \sum_{i=0}^{n} r_i t^i$$

over a ring R we define

$$\mathrm{D}f = \sum_{i=0}^{n} i r_i t^{i-1}$$

called for obvious reasons the *formal derivative* of f. (The term for $i = 0$ vanishes, so t^{-1} does not occur.) It is not hard to check directly that

$$\begin{aligned} \mathrm{D}(f + g) &= \mathrm{D}f + \mathrm{D}g \\ \mathrm{D}(fg) &= (\mathrm{D}f)g + f(\mathrm{D}g) \end{aligned}$$

This enables us to check for repeated factors.

Theorem 1.6. *Let K be a field of characteristic zero. A nonzero polynomial f over K is divisible by the square of a polynomial of degree > 0 if and only if f and $\mathrm{D}f$ have a common factor of degree > 0.*

Proof: First suppose $f = g^2h$. Then $\mathrm{D}f = g^2\mathrm{D}h + 2g(\mathrm{D}g)h$, so f and $\mathrm{D}f$ have g as a common factor.

Conversely, suppose that f and $\mathrm{D}f$ have a common factor g, so $g|f$ and $g|\mathrm{D}f$. We may assume g irreducible (by passing to any of its irreducible factors). Then $f = gh$ for some h, so $\mathrm{D}f = \mathrm{D}g.h + g.\mathrm{D}h$. Therefore $g|\mathrm{D}g.h$.

But g is irreducible and $\partial \mathrm{D}g < \partial g$, so $g|h$. Thus $h = gk$ for some k, and $f = gh = g^2k$, so $g^2|f$. □

If the field has characteristic $p > 0$, then the first part of Theorem 1.6, that f having a squared factor implies f and $\mathrm{D}f$ have a common factor, is still true, and the proof is the same as above. The second part may fail: see Exercise 1.5.

An important result, needed later, is:

Corollary 1.7. *An irreducible polynomial over a subfield K of $\mathbb{C}$ has no repeated zeros in $\mathbb{C}$.*

Proof: Suppose f is irreducible over K. Then f and $\mathrm{D}f$ must be coprime, because a common factor would be a squared factor of f by (1.6), but f is irreducible. Thus there exist polynomials a, b over K such that $af + b\mathrm{D}f = 1$, and the same equation interpreted over $\mathbb{C}$ shows f and $\mathrm{D}f$ to be coprime over $\mathbb{C}$. By Theorem 1.6, f has no repeated zeros. □

Gauss's Lemma and Eisenstein's Criterion

We often consider factorisation of polynomials over $\mathbb{Q}$. When such a polynomial has integer coefficients, it turns out that we need to consider only factors with integer coefficients. This fact is enshrined in a result of Gauss:

Lemma 1.8. (Gauss's Lemma) *Let $p \in \mathbb{Z}[t]$ and suppose that $p = gh$ where $g, h \in \mathbb{Q}[t]$. Then there exists $\lambda \in \mathbb{Q}, \lambda \neq 0$, such that $\lambda g, \lambda^{-1}h \in \mathbb{Z}[t]$.*

Proof: Multiplying by the product of the denominators of the coefficients of g, h we rewrite $p = gh$ as

$$np = g'h'$$

where g', h' are rational multiples of g, h respectively, $n \in \mathbb{Z}$ and $g', h' \in \mathbb{Z}[t]$. Therefore n divides the coefficients of the product $g'h'$. We now divide the equation successively by the prime factors of n. We shall establish that if k is a prime factor of n, then k divides all the coefficients of g' or all those of h'. Whichever it is, we can divide that particular polynomial by k to give another polynomial with integer coefficients. After dividing in this way by all the prime factors of n, we are left with

$$p = \overline{g}\overline{h}$$

where $\overline{g}, \overline{h} \in \mathbb{Z}[t]$ are rational multiples of g, h respectively. Putting $\overline{g} = \lambda g$ for $\lambda \in \mathbb{Q}$, we obtain $\overline{h} = \lambda^{-1} h$ and the result follows.

It remains to prove that if

$$\begin{aligned} g' &= g_0 + g_1 + \cdots + g_r t^r \\ h' &= h_0 + h_1 + \cdots + h_s t^s \end{aligned}$$

and a prime k divides all the coefficients of $g'h'$, then k must divide all the g_i or all the h_j. But if a prime k does not divide all the g_i and all the h_j, we can choose the *first* of each set of coefficients, say g_m, h_q, that are not divisible by k. Then the coefficient of t^{m+q} in $g'h'$ is

$$g_0 h_{m+q} + g_1 h_{m+q-1} + \cdots + g_m h_g + \cdots g_{m+q} h_0$$

Since every term in this expression is divisible by k except $h_q g_m$, the whole coefficient is not divisible by k, a contradiction. □

For later use, we record an easy corollary:

Corollary 1.9. *Let p be a monic polynomial in $\mathbb{Z}[t]$, and suppose that $p = gh$ where $g \in \mathbb{Q}[t]$ is monic. Then g and h are in $\mathbb{Z}[t]$.*

Proof: Clearly h is monic. By Theorem 1.8 there exists $\lambda \in \mathbb{Q}, \lambda \neq 0$, such that $\lambda g, \lambda^{-1} h \in \mathbb{Z}[t]$. Since g is monic, $\lambda = \pm 1$, and changing λ to $-\lambda$ is necessary we may assume that $\lambda = 1$. Thus $h \in \mathbb{Z}[t]$. □

Corollary 1.10. *Let $f \in \mathbb{Z}[t]$ be monic. Then every rational zero of f is an integer.* □

Some standard tricks are useful to prove irreducibility for various specific polynomials over $\mathbb{Z}$. The first of these is *Eisenstein's Criterion*, named after Gotthold Eisenstein:

Theorem 1.11. (Eisenstein's Criterion) *Let*

$$f = a_0 + a_1 t + \cdots + a_n t^n$$

be a polynomial over $\mathbb{Z}$. Suppose there is a prime q such that

$$\begin{aligned} &\text{(a) } q \nmid a_n \\ &\text{(b) } q \mid a_i \qquad (i = 0, 1, \ldots, n-1) \\ &\text{(c) } q^2 \nmid a_0 \end{aligned}$$

Then, apart from constant factors, f is irreducible over $\mathbb{Z}$, hence irreducible over $\mathbb{Q}$.

Proof: By Lemma 1.8 it is enough to show that f can have only constant factors over $\mathbb{Z}$.

If not, then $f = gh$ where

$$g = g_0 + g_1 t + \cdots + g_r t^r \qquad h = h_0 + h_1 t + \cdots + h_s t^s$$

with all $g_i, h_j \in \mathbb{Z}$ and $r, s > 1, r + s = n$.

Now $g_0 h_0 = a_0$, so (b) implies q divides one of g_0, h_0 while (c) implies that it cannot divide both. Without loss of generality, suppose q divides g_0 but not h_0. Not all g_i are divisible by q because this would imply that q divides a_n, contrary to (a). Let g_m be the first coefficient of g not divisible by q. Then

$$a_m = g_0 h_m + \cdots + g_m h_0$$

where $m \leq r < n$. All the summands on the right are divisible by q except the last, which means that a_m is not divisible by q, contradicting (b). □

Reduction

A second useful method for proving irreducibility is *reduction modulo n*, as follows. Suppose $0 \neq p \in \mathbb{Z}[t]$, with p reducible: say $p = qr$. The natural homomorphism $\mathbb{Z} \to \mathbb{Z}_n$ gives rise to a homomorphism $\mathbb{Z}[t] \to \mathbb{Z}_n[t]$. Using bars to denote images under this map, $\overline{p} = \overline{q}\,\overline{r}$. If $\partial\overline{p} = \partial p$, then clearly $\partial\overline{q} = \partial q$, $\partial\overline{r} = \partial r$, and $\overline{p}$ is also reducible. This proves:

Theorem 1.12. *If $p \in \mathbb{Z}[t]$ and its image $\overline{p} \in \mathbb{Z}_n[t]$ is irreducible, with $\partial\overline{p} = \partial p$, then p is irreducible as an element of $\mathbb{Z}[t]$.* □

In practice we take n to be prime, though this is not essential. The point of reduction modulo n is that $\mathbb{Z}_n$ is finite, so only a finite number of possible factors of $\overline{p}$ need to be considered.

Examples.

(a) The polynomial $t^2 - 2$ satisfies Eisenstein's Criterion with $q = 2$.

(b) The polynomial $t^{11} - 7t^6 + 21t^5 + 49t - 56$ satisfies Eisenstein's Criterion with $q = 7$.

(c) The polynomial $t^5 - t + 1$ does not satisfy Eisenstein's Criterion for any q. Instead we try reduction modulo 5. There is no linear factor since none of 0, 1, 2, 3, 4 yields 0 when substituted for t, so the only possible way to factorise is

$$t^5 - t + 1 = (t^2 + \alpha t + \beta)(t^3 + \gamma t^2 + \delta t + \varepsilon)$$

where $\alpha, \beta, \gamma, \delta, \varepsilon$ take values 0, 1, 2, 3 or 4 (mod 5). This gives a system of equations on comparing coefficients: there is only a finite number of possibilities, all of which are easily eliminated. Hence the polynomial is irreducible mod 5, so irreducible over $\mathbb{Z}$.

1.3 Field Extensions

When finding the zeros of a polynomial p over a field K, it is often necessary to pass to a larger field L that contains K. This leads to:

Definition 1.13. If a field K is a subfield of a field L, then L is a *field extension* of K (more simply, an *extension*). We use the notation L/K to show that L is an extension of K.

If several extensions occur in sequence, such as $H \subseteq K \subseteq L$, we often write this as $L/K/H$.

Here the symbol / does not indicate a quotient, as it would in group theory. 'Quotient field' makes very little sense.

For example, $p(t) = t^2 + 1$ has no zeros in $\mathbb{R}$, but considering p as a polynomial over $\mathbb{C}$, it has zeros $\pm\mathrm{i}$ and a factorisation

$$p(t) = (t + \mathrm{i})(t - \mathrm{i})$$

Thus we can factorise $t^2 + 1$ into linear factors in the extension $\mathbb{C}/\mathbb{R}$, and all of its zeros lie in the larger field.

Field extensions often arise in a slightly more general context as a monomorphism $j : K \to L$ where K and L are fields. It is customary in these cases to identify K with its image $j(K)$, which is a subfield of L; then a field extension is a pair of fields (K, L) where K is a subfield of L. In most field extensions we encounter, K and L are subfields of $\mathbb{C}$.

If L/K is a field extension, then L has a natural structure as a vector space over K (where vector addition is addition in L and scalar multiplication of $\lambda \in K$ on $v \in L$ is just $\lambda v \in L$).

Definition 1.14. The dimension of L considered as a vector space over K is called the *degree* of the extension L/K, or the *degree of L over K*, written

$$[L : K] = \dim_K L$$

(Some sources write this as $[L/K]$, which is consistent but less traditional.)

For example, $[\mathbb{C} : \mathbb{R}] = 2$ and $[\mathbb{R} : \mathbb{Q}] = \infty$.

The degree has an important multiplicative property:

Theorem 1.15. (Tower Law) *If $H \subseteq K \subseteq L$ are fields, then*

$$[L : H] = [L : K][K : H]$$

Proof: The idea is simple. Let $\{u_i\}(i \in I)$ be a basis for L over K, and let $\{v_j\}(j \in J)$ be a basis for K over H. Then $\{u_i v_j\}$ $(i, j \in I \times J)$ is a basis for L over H.

In more detail, every $x \in L$ can be expressed uniquely as $x = \sum_{j \in J} k_j v_j$ for $k_j \in K$, and every k_j can be expressed uniquely as $k_j = \sum_{i \in I} h_{ij} u_i$ for $h_{ij} \in H$. Now

$$x = \sum_{j \in J} \left(\sum_{i \in I} h_{ij} u_i \right) v_j = \sum_{(i,j) \in I \times J} h_{ij} u_i v_j$$

so the $u_j v_i$ span L over H. Linear independence is proved in a similar manner by grouping terms and using uniqueness in each extension. □

Definition 1.16. If $[L : K]$ is finite then L is a *finite extension* of K.

Given a field extension L/K and an element $\alpha \in L$, there may or may not exist a nonzero polynomial $p \in K[t]$ such that $p(\alpha) = 0$. If not, we say that α is *transcendental* over K. If such a p exists, we say that α is *algebraic* over K. If α is algebraic over K, then there exists a unique monic polynomial q of minimal degree subject to $q(\alpha) = 0$, and q is called the *minimal polynomial* of α over K. The next result is easy but important.

Proposition 1.17. *The minimal polynomial of an algebraic element α is irreducible over K.*

Proof: Let $m \in K[t]$ be the minimal polynomial of α. If $m(t) = p(t)q(t)$ where $p, q \in K[t]$ then $0 = m(\alpha) = p(\alpha)q(\alpha)$. Therefore either $p(\alpha) = 0$ or $q(\alpha) = 0$. By minimality of the degree, one of p, q has degree 0, so m is irreducible. □

If $\alpha_1, \ldots \alpha_n \in L$, we write

$$K(\alpha_1, \ldots, \alpha_n)$$

for the smallest subfield of L containing K and the elements $\alpha_1, \ldots, \alpha_n$. We say that L is *finitely generated* over K, with *generators* α_j.

For example, $\mathbb{C} = \mathbb{R}(\mathrm{i})$ with one generator i.

In an analogous way, if S is a subring of a ring R and $\alpha_1, \ldots, \alpha_n \in R$, we write

$$S[\alpha_1, \ldots, \alpha_n]$$

for the smallest subring of R containing S and the elements $\alpha_1, \ldots, \alpha_n$. Clearly $S[\alpha_1, \ldots, \alpha_n]$ consists of all polynomials in $\alpha_1, \ldots, \alpha_n$ with coefficients in S. For instance $S[\alpha]$ consists of all polynomials

$$s_0 + s_1\alpha + \cdots + s_m\alpha^m \qquad (s_i \in S)$$

The structure of the field $K(\alpha)$ depends on α in an interesting way. If α is transcendental over K, then for $k_m \neq 0$ we have

$$k_0 + k_1\alpha + \cdots + k_m\alpha^m \neq 0 \qquad (k_i \in K)$$

In this case $K(\alpha)$ must include all rational expressions

$$\frac{s_0 + s_1\alpha + \cdots + s_n\alpha^n}{k_0 + k_1\alpha + \cdots + k_m\alpha^m} \qquad (s_j, k_i \in K, k_m \neq 0)$$

and clearly consists precisely of these elements.

However, for α algebraic, we have:

Theorem 1.18. *If L/K is a field extension and $\alpha \in L$, then α is algebraic over K if and only if $K(\alpha)$ is a finite extension of K. In this case, $[K(\alpha)/K] = \partial p$ where p is the minimal polynomial of α over K, and $K(\alpha) = K[\alpha]$.*

Proof: If $[K(\alpha)/K] = n < \infty$ then the powers, $1, \alpha, \alpha^2, \ldots, \alpha^n$ are linearly dependent over K, whence α is algebraic. Conversely, suppose that α is algebraic with minimal polynomial p of degree m. We claim that $K(\alpha)$ is the vector space over K spanned by $1, \alpha, \ldots, \alpha^{m-1}$. This space, call it V, is certainly closed under addition, subtraction, and multiplication by α; for the last statement note that $\alpha^m = -p(\alpha) + \alpha^m = q(\alpha)$ where $\partial q < m$. Hence V is closed under multiplication, and so it forms a ring. It remains to prove that if $0 \neq v \in V$ then $1/v \in V$. Now $v = h(\alpha)$ where $h \in K[t]$ and $\partial h < m$. Since p is irreducible, p and h are coprime, so there exist $f, g \in K[t]$ such that

$$f(t)p(t) + g(t)h(t) = 1$$

Then

$$1 = f(\alpha)p(\alpha) + g(\alpha)h(\alpha) = g(\alpha)h(\alpha)$$

so that $1/v = g(\alpha) \in V$ as required. Now $[K(\alpha)/K] = \dim_K V = m$. $\square$

A more abstract version of the proof of Theorem 1.18 leads to:

Theorem 1.19. *Let K be a field and let $p(t) \in K[t]$ be an irreducible monic polynomial. Then there exists up to isomorphism a unique extension L such that L contains an element α with minimal polynomial p and $L = K(\alpha)$. This isomorphism can be chosen to be the identity on K.*

Proof: Assume L is such an extension. Define a map $\phi : K[t] \to L$ by $\phi(f(t)) = f(\alpha)$. It is clear that ϕ is a ring homomorphism, and it is onto since $L = K(\alpha)$. The kernel is

$$\ker \phi = \{f \in K[t] : f(\alpha) = 0\}$$

But this is the ideal $\langle p \rangle$ since p is the minimal polynomial.

Therefore $L \cong K[t]/\langle p \rangle$, so L is unique up to isomorphism. Clearly ϕ is the identity on K. □

Corollary 1.20. *Suppose that $K(\alpha)/K$ and $K(\beta)/K$ are simple algebraic extensions such that α and β have the same minimal polynomial m over K. Then the two extensions are isomorphic, and the isomorphism of the large fields can be taken to map α to β and to be the identity on K.*

Proof: Both extensions are isomorphic to $K[t]/\langle m \rangle$. The isomorphisms concerned map t to α and t to β, respectively. Call them ι, j, respectively. Then $j\iota^{-1}$ is an isomorphism from $K(\alpha)$ to $K(\beta)$ that is the identity on K and maps α to β. □

It is customary to express this construction by the phrase 'adjoin to K an element α with $p(\alpha) = 0$' and to write $K(\alpha)$ for the resulting field. This, and much else, is discussed in Stewart [128] chapter 17.

1.4 Symmetric Polynomials

Let $R[t_1, t_2, \dots, t_n]$ denote the ring of polynomials in indeterminates $t_1, t_2, \dots, t_n$ with coefficients in a ring R. Let $\mathbb{S}_n$ denote the symmetric group of permutations on $\{1, 2, \dots, n\}$. For any permutation $\pi \in \mathbb{S}_n$ and any polynomial $f \in R[t_1, \dots, t_n]$ define the polynomial f^π by

$$f^\pi(t_1, \dots, t_n) = f(t_{\pi(1)}, \dots, t_{\pi(n)})$$

For example if $f = t_1 + t_2 t_3$ and π is the cycle (123) then $f^\pi = t_2 + t_3 t_1$. The polynomial f is *symmetric* if $f^\pi = f$ for all $\pi \in \mathbb{S}_n$. For example $t_1 + \cdots + t_n$

is symmetric. More generally the *elementary symmetric polynomials*

$$s_r(t_1, \ldots, t_n) \qquad (1 \le r \le n)$$

are defined to be the sums of all possible distinct products of r distinct t_i's. Thus

$$\begin{aligned} s_1(t_1, \ldots, t_n) &= t_1 + t_2 + \cdots + t_n \\ s_2(t_1, \ldots, t_n) &= t_1t_2 + t_1t_3 + \cdots + t_2t_3 + \cdots + t_{n-1}t_n \\ &\vdots \\ s_n(t_1, \ldots, t_n) &= t_1t_2 \ldots t_n \end{aligned}$$

These polynomials arise in the following circumstances. Consider a polynomial of degree n over a subfield K of $\mathbb{C}$:

$$f = a_nt^n + \cdots + a_0$$

Resolve it into linear factors over $\mathbb{C}$:

$$f = a_n(t - \alpha_1) \ldots (t - \alpha_n)$$

Expanding the product leads to:

$$f = a_n(t^n - s_1t^{n-1} + \cdots + (-1)^n s_n)$$

where s_i denotes $s_i(\alpha_1, \ldots, \alpha_n)$.

A polynomial in $s_1, \ldots, s_n$ can clearly be rewritten as a symmetric polynomial in $t_1, \ldots, t_n$. The converse is also true, a fact first proved by Isaac Newton:

Theorem 1.21. *Let R be a ring. Then every symmetric polynomial in $R[t_1, \ldots, t_n]$ is expressible as a polynomial with coefficients in R in the elementary symmetric polynomials $s_1, \ldots, s_n$.*

Proof: We demonstrate a specific technique for reducing a symmetric polynomial to a combination of elementary ones. First we order the monomials $t_1^{\alpha_1} \ldots t_n^{\alpha_n}$ by a 'lexicographic' order in which $t_1^{\alpha_1} \ldots t_n^{\alpha_n}$ precedes $t_1^{\beta_1} \ldots t_n^{\beta_n}$ if the first nonzero $\alpha_i - \beta_i$ is positive. Given a polynomial $p \in R[t_1, \ldots, t_n]$ we order its terms lexicographically. If p is symmetric, then for every monomial $at_1^{\alpha_1} \ldots t_n^{\alpha_n}$ occurring in p, there occurs a similar monomial with the exponents permuted. Let α_1 be the highest exponent occurring in monomials of p: then there is a term containing $t_1^{\alpha_1}$. The leading term of p

in lexicographic ordering contains $t_1^{\alpha_1}$, and among all such monomials we select the one with the highest occurring power of t_2 and so on. In particular, the leading term of a symmetric polynomial is of the form $at_1^{\alpha_1} \dots t_n^{\alpha_n}$ where $\alpha_1 \geq \dots \geq \alpha_n$. For example, the leading term of

$$s_1^{k_1} \dots s_n^{k_n} = (t_1 + \dots + t_n)^{k_1} \dots (t_1 \dots t_n)^{k_n}$$

is

$$t_1^{k_1+\dots+k_n} t_2^{k_2+\dots+k_n} \dots t_n^{k_n}$$

By choosing $k_1 = \alpha_1 - \alpha_2, \dots, k_{n-1} = \alpha_{n-1} - \alpha_n$, $k_n = \alpha_n$ (which is possible because $\alpha_1 \geq \dots \geq \alpha_n$) we can make this the same as the leading term of p. Then

$$p - as_1^{\alpha_1-\alpha_2} \dots s_{n-1}^{\alpha_{n-1}-\alpha_n} s_n^{\alpha_n}$$

has a lexicographic leading term

$$bt_1^{\beta_1} \dots t_n^{\beta_n} \qquad (\beta_1 \geq \dots \geq \beta_n)$$

which follows $at_1^{\alpha_1} \dots t_n^{\alpha_n}$ in the ordering. But only a finite number of monomials $t_1^{\gamma_1} \dots t_n^{\gamma_n}$ satisfying $\gamma_1 \geq \dots \geq \gamma_n$ follow $t_1^{\alpha_1} \dots t_n^{\alpha_n}$ lexicographically, and so a finite number of repetitions of the given process reduce p to a polynomial in $s_1, \dots, s_n$. □

Example 1.22. The symmetric polynomial

$$p = t_1^2 t_2 + t_1^2 t_3 + t_1 t_2^2 + t_1 t_3^2 + t_2^2 t_3 + t_2 t_3^2$$

is written lexicographically. Here $n = 3$, $\alpha_1 = 2$, $\alpha_2 = 1$, $\alpha_3 = 0$ and the method tells us to consider

$$p - s_1 s_2$$

This simplifies to give

$$p - s_1 s_2 = 3t_1 t_2 t_3$$

The polynomial $3t_1t_2t_3$ is visibly $3s_3$, but the method, using $\alpha_1 = \alpha_2 = \alpha_3 = 1$, leads to the same conclusion.

This result about symmetric functions proves to be extremely useful in the following instance:

Corollary 1.23. *Suppose that L is an extension of the field K, $p \in K[t]$, $\partial p = n$ and the zeros of p are $\theta_1, \dots, \theta_n \in L$. If $h(t_1, \dots, t_n) \in K[t_1, \dots, t_n]$ is symmetric, then $h(\theta_1, \dots, \theta_n) \in K$.* □

1.5 Modules

A module over a ring is the analogue of a vector space over a field. Modules are vital in algebraic number theory, because they synthesise numerous results that classically required separate proofs. They also provide a clean abstract language to express key ideas.

Definition 1.24. Let R be a ring (recall we assume it is commutative with 1). An *R-module* is an abelian group M (written additively), together with a function $\alpha : R \times M \to M$, for which we write $\alpha(r, m) = rm$ $(r \in R, m \in M)$, satisfying

$$\begin{array}{lrcl} \text{(a)} & (r+s)m & = & rm + sm \\ \text{(b)} & r(m+n) & = & rm + rn \\ \text{(c)} & r(sm) & = & (rs)m \\ \text{(d)} & 1m & = & m \end{array}$$

for all $r, s \in R$, $m, n \in M$.

Although condition (d) is obligatory in this text, be warned that in other parts of mathematics it may not be.

The function α is called an *R-action* on M.

If R is a field K, then an R-module is the same thing as a vector space over K. In this sense we can think of an R-module as a generalisation of a vector space, but because division need not be possible in R, many of the results and techniques of vector space theory do not carry over unchanged to R-modules. The basic theory of modules may be found in Fraleigh [41] section 37.2, p. 338. In particular we define an *R-submodule* of M to be a subgroup N of M (under addition) such that if $n \in N$, $r \in R$, then $rn \in N$. The *quotient module* M/N is the corresponding quotient group, with R-action

$$r(N + m) = N + rm \qquad (r \in R, m \in M)$$

If $X \subseteq M, Y \subseteq R$, we define YX to be the set of all finite sums $\sum_i y_i x_i$ where $y_i \in Y, x_i \in X$.

The submodule of M *generated by* X, which we write

$$\langle X \rangle_R$$

is the smallest submodule containing X. This is equal to RX. If $N = \langle x_1, \ldots, x_n \rangle_R$ then N is a *finitely generated* R-module.

A $\mathbb{Z}$-module is nothing more than an abelian group M (written additively), and conversely, given an additive abelian group M we can make it into a $\mathbb{Z}$-module by defining

$$\begin{array}{rlll} & 0m = 0 \quad 1m = m & (m \in M) & \\ \text{then inductively} & (n+1)m = nm + m & (n \in \mathbb{Z}, n > 0) & (1.2) \\ \text{and} & (-n)m = -(nm) & (n \in \mathbb{Z}, n > 0) & \end{array}$$

We discuss this case further in the next section.

More generally there are several natural ways in which R-modules can arise, of which we distinguish three:

(1) Suppose that R is a subring of a ring S. Then S is an R-module with action

$$\alpha(r, s) = rs \quad (r \in R, s \in S)$$

where the product is just that of elements in S.

(2) Suppose that I is an ideal of the ring R. Then I is an R-module under

$$\alpha(r, i) = ri \quad (r \in R, i \in I)$$

where the product is that in R.

(3) Suppose that $J \subseteq I$ is another ideal. Then J is also an R-module. The quotient module I/J has the action

$$r(J + i) = J + ri \quad (r \in R, i \in I)$$

Unlike vector spaces, R-modules need not possess a basis. However, there is a special class of R-modules that behaves like a vector space in this respect.

Definition 1.25. A finitely generated R-module M is *free* on the generators $\{x_1, \ldots, x_n\} \subseteq M$ if every element $x \in M$ can be expressed uniquely as

$$x = r_1 x_1 + \cdots + r_n x_n \qquad r_j \in R, \ 1 \leq j \leq n$$

In this case we call $\{x_1, \ldots, x_n\}$ a *basis* for M over R.

The *direct product* $M_1 \times M_2$ (or *direct sum* $M_1 \oplus M_2$) of two R-modules M_1, M_2 is the Cartesian product $M_1 \times M_2$ with action

$$r(m_1, m_2) = (rm_1, rm_2)$$

Proposition 1.26. *If M is a free R-module with basis $\{x_1, \ldots, x_n\}$ then $M \cong R \times \cdots \times R \equiv R^n$ with n direct factors.*

Proof: The map $\phi : R^n \to M$ defined by

$$\phi(r_1, \dots, r_n) = r_1x_1 + \cdots + r_nx_n$$

is an isomorphism. □

In additive notation $R^n = R \oplus \cdots \oplus R$.

1.6 Free Abelian Groups

The study of algebraic numbers involves subfields and subrings of $\mathbb{C}$. A typical instance is the subring

$$\mathbb{Z}[\mathrm{i}] = \{a + b\mathrm{i} \in \mathbb{C} \mid a, b \in \mathbb{Z}\}$$

The additive group of $\mathbb{Z}[\mathrm{i}]$ is isomorphic to $\mathbb{Z} \times \mathbb{Z}$. More generally the additive groups of those subrings of $\mathbb{C}$ that we study are usually isomorphic to the direct product of a *finite* number of copies of $\mathbb{Z}$. In this section we study such abelian groups, for later use.

Let G be an abelian group. In this section we use additive notation for G, so the group operation is denoted by $+$, the identity by 0, the inverse of g by $-g$ and powers of g by $2g, 3g, \dots$. In later chapters we encounter cases where multiplicative notation is more appropriate.

We saw in (1.2) that every abelian group G can be considered as a $\mathbb{Z}$-module, and conversely. Suppose that G is finitely generated as a $\mathbb{Z}$-module, so there exist $g_1, \dots, g_n \in G$ such that every $g \in G$ is a sum

$$g = m_1g_1 + \cdots + m_ng_n \qquad (m_i \in \mathbb{Z})$$

then G is a *finitely generated abelian group.*

Generalising the notion of linear independence in a vector space, we say that elements $g_1, \dots, g_n$ in an abelian group G are *linearly independent* (over $\mathbb{Z}$) if any equation

$$m_1g_1 + \cdots + m_ng_n = 0$$

with $m_1, \dots, m_n \in \mathbb{Z}$ implies $m_1 = \dots = m_n = 0$. Borrowing module language, a linearly independent set that generates G is a *basis* ($\mathbb{Z}$-basis for emphasis). If $\{g_1, \dots, g_n\}$ is a basis, every $g \in G$ has a unique representation in the form

$$g = m_1g_1 + \cdots + m_ng_n \qquad (m_i \in \mathbb{Z})$$

because an alternative expression

$$g = k_1 g_1 + \cdots + k_n g_n \qquad (k_i \in \mathbb{Z})$$

implies

$$(m_1 - k_1)g_1 + \cdots + (m_n - k_n)g_n = 0$$

and linear independence implies that $m_i - k_i = 0$, that is, $m_i = k_i$ $(1 \leq i \leq n)$.

Denote the direct product of n copies of the additive group of integers by $\mathbb{Z}^n$. Then an abelian group with a basis of n elements is isomorphic to $\mathbb{Z}^n$.

Theorem 1.27. *Let G be a finitely generated free abelian group. Then all bases of G have the same number of elements.*

Proof: Let $2G$ be the subgroup of G consisting of all elements of the form $g + g$ $(g \in G)$. If G has a basis of n elements, then $G/2G$ is a group of order 2^n. Since the definition of $2G$ does not depend on any particular basis, every basis must have the same number of elements. □

An abelian group with a basis of n elements is called a *free abelian group* of *rank* n. If G is free abelian of rank n and $\{x_1, \ldots, x_n\}$, $\{y_1, \ldots, y_n\}$ are both bases, then there exist integers a_{ij}, b_{ij} such that

$$y_i = \sum_{j=1}^{n} a_{ij} x_j \qquad x_i = \sum_{j=1}^{n} b_{ij} y_j$$

If we consider the matrices

$$A = (a_{ij}) \qquad B = (b_{ij})$$

then $AB = I_n$, the identity matrix. Hence

$$\det(A)\det(B) = 1$$

where det indicates the determinant. Since $\det(A)$ and $\det(B)$ are integers,

$$\det(A) = \det(B) = \pm 1$$

A square matrix over $\mathbb{Z}$ with determinant ± 1 is *unimodular*. We have:

Lemma 1.28. *Let G be a free abelian group of rank n with basis $\{x_1, \ldots, x_n\}$. Let (a_{ij}) be an $n \times n$ matrix with integer entries. Then the elements*

$$y_i = \sum_{j=1}^{n} a_{ij} x_j$$

form a basis of G if and only if (a_{ij}) is unimodular.

Proof: The 'only if' part has already been dealt with. Now suppose $A = (a_{ij})$ is unimodular. Since $\det(A) \neq 0$ the y_j are linearly independent. We have

$$A^{-1} = (\det(A))^{-1}\tilde{A}$$

where $\tilde{A}$ is the adjoint matrix and has integer entries. Hence $A^{-1} = \pm\tilde{A}$ has integer entries. Putting $B = A^{-1} = (b_{ij})$ we obtain $x_i = \sum_j b_{ij}y_j$, demonstrating that the y_j generate G. Thus they form a basis. □

The central result in the theory of finitely generated free abelian groups concerns the structure of subgroups:

Theorem 1.29. *Every subgroup H of a free abelian group G of rank n is free of rank $s \leq n$. Moreover there exists a basis $u_1, \ldots, u_n$ for G and positive integers $\alpha_1, \ldots, \alpha_s$ such that $\alpha_1 u_1, \ldots, \alpha_s u_s$ is a basis for H.*

Proof: We use induction on the rank n of G. For $n = 1$, G is infinite cyclic and the result is a consequence of the subgroup structure of the cyclic group. If G has rank n, pick any basis $w_1, \ldots, w_n$ of G. Every $h \in H$ is of the form

$$h = h_1 w_1 + \cdots + h_n w_n$$

Either $H = \{0\}$, in which case the theorem is trivial, or there exist nonzero coefficients h_i for some $h \in H$. From all such coefficients, let $\lambda(w_1, \ldots, w_n)$ be the least positive integer occurring. Now choose the basis $w_1, \ldots, w_n$ to make $\lambda(w_1, \ldots, w_n)$ minimal. Let α_1 be this minimal value, and number the w_i in such a way that

$$v_1 = \alpha_1 w_1 + \beta_2 w_2 + \cdots + \beta_n w_n$$

is an element of H in which α_1 occurs as a coefficient. Let

$$\beta_i = \alpha_1 q_i + r_i \qquad (2 \leq i \leq n)$$

where $0 \leq r_i < \alpha_1$, so that r_i is the remainder on dividing β_i by α_1. Define

$$u_1 = w_1 + q_2 w_2 + \cdots + q_n w_n$$

Then it is easy to verify that $u_1, w_2, \ldots, w_n$ is another basis for G. (The appropriate matrix is clearly unimodular.) With respect to the new basis,

$$v_1 = \alpha_1 u_1 + r_2 w_2 + \cdots + r_n w_n$$

By the minimality of $\alpha_1 = \lambda(w_1, \dots, w_n)$ for *all* bases we have

$$r_2 = \cdots = r_n = 0$$

Hence $v_1 = \alpha_1 u_1$. With respect to the new basis, let

$$H' = \{m_1 u_1 + m_2 w_2 + \cdots + m_n w_n \mid m_1 = 0\}$$

Clearly $H' \cap V_1 = \{0\}$, where V_1 is the subgroup generated by v_1. We claim that $H = H' + V_1$. For if $h \in H$ then

$$h = \gamma_1 u_1 + \gamma_2 w_2 + \cdots + \gamma_n w_n$$

and putting

$$\gamma_1 = \alpha_1 q + r_1 \qquad (0 \leq r_1 < \alpha_1)$$

we see that H contains

$$h - qv_1 = r_1 u_1 + \gamma_2 w_2 + \cdots + \gamma_n w_n$$

Minimality of α_1 once more implies that $r_1 = 0$, so $h - qv_1 \in H'$. Therefore H is isomorphic to $H' \times V_1$ and H' is a subgroup of the group G' with generators $w_2, \dots, w_n$. Clearly G' is free abelian of rank $n - 1$. By induction, H' is free of rank $\leq n - 1$, and there exist bases $u_2, \dots, u_n$ of G' and $v_2, \dots, v_s$ of H' such that $v_i = \alpha_i u_i$ for positive integers α_i. □

From the above two results we deduce a useful theorem about orders of quotient groups. In its statement we use $|X|$ to denote the cardinality of the set X, and $|x|$ to denote the absolute value of the real number x. No confusion need arise.

Theorem 1.30. *Let G be a free abelian group of rank r, and H a subgroup of G. Then G/H is finite if and only if the ranks of G and H are equal. If this is the case, and if G and H have $\mathbb{Z}$-bases $x_1, \dots, x_r$ and $y_1, \dots, y_r$ respectively, with $y_i = \sum_j a_{ij} x_j$, then*

$$|G/H| = |\det(a_{ij})| \tag{1.3}$$

Proof: Let H have rank s. Use Theorem 1.29 to choose $\mathbb{Z}$-bases $u_1, \dots, u_r$ of G and $v_1, \dots, v_s$ of H with $v_i = \alpha_i u_i$ for $1 \leq i \leq s$. Clearly G/H is the direct product of finite cyclic groups of orders $\alpha_1, \dots, \alpha_s$ and $r - s$ infinite cyclic groups. Hence $|G/H|$ is finite if and only if $r = s$, and in that case

$$|G/H| = \alpha_1 \dots \alpha_r$$

Now

$$u_i = \sum_j b_{ij} x_j \qquad v_i = \sum_j c_{ij} u_j \qquad y_i = \sum_j d_{ij} v_j$$

where the matrices $(b_{ij}) = B$ and $(d_{ij}) = D$ are unimodular by Lemma 1.28, and

$$C = (c_{ij}) = \begin{bmatrix} \alpha_1 & & & \\ & \alpha_2 & \mathbf{0} & \\ & \mathbf{0} & \ddots & \\ & & & \alpha_r \end{bmatrix}$$

Clearly if $A = (a_{ij})$ then $A = BCD$, so

$$\det(A) = \det(B)\det(C)\det(D)$$

Therefore

$$|\det(A)| = |\pm 1||\det(C)||\pm 1| = |\alpha_1 \ldots \alpha_r| = |G/H|$$

as claimed. □

For example, if G has rank 3 and $\mathbb{Z}$-basis x, y, z, and if H has $\mathbb{Z}$-basis

$$\{3x + y - 2z, 4x - 5y + z, x + 7z\}$$

then $|G/H|$ is the absolute value of

$$\begin{vmatrix} 3 & 1 & -2 \\ 4 & -5 & 1 \\ 1 & 0 & 7 \end{vmatrix}$$

namely 142.

Proposition 1.31. *Every finitely generated abelian group with n generators is the direct product of a finite abelian group and a free group on k generators where $k \leq n$.*

Proof: Let G be a finitely generated abelian group; let the generators be $w_1, \ldots, w_n$, which need not be independent. Define a map $f : \mathbb{Z}^n \to G$ by:

$$f(m_1, \ldots, m_n) = m_1 w_1 + \cdots + m_n w_n$$

This is surjective, so G is isomorphic to $\mathbb{Z}^n/H$ where H is the kernel of f.

We can use Theorem 1.29 to choose a new basis $u_1, \ldots, u_n$ of $\mathbb{Z}^n$ so that $\alpha_1 u_1, \ldots, \alpha_s u_s$ is a basis for H. Let A be the subgroup of $\mathbb{Z}^n$ generated by

$u_1, \ldots, u_s$ and B be the subgroup generated by $u_{s+1}, \ldots, u_n$. Clearly G is isomorphic to $(A/H) \times B$, so is the direct product of a finite abelian group A/H and a free group B on $n - s$ generators. Now set $k = n - s$. □

Proposition 1.32. *A subgroup of a finitely generated abelian group is finitely generated.*

Proof: Let K be a subgroup of a finitely generated abelian group G. If $G = F \times B$ where F is finite and B is finitely generated and free, then $K \cong (F \cap K) \times H$ where $H \subseteq B$. Then $F \cap K$ is finite and (by Theorem 1.29) H is finitely generated and free, so K is finitely generated. □

The results in this section are not the best possible statements about the structure of finitely generated abelian groups. Refinements are in Fraleigh [60] chapter 9 pp. 86–93. The results we have just established are ample for our needs.

1.7 Finite Fields

Most of the fields we encounter in this book are subfields of the complex numbers, so they have characteristic zero, with prime subfield $\mathbb{Q}$. However, some constructions, particularly quotient rings, lead to fields of prime characteristic p with prime subfield $\mathbb{Z}_p$. Prominent among these fields are the *finite fields*: fields with finitely many elements. The $\mathbb{Z}_p$ are examples, but there are others. Galois realised that every finite field has p^n elements for some prime p and $n > 0$; moreover, up to isomorphism there is a unique finite field of that order. We denote it by

$$\mathbb{F}_{p^n}$$

so $\mathbb{F}_p$ is the same as $\mathbb{Z}_p$. However, when $n \geq 2$ the ring $\mathbb{Z}_{p^n}$ is not a field, because $p.p^{n-1} = 0$ but $p, p^{n-1} \neq 0$.

The field $\mathbb{F}_q$ where $q = p^n$ can be constructed as the splitting field of the polynomial $f(t) = t^q - t$ over $\mathbb{Z}_p$ (see Stewart [128] theorem 19.2). Here 'splitting field' means that $\mathbb{F}_q$ is generated by the zeros of f as an extension of the prime subfield $\mathbb{Z}_p$. We do not give details here, but we prove one important result:

Theorem 1.33. *The multiplicative group $\mathbb{F}_q^*$ of a finite field $\mathbb{F}_q$ of order $q = p^n$ is cyclic of order $q - 1$.*

To prove this, we need some simple properties of finite abelian groups.

Definition 1.34. The *exponent* $e(G)$ of a finite group G is the least common multiple of the orders of the elements of G.

The order of any element of G divides the order $|G|$, so $e(G)$ divides $|G|$. In general, G need not possess an element of order $e(G)$. For example if $G = \mathbb{S}_3$ then $e(G) = 6$, but G has no element of order 6. Abelian groups are better behaved:

Lemma 1.35. *Any finite abelian group G contains an element of order $e(G)$.*

Proof: Let $e = e(G) = p_1^{\alpha_1} \dots p_n^{\alpha_n}$ where the p_j are distinct primes and $\alpha_j \geq 1$. The definition of $e(G)$ implies that for each j, the group G must possess an element g_j whose order is divisible by $p_j^{\alpha_j}$. Then a suitable power a_j of g_j has order $p_j^{\alpha_j}$. Define

$$g = a_1 a_2 \dots a_n \tag{1.4}$$

Suppose that $g^m = 1$ where $m \geq 1$. Then

$$a_j^m = a_1^{-m} \dots a_{j-1}^{-m} a_{j+1}^{-m} \dots a_n^{-m}$$

So if

$$q = p_1^{\alpha_1} \dots p_{j-1}^{\alpha_{j-1}} p_{j+1}^{\alpha_{j+1}} \dots p_n^{\alpha_n}$$

then $a_j^{mq} = 1$. But q is prime to the order of a_j, so $p_j^{\alpha_j}$ divides m. Hence e divides m. But clearly $g^e = 1$. Hence g has order e, which is what we want. □

Corollary 1.36. *If G is a finite abelian group such that $e(G) = |G|$, then G is cyclic.*

Proof: The element g in (1.4) generates G. □

We can apply this corollary immediately.

Theorem 1.37. *If G is a finite subgroup of the multiplicative group K^* of a field K, then G is cyclic.*

Proof: Since multiplication in K is commutative, G is an abelian group. Let $e = e(G)$. For any $x \in G$ we have $x^e = 1$, so that x is a zero of the polynomial $t^e - 1$ over K. This polynomial has at most e zeros, so $|G| \leq e$. But $e \leq |G|$, hence $e = |G|$; by Corollary 1.36, G is cyclic. $\square$

Theorem 1.33 is a special case of this result.

Example 1.38. Let K be the finite field $\mathbb{F}_{25}$. Since $t^2 - 2$ is irreducible over $\mathbb{Z}_5$, the quotient $\mathbb{Z}_5[t]/\langle t^2 - 2\rangle$ is a field. It contains 25 elements, so we can use it to define $\mathbb{F}_{25}$. The elements of $\mathbb{F}_{25}$ can be represented in the form $a + b\alpha$ where $\alpha^2 = 2$ and $a, b \in \mathbb{Z}_5$. There is no harm in writing $\alpha = \sqrt{2}$. By trial and error we are led to consider the element $2 + \sqrt{2}$, whose successive powers are:

$$\begin{array}{ccccccc} 1 & 2+\sqrt{2} & 1+4\sqrt{2} & 4\sqrt{2} & 3+3\sqrt{2} & 2+4\sqrt{2} & 2 \\ 4+2\sqrt{2} & 2+3\sqrt{2} & 3\sqrt{2} & 1+\sqrt{2} & 4+3\sqrt{2} & 4 & \\ 3+4\sqrt{2} & 4+\sqrt{2} & \sqrt{2} & 2+2\sqrt{2} & 3+\sqrt{2} & 3 & \\ 1+3\sqrt{2} & 3+2\sqrt{2} & 2\sqrt{2} & 4+4\sqrt{2} & 1+2\sqrt{2} & 1 & \end{array}$$

Hence $2 + \sqrt{2}$ generates the multiplicative group.

1.8 Exercises

1.1 Show that Theorem 1.4 becomes false if the word 'finite' is omitted from the hypotheses.

1.2 Which of the following polynomials over $\mathbb{Z}$ are irreducible?

(a) $t^2 + 3$

(b) $t^2 - 169$

(c) $t^3 + t^2 + t + 1$

(d) $xt^3 + 2t^2 + 3t + 4$

1.3 Use a computer algebra package to factorise the following polynomials over $\mathbb{Z}$:

(a) $t^{10} + 3t^9 + 2t^8 + 4t^7 + 10t^6 + 12t^5 + 10t^4 + 7t^3 + 12t^2 + 9t + 5$

(b) $t^{12} + t^7 - t^6 + t^5 + t^4 - t^2 + 1$

(c) $t^7 + t^5 + t^3 + t^2 + t + 1$

(d) $t^{11} - 12t^8 + 4t^6 + 35t^5 - 36t^3 - 24t^2 + 32$

(e) $t^{13} + t^{11} + t^7 + t^5 + t^3 + t^2 + t + 1$

(f) $t^{13} + 7t^6 - 6$.

1.4 Write down some polynomials over $\mathbb{Z}$ and factorise them into irreducibles.

1.5 Does Theorem 1.6 remain true over a field of characteristic $p > 0$?

1.6 Find the minimal polynomial over $\mathbb{Q}$ of

(a) $(1 + \mathrm{i})/\sqrt{2}$

(b) $\mathrm{i} + \sqrt{2}$

(c) $e^{2\pi\mathrm{i}/3} + 2$

1.7 Find the degrees of the following field extensions:

(a) $\mathbb{Q}(\sqrt{7})/\mathbb{Q}$

(b) $\mathbb{C}(\sqrt{7})/\mathbb{C}$

(c) $\mathbb{Q}(\sqrt{5}, \sqrt{7}, \sqrt{35})/\mathbb{Q}$

(d) $\mathbb{R}(\theta)/\mathbb{R}$ where $\theta^3 - 7\theta + 6 = 0$ and $\theta \notin \mathbb{R}$

(e) $\mathbb{Q}(\pi)/\mathbb{Q}$ (*Hint:* You may assume that π is transcendental.)

1.8 Let K be the field generated by the elements $e^{2\pi\mathrm{i}/n}$ $(n = 1, 2, \ldots)$. Show that K is an algebraic extension of $\mathbb{Q}$, but that $[K : \mathbb{Q}]$ is not finite. (*Hint:* It may help to show that the minimal polynomial of $e^{2\pi\mathrm{i}/p}$ for p prime is $t^{p-1} + t^{p-2} + \ldots + 1$.)

1.9 Express the following polynomials in terms of elementary symmetric polynomials, where possible.

(a) $t_1^2 + t_2^2 + t_3^2$ $(n = 3)$
(b) $t_1^3 + t_2^3$ $(n = 2)$
(c) $t_1t_2^2 + t_2t_3^2 + t_3t_1^2$ $(n = 3)$
(d) $t_1 + t_2^2 + t_3^3$ $(n = 3)$

1.10 A polynomial belonging to $\mathbb{Z}[t_1, \ldots, t_n]$ is *antisymmetric* if it is invariant under even permutations of the variables, but changes sign under odd permutations. Let

$$\Delta = \prod_{i<j} (t_i - t_j) .$$

Show that Δ is antisymmetric. If f is any antisymmetric polynomial, prove that f is expressible as a polynomial in the elementary symmetric polynomials, together with Δ. (*Hint:* Show that Δ divides f and consider f/Δ.)

1.11 Find the orders of the groups G/H where G is free abelian with $\mathbb{Z}$-basis x, y, z and H is generated by:

(a) $2x,\ 3y,\ 7z$

(b) $x + 3y - 5z,\ 2x - 4y,\ 7x + 2y - 9z$

(c) x

(d) $41x + 32y - 999z,\ 16y + 3z,\ 2y + 111z$

(e) $41x + 32y - 999z$

1.12 Let K be a field. Show that M is a K-module if and only if it is a vector space over K. Show that the submodules of M are precisely the vector subspaces. Do these statements remain true if we do not use condition (d) of Definition 1.24 for modules?

1.13 Let $\mathbb{Z}$ be a $\mathbb{Z}$-module with the obvious action. Find all the submodules.

1.14 Let R be a ring, and let M be a finitely generated R-module. Is it true that M necessarily has only finitely many distinct R-submodules? If not, is there an extra condition on R that leads to this conclusion?

1.15 An abelian group G is *torsion-free* if $g \in G$, $g \neq 0$ and $kg = 0$ for $k \in \mathbb{Z}$ implies $k = 0$. Prove that a finitely generated torsion-free abelian group is a finitely generated free group.

1.16 Find a polynomial in $\mathbb{Z}[t]$ that is reducible modulo 2, 3, and 5, but irreducible modulo 7. (*Hint*: Consider quadratic polynomials. The answer is not unique.)

1.17 Prove that if $\{p_1, \ldots, p_r\}$ is a finite set of primes, and q is a prime different from all p_i, then there is a polynomial in $\mathbb{Z}[t]$ that is reducible modulo all p_i, but irreducible modulo q. (*Hint*: Generalise the previous example.)

1.18 Use reduction modulo 3 to show that the Diophantine equation $x^2 + y^2 = 3z^2$ has no nontrivial solutions in integers. What about the Diophantine equations $x^2 + y^2 = 7z^2$ and $x^2 + y^2 = 11z^2$? Justify your answers.

Show that the Diophantine equations $x^2 + y^2 = 5z^2$ and $x^2 + y^2 = 13z^2$ have nontrivial integer solutions.

State a conjecture about the existence (or not) of nontrivial integer solutions to the Diophantine equation $x^2 + y^2 = pz^2$ where p is an odd prime.

1.19 Find the irreducible factors of $t^3 - t + 1$ modulo 23.

Find the irreducible factors of $t^3 + t + 1$ modulo 31.

1.20 If $\mathfrak{a}, \mathfrak{b}$ are ideals of a ring R, show that $\mathfrak{ab} \subseteq \mathfrak{a} \cap \mathfrak{b}$.

Find a ring R that has two ideals $\mathfrak{a}, \mathfrak{b}$ such that $\mathfrak{ab} \neq \mathfrak{a} \cap \mathfrak{b}$.

1.21 Prove that if $f, g \in K[t]$ for a field K, and $h = \gcd(f, g)$, then there exist $a, b \in K[t]$ such that $h = af + bg$, by considering an element of minimal degree in the set $\{af + bg : a, b \in K[t]\}$.

1.22 Let $f(t) \in \mathbb{Z}[t]$ be monic, with reduction $\bar{f}(t) \in \mathbb{Z}_p[t]$ modulo a prime p. Which of the following statements are true? Justify your answer.

(a) If f is reducible over $\mathbb{Z}$ then $\bar{f}$ is reducible over $\mathbb{Z}_p$.

(b) If f is irreducible over $\mathbb{Z}$ then $\bar{f}$ is irreducible over $\mathbb{Z}_p$.

(c) If $\bar{f}$ is reducible over $\mathbb{Z}_p$ then f is reducible over $\mathbb{Z}$.

(d) If $\bar{f}$ is irreducible over $\mathbb{Z}_p$ then f is irreducible over $\mathbb{Z}$.

1.23 Prove that $f(t) = t^4 + 1 \in \mathbb{Z}[t]$ is irreducible over $\mathbb{Z}$, but $\bar{f}(t) \in \mathbb{Z}_p[t]$ is reducible over $\mathbb{Z}_p$ for all primes p.

(*Hint*: You may use the following results from the Law of Quadratic Reciprocity, Theorem 22.9 below:

For odd primes p,

The number 2 is a square modulo p if and only if $p \equiv \pm 1 \pmod{8}$.

The number -2 is a square modulo p if and only if $p \equiv 1, 3 \pmod{8}$.

The number -1 is a square modulo p if and only if $p \equiv 1 \pmod{4}$.)

2

Algebraic Numbers

In this chapter we introduce the algebraic numbers as solutions of polynomial equations with integer coefficients. Among these numbers, the major players are the solutions of equations with integer coefficients whose leading coefficient is 1. These are the algebraic integers. We develop a theory of factorisation of algebraic integers, analogous to factorisation of whole numbers. In many ways the theories are alike, but in at least one essential way—uniqueness of factorisation—there are differences.

The most important of these differences is that in many rings of algebraic integers, factorisation into irreducibles is not unique. We postpone discussion of this issue until Chapter 5.

Here we observe a simpler issue: factorisation into irreducibles depends on the ring in which factorisation is performed. In $\mathbb{Z}$ the number 5 is irreducible. The only ways to write it as a product are trivial: multiply ± 5 and ± 1. However, in $\mathbb{Z}[\sqrt{5}]$ it can be written as the nontrivial product $5 = \sqrt{5} \cdot \sqrt{5}$; moreover, it turns out that $\sqrt{5}$ cannot be further factorised in this ring. Thus 5 is irreducible in $\mathbb{Z}$, yet reducible in $\mathbb{Z}[\sqrt{5}]$. It is therefore essential to specify the ring in which factorisation is carried out.

The natural context is a ring of algebraic integers, contained in its associated algebraic number field. We begin with algebraic number fields that obey a finiteness condition: they are finite-dimensional as vector spaces over the rationals. We prove that such a field is of the form $\mathbb{Q}[\theta]$ for a single algebraic number θ.

We introduce the conjugates of an algebraic number and the discriminant of a basis for $\mathbb{Q}[\theta]$ over $\mathbb{Q}$, using the conjugates of θ to show that the discriminant is always a nonzero rational number. Algebraic integers are defined and shown to form a ring. The ring of algebraic integers in a number field is shown to have an integral basis whose discriminant is an integer. This integer is independent of the choice of integral basis and is called the discriminant of the number field.

We introduce the norm and trace of an algebraic number, which are ordinary integers when the algebraic number is an algebraic integer. Using the norm and trace in later chapters we can translate some statements about algebraic integers into statements about ordinary integers, which are easier to handle. We end with some techniques for computing rings of integers.

2.1 Algebraic Numbers

Definition 2.1. A complex number α is *algebraic* if it is algebraic over $\mathbb{Q}$, that is, it satisfies a nonzero polynomial equation with coefficients in $\mathbb{Q}$. Equivalently, clearing out denominators, we may assume the coefficients are in $\mathbb{Z}$.

Let $\mathbb{A}$ denote the set of algebraic numbers. In fact $\mathbb{A}$ is a field:

Theorem 2.2. *The set $\mathbb{A}$ of algebraic numbers is a subfield of the complex field $\mathbb{C}$.*

Proof: We use Theorem 1.18, which in this case says that α is algebraic if and only if $[\mathbb{Q}(\alpha) : \mathbb{Q}]$ is finite. Suppose that α, β are algebraic. Then

$$[\mathbb{Q}(\alpha, \beta) : \mathbb{Q}] = [\mathbb{Q}(\alpha, \beta) : \mathbb{Q}(\alpha)]\,[\mathbb{Q}(\alpha) : \mathbb{Q}]$$

Since β is algebraic over $\mathbb{Q}$ it is certainly algebraic over $\mathbb{Q}(\alpha)$, so the first factor on the right is finite; and the second factor is also finite. Hence $[\mathbb{Q}(\alpha, \beta) : \mathbb{Q}]$ is finite. But each of $\alpha + \beta$, $\alpha - \beta$, $\alpha\beta$, and (for $\beta \neq 0$) α/β belongs to $\mathbb{Q}(\alpha, \beta)$. So all of these are in $\mathbb{A}$. □

The whole field $\mathbb{A}$ is not as interesting, for us, as certain of its subfields. The trouble with $\mathbb{A}$ is that $[\mathbb{A} : \mathbb{Q}]$ is not finite; see Chapter 1, Exercise 1.7. (However, at a more advanced stage it is usual to work with all of $\mathbb{A}$ simultaneously, but this involves extra technical concepts, such as profinite groups.)

Definition 2.3. An *algebraic number field* is a subfield K of $\mathbb{C}$ such that $[K : \mathbb{Q}]$ is finite.

We shorten this term to *number field* from now on.

Finiteness of the degree $[K : \mathbb{Q}]$ implies that every element of K is algebraic, so $K \subseteq \mathbb{A}$.

Proposition 2.4. *Let $K \subseteq \mathbb{C}$ be a subfield. Then the following are equivalent:*
(a) *K is a number field.*
(b) *$[K : \mathbb{Q}]$ is finite.*
(c) *$K \subseteq \mathbb{A}$ and K is finitely generated over $\mathbb{Q}$.*

Proof: By induction on n and the Tower Law, a subfield K of $\mathbb{C}$ is a finite extension of $\mathbb{Q}$ if and only if $K = \mathbb{Q}(\alpha_1, \ldots, \alpha_n)$ is finitely generated over $\mathbb{Q}$ with each α algebraic. □

By condition (c) of Proposition 2.4, any number field K has the form $K = \mathbb{Q}(\alpha_i, \ldots, \alpha_n)$ for finitely many algebraic numbers $\alpha_1, \ldots, \alpha_n$.

We can strengthen this observation considerably. For simplicity we prove only the number field case, but the same method works for any infinite field K with some routine technicalities about splitting fields.

Theorem 2.5. (Primitive Element Theorem for Number Field Extensions) *If L/K is an extension of number fields, then $L = K(\theta)$ for some $\theta \in L$.*

Proof: Arguing by induction, it is sufficient to prove that if $L = L_1(\alpha, \beta)$ where L_1 is a subfield of L containing K, then $L = L_1(\theta)$ for some $\theta \in L$. Let p and q respectively be the minimal polynomials of α, β over L_1, and suppose that over $\mathbb{C}$ these factorise as

$$\begin{aligned} p(t) &= (t - \alpha_1) \ldots (t - \alpha_n) \\ q(t) &= (t - \beta_1) \ldots (t - \beta_m) \end{aligned}$$

where we choose the numbering so that $\alpha_1 = \alpha$, $\beta_1 = \beta$. By Corollary 1.7 the α_i are distinct, as are the β_j. Hence for each i and each $k \neq 1$ there is at most one element $x \in L_1$ such that

$$\alpha_i + x\beta_k = \alpha_1 + x\beta_1$$

Namely, those x of the form

$$x = \frac{\alpha_1 - \alpha_i}{\beta_k - \beta_1}$$

Since there are only finitely many such equations and L_1 is infinite, we may choose $c \neq 0$ in L_1, not equal to any of these x's, and then

$$\alpha_i + c\beta_k \neq \alpha_1 + c\beta_1$$

for $1 \le i \le n$, $2 \le k \le m$. Define

$$\theta = \alpha + c\beta$$

We prove that $L_1(\theta) = L_1(\alpha, \beta)$. Obviously $L_1(\theta) \subseteq L_1(\alpha, \beta)$, and it suffices to prove that $\beta \in L_1(\theta)$ since $\alpha = \theta - c\beta$.

To do so, observe that

$$p(\theta - c\beta) = p(\alpha) = 0$$

which suggests defining the polynomial

$$r(t) = p(\theta - ct) \in L_1(\theta)[t]$$

Now β is a zero of both $q(t)$ and $r(t)$ as polynomials over $L_1(\theta)$. These polynomials have only one common zero, for if $q(\xi) = r(\xi) = 0$ then ξ is one of $\beta_1, \ldots, \beta_m$ and also $\theta - c\xi$ is one of $\alpha_1, \ldots, \alpha_n$. Our choice of c forces $\xi = \beta$. Let $h(t)$ be the minimal polynomial of β over $L_1(\theta)$. Then $h(t)|q(t)$ and $h(t)|r(t)$. Since q and r have just one common zero in $\mathbb{C}$ we have $\partial h = 1$, so

$$h(t) = t + \mu$$

for $\mu \in L_1(\theta)$. Now $0 = h(\beta) = \beta + \mu$ so that $\beta = -\mu \in L_1(\theta)$ as required. $\square$

Example 2.6. $\mathbb{Q}(\sqrt{2}, \sqrt[3]{5})$. We have

$$\alpha_1 = \sqrt{2} \qquad \alpha_2 = -\sqrt{2}$$

$$\beta_1 = \sqrt[3]{5} \qquad \beta_2 = \omega\sqrt[3]{5} \qquad \beta_3 = \omega^2\sqrt[3]{5}$$

where

$$\omega = \tfrac{1}{2}(-1 + \mathrm{i}\sqrt{3})$$

is a complex cube root of 1. The number $c = 1$ satisfies

$$\alpha_i + c\beta_k \neq \alpha + c\beta$$

for $i = 1, 2$, $k = 2, 3$; since the number on the left is not real in any of the four cases, whereas that on the right is. Hence $\mathbb{Q}(\sqrt{2}, \sqrt[3]{5}) = \mathbb{Q}(\sqrt{2} + \sqrt[3]{5})$.

The expression of K as $\mathbb{Q}(\theta)$ is, of course, not unique; for $\mathbb{Q}(\theta) = \mathbb{Q}(-\theta) = \mathbb{Q}(\theta + 1) = \ldots$ and so on.

2.2 Conjugates

Recall from Section 1.1 that a ring monomorphism is an injective homomorphism $\phi : R \to S$, where R and S are rings. Injectivity is equivalent to $\ker \phi = 0$. We are especially interested in the case where R and S are fields. In this case, since $\ker \phi$ is an ideal of R, it is either the whole of R or zero. Thus a field monomorphism is a ring homomorphism that is not identically zero.

If $K = \mathbb{Q}(\theta)$ is a number field there are, in general, several distinct monomorphisms $\sigma : K \to \mathbb{C}$. For instance, if $K = \mathbb{Q}(\mathrm{i})$ the possibilities are

$$\begin{aligned} \sigma_1(x + \mathrm{i}y) &= x + \mathrm{i}y \\ \sigma_2(x + \mathrm{i}y) &= x - \mathrm{i}y \end{aligned}$$

for $x, y \in \mathbb{Q}$. The full set of such monomorphisms plays a fundamental role in the theory, so we begin with a description.

Theorem 2.7. *Let $K = \mathbb{Q}(\theta)$ be a number field of degree n over $\mathbb{Q}$. Then there are exactly n distinct monomorphisms $\sigma_i : K \to \mathbb{C}$ $(i = 1, \ldots, n)$. The elements $\sigma_i(\theta) = \theta_i$ are the distinct zeros in $\mathbb{C}$ of the minimal polynomial of θ over $\mathbb{Q}$.*

Proof: Let $\theta_1, \ldots, \theta_n$ be the zeros of the minimal polynomial p of θ. By Corollary 1.4 the θ_i are distinct. Each θ_i also has minimal polynomial p because it must divide p, and p is irreducible.

By Corollary 1.20, there is a unique field isomorphism $\sigma_i : \mathbb{Q}(\theta) \to \mathbb{Q}(\theta_i)$ such that $\sigma_i(\theta) = \theta_i$. In fact, if $\alpha \in \mathbb{Q}(\theta)$ then $\alpha = r(\theta)$ for a unique $r \in \mathbb{Q}[t]$ with $\partial r < n$, and

$$\sigma_i(\alpha) = r(\theta_i) \tag{2.1}$$

Conversely if $\sigma : K \to \mathbb{C}$ is a monomorphism then σ is the identity on $\mathbb{Q}$. Now

$$0 = \sigma(p(\theta)) = p(\sigma(\theta))$$

so $\sigma(\theta)$ is one of the θ_i, hence σ is one of the σ_i. □

Definition 2.8. With the above notation let $\alpha \in K = \mathbb{Q}(\theta)$. The *field polynomial* of α over K is

$$c_\alpha(t) = \prod_{i=1}^{n} (t - \sigma_i(\alpha)) \tag{2.2}$$

(Another term is *characteristic polynomial.*)

As it stands, the field polynomial is in $K[t]$. In fact more is true:

Theorem 2.9. *The coefficients of the field polynomial are rational numbers, so that* $c_\alpha(t) \in \mathbb{Q}[t]$.

Proof: We have $\alpha = r(\theta)$ for $r \in \mathbb{Q}[t]$, $\partial r < n$. By (2.1) the field polynomial (2.2) takes the form

$$c_\alpha(t) = \prod_i (t - r(\theta_i))$$

where the θ_i run through all zeros of the minimal polynomial p of θ, whose coefficients are in $\mathbb{Q}$. It is easy to see that the coefficients of $c_\alpha(t)$ are of the form $h(\theta_1, \ldots, \theta_n)$ where $h(t_1, \ldots, t_n)$ is a symmetric polynomial in $\mathbb{Q}[t_1, \ldots, t_n]$. Now use Corollary 1.23. □

Definition 2.10. The elements $\sigma_i(\alpha)$ for $i = 1, \ldots, n$ are the *K-conjugates* of α.

Although the θ_i are distinct (and are the K-conjugates of θ) it is not always the case that the K-conjugates of α are distinct: for instance $\sigma_i(1) = 1$ for all i. The precise situation is given by:

Theorem 2.11. *With the above notation,*

(a) *The field polynomial c_α is a power of the minimal polynomial p_α.*

(b) *The K-conjugates of α are the zeros of p_α in $\mathbb{C}$, each repeated n/m times where $m = \partial p_\alpha$ is a divisor of n.*

(c) *The element $\alpha \in \mathbb{Q}$ if and only if all of its K-conjugates are equal.*

(d) $\mathbb{Q}(\alpha) = \mathbb{Q}(\theta)$ *if and only if all K-conjugates of α are distinct.*

Proof: The main point is (a). Now $q = p_\alpha$ is irreducible and α is a zero of $c = c_\alpha$, so $c = q^s h$ where q and h are coprime and both are monic. (This follows by factorising c into irreducibles.) We claim that h is constant. If not, some $\alpha_i = \sigma_i(\alpha) = r(\theta_i)$ is a zero of h, where $\alpha = r(\theta)$. Therefore if $g(t) = h(r(t))$ then $g(\theta_i) = 0$. Let p be the minimal polynomial of θ over $\mathbb{Q}$, hence also of each θ_i. Then $p|g$, so that $g(\theta_j) = 0$ for all j, and in particular $g(\theta) = 0$. Therefore, $h(\alpha) = h(r(\theta)) = g(\theta) = 0$, so q divides h, a contradiction. Hence h is constant and monic, so $h = 1$ and $f = q^s$.

Part (b) is an immediate consequence of (a) by the definition of the field polynomial.

To prove (c), it is clear that $\alpha \in \mathbb{Q}$ implies $\sigma_i(\alpha) \in \mathbb{Q}$. Conversely, if all $\sigma_i(\alpha)$ are equal then since the zeros of $q = p_\alpha$ are distinct and $c_\alpha = q^s$, we have $\partial q = 1$ so $\alpha \in \mathbb{Q}$.

Finally for (d): if all $\sigma_i(\alpha)$ are distinct then $\partial p_\alpha = n$, so $[\mathbb{Q}(\alpha) : \mathbb{Q}] = n = [\mathbb{Q}(\theta) : \mathbb{Q}]$. Thus $\mathbb{Q}(\alpha) = \mathbb{Q}(\theta)$. Conversely if $\mathbb{Q}(\alpha) = \mathbb{Q}(\theta)$ then $\partial p_\alpha = n$ so the $\sigma_i(\alpha)$ are distinct. □

Warning. The K-conjugates of α need not be elements of K. Even the θ_i need not be elements of K. For example, let θ be the real cube root of 2. Then $\mathbb{Q}(\theta)$ is a subfield of $\mathbb{R}$. The K-conjugates of θ, however, are θ, $\omega\theta$, $\omega^2\theta$, where $\omega = \frac{1}{2}(-1+i\sqrt{3})$. The last two of these are nonreal, hence they do not lie in $\mathbb{Q}(\theta)$.

2.3 Discriminants

We now discuss an important concept in algebraic number theory: the discriminant. It is a numerical invariant of any number field, and its significance becomes clear as we proceed. We work up to its definition in two stages, starting with the discriminant of a $\mathbb{Q}$-basis. We return to the topic in Section 2.4, where we define the discriminant of a number field.

Definition 2.12. Let $K = \mathbb{Q}(\theta)$ have degree n, and let $\{\alpha_i, \ldots, \alpha_n\}$ be a basis of K over $\mathbb{Q}$. The *discriminant* of this basis is

$$\Delta[\alpha_1, \ldots, \alpha_n] = (\det[\sigma_i(\alpha_j)])^2 \tag{2.3}$$

We examine how the discriminant behaves under a change of basis. If $\{\beta_1, \ldots, \beta_n\}$ is another basis then

$$\beta_k = \sum_{i=1}^{n} c_{ik}\alpha_i \qquad (c_{ik} \in \mathbb{Q})$$

for $k = 1, \ldots, n$, and $\det(c_{ik}) \neq 0$. The σ_i are monomorphisms, hence the identity on $\mathbb{Q}$, so the product formula for determinants shows that

$$\Delta[\beta_1, \ldots, \beta_n] = [\det(c_{ik})]^2 \Delta[\alpha_1, \ldots, \alpha_n] \tag{2.4}$$

For immediate use, we recall a well-known concept from basic algebra:

Definition 2.13. A determinant of the form $D = \det(t_i^j)$ is a *Vandermonde* determinant.

Proposition 2.14. *The Vandermonde determinant* $D = \det(t_i^j)$ *has value*

$$D = \prod_{1 \le i < j \le n} (t_i - t_j) \tag{2.5}$$

Proof: To see this, think of everything as lying inside $\mathbb{Q}[t_1, \ldots, t_n]$. Then for $t_i = t_j$ the determinant has two equal rows, so it vanishes. Hence D is divisible by each $(t_i - t_j)$. To avoid repeating such a factor twice, we take $i < j$. Then comparison of degrees easily shows that D has no other nonconstant factors; comparing coefficients of $t_1 t_2^2 \ldots t_n^n$ gives (2.5). □

Historical Curiosity

What today we call the Vandermonde determinant does not appear in any of Vandermonde's papers. Lebesgue [77] pp. 30–31 writes, 'D'ou vient donc la dénomination: déterminant de Vandermonde?' (where does the name 'Vandermonde determinant' come from?) and traces this to a misreading of a system of linear equations in unknowns $\zeta 1, \zeta 2, \zeta 3 \ldots$ occurring in [139]. Today we would write $\zeta_1, \zeta_2, \zeta_3 \ldots$, which are just n distinct unknowns, but the symbol ζn was misinterpreted as ζ^n instead of ζ_n. Lebesgue concludes that '... le determinant de Vandermonde n'est pas de Vandermonde'.

Theorem 2.15. *The discriminant of any basis for $K = \mathbb{Q}(\theta)$ is rational and nonzero. If all K-conjugates of θ are real then the discriminant of any basis is positive.*

Proof: Pick a basis that makes computations straightforward: the obvious one is $\{1, \theta, \ldots, \theta^{n-1}\}$. If the conjugates of θ are $\theta_1, \ldots, \theta_n$ then

$$\Delta[1, \theta, \ldots, \theta^{n-1}] = (\det \theta_i^j)^2$$

By Proposition 2.14,

$$\Delta = \Delta[1, \theta, \ldots, \theta^{n-1}] = \left(\prod(\theta_i - \theta_j)\right)^2$$

Now D is antisymmetric in the t_i, so D^2 is symmetric. By the usual argument about symmetric polynomials, Corollary 1.23, Δ is rational. Since the θ_i are distinct, $\Delta \neq 0$.

Now let $\{\beta_1, \ldots, \beta_n\}$ be any basis. Then

$$\Delta[\beta_1, \ldots, \beta_n] = (\det c_{ik})^2 \Delta$$

for certain rational numbers c_{ik}, and $\det(c_{ik}) \neq 0$, so

$$\Delta[\beta_1, \ldots \beta_n] \neq 0$$

and is rational. Clearly if all θ_i are real then Δ is a positive real number, hence so is $\Delta[\beta_1, \ldots, \beta_n]$. □

With the above notation, Δ vanishes if and only if some θ_i is equal to another θ_j. Hence the nonvanishing of Δ lets us 'discriminate' among the θ_i, which motivates calling Δ the discriminant.

2.4 Algebraic Integers

Just as the rational numbers $\mathbb{Q}$ are intimately related to the integers $\mathbb{Z}$, so every number field K has a distinguished subring of algebraic integers. Although $K = \mathbb{Q}(\theta)$ for suitable θ, this ring need not equal $\mathbb{Z}[\theta]$. The definition is subtler:

Definition 2.16. A complex number θ is an *algebraic integer* if there is a *monic* polynomial $p(t)$ with integer coefficients such that $p(\theta) = 0$. In other words,

$$\theta^n + a_{n-1}\theta^{n-1} + \cdots + a_0 = 0$$

where $a_i \in \mathbb{Z}$ for all i.

For example, $\theta = \sqrt{-2}$ is an algebraic integer, since $\theta^2 + 2 = 0$. More surprisingly, $\tau = \frac{1}{2}(1 + \sqrt{5})$ is also an algebraic integer, since $\tau^2 - \tau - 1 = 0$. On the other hand, $\phi = 22/7$ is not an algebraic integer. It satisfies equations over $\mathbb{Z}$ like $7\phi - 22 = 0$, but this is not monic; or like $\phi - 22/7 = 0$, whose coefficients are not integers; but it can be shown without difficulty that ϕ does not satisfy any monic polynomial equation with integer coefficients.

Definition 2.17. We write $\mathbb{B}$ for the set of algebraic integers.

One of our aims is to prove that $\mathbb{B}$ is a subring of $\mathbb{A}$. We prepare for this by proving:

Lemma 2.18. *A complex number θ is an algebraic integer if and only if the additive group generated by all powers $1, \theta, \theta^2, \ldots$ is finitely generated.*

Proof: If θ is an algebraic integer, then for some n

$$\theta^n + a_{n-1}\theta^{n-1} + \cdots + a_0 = 0 \tag{2.6}$$

where the $a_i \in \mathbb{Z}$. We claim that every power of θ lies in the additive group generated by $1, \theta, \ldots, \theta^{n-1}$. Call this group Γ. Then (2.6) shows that $\theta^n \in \Gamma$. Inductively, if $m \geq n$ and $\theta^m \in \Gamma$ then

$$\theta^{m+1} = \theta^{m+1-n}\theta^n = \theta^{m+1-n}(-a_{n-1}\theta^{n-1} - \cdots - a_0) \in \Gamma$$

This proves that every power of θ lies in Γ, which gives one implication.

For the converse, suppose that every power of θ lies in a finitely generated additive group G. The subgroup Γ of G generated by the powers $1, \theta, \theta^2, \ldots$ must also be finitely generated by Proposition 1.32. Let $v_1, \ldots, v_n$ be generators. Each v_i is a polynomial in θ with integer coefficients, so θv_i is also such a polynomial. Hence there exist integers b_{ij} such that

$$\theta v_i = \sum_{j=1}^{n} b_{ij} v_j$$

This leads to a system of homogeneous equations for the v_i of the form

$$\begin{aligned}
(b_{11} - \theta)v_1 + b_{12}v_2 + \cdots + b_{1n}v_n &= 0 \\
b_{21}v_1 + (b_{22} - \theta)v_2 + \cdots + b_{2n}v_n &= 0 \\
&\vdots \\
b_{n1}v_1 + b_{n2}v_2 + \cdots + (b_{nn} - \theta)v_n &= 0
\end{aligned}$$

Since there exists a solution $v_1, \ldots, v_n \in \mathbb{C}$, not all zero, the determinant

$$\begin{vmatrix}
b_{11} - \theta & b_{12} & \ldots & b_{1n} \\
b_{21} & b_{22} - \theta & \ldots & b_{2n} \\
\vdots & \vdots & \ddots & \vdots \\
b_{n1} & b_{n2} & \ldots & b_{nn} - \theta
\end{vmatrix}$$

is zero. Expand to see that θ satisfies a monic polynomial equation with integer coefficients. □

Theorem 2.19. *The algebraic integers form a subring of the field of algebraic numbers.*

Proof: Let $\theta, \phi \in \mathbb{B}$. We have to show that $\phi + \theta$ and $\theta\phi \in \mathbb{B}$. By Lemma 2.18 all powers of θ lie in a finitely generated additive subgroup Γ_θ of $\mathbb{C}$, and all powers of ϕ lie in a finitely generated additive subgroup Γ_ϕ. But now all powers of $\theta+\phi$ and of $\theta\phi$ are integer linear combinations of elements $\theta^i\phi^j$ which lie in $\Gamma_\theta\Gamma_\phi \subseteq \mathbb{C}$. If Γ_θ has generators $v_1, \ldots, v_n$ and Γ_ϕ has generators $w_1, \ldots, w_m$, then $\Gamma_\theta\Gamma_\phi$ is the additive group generated by all $v_i w_j$ for $1 \leq i \leq n$, $1 \leq j \leq m$. Hence all powers of $\theta + \phi$ and of $\theta\phi$ lie in a finitely generated additive subgroup of $\mathbb{C}$, so by Lemma 2.18 $\theta + \phi$ and $\theta\phi$ are algebraic integers. Hence $\mathbb{B}$ is a subring of $\mathbb{A}$. □

A simple extension of this technique lets us prove a useful theorem:

Theorem 2.20. *Let θ be a complex number satisfying a monic polynomial equation whose coefficients are algebraic integers. Then θ is an algebraic integer.*

Proof: Suppose that

$$\theta^n + \psi_{n-1}\theta^{n-1} + \cdots + \psi_0 = 0$$

where $\psi_0, \ldots, \psi_{n-1} \in \mathbb{B}$. The ψ_i generate a subring Ψ of $\mathbb{B}$. The argument of Lemma 2.18 shows that all powers of θ lie inside a finitely generated Ψ-submodule M of $\mathbb{C}$, spanned by $1, \theta, \ldots, \theta^{n-1}$. By Theorem 2.19, the ψ_i and all of their powers lie inside a finitely generated additive group Γ_i with generators γ_{ij} $(1 \leq j \leq n_i)$. Therefore M lies inside the additive group generated by all elements

$$\gamma_{1j_1}, \gamma_{2j_2}, \ldots, \gamma_{n-1,j_{n-1}}\theta^k$$

$(1 \leq j_i \leq n_i,\ 0 \leq i \leq n-1,\ 0 \leq k \leq n-1)$, which is a finite set. So M is finitely generated as an additive group. □

Theorems 2.19 and 2.20 let us construct many new algebraic integers from known ones. For instance, $\sqrt{2}$ and $\sqrt{3}$ are clearly algebraic integers. Then Theorem 2.19 says that numbers such as $\sqrt{2} + \sqrt{3}$, $7\sqrt{2} - 41\sqrt{3}$, $(\sqrt{2})^5(1 + \sqrt{3})^2$ are also algebraic integers. And Theorem 2.20 says that zeros of polynomials such as

$$t^{23} - (14 + \sqrt[5]{3})t^9 + (\sqrt[3]{2})t^5 - 19\sqrt{3}$$

are algebraic integers. It would not be easy, particularly in the last instance, to compute explicit polynomials over $\mathbb{Z}$ of which these algebraic integers are zeros; although it can in principle be done by using symmetric polynomials. In fact Theorems 2.19 and 2.20 can be proved this way.

2.5 Ring of Integers of a Number Field

The ideas that we now discuss are absolutely central to the entire book.

Definition 2.21. For any number field K, the ring

$$\mathfrak{O} = K \cap \mathbb{B}$$

is the *ring of algebraic integers* of K, or more briefly the *ring of integers* of K.

If it is not immediately clear which number field is involved, we write this ring more explicitly as $\mathfrak{O}_K$.

The symbol '$\mathfrak{O}$' is traditional but confusing. It looks a bit like a letter D, but actually it is a Gothic (often called 'Fraktur') capital O (for 'order', the old terminology, in German). Since K and $\mathbb{B}$ are subrings of $\mathbb{C}$ it follows that $\mathfrak{O}$ is a subring of K. Further $\mathbb{Z} \subseteq \mathbb{Q} \subseteq K$ and $\mathbb{Z} \subseteq \mathbb{B}$ so $\mathbb{Z} \subseteq \mathfrak{O}$.

The next lemma is easy to prove:

Lemma 2.22. *If $\alpha \in K$ then $c\alpha \in \mathfrak{O}$ for some nonzero $c \in \mathbb{Z}$.*

Proof: Let m be the minimal polynomial of α over $\mathbb{Q}$. Then

$$\alpha^m + a_{m-1}\alpha^{m-1} + \cdots + a_0 = 0$$

with all $a_i \in \mathbb{Q}$. Now

$$(c\alpha)^m + ca_{m-1}(c\alpha)^{m-1} + \cdots + c^m a_0 = 0$$

Choose c to be the product of the denominators of the a_i. Then $c \in \mathbb{Z}$ and $c^i a_{m-i} \in \mathbb{Z}$ for $0 \leq i \leq m-1$, so $c\alpha$ is a zero of a monic polynomial over $\mathbb{Z}$. □

Corollary 2.23. *If K is a number field then $K = \mathbb{Q}(\theta)$ for an algebraic integer θ.*

Proof: By Corollary 2.23 and Theorem 2.5, $K = \mathbb{Q}(\phi)$ for an algebraic number ϕ. By Lemma 2.22, $\theta = c\phi$ is an algebraic integer for some $0 \neq c \in \mathbb{Z}$. Clearly $\mathbb{Q}(\phi) = \mathbb{Q}(\theta)$. □

Warning For $\theta \in \mathbb{C}$, write $\mathbb{Z}[\theta]$ for the set of elements $p(\theta)$ for polynomials $p \in \mathbb{Z}[t]$. If $K = \mathbb{Q}(\theta)$ where θ is an algebraic integer then certainly $\mathfrak{O}$ contains $\mathbb{Z}[\theta]$ since $\mathfrak{O}$ is a ring containing θ. However, $\mathfrak{O}$ need not equal $\mathbb{Z}[\theta]$. For example, $\mathbb{Q}(\sqrt{5})$ is a number field and $\sqrt{5}$ an algebraic integer. But, as we have seen,

$$\tau = \frac{1+\sqrt{5}}{2}$$

is a zero of $t^2 - t - 1$, hence an algebraic integer. Moreover, τ lies in $\mathbb{Q}(\sqrt{5})$ so it belongs to $\mathfrak{O}$. It does not belong to $\mathbb{Z}[\sqrt{5}]$.

There is a useful criterion, in terms of the minimal polynomial, for a number to be an algebraic integer:

Lemma 2.24. *An algebraic number α is an algebraic integer if and only if its minimal polynomial over $\mathbb{Q}$ has coefficients in $\mathbb{Z}$.*

Proof: Let p be the minimal polynomial of α over $\mathbb{Q}$, and recall that this is monic and irreducible in $\mathbb{Q}[t]$. If $p \in \mathbb{Z}[t]$ then α is an algebraic integer. Conversely, if α is an algebraic integer then $q(\alpha) = 0$ for some monic $q \in \mathbb{Z}[t]$, and $p|q$. Now $p \in \mathbb{Z}[t]$ by Corollary 1.9. □

When confusion about the word 'integer' might occur, we adopt the following convention: a *rational integer* is an element of $\mathbb{Z}$, and a plain *integer* is an algebraic integer. (The aim is to reserve the shorter term for the concept most often encountered.) Any remaining possibility of confusion is eliminated by:

Lemma 2.25. *An algebraic integer is a rational number if and only if it is a rational integer. Equivalently, $\mathbb{B} \cap \mathbb{Q} = \mathbb{Z}$.*

Proof: Clearly $\mathbb{Z} \subseteq \mathbb{B} \cap \mathbb{Q}$. Let $\alpha \in \mathbb{B} \cap \mathbb{Q}$; since $\alpha \in \mathbb{Q}$ its minimal polynomial over $\mathbb{Q}$ is $t - \alpha$. By Lemma 2.24 the coefficients of this are in $\mathbb{Z}$, hence $-\alpha \in \mathbb{Z}$, hence $\alpha \in \mathbb{Z}$. □

2.6 Integral Bases

Let K be a number field of degree n (over $\mathbb{Q}$). A *basis* (or $\mathbb{Q}$-*basis* for emphasis) of K is a basis for K as a vector space over $\mathbb{Q}$. Theorem 2.5 shows that $K = \mathbb{Q}(\theta)$ where θ is an algebraic integer, so the minimal polynomial p of θ has degree n and $\{1, \theta, \ldots, \theta^{n-1}\}$ is a basis for K.

Definition 2.26. The ring $\mathfrak{O}$ of integers of K is an abelian group under addition; for clarity we denote this group by $(\mathfrak{O}, +)$.

An *integral basis* for K (or for $\mathfrak{O}$) is a $\mathbb{Z}$-basis for $(\mathfrak{O}, +)$.

Thus $\{\alpha_1, \ldots, \alpha_s,\}$ is an integral basis if and only if all $\alpha_i \in \mathfrak{O}$ and every element of $\mathfrak{O}$ is *uniquely* expressible in the form

$$a_1\alpha_1 + \cdots + a_s\alpha_s$$

for rational integers $a_1, \ldots, a_s$. It is obvious from Lemma 2.22 that any integral basis for K is a $\mathbb{Q}$-basis, so $s = n$.

We must verify that integral bases exist. In fact they do, but they are not always what naively we might expect them to be. In particular, a basis consisting of (algebraic) integers need not be an integral basis.

For instance, $K = \mathbb{Q}[\theta]$ (which equals $\mathbb{Q}(\theta)$) for an algebraic integer θ by Corollary 2.23, so $\{1, \theta, \ldots, \theta^{n-1}\}$ is a $\mathbb{Q}$-basis for K which consists of integers. However, it does *not* follow that $\{1, \theta, \ldots, \theta^{n-1}\}$ is an integral basis, because some elements in $\mathbb{Q}[\theta]$ with non-integer coefficients may also be (algebraic) integers. As an example, we again consider $K = \mathbb{Q}(\sqrt{5})$. We saw that $\tau = \frac{1}{2} + \frac{1}{2}\sqrt{5}$ satisfies the equation $t^2 - t + 1 = 0$, so τ is an integer in $\mathbb{Q}(\sqrt{5})$, but it is not an element of $\mathbb{Z}[\sqrt{5}]$.

Our first problem, therefore, is to show that integral bases exist. That they do is equivalent to the statement that $(\mathfrak{O}, +)$ is a free abelian group of rank n. To prove this we first establish:

Lemma 2.27. *If $\{\alpha_1, \ldots, \alpha_n\}$ is a basis of K consisting of integers, then the discriminant $\Delta[\alpha_1, \ldots, \alpha_n]$ is a rational integer, not equal to zero.*

Proof: By Theorem 2.15, $\Delta = \Delta[\alpha_1, \ldots, \alpha_n]$ is rational. It is an integer since the α_i are. Hence by Lemma 2.25 it is a rational integer. By Theorem 2.15, $\Delta \neq 0$. □

Theorem 2.28. *Every number field K possesses an integral basis, and the additive group of $\mathfrak{O}$ is free abelian of rank n equal to the degree of K.*

Proof: By the Primitive Element Theorem 2.5, $K = \mathbb{Q}(\theta)$ for θ is an integer. Hence there exist bases for K consisting of integers, for example $\{1, \theta, \ldots, \theta^{n-1}\}$. We have already seen that such $\mathbb{Q}$-bases need not be integral bases. However, the discriminant of a $\mathbb{Q}$-basis consisting of integers is always a rational integer (Lemma 2.27), so what we do is to select a basis $\{\omega_1, \ldots, \omega_n\}$ of integers for which

$$|\Delta[\omega_1, \ldots, \omega_n]|$$

is least. We claim that this is in fact an integral basis. If not, there is an integer ω of K such that

$$\omega = a_1\omega_1 + \cdots + a_n\omega_n$$

for $a_i \in \mathbb{Q}$, *not all in* $\mathbb{Z}$. Choose the numbering so that $a_1 \notin \mathbb{Z}$. Then $a_1 = a + r$ where $a \in \mathbb{Z}$ and $0 < r < 1$. Define

$$\psi_1 = \omega - a\omega_1 \qquad \psi_i = \omega_i \qquad (i = 2, \ldots, n)$$

Then $\{\psi_1, \ldots, \psi_n\}$ is a basis consisting of integers. The determinant relevant to the change of basis from the ω's to the ψ's is

$$\begin{vmatrix} a_1 - a & a_2 & a_3 & \dots & a_n \\ 0 & 1 & 0 & \dots & 0 \\ 0 & 0 & 1 & \dots & 0 \\ \vdots & \vdots & \vdots & \ddots & \vdots \\ 0 & 0 & 0 & \dots & 1 \end{vmatrix} = r$$

so

$$\Delta[\psi_1, \dots, \psi_n] = r^2 \Delta[\omega_1, \dots, \omega_n]$$

Since $0 < r < 1$ this contradicts the choice of $\{\omega_1, \dots, \omega_n\}$ making $|\Delta[\omega_1, \dots, \omega_n]|$ minimal.

Therefore $\{\omega_1, \dots, \omega_n\}$ is an integral basis, so $(\mathfrak{O}, +)$ is free abelian of rank n. □

This raises the question of *finding* integral bases in cases such as $\mathbb{Q}(\sqrt{5})$ where the $\mathbb{Q}$-basis $\{1, \sqrt{5}\}$ is not an integral basis. We consider a more general case in Chapter 3, but this particular example is worth a brief discussion.

A general element of $\mathbb{Q}(\sqrt{5})$ has the form $p + q\sqrt{5}$ for $p, q \in \mathbb{Q}$, with minimal polynomial

$$(t - p - q\sqrt{5})(t - p + q\sqrt{5}) = t^2 - 2pt + (p^2 - 5q^2)$$

Then $p + q\sqrt{5}$ is an integer if and only if the coefficients $2p, p^2 - 5q^2$ are rational integers. Thus $p = \frac{1}{2}P$ where P is a rational integer. For P even, we have p^2 a rational integer, so $5q^2$ is a rational integer also, implying q is a rational integer. For P odd, a straightforward calculation (performed in the next chapter in greater generality) shows $q = \frac{1}{2}Q$ where Q is also an odd rational integer.

This implies that $\mathfrak{O} = \mathbb{Z}[\frac{1}{2} + \frac{1}{2}\sqrt{5}]$, so an integral basis is $\{1, \frac{1}{2} + \frac{1}{2}\sqrt{5}\}$.

We can prove this by another route using the discriminant. First, we need a definition and a theorem:

Definition 2.29. A rational integer is *squarefree* if it is not divisible by the square of a prime.

For example, 5 is squarefree, as are 6, 7, but not 8 or 9. Given a $\mathbb{Q}$-basis of K consisting of integers, we compute the discriminant and then we have:

Theorem 2.30. *Suppose that $\alpha_1, \dots, \alpha_n \in \mathfrak{O}$ form a $\mathbb{Q}$-basis for K. If $\Delta[\alpha_1, \dots, \alpha_n]$ is squarefree then $\{\alpha_1, \dots, \alpha_n\}$ is an integral basis.*

Proof: Let $\{\beta_1, \ldots, \beta_n\}$ be an integral basis. Then there exist rational integers c_{ij} such that $\alpha_i = \Sigma c_{ij}\beta_j$, and (2.4) implies that

$$\Delta[\alpha_1, \ldots, \alpha_n] = (\det c_{ij})^2 \Delta[\beta_1, \ldots, \beta_n]$$

Since the left-hand side is squarefree, $\det c_{ij} = \pm 1$, so (c_{ij}) is unimodular. Hence by Lemma 1.28 $\{\alpha_1, \ldots, \alpha_n\}$ is a $\mathbb{Z}$-basis for $\mathfrak{O}$, that is, an integral basis for K. □

The two monomorphisms $\mathbb{Q}(\sqrt{5}) \to \mathbb{C}$ are

$$\begin{aligned} \sigma_1(p + q\sqrt{5}) &= p + q\sqrt{5} \\ \sigma_2(p + q\sqrt{5}) &= p - q\sqrt{5} \end{aligned}$$

Hence the discriminant $\Delta[1, \frac{1}{2} + \frac{1}{2}\sqrt{5}]$ is

$$\begin{vmatrix} 1 & \frac{1}{2} + \frac{1}{2}\sqrt{5} \\ 1 & \frac{1}{2} - \frac{1}{2}\sqrt{5} \end{vmatrix}^2 = 5$$

Since 5 is squarefree, the $\mathbb{Q}$-basis $\{1, \frac{1}{2} + \frac{1}{2}\sqrt{5}\}$ for $\mathbb{Q}(\sqrt{5})$ is an integral basis.

Later we show that there exist integral bases whose discriminants are not squarefree, so the converse of Theorem 2.30 is false.

For two integral bases $\{\alpha_1, \ldots, \alpha_n\}$, $\{\beta_1, \ldots, \beta_n\}$ of an algebraic number field K, we have

$$\Delta[\alpha_1, \ldots, \alpha_n] = (\pm 1)^2 \Delta[\beta_1, \ldots, \beta_n] = \Delta[\beta_1, \ldots, \beta_n]$$

because the matrix corresponding to the change of basis is unimodular. Hence the discriminant of an integral basis is independent of which integral basis we choose.

Definition 2.31. The discriminant of any integral basis for K, which is independent of the basis, is the *discriminant of* K (or of $\mathfrak{O}$). It is always a nonzero rational integer. We denote it by d_K.

Obviously, isomorphic number fields have the same discriminant: it is an invariant of the field K and its ring of integers $\mathfrak{O}$. The vital role played by the discriminant will become apparent as the drama unfolds.

Two Advanced Results

We mention two major results about the discriminant whose proofs are outside the scope of this book. The first was proved by Ludwig Stickelberger [131].

Theorem 2.32. (Stickelberger's Theorem) *The discriminant of any number field is congruent to either* 0 *or* 1 (mod 4). □

There is another theorem with the same name, which asserts that the Stickelberger ideal of an abelian number field annihilates the class group of the field. The two are confused in some sources.

The next theorem is generally credited to Charles Hermite and Hermann Minkowski. However, Hermite's result [63] was published in 1857, and he proved only a special (and much simpler) theorem in which the degree of the number field is fixed. Later Minkowski proved that the degree is bounded by a function of the discriminant, which completes the proof.

Theorem 2.33. (Hermite–Minkowski Theorem) *For any positive integer N there are only finitely many number fields K such that the absolute value $|d_K|$ of the discriminant is less than or equal to N.* □

2.7 Norms and Traces

These important concepts often let us transform a problem about algebraic integers into one about rational integers. As usual, let $K = \mathbb{Q}(\theta)$ be a number field of degree n and let $\sigma_1, \dots, \sigma_n$ be the monomorphisms $K \to \mathbb{C}$.

Definition 2.34. For any $\alpha \in K$, the *norm* of α is

$$\mathrm{N}_K(\alpha) = \prod_{i=1}^{n} \sigma_i(\alpha)$$

and the *trace* of α is

$$\mathrm{T}_K(\alpha) = \sum_{i=1}^{n} \sigma_i(\alpha)$$

When the field K is clear from the context, we abbreviate the norm and trace of α to $\mathrm{N}(\alpha)$ and $\mathrm{T}(\alpha)$.

Proposition 2.35. (a) *If α is an integer then the norm and trace of α are rational integers.*

(b) *An element $\alpha \in K$ is an integer if and only if the field polynomial has rational integer coefficients.*

Proof: (a) The field polynomial is

$$c_\alpha(t) = \prod_{i=1}^{n}(t - \sigma_i(\alpha))$$

so the norm and trace are, up to sign, the coefficient of t^{n-1} and the constant term in c_α.

(b) The field polynomial is a power of the minimal polynomial by Theorem 2.11(a). Now use Lemma 2.24 and Gauss's Lemma 1.8. □

Since the σ_i are monomorphisms it is clear that

$$\mathrm{N}(\alpha\beta) = \mathrm{N}(\alpha)\mathrm{N}(\beta) \tag{2.7}$$

and if $\alpha \neq 0$ then $\mathrm{N}(\alpha) \neq 0$. If p, q are rational numbers then

$$\mathrm{T}(p\alpha + q\beta) = p\mathrm{T}(\alpha) + q\mathrm{T}(\beta). \tag{2.8}$$

For instance, if $K = \mathbb{Q}(\sqrt{7})$ then the integers of K are $\mathfrak{O} = \mathbb{Z}[\sqrt{7}]$, see Theorem 3.3. The maps σ_i are

$$\begin{aligned} \sigma_1(p + q\sqrt{7}) &= p + q\sqrt{7} \\ \sigma_2(p + q\sqrt{7}) &= p - q\sqrt{7} \end{aligned}$$

Hence

$$\begin{aligned} \mathrm{N}(p + q\sqrt{7}) &= p^2 - 7q^2 \\ \mathrm{T}(p + q\sqrt{7}) &= 2p \end{aligned}$$

Since norms are not too hard to compute (they can always be found from symmetric polynomial considerations, often with shortcuts) whereas discriminants involve complicated work with determinants, the following result is sometimes useful:

Proposition 2.36. *Let $K = \mathbb{Q}(\theta)$ be a number field where θ has minimal polynomial p of degree n. The $\mathbb{Q}$-basis $\{1, \theta, \ldots, \theta^{n-1}\}$ has discriminant*

$$\Delta\left[1, \ldots, \theta^{n-1}\right] = (-1)^{n(n-1)/2}\mathrm{N}(\mathrm{D}p(\theta))$$

where $\mathrm{D}p$ is the formal derivative of p.

Proof: The proof of Theorem 2.15 yields:

$$\Delta = \Delta[1, \theta, \ldots, \theta^{n-1}] = \prod_{1 \leq i < j \leq n} (\theta_i - \theta_j)^2$$

where $\theta_1, \ldots, \theta_n$ are the conjugates of θ. Now

$$p(t) = \prod_{i=1}^{n}(t - \theta_i)$$

so

$$\mathrm{D}p(t) = \sum_{j=1}^{n}\prod_{\substack{i=1 \\ i \neq j}}^{n}(t - \theta_i)$$

and therefore

$$\mathrm{D}p(\theta_j) = \prod_{\substack{i=1 \\ i \neq j}}^{n}(\theta_j - \theta_i)$$

Multiply all these equations together for $j = 1, \ldots, n$:

$$\prod_{j=1}^{n}\mathrm{D}p(\theta_j) = \prod_{\substack{i,j=1 \\ i \neq j}}^{n}(\theta_j - \theta_i)$$

The left-hand side is $\mathrm{N}(\mathrm{D}p(\theta))$. On the right, each factor $(\theta_i - \theta_j)$ for $i < j$ appears twice, once as $(\theta_i - \theta_j)$ and once as $(\theta_j - \theta_i)$. The product of these two factors is $-(\theta_i - \theta_j)^2$. Multiplying up, we get Δ multiplied by $(-1)^s$ where s is the number of pairs (i, j) with $1 \leq i < j \leq n$, namely $s = n(n-1)/2$. □

We close this section with a simple identity linking the discriminant and trace. It can be used as an alternative definition of the discriminant.

Proposition 2.37. *If $\{\alpha_1, \ldots, \alpha_n\}$ is any $\mathbb{Q}$-basis of K, then*

$$\Delta[\alpha_1, \ldots, \alpha_n] = \det(\mathrm{T}(\alpha_i\alpha_j))$$

Proof: $\mathrm{T}(\alpha_i\alpha_j) = \sum_{r=1}^{n}\sigma_r(\alpha_i\alpha_j) = \sum_{r=1}^{n}\sigma_r(\alpha_i)\sigma_r(\alpha_j)$. Hence

$$\begin{aligned}
\Delta[\alpha_1, \ldots, \alpha_n] &= (\det(\sigma_i(\alpha_j)))^2 \\
&= (\det(\sigma_j(\alpha_i)))(\det(\sigma_i(\alpha_j))) \\
&= \det\left(\sum_{r=1}^{n}\sigma_r(\alpha_i)\sigma_r(\alpha_j)\right) \\
&= \det(\mathrm{T}(\alpha_i\alpha_j))
\end{aligned}$$

□

2.8 Computing Rings of Integers

We now discuss how to find the ring of integers of a given number field. With the methods available to us, this involves moderately heavy calculations, but by taking advantage of shortcuts the technique can be made reasonably efficient. In particular Example 2.42 below shows that not every number field has an integral basis of the form $\{1, \theta, \ldots, \theta^{n-1}\}$.

The method is based on the following result:

Theorem 2.38. *Let G be an additive subgroup of $\mathfrak{O}$ of rank equal to the degree of K, with $\mathbb{Z}$-basis $\{\alpha_1, \ldots, \alpha_n\}$. Then $|\mathfrak{O}/G|^2$ divides $\Delta[\alpha_1, \ldots, \alpha_n]$.*

Proof: By Theorem 1.29 there exists a $\mathbb{Z}$-basis for $\mathfrak{O}$ of the form $\{\beta_1, \ldots, \beta_n\}$ such that G has a $\mathbb{Z}$-basis $\{\mu_1\beta_1, \ldots, \mu_n\beta_n\}$ for suitable $\mu_i \in \mathbb{Z}$. Now

$$\Delta[\alpha_1, \ldots, \alpha_n] = \Delta[\mu_1\beta_1, \ldots, \mu_n\beta_n]$$

since by Lemma 1.28 a basis change has a unimodular matrix. The right-hand side is

$$(\mu_1 \ldots \mu_n)^2 \Delta[\beta_1, \ldots, \beta_n] = (\mu_1 \ldots \mu_n)^2 \Delta$$

where Δ is the discriminant of K and so lies in $\mathbb{Z}$. But

$$|\mu_1 \ldots \mu_n| = |\mathfrak{O}/G|$$

by (1.3). Therefore

$$|\mathfrak{O}/G|^2 \text{ divides } \Delta\left[\alpha_1, \ldots, \alpha_n\right]$$

□

In the above situation we use the notation

$$\Delta_G = \Delta[\alpha_1, \ldots, \alpha_n]$$

We then have a generalisation of Theorem 2.30:

Proposition 2.39. *Suppose that $G \neq \mathfrak{O}$. Then there exists an algebraic integer $u \neq 0$ of the form*

$$u = \frac{1}{p}(\lambda_1\alpha_1 + \cdots + \lambda_n\alpha_n) \tag{2.9}$$

where $0 \le \lambda_i \le p-1$, $\lambda_i \in \mathbb{Z}$, and p is a prime such that p^2 divides Δ_G.

Proof: If $G \neq \mathfrak{O}$ then $|\mathfrak{O}/G| > 1$. Therefore (by the structure theory for finite abelian groups) there exists a prime p dividing $|\mathfrak{O}/G|$ and an element

$u \in \mathfrak{O}/G$ such that $g = pu \in G$. By Theorem 2.38, p^2 divides Δ_G. Further,

$$u = \frac{1}{p}g = \frac{1}{p}(\lambda_1\alpha_1 + \cdots + \lambda_n\alpha_n)$$

since $\{\alpha_i\}$ forms a $\mathbb{Z}$-basis for G. □

This really *is* a generalisation of Theorem 2.30: if Δ_G is squarefree then no such p exists, so $G = \mathfrak{O}$.

With a little care we can avoid recomputing the discriminant for a new basis, and just divide by p^2:

Lemma 2.40. *Let u be as in (2.9), and let i be such that $\lambda_i \neq 0$. Then there exists $v \in \mathfrak{O}$ such that*

$$\Delta[\alpha_1, \ldots, \alpha_{i-1}, v, \alpha_{i+1}, \ldots, \alpha_n] = \frac{1}{p^2}\Delta[\alpha_1, \ldots, \alpha_n] \tag{2.10}$$

Proof: Let u be as in (2.9). Without loss of generality we can renumber the indices so that $i = 1$ and $\lambda_1 \neq 0$. Then there exists $\mu \in \mathbb{Z}$ such that $\mu\lambda_1 \equiv 1 \pmod{p}$. Now let

$$v = \frac{1}{p}\alpha_1 + \frac{\mu\lambda_2}{p}\alpha_2 + \cdots + \frac{\mu\lambda_n}{p}\alpha_n \in \mathfrak{O}$$

The basis-change matrix from $\{\alpha_1, \ldots, \alpha_n\}$ to $\{\alpha_1, \ldots, \alpha_{i-1}, v, \alpha_{i+1}, \ldots, \alpha_n\}$ has the form

$$\begin{bmatrix} \frac{1}{p} & \frac{\mu\lambda_2}{p} & \ldots & \frac{\mu\lambda_n}{p} \\ 0 & 1 & \ldots & 0 \\ \vdots & \vdots & \ddots & \vdots \\ 0 & 0 & \ldots & 1 \end{bmatrix}$$

with determinant $\frac{1}{p}$. Now (2.4) implies (2.10). □

Reducing the coefficients $\mu\lambda_i$ modulo p we can also modify v, so that when $2 \leq i \leq n$ we have $\frac{\mu\lambda_i}{p} = \frac{\rho_i}{p}$, where $0 \leq \rho_i \leq p - 1$.

We may use Proposition 2.39 and Lemma 2.40 as the basis of a trial-and-error search for algebraic integers in $\mathfrak{O}$ but not in G, because there are only finitely many possibilities. The idea is:

(a) Start with an initial guess G for $\mathfrak{O}$.

(b) Compute Δ_G.

(c) For each prime p whose square divides Δ_G, test all numbers of the form (2.6) to see which are algebraic integers.

(d) If any new integers arise, enlarge G to a new G' by adding in a new integer as in Lemma 2.40; then divide $\Delta_G$by p^2 to obtain $\Delta_{G'}$.

(e) Repeat until no new algebraic integers are found.

Example 2.41. Find the ring of integers of $\mathbb{Q}(\sqrt[3]{2})$. Let $\theta \in \mathbb{R}$, $\theta^3 = 2$. The natural first guess is that $\mathfrak{O}$ has $\mathbb{Z}$-basis $\{1, \theta, \theta^2\}$. Let G be the abelian group generated by this set. Certainly $G \subseteq \mathfrak{O}$.

Let $\omega = e^{2\pi i/3}$ be a cube root of unity. A general element of G has the form

$$x_1 = a + b\theta + c\theta^2$$

with conjugates

$$\begin{aligned} x_2 &= a + b\omega\theta + c\omega^2\theta^2 \\ x_3 &= a + b\omega^2\theta + c\omega\theta^2 \end{aligned}$$

The trace of x_1 is

$$\mathrm{T}(x_1) = x_1 + x_2 + x_3 = 3a$$

and the norm (after some manipulation) computes as

$$\mathrm{N}(x_1) = x_1x_2x_3 = a^3 + 2b^3 + 4d^3 - 6abcd$$

If x_1 is an (algebraic) integer then both $\mathrm{T}(x_1)$ and $\mathrm{N}(x_1)$ must be rational integers.

The ring $\mathfrak{O}$ must also include

$$\begin{aligned} \theta x_1 &= 2c + a\theta + b\theta^2 \\ \theta^2 x_1 &= 2b + 2c\theta + a\theta^2 \end{aligned}$$

whose traces are $\mathrm{T}(x_2) = 6c$ and $\mathrm{T}(x_3) = 6b$. These must also be rational integers. Therefore $3a, 6b, 6c$ are rational integers. That is, there are rational integers p, q, r such that

$$a = \frac{p}{3} \qquad b = \frac{q}{6} \qquad c = \frac{r}{6}$$

Subtracting suitable integer multiples of $1, \theta$, and. θ^2 we may assume that $0 \leq p \leq 2$, $0 \leq q \leq 5$, and $0 \leq r \leq 5$. We therefore have at most $3.6.6 = 108$ cases to check. This number can be reduced: if x_1 is an integer then so is $2x_1$. So we can first check whether there are any new integers with p, q even. In other words, we can consider only the numbers

$$\frac{p}{3} + \frac{q}{3}\theta + \frac{r}{3}\theta^2$$

where p, q, r are integers equal to $0, 1$, or 2.

Now calculate the 27 possible norms (easy with computer algebra). It turns out that the only case where the norm is a rational integer is $p = q = r = 0$. That is, there are no new elements of $\mathfrak{O}$. Thus $\mathfrak{O} = G$, and $\mathfrak{O} = \mathbb{Z}[\sqrt[3]{2}]$.

Example 2.42. Find the ring of integers of $\mathbb{Q}(\sqrt[3]{5})$.

Let $\theta \in \mathbb{R}$, $\theta^3 = 5$. The natural first guess is that $\mathfrak{O}$ has $\mathbb{Z}$-basis $\{1, \theta, \theta^2\}$. Let G be the abelian group generated by this set. Let $\omega = e^{2\pi i/3}$ be a cube root of unity. Compute

$$\begin{aligned}
\Delta_G &= \begin{vmatrix} 1 & \theta & \theta^2 \\ 1 & \omega\theta & \omega^2\theta^2 \\ 1 & \omega^2\theta & \omega\theta^2 \end{vmatrix}^2 \\
&= \theta^6 \begin{vmatrix} 1 & 1 & 1 \\ 1 & \omega & \omega^2 \\ 1 & \omega^2 & \omega \end{vmatrix}^2 \\
&= 5^2 \cdot (\omega^2 + \omega^2 + \omega^2 - \omega - \omega - \omega)^2 \\
&= 5^2 \cdot 3^2 \cdot (\omega^2 - \omega)^2 \\
&= 3^2 \cdot 5^2 \cdot (-3) \\
&= -3^3 \cdot 5^2.
\end{aligned}$$

By Proposition 2.39 we must consider two possibilities:

(a) Can $\alpha = \frac{1}{3}(\lambda_1 + \lambda_2\theta + \lambda_3\theta^2)$ be an algebraic integer, for $0 \leq \lambda_i \leq 2$?

(b) Can $\alpha = \frac{1}{5}(\lambda_1 + \lambda_2\theta + \lambda_3\theta^2)$ be an algebraic integer, for $0 \leq \lambda_i \leq 4$?

Consider case (b), which is harder. First use the trace: we have

$$\mathrm{T}(\alpha) = 3\lambda_1/5 \in \mathbb{Z}$$

so that $\lambda_1 \in 5\mathbb{Z}$. Then $\alpha' = (\lambda_2\theta + \lambda_3\theta^2)/5$ is also an algebraic integer.

Now compute the norm of α'. (It is easier to do this for α' than for α because there are fewer terms, which is why we use the trace first.) We have

$$\begin{aligned}
\mathrm{N}(a\theta + b\theta^2) &= (a\theta + b\theta^2)(a\omega\theta + b\omega^2\theta^2)(a\omega^2\theta + b\omega\theta^2) \\
&= \omega \cdot \omega^2(a\theta + b\theta^2)(a\theta + \omega b\theta^2)(a\theta + \omega^2 b\theta^2) \\
&= (a\theta)^3 + (b\theta^2)^3 \\
&= 5a^3 + 25b^3
\end{aligned}$$

Thus in order for α to be an algebraic integer, we must have $N(\alpha') \in \mathbb{Z}$. But $N(\alpha') = (5\lambda_2^3 + 25\lambda_3^3)/125 = (\lambda_2^3 + 5\lambda_3^3)/25$. One way to finish the calculation is just to try all cases, see Table 2.1.

Table 2.1. Try all cases.

λ_2	λ_3	$\lambda_2^3 + 5\lambda_3^3$	Divisible by 25?
0	1	5	No
0	2	40	No
0	3	135	No
0	4	320	No
1	0	1	No
1	1	6	No
1	2	41	No
1	3	136	No
1	4	321	No
2	0	8	No
2	1	13	No
2	2	48	No
2	3	143	No
2	4	328	No
3	0	27	No
3	1	32	No
3	2	67	No
3	3	162	No
3	4	347	No
4	0	64	No
4	1	69	No
4	2	104	No
4	3	199	No
4	4	384	No

However, a little thought leads to a better idea. Suppose that $\lambda_2^3+5\lambda_3^3 \equiv 0 \pmod{25}$. If $\lambda_3 \equiv 0 \pmod 5$, then we must also have $\lambda_2 \equiv 0 \pmod 5$. If not, we have $5 \equiv (-\lambda_2/\lambda_3)^3 \pmod{25}$. Therefore 5 is a *cubic residue* (mod 25), that is, is congruent to a cube. The factor 5 shows that we must have $5 \equiv (5k)^3 \pmod{25}$, but then $5 \equiv 0 \pmod{25}$, an impossibility.

Whichever argument we use, we have shown that no new α' occurs in case (b). The analysis in case (a) is similar, and left as Exercise 2.6.

In order for α to be an algebraic integer, it is *necessary* for $\mathrm{N}(\alpha)$ and $\mathrm{T}(\alpha)$ to be rational integers, but these conditions need not be *sufficient.* If the use of norms and traces produces a candidate for a new algebraic integer, we still have to check that it is one—for example, by finding its minimal polynomial. However, our main use of $\mathrm{N}(\alpha)$ and $\mathrm{T}(\alpha)$ is to rule out possible candidates, so this step is not always needed.

Example 2.43. (a) Find the ring of integers of $\mathbb{Q}(\sqrt[3]{175})$.

(b) Show that it has no $\mathbb{Z}$-basis of the form $\{1, \theta, \theta^2\}$.

(a) Let $t = \sqrt[3]{175} = \sqrt[3]{(5^2 \cdot 7)}$. Consider also $u = \sqrt[3]{5 \cdot 7^2} = \sqrt[3]{245}$. Now

$$ut = 35 \qquad u^2 = 7t \qquad t^2 = 5u$$

Let $\mathfrak{O}$ be the ring of integers of $K = \mathbb{Q}(\sqrt[3]{175})$. We have $u = 35/t \in K$. But $u^3 - 245 = 0$ so $u \in \mathbb{B}$. Therefore $u \in \mathbb{B} \cap K = \mathfrak{O}$. A good initial guess is that $\mathfrak{O} = G$, where G is the abelian group generated by $\{1, t, u\}$.

To see if this is correct, compute Δ_G. The monomorphisms $K \to \mathbb{C}$ are $\sigma_1, \sigma_2, \sigma_3$ where $\sigma_1(t) = t$, $\sigma_2(t) = \omega t$, $\sigma_3(t) = \omega^3 t$. Since $tu = 35$, which must be fixed by each σ_i, we have $\sigma_1(u) = u$, $\sigma_2(u) = \omega^2 u$, $\sigma_3(u) = \omega u$. Therefore

$$\Delta_G = \begin{vmatrix} 1 & t & u \\ 1 & \omega t & \omega^2 u \\ 1 & \omega^2 t & \omega u \end{vmatrix}^2$$

which works out as $-3^3 \cdot 5^2 \cdot 7^2$.

There are now three primes to try: $p = 3, 5$, or 7.

If $p = 5$ or 7 then, as in Example 2.42, use of the trace lets us assume that our putative integer is $\frac{1}{p}(at + bu)$ for $a, b, \in \mathbb{Z}$. Now

$$\mathrm{N}(at + bu) = 175a^3 + 245b^3$$

and we must see whether this can be congruent to $0 \pmod{5^3 \text{ or } 7^3}$ for a, b not congruent to zero.

Suppose that $175a^3 + 245b^3 \equiv 0 \pmod{125}$, that is, $35a^3 + 49b^3 \equiv 0 \pmod{25}$. Write this as $10a^3 - b^3 \equiv 0 \pmod{25}$. If $a \equiv 0 \pmod 5$ then also $b \equiv 0 \pmod 5$. If not, $10 \equiv (b/a)^3 \pmod{25}$ is a cube $\pmod{25}$, but then $10 \equiv (5k)^3 \pmod{25}$, hence $10 \equiv 0 \pmod{25}$ which is absurd. The case $p = 7$ is dealt with in the same way.

When $p = 3$ the trace is no help, and we must compute the norm of

$$\frac{1}{3}(a + bt + cu)$$

for $a, b, c \in \mathbb{Z}$. The calculation is more complicated, but not too bad since we have to consider only $a, b, c = 0, 1, 2$. No new integers occur.

Therefore $\mathfrak{O} = G$ as hoped.

(b) Now we have to show that there is no $\mathbb{Z}$-basis of the form $\{1, \theta, \theta^2\}$, where $\theta = a + bt + cu$. The set $\{1, \theta, \theta^2\}$ is a $\mathbb{Z}$-basis if and only if $\{1, \theta + 1, (\theta + 1)^2\}$ is a $\mathbb{Z}$-basis, so without loss of generality we may assume that $a = 0$. Now

$$(bt + cu)^2 = b^2t^2 + 2bctu + c^2u^2 = 5b^2u + 70bc + 7c^2t$$

Therefore $\{1, bt + cu, (bt + cu)^2\}$ is a $\mathbb{Z}$-basis if and only if the matrix

$$\begin{vmatrix} 1 & 0 & 0 \\ 0 & b & c \\ 70bc & 7c^2 & 5b^2 \end{vmatrix}$$

is unimodular; that is,

$$5b^3 - 7c^3 = \pm\, 1.$$

Consider this modulo 7. Cubes are congruent to 0, 1, or -1 (mod 7), so $5(-1,\ 0,\ \text{or}\ 1) \equiv \pm\, 1 \pmod 7$, a contradiction. Hence no such $\mathbb{Z}$-basis exists.

Example 2.44. Find the ring of integers of $\mathbb{Q}(\sqrt{2}, \mathrm{i})$.

In this example our initial guess turns out not to be good enough, illustrating how to continue the analysis when this unfortunate event occurs.

The obvious guess is $\{1, \sqrt{2}, \mathrm{i}, \mathrm{i}\sqrt{2}\}$. Let G be the group these generate. We have $\Delta_G = -64$, so $\mathfrak{O}$ may contain elements of the form $\frac{1}{2}g$ (and then possibly $\frac{1}{4}g$ or $\frac{1}{8}g$) for $g \in G$. The norm is

$$\mathrm{N}(a + b\sqrt{2} + c\mathrm{i} + d\mathrm{i}\sqrt{2}) = (a^2 - c^2 - 2b^2 + 2d^2)^2 + 4(ac - 2bd)^2$$

We must find whether this is divisible by 16 for $a, b, c, d = 0$ or 1, and not all zero. By trial and error the only case where this occurs is $b = d = 1$, $a = c = 0$. So

$$\alpha = \tfrac{1}{2}(\theta + \theta\mathrm{i})$$

might be an integer (where $\theta = \sqrt{2}$). In fact $\alpha^2 = \mathrm{i}$, so $\alpha^4 + 1 = 0$, and α *is* an integer.

We therefore revise our initial guess to

$$G' = \{1, \theta, \mathrm{i}, \theta\mathrm{i}, \tfrac{1}{2}\theta(1 + \mathrm{i})\}$$

Since $2 \cdot \frac{1}{2}\theta(1 + \mathrm{i}) = \theta + \theta\mathrm{i}$ this has a $\mathbb{Z}$-basis

$$\{1, \theta, \mathrm{i}, \tfrac{1}{2}\theta(1 + \mathrm{i})\}$$

By Lemma 2.40,

$$\Delta_{G'} = -64/2^2 = -16$$

This still has a square factor, but a recalculation of the usual kind shows that nothing of the form $\frac{1}{2}g$ (where we may now assume that the term in $\frac{1}{2}\theta(1+\mathrm{i})$ occurs with nonzero coefficient) has integer norm. So no new integers arise and $\mathfrak{O} = G'$.

2.9 Exercises

2.1 Which of the following complex numbers are algebraic? Which are algebraic integers?

(a) $355/113$

(b) $e^{2\pi\mathrm{i}/23}$

(c) $e^{\pi\mathrm{i}/23}$

(d) $\sqrt{17} + \sqrt{19}$

(e) $(1+\sqrt{17})/(2\sqrt{-19})$

(f) $\sqrt{(1+\sqrt{2})} + \sqrt{(1-\sqrt{2})}$

2.2 Express $\mathbb{Q}(\sqrt{3}, \sqrt[3]{5})$ in the form $\mathbb{Q}(\theta)$.

2.3 Find all monomorphisms $\mathbb{Q}(\sqrt[3]{7}) \to \mathbb{C}$.

2.4 Let $K = \mathbb{Q}(\sqrt{3}, \sqrt{5})$. Find the discriminants of the $\mathbb{Q}$-bases $[1, \sqrt{3}, \sqrt{5}, \sqrt{15}]$ and $[1, \sqrt{3}, \frac{1+\sqrt{5}}{2}, \frac{\sqrt{3}+\sqrt{15}}{2}]$.

2.5 Let $K = \mathbb{Q}(\sqrt[4]{2})$. Find all monomorphisms $\sigma : K \to \mathbb{C}$ and the minimal polynomials (over $\mathbb{Q}$) and field polynomials (over K) of (a) $\sqrt[4]{2}$ (b) $\sqrt{2}$ (c) 2 (d) $\sqrt{2}+1$. Compare with Theorem 2.7.

2.6 Complete Example 2.42 by discussing the case $p = 3$.

2.7 Complete Example 2.43 by discussing the case $p = 3$.

2.8 Compute integral bases and discriminants of

(a) $\mathbb{Q}(\sqrt{2}, \sqrt{3})$

(b) $\mathbb{Q}(\sqrt{2}, \mathrm{i})$

(c) $\mathbb{Q}(\sqrt[3]{2})$

(d) $\mathbb{Q}(\sqrt[4]{2})$

2.9 If $\alpha_1, \ldots, \alpha_n$ are $\mathbb{Q}$-linearly independent algebraic integers in $\mathbb{Q}(\theta)$, and if

$$\Delta[\alpha_1, \ldots, \alpha_n] = d$$

where d is the discriminant of $\mathbb{Q}(\theta)$, show that $\{\alpha_i, \ldots, \alpha_n\}$ is an integral basis for $\mathbb{Q}(\theta)$.

2.10 If $[K : \mathbb{Q}] = n$, $\alpha \in \mathbb{Q}$, show that

$$\mathrm{N}_K(\alpha) = \alpha^n \qquad \mathrm{T}_K(\alpha) = n\alpha$$

2.11 Give examples to show that for fixed α, $\mathrm{N}_K(\alpha)$ and $\mathrm{T}_K(\alpha)$ depend on K. (This is to emphasise that the norm and trace must always be defined in the context of a specific field K; there is no such thing as the norm or trace of α without a specified field.)

2.12 The norm and trace may be generalised by considering number fields $K \supseteq L$. Suppose $K = L(\theta)$ and $[K : L] = n$. Consider monomorphisms $\sigma : K \to \mathbb{C}$ such that $\sigma(x) = x$ for all $x \in L$. Show that there are precisely n such monomorphisms $\sigma_1, \ldots, \sigma_n$ and describe them. For $\alpha \in K$, define

$$\mathrm{N}_{K/L}(\alpha) = \prod_{i=1}^{n} \sigma_i(\alpha)$$

$$\mathrm{T}_{K/L}(\alpha) = \sum_{i=1}^{n} \sigma_i(\alpha)$$

(Compared with our earlier notation, we have $\mathrm{N}_K = \mathrm{N}_{K/\mathbb{Q}}$, $\mathrm{T}_K = \mathrm{T}_{K/\mathbb{Q}}$.) Prove that

$$\mathrm{N}_{K/L}(\alpha_1\alpha_2) = \mathrm{N}_{K/L}(\alpha_1)\mathrm{N}_{K/L}(\alpha_2)$$

$$\mathrm{T}_{K/L}(\alpha_1 + \alpha_2) = \mathrm{T}_{K/L}(\alpha_1) + \mathrm{T}_{K/L}(\alpha_2)$$

Let $K = \mathbb{Q}(\sqrt[4]{3})$, $L = \mathbb{Q}(\sqrt{3})$. Calculate $\mathrm{N}_{K/L}(\sqrt{\alpha})$, $\mathrm{T}_{K/L}(\alpha)$ for $\alpha = \sqrt[4]{3}$ and $\alpha = \sqrt[4]{3} + \sqrt{3}$.

2.13 In the field $\mathbb{Q}(\theta)$ where $\theta = \sqrt[3]{d}$, let $x = a + b\theta + c\theta^2$ for $a, b, c \in \mathbb{Q}$. Let the field polynomial of x be $t^3 + a_2t^2 + a_1t + a_0$. Calculate a_0, a_1, a_2.

2.14 Find the zeros of $f(t) = t^4 - 2t^2 - 1$ in $\mathbb{C}$. How many are real? Show that the number field generated by the zeros contains $\mathbb{Q}(\sqrt{2})$. Prove that $f(t)$ is irreducible over $\mathbb{Q}$ and has no linear factors in $\mathbb{Q}(\sqrt{2})$. Is it irreducible over $\mathbb{Q}(\sqrt{2})$? Justify your answer.

2.15 Let $\theta = \sqrt[3]{d} \in \mathbb{R}$ and let $K = \mathbb{Q}(\theta)$. Compute the discriminant of the $\mathbb{Q}$-basis $\{1, \theta, \theta^2\}$ of K.

2.16 Prove that the ring of integers of $\mathbb{Q}(\sqrt{2}, \sqrt{-5})$ is $\mathbb{Z}[\sqrt{2}, \sqrt{-5}]$. (Computer assistance recommended.)

2.17 Let $f(t) = t^3 + at^2 + bt + c$ where $a, b, c \in \mathbb{Z}$. Let θ be a zero of f. Show that $\Delta[1, \theta, \theta^2] = a^2b^2 - 4b^3 - 4a^3c - 27c^2 + 18abc$. Deduce that if $a = 0$ then $\Delta[1, \theta, \theta^2] = -4b^3 - 27c^2$. (Computer assistance recommended.)

2.18 Let θ be a zero of $t^3 + pt + q$ where $p, q \in \mathbb{Z}$. Prove that $\Delta[1, \theta, \theta^2]$ is not squarefree if q is even or p is a multiple of 3. For all other cases when $1 \leq q \leq 9$ and $-5 \leq p \leq 5$, list the values for which f is irreducible over $\mathbb{Q}$ and $\Delta[1, \theta, \theta^2]$ is squarefree. (Computer assistance recommended.)

Deduce that the corresponding number fields $\mathbb{Q}[\theta]$ have integral basis $\{1, \theta, \theta^2\}$.

2.19 Using computer algebra, show that each of the following quartic polynomials in $\mathbb{Z}[t]$ is irreducible. Calculate the corresponding discriminants $\Delta[1, \theta, \theta^2, \theta^3]$, where θ is any zero of the polynomial. Deduce that $\{1, \theta, \theta^2, \theta^3\}$ is an integral basis in each case.

(a) $t^4 + t^3 - 3t^2 + 2t + 1$

(b) $t^4 + t^3 - 3t^2 + 2t + 7$

(c) $t^4 + 3t^3 - 3t^2 + 2t + 5$

(d) $t^4 - t^3 + 3t^2 - t + 3$

3

Quadratic and Cyclotomic Fields

In this chapter we investigate two special types of number field. Quadratic fields are those of degree 2 and are especially important in the study of quadratic forms. Cyclotomic fields are generated by pth roots of unity, and in this chapter we consider only the case p prime; these fields are central to Kummer's approach to Fermat's Last Theorem and play a substantial role in subsequent work. We return to both types of field at later stages. For the moment we content ourselves with finding the rings of integers, integral bases, and discriminants.

3.1 Quadratic Fields

Definition 3.1. A *quadratic field* is a number field K of degree 2 over $\mathbb{Q}$.

Recall from Definition 2.29 that a nonzero integer is squarefree if it has no factor p^2 where p is prime. Prime factorisation in $\mathbb{Z}$ implies that any integer can be written as $r^2 d$ where $r, d \in \mathbb{Z}$ and d is squarefree.

Theorem 3.2. *The quadratic fields are precisely those of the form* $\mathbb{Q}(\sqrt{d})$ *for a squarefree rational integer* d.

Proof: If K is a quadratic field then $K = \mathbb{Q}(\theta)$, where θ is an algebraic integer that is a zero of

$$t^2 + at + b \qquad (a, b \in \mathbb{Z})$$

Thus

$$\theta = \frac{-a \pm \sqrt{(a^2 - 4b)}}{2}$$

Let $a^2 - 4b = r^2 d$ where $r, d \in \mathbb{Z}$ and d is squarefree. Then

$$\theta = \frac{-a \pm r\sqrt{d}}{2}$$

so $\mathbb{Q}(\theta) = \mathbb{Q}(\sqrt{d})$. □

Next we determine the ring of integers of $\mathbb{Q}(\sqrt{d})$, for squarefree d. The answer, it turns out, depends on an arithmetical property of d.

Theorem 3.3. *Let d be a squarefree rational integer. Then the integers of $\mathbb{Q}(\sqrt{d})$ are:*
(a) $\mathbb{Z}[\sqrt{d}]$ *if* $d \not\equiv 1 \pmod 4$
(b) $\mathbb{Z}[\frac{1}{2} + \frac{1}{2}\sqrt{d}]$ *if* $d \equiv 1 \pmod 4$.

Proof: Every element $\alpha \in \mathbb{Q}(\sqrt{d})$ is of the form $\alpha = r + s\sqrt{d}$ for $r, s \in \mathbb{Q}$. Hence

$$\alpha = \frac{a + b\sqrt{d}}{c}$$

where $a, b, c \in \mathbb{Z}$, $c > 0$, and no prime divides all of a, b, c. Now α is an integer if and only if the coefficients of the minimal polynomial

$$\left(t - \left(\frac{a + b\sqrt{d}}{c}\right)\right)\left(t - \left(\frac{a - b\sqrt{d}}{c}\right)\right)$$

are integers. Thus

$$\frac{a^2 - b^2 d}{c^2} \in \mathbb{Z} \tag{3.1}$$

$$\frac{2a}{c} \in \mathbb{Z} \tag{3.2}$$

If c and a have a common prime factor p then (3.1) implies that p divides b (since d is squarefree) which contradicts our previous assumption. Hence from (3.2) $c = 1$ or 2. If $c = 1$ then α is an integer of K in any case, so we may concentrate on the case $c = 2$. Now a and b must both be odd, and $(a^2 - b^2 d)/4 \in \mathbb{Z}$. Hence

$$a^2 - b^2 d \equiv 0 \pmod 4$$

Now an odd number $2k + 1$ has square $4k^2 + 4k + 1 \equiv 1 \pmod 4$, hence $a^2 \equiv 1 \equiv b^2 \pmod 4$, and this implies that $d \equiv 1 \pmod 4$. Conversely, if $d \equiv 1 \pmod 4$ then, for odd a and b, equations (3.1) and (3.2) imply that α is an integer.

To sum up: if $d \equiv 1 \pmod 4$ then $c = 1$ and so (a) holds; whereas if $d \equiv 1 \pmod 4$ we can also have $c = 2$ and a, b odd, whence easily (b) holds. □

The monomorphisms $K \to \mathbb{C}$ are

$$\begin{aligned} \sigma_1(r + s\sqrt{d}) &= r + s\sqrt{d} \\ \sigma_2(r + s\sqrt{d}) &= r - s\sqrt{d} \end{aligned}$$

We can therefore compute discriminants:

Theorem 3.4. (a) *If* $d \not\equiv 1 \pmod 4$ *then* $\mathbb{Q}(\sqrt{d})$ *has an integral basis* $\{1, \sqrt{d}\}$, *and the discriminant is* $4d$.

(b) *If* $d \equiv 1 \pmod 4$ *then* $\mathbb{Q}(\sqrt{d})$ *has an integral basis* $\{1, \frac{1}{2} + \frac{1}{2}\sqrt{d}\}$, *and the discriminant is* d.

Proof: The assertions regarding bases are clear from Theorem 3.3. Compute discriminants:

$$\begin{aligned} \left|\begin{array}{cc} 1 & \sqrt{d} \\ 1 & -\sqrt{d} \end{array}\right|^2 &= (-2\sqrt{d})^2 = 4d \\ \left|\begin{array}{cc} 1 & \frac{1}{2} + \frac{1}{2}\sqrt{d} \\ 1 & \frac{1}{2} - \frac{1}{2}\sqrt{d} \end{array}\right|^2 &= (-\sqrt{d})^2 = d \end{aligned}$$

□

Since the discriminants of isomorphic fields are equal, the fields $\mathbb{Q}(\sqrt{d})$ are not isomorphic for distinct squarefree d. This completes the classification of quadratic fields.

A special case, of historical interest as the first number field to be studied as such, is:

Definition 3.5. The *Gaussian field* is $\mathbb{Q}(\mathrm{i})$. The ring of *Gaussian integers* is $\mathbb{Z}[\mathrm{i}]$.

Since $-1 \not\equiv 1 \pmod 4$ the ring $\mathbb{Z}[\mathrm{i}]$ is the ring of integers of $\mathbb{Q}(\mathrm{i})$ by Theorem 3.3. The discriminant is -4.

Incidentally, these results show that Theorem 2.30 is not always applicable: an integral basis *can* have a discriminant that is not squarefree, for instance, the Gaussian integers themselves.

For future use we record the norms and traces:

$$\begin{aligned} \mathrm{N}(r+s\sqrt{d}) &= r^2 - ds^2 \\ \mathrm{T}(r+s\sqrt{d}) &= 2r \end{aligned}$$

Definition 3.6. A quadratic field $\mathbb{Q}(\sqrt{d})$ is *real* if d is positive. It is *imaginary* if d is negative.

A real quadratic field contains only real numbers, but an imaginary quadratic field contains proper complex numbers as well as imaginary numbers.

3.2 Cyclotomic Fields

Definition 3.7. A *cyclotomic field* is one of the form $\mathbb{Q}(\zeta)$ where $\zeta = e^{2\pi\mathrm{i}/n}$ is a primitive complex nth root of unity.

The name means 'circle-cutting' and refers to the equal spacing of powers of ζ around the unit circle in the complex plane. In this book we focus initially on the case $n = p$, a prime number. In Chapter 19 we extend the discussion to composite n. Further, if $p = 2$ then $\zeta = -1$ so that $\mathbb{Q}(\zeta) = \mathbb{Q}$, so we ignore this case and assume p odd.

Lemma 3.8. *The minimal polynomial of $\zeta = e^{2\pi\mathrm{i}/p}$ over $\mathbb{Q}$, for an odd prime p, is*

$$f(t) = t^{p-1} + t^{p-2} + \cdots + t + 1$$

The degree of $\mathbb{Q}(\zeta)$ is $p-1$.

Proof: We have

$$f(t) = \frac{t^p - 1}{t - 1}$$

Now $\zeta - 1 \neq 0$ and $\zeta^p = 1$, so $f(\zeta) = 0$. Therefore it is enough to prove that f is irreducible. This we do by a standard piece of trickery. We have

$$f(t+1) = \frac{(t+1)^p - 1}{t} = \sum_{r=1}^{p} \binom{p}{r} t^{r-1}$$

The binomial coefficient $\binom{p}{r}$ is divisible by p if $1 \leq r \leq p-1$, and $\binom{p}{1} = p$ is not divisible by p^2. Hence by Eisenstein's Criterion (Theorem 1.11) $f(t+1)$ is irreducible. Therefore $f(t)$ is irreducible, and is the minimal polynomial of ζ. Since $\partial f = p-1$ we have $[\mathbb{Q}(\zeta) : \mathbb{Q}] = p-1$ by Theorem 1.18. □

The powers $\zeta, \zeta^2, \ldots, \zeta^{p-1}$ are also pth roots of unity, not equal to 1, so by the same argument they also have $f(t)$ as minimal polynomial. Clearly

$$f(t) = (t-\zeta)(t-\zeta^2)\ldots(t-\zeta^{p-1}) \tag{3.3}$$

so the conjugates of ζ are $\zeta, \zeta^2, \ldots, \zeta^{p-1}$. Therefore the monomorphisms from $\mathbb{Q}(\zeta)$ to $\mathbb{C}$ are the σ_i determined by

$$\sigma_i(\zeta) = \zeta^i \qquad (1 \leq i \leq p-1)$$

Because the minimal polynomial $f(t)$ has degree $p-1$, a basis for $\mathbb{Q}(\zeta)$ over $\mathbb{Q}$ is $1, \zeta, \ldots, \zeta^{p-2}$, so for a general element

$$\alpha = a_0 + a_1\zeta + \cdots + a_{p-2}\zeta^{p-2} \qquad (a_i \in \mathbb{Q})$$

we have

$$\sigma_i(a_0 + \zeta + \cdots + a_{p-2}\zeta^{p-2}) = a_0 + \zeta^i + \cdots + a_{p-2}\zeta^{i(p-2)}$$

From this formula the norm and trace may be calculated using Definition 2.34. In particular

$$\mathrm{N}(\zeta) = \zeta \cdot \zeta^2 \ldots \zeta^{p-1} = \zeta^{1+2+\cdots+(p-1)} = \zeta^{p(p-1)/2} = (\zeta^p)^{(p-1)/2}$$

and since p is odd,

$$\mathrm{N}(\zeta^i) = 1 \quad (1 \leq i \leq p-1) \tag{3.4}$$

The trace of ζ^i can be found by a similar argument. We have

$$\mathrm{T}(\zeta^i) = \mathrm{T}(\zeta) = \zeta + \zeta^2 + \cdots + \zeta^{p-1} = -1 \quad (1 \leq i \leq p-1) \tag{3.5}$$

For $a \in \mathbb{Q}$ we trivially have

$$\mathrm{N}(a) = a^{p-1} \qquad \mathrm{T}(a) = (p-1)a \tag{3.6}$$

Since $\zeta^p = 1$, we can use these formulas to extend (3.4) and (3.5) to

$$\mathrm{N}(\zeta^s) = 1 \qquad \text{for all } s \in \mathbb{Z} \tag{3.7}$$

and

$$\mathrm{T}(\zeta^s) = \begin{cases} -1 & \text{if } s \not\equiv 0 \pmod{p} \\ p-1 & \text{if } s \equiv 0 \pmod{p} \end{cases} \tag{3.8}$$

For a general element of $\mathbb{Q}(\zeta)$, the trace is easily calculated:

$$\begin{aligned}
\mathrm{T}\left(\sum_{i=0}^{p-2} a_i\zeta^i\right) &= \sum_{i=0}^{p-2}\mathrm{T}(a_i\zeta^i) \\
&= \mathrm{T}(a_0) + \sum_{i=1}^{p-2}\mathrm{T}(a_i\zeta^i) \\
&= (p-1)a_0 - \sum_{i=0}^{p-2} a_i
\end{aligned}$$

so

$$\mathrm{T}\left(\sum_{i=0}^{p-2} a_i\zeta^i\right) = pa_0 - \sum_{i=1}^{p-2} a_i \tag{3.9}$$

The norm is more complicated in general, but a useful special case is

$$\mathrm{N}(1-\zeta) = \prod_{i=1}^{p-1}(1-\zeta^i)$$

which can be calculated by putting $t = 1$ in (3.3) to obtain

$$\prod_{i=1}^{p-1}(1-\zeta^i) = p \tag{3.10}$$

so

$$\mathrm{N}(1-\zeta) = p \tag{3.11}$$

and the same goes for its conjugates: $\mathrm{N}(1-\zeta^i) = p$ for $1 \leq i \leq p-1$.

Similarly, putting $t = -1$, we obtain

$$\mathrm{N}(1+\zeta) = 1 \tag{3.12}$$

Therefore $1+\zeta$ is a unit, and so are its conjugates $1+\zeta^i$ for $1 \leq i \leq p-1$.

We can put these computations to good use, first by showing that the integers of $\mathbb{Q}(\zeta)$ are what one naively might expect:

Theorem 3.9. *The ring* $\mathfrak{O}$ *of integers of* $\mathbb{Q}(\zeta)$ *is* $\mathbb{Z}[\zeta]$.

Proof: Suppose that $\alpha = a_0 + a_1\zeta + \cdots + a_{p-2}\zeta^{p-2}$ is an integer in $\mathbb{Q}(\zeta)$. We must demonstrate that the rational numbers a_i are actually rational integers.

For $0 \leq k \leq p-2$ the element

$$\alpha\zeta^{-k} - \alpha\zeta$$

is an integer, so its trace is a rational integer. But

$$\begin{aligned} & \mathrm{T}(\alpha\zeta^{-k} - \alpha\zeta) \\ = & \mathrm{T}(a_0\zeta^{-k} + \cdots + a_k + \cdots + a_{p-2}\zeta^{p-k-2} - a_0\zeta - \ldots - a_{p-2}\zeta^{p-1}) \\ = & pa_k - (a_0 + \cdots + a_{p-2}) - (-a_0 - \ldots - a_{p-2}) \\ = & pa_k \end{aligned}$$

Hence $b_k = pa_k$ is a rational integer.

Put $\lambda = 1 - \zeta$. Then

$$\begin{aligned} p\alpha &= b_0 + b_1\zeta + \ldots + b_{p-2}\zeta^{p-2} \\ &= c_0 + c_1\lambda + \cdots + c_{p-2}\lambda^{p-2} \end{aligned} \tag{3.13}$$

where, substituting $\zeta = 1 - \lambda$ and expanding,

$$c_i = \sum_{j=i}^{p-2} (-1)^i \binom{j}{i} b_j \in \mathbb{Z}$$

Since $\lambda = 1 - \zeta$ we also have, symmetrically,

$$b_i = \sum_{j=i}^{p-2} (-1)^i \binom{j}{i} c_j \tag{3.14}$$

We claim that all c_i, are divisible by p. By induction we may assume this for all c_i with $i \leq k-1$, where $0 \leq k \leq p-2$. Since $c_0 = b_0 + \cdots + b_{p-2} = p(-\mathrm{T}(\alpha) + b_0)$, we have $p|c_0$, so it is true for $k = 0$. Now by (3.10)

$$\begin{aligned} p &= \prod_{i=1}^{p-1} (1 - \zeta^i) \\ &= (1 - \zeta)^{p-1} \prod_{i=1}^{p-1} (1 + \zeta + \cdots + \zeta^{i-1}) \\ &= \lambda^{p-1}\kappa \end{aligned} \tag{3.15}$$

where $\kappa \in \mathbb{Z}[\zeta] \subseteq \mathfrak{O}$. Consider (3.13) as a congruence modulo the ideal $\langle \lambda^{k+1} \rangle$ of $\mathfrak{O}$. By (3.15)

$$p \equiv 0 \pmod{\langle \lambda^{k+1} \rangle}$$

so the left-hand side of (3.13) and the terms up to $c_{k-1}\lambda^{k-1}$ vanish. Further, the terms from $c_{k+1}\lambda^{k+1}$ onwards are multiples of λ^{k+1} and also vanish. There remains:

$$c_k\lambda^k \equiv 0 \pmod{\langle \lambda^{k+1} \rangle}$$

This is equivalent to $c_k\lambda^k = \mu\lambda^{k+1}$ for some $\mu \in \mathfrak{O}$, from which we obtain $c_k = \mu\lambda$. Take norms:

$$c_k^{p-1} = \mathrm{N}(c_k) = \mathrm{N}(\mu)\mathrm{N}(\lambda) = p\mathrm{N}(\mu)$$

since $\mathrm{N}(\lambda) = p$ by (3.11). Hence $p|c_k^{p-1}$, so $p|c_k$. By induction $p|c_k$ for all k, and then (3.14) shows that $p|b_k$ for all k. Therefore $a_k \in \mathbb{Z}$ for all k. □

Now we can compute the discriminant.

Theorem 3.10. *Let p be an odd prime and $\zeta = e^{2\pi i/p}$. The discriminant of $\mathbb{Q}(\zeta)$ is*

$$(-1)^{(p-1)/2}p^{p-2}$$

Proof: By Theorem 3.9 an integral basis is $\{1, \zeta, \ldots, \zeta^{p-2}\}$. By Proposition 2.36 the discriminant is

$$(-1)^{(p-1)(p-2)/2} \cdot \mathrm{N}(\mathrm{D}f(\zeta))$$

with $f(t)$ as above. Since p is odd the first factor reduces to $(-1)^{(p-1)/2}$. To evaluate the second, recall that

$$f(t) = \frac{t^p - 1}{t - 1}$$

so

$$\mathrm{D}f(t) = \frac{(t-1)pt^{p-1} - (t^p - 1)}{(t-1)^2} \tag{3.16}$$

whence

$$\mathrm{D}f(\zeta) = \frac{-p\zeta^{p-1}}{\lambda}$$

where $\lambda = 1 - \zeta$ as before. Hence

$$\mathrm{N}(\mathrm{D}f(\zeta)) = \frac{\mathrm{N}(p)\mathrm{N}(\zeta)^{p-1}}{\mathrm{N}(\lambda)} = \frac{(-p)^{p-1}1^{p-1}}{p} = p^{p-2}$$

□

The case $p = 3$ deserves special mention, for $\mathbb{Q}(\zeta)$ then has degree $p - 1 = 2$, so it is a quadratic field. Since

$$e^{2\pi i/3} = \frac{-1 + \sqrt{-3}}{2}$$

it is equal to $\mathbb{Q}(\sqrt{-3})$. As a check on our discriminant calculations: Theorem 3.4 gives -3 (since $-3 \equiv 1 \pmod 4$), and Theorem 3.10 gives $(-1)^{2/2}3^1 = -3$ as well.

3.3 Exercises

3.1 Find integral bases and discriminants for:

(a) $\mathbb{Q}(\sqrt{3})$

(b) $\mathbb{Q}(\sqrt{-7})$

(c) $\mathbb{Q}(\sqrt{11})$

(d) $\mathbb{Q}(\sqrt{-11})$

(e) $\mathbb{Q}(\sqrt{6})$

(f) $\mathbb{Q}(\sqrt{-6})$

3.2 Let $\tau = (1+\sqrt{5})/2$. Prove that if $n \in \mathbb{Z}$ and $n \geq 0$ then $\tau^n \in \mathbb{Z}[\sqrt{5}]$ if and only if $3|n$. More precisely, show that

$$\begin{aligned} \tau^{3k} &\in \mathbb{Z}[\sqrt{5}] \\ \tau^{3k+1} - \tau &\in \mathbb{Z}[\sqrt{5}] \\ \tau^{3k+2} - \tau &\in \mathbb{Z}[\sqrt{5}] \end{aligned}$$

(*Hint*: Calculate the first few powers of τ and look for patterns; then use induction. Use computer algebra if you wish.)

3.3 Use computer algebra to explore what happens to positive powers ρ^n where $\rho = (1+\sqrt{d})/2$ for other squarefree integers d. Formulate a plausible conjecture. Prove it.

(*Hint*: There are four obvious types of behaviour: (1) Like the case $d = 5$. (2) No powers lie in $\mathbb{Z}[\sqrt{d}]$ but they all lie in $\frac{1}{2}\mathbb{Z}[\sqrt{d}]$. (3) Denominators in lowest terms are 2^n. (4) Denominators are $2^{\lceil n/2 \rceil + 1}$. A simple condition determines which occurs.)

3.4 Let $K = \mathbb{Q}(\zeta)$ where $\zeta = e^{2\pi i/5}$. Calculate $\mathrm{N}_K(\alpha)$ and $\mathrm{T}_K(\alpha)$ for the following values of α:

(a) ζ^2

(b) $\zeta + \zeta^2$

(c) $1 + \zeta + \zeta^2 + \zeta^3 + \zeta^4$

3.5 Let $K = \mathbb{Q}(\zeta)$ where $\zeta = e^{2\pi i/p}$ for a rational prime p. In the ring of integers $\mathbb{Z}[\zeta]$, show that $\alpha \in \mathbb{Z}[\zeta]$ is a unit if and only if $\mathrm{N}_K(\alpha) = \pm 1$.

3.6 If $\zeta = e^{2\pi i/3}$, $K = \mathbb{Q}(\zeta)$, prove that the norm of $\alpha \in \mathbb{Z}[\zeta]$ has the form $\frac{1}{4}(a^2 + 3b^2)$ where a, b are rational integers that are either both even or both odd. Using the result of Exercise 3.5, deduce that there are precisely six units in $\mathbb{Z}[\zeta]$ and find them all.

3.7 If $\zeta = e^{2\pi i/5}$, $K = \mathbb{Q}(\zeta)$, prove that the norm of $\alpha \in \mathbb{Z}[\zeta]$ has the form $\frac{1}{4}(a^2 - 5b^2)$ where a, b are rational integers. (*Hint:* When calculating $\mathrm{N}(\alpha)$, first calculate $\sigma_1(\alpha)\sigma_4(\alpha)$ where $\sigma_i(\zeta) = \zeta^i$. Show that this is of the form $q + r\theta + s\phi$ where q, r, s are rational integers, $\theta = \zeta + \zeta^4$, $\phi = \zeta^2 + \zeta^3$. In the same way, establish $\sigma_2(\alpha)\sigma_3(\alpha) = q + s\theta + r\phi$.) Use multiplicativity of the norm to prove that $\mathbb{Z}[\zeta]$ has an infinite number of units.

3.8 Let $\zeta = \mathrm{e}^{2\pi i/5}$. For $K = \mathbb{Q}(\zeta)$, use the formula

$$\mathrm{N}_K(a + b\zeta) = (a^5 + b^5)/(a + b)$$

to calculate the following norms:

(a) $\mathrm{N}_K(\zeta + 2)$ (b) $\mathrm{N}_K(\zeta - 2)$ (c) $\mathrm{N}_K(\zeta + 3)$.

Using the identity $\mathrm{N}_K(\alpha\beta) = \mathrm{N}_K(\alpha)\mathrm{N}_K(\beta)$, deduce that $\zeta + 2$, $\zeta - 2$, $\zeta + 3$ have no proper factors (that is, factors that are not units) in $\mathbb{Z}[\zeta]$.

Factorise 11, 31, 61 in $\mathbb{Z}[\zeta]$.

3.9 If $\zeta = e^{2\pi i/5}$, as in Exercise 3.6, calculate

(a) $\mathrm{N}_K(\zeta + 4)$

(b) $\mathrm{N}_K(\zeta - 3)$

Deduce that any proper factor of $\zeta + 4$ in $\mathbb{Z}[\zeta]$ has norm 5 or 41. Given that $\zeta - 1$ is a factor of $\zeta + 4$, find another factor. Verify that $\zeta - 3$ is a unit times $(\zeta^2 + 2)^2$ in $\mathbb{Z}[\zeta]$.

3.10 Show that the multiplicative group of nonzero elements of $\mathbb{Z}_7$ is cyclic with generator the residue class of 3. If $\zeta = \mathrm{e}^{2\pi i/7}$, define the monomorphism $\sigma : \mathbb{Q}(\zeta) \to \mathbb{C}$ by $\sigma(\zeta) = \zeta^3$. Show that all other monomorphisms from $\mathbb{Q}(\zeta)$ to $\mathbb{C}$ are of the form $\sigma^i (1 \leq i \leq 6)$ where $\sigma^6 = 1$. For any $\alpha \in \mathbb{Q}(\zeta)$, define $c(\alpha) = \alpha\sigma^2(\alpha)\sigma^4(\alpha)$, and show $\mathrm{N}(\alpha) = c(\alpha) \cdot \sigma c(\alpha)$. Demonstrate that $c(\alpha) = \sigma^2 c(\alpha) = \sigma^4 c(\alpha)$. Using the relation $1 + \zeta + \cdots + \zeta^6 = 0$, show that every element $\alpha \in \mathbb{Q}(\zeta)$ can be written uniquely as $\sum_{i=1}^{6} a_i\zeta^i (a_i \in \mathbb{Q})$. Deduce that $c(\alpha) = a_1\theta_1 + a_3\theta_2$ where $\theta_1 = \zeta + \zeta^2 + \zeta^4$, $\theta_2 = \zeta^3 + \zeta^5 + \zeta^6$. Show that $\theta_1 + \theta_2 = -1$ and calculate $\theta_1\theta_2$. Verify that $c(\alpha)$ may be written in the form $b_0 + b_1\theta_1$ where b_0, $b_1 \in \mathbb{Q}$, and show that $\sigma c(\alpha) = b_0 + b_1\theta_2$. Deduce that

$$\mathrm{N}(\alpha) = b_0^2 - b_0 b_1 + 2b_1^2$$

Now calculate $\mathrm{N}(\zeta + 5\zeta^6)$.

3.11 Suppose that p is a rational prime and $\zeta = e^{2\pi i/p}$. By Theorem 1.33, the group of nonzero elements of $\mathbb{Z}_p$ is cyclic. Show that there is a monomorphism $\sigma : \mathbb{Q}(\zeta) \to \mathbb{C}$ such that σ^{p-1} is the identity and all monomorphisms from $\mathbb{Q}(\zeta)$ to $\mathbb{C}$ are of the form $\sigma^i (1 \leq i \leq p-1)$. If $p - 1 = kr$, define $c_k(\alpha) = \alpha\sigma^r(\alpha)\sigma^{2r}(\alpha)\ldots\sigma^{(k-1)r}(\alpha)$. Show that

$$\mathrm{N}(\alpha) = c_k(\alpha) \cdot \sigma c_k(\alpha) \ldots \sigma^{r-1} c_k(\alpha)$$

Prove that every element of $\mathbb{Q}(\zeta)$ is uniquely of the form $\sum_{i=1}^{p-1} a_i\zeta^i$ (with no term a_0). By demonstrating that $\sigma^r(c_k(\alpha)) = c_k(\alpha)$, deduce that $c_k(\alpha) = b_1\eta_1 + \cdots + b_k\eta_k$, where

$$\eta_1 = \zeta + \sigma^r(\zeta) + \sigma^{2r}(\zeta) + \cdots + \sigma^{(k-1)r}(\zeta)$$

and $\eta_{i+1} = \sigma^i(\eta_1)$.

3.12 Interpret the results of Exercise 3.9 when $p = 5$, $k = r = 2$, by showing that the residue class of 2 is a generator of the multiplicative group of nonzero elements of $\mathbb{Z}_5$. Demonstrate that $c_2(\alpha)$ is of the form $b_1\eta_1 + b_2\eta_2$ where $\eta_1 = \zeta + \zeta^4$, $\eta_2 = \zeta^2 + \zeta^3$.

Calculate the norms of the following elements in $\mathbb{Q}(\zeta)$:

(a) $\zeta + 2\zeta^2$

(b) $\zeta + \zeta^4$

(c) $15\zeta + 15\zeta^4$

(d) $\zeta + \zeta^2 + \zeta^3 + \zeta^4$

3.13 This exercise previews Chapter 5.

In $\mathbb{Z}[\sqrt{-5}]$, prove that 6 factorises in two ways:

$$6 = 2 \cdot 3 = (1 + \sqrt{-5})(1 - \sqrt{-5})$$

Verify that $2, 3, 1+\sqrt{-5}, 1-\sqrt{-5}$ have no proper factors in $\mathbb{Z}[\sqrt{-5}]$. (*Hint:* Take norms and note that if γ factorises as $\gamma = \alpha\beta$, then $\mathrm{N}(\gamma) = \mathrm{N}(\alpha)\mathrm{N}(\beta)$ is a factorisation of rational integers.) Deduce that it is possible in $\mathbb{Z}[\sqrt{-5}]$ for 2 to have no proper factors, yet 2 divides a product $\alpha\beta$ without dividing either α or β.

3.14 Prove that the set $\mathbb{Z}[\zeta_n]$ is dense in $\mathbb{C}$ (with its usual topology) except when $n = 1, 2, 3, 4$, or 6. You may use the result of Lehmer [79] that if r is rational then $\cos(2\pi r)$ is rational of and only if r is an integer multiple of $\frac{1}{6}$ or $\frac{1}{4}$.

4

Pell's Equation

This chapter can be omitted without affecting the proof of Kummer's special case of Fermat's Last Theorem.

Already we can obtain payoff in number theory by applying algebraic number theory to solve a classic Diophantine equation known as *Pell's Equation*:

$$x^2 - dy^2 = 1 \tag{4.1}$$

where d is a specified positive integer that is not a square. We seek solutions where x, y are integers. (If d is square there are no such solutions.) If (x, y) is a solution then so are $(\pm x, \pm y)$, so we can usually restrict attention to *non-negative* solutions with $x, y \geq 0$.

If $y = 0$ then $x = \pm 1$, so $y \geq 1$ for a nontrivial non-negative solution. Then $x^2 = 1 + dy^2 > 0$, so both x and y are positive. In this case we call (x, y) a *positive* solution. The obvious solutions $x = \pm 1, y = 0$ are the *trivial solutions*, and we want to find nontrivial ones. For example, when $d = 2$ there is a nontrivial solution $x = 3, y = 2$, since $3^2 - 2.2^2 = 9 - 8 = 1$. Similarly when $d = 3$ there is a nontrivial solution $x = 2, y = 1$.

The main result of this chapter is that for all such d there exists a unique 'minimal' solution in positive integers, from which all other solutions can be derived. Similar methods also apply to the closely related *negative Pell's Equation*

$$x^2 - dy^2 = -1 \tag{4.2}$$

and indeed to

$$x^2 - dy^2 = k \tag{4.3}$$

for integers k other than ± 1. See Exercise 4.2.

We begin with a quick trip through the history of this ancient problem. Next we relate it to the norm in a quadratic number field. By examining the geometry of the curve $x^2 - dy^2 = 1$ in the plane, which is a hyperbola, we show that any nontrivial solution, if it exists, generates all solutions by a simple algebraic process. We then develop part of the theory of continued fractions, to prove existence of a nontrivial solution. (In Section 10.5 we read off another existence proof as a simple corollary of Dirichlet's Units Theorem.)

4.1 Brief History of Pell's Equation

The equation's name is traditional, but it misrepresents history. In 1668 John Pell revised an English translation (*Translation of Rhonius's Algebra*) by Thomas Branker of Johann Rahn's 1659 book *Teutsche Algebra* ('German Algebra' in Dutch). This discussed a solution of (4.1) by William Brouncker. Euler thought that this solution was Pell's own work, and coined the name 'Pell's Equation'. Many scholars believe that Pell wrote most, if not all, of Rahn's book, so Euler may not have been as far off the mark as is often assumed.

The story goes back much further. The ancient Greeks were interested in solutions when $d = 2$, which give good rational approximations to $\sqrt{2}$. The same goes for solutions of (4.2). Indeed,

$$x^2 - 2y^2 = \pm 1 \Rightarrow \left(\frac{x}{y}\right)^2 = 2 \mp \frac{1}{y^2}$$

so if y is large, x/y is very close to $\sqrt{2}$. An example is $x = 17, y = 12$, where $17^2 - 2.12^2 = 289 - 2.144 = 289 - 288 = 1$. The rational number $17/12 = 1.4167$ to four decimal places, whereas $\sqrt{2} = 1.4142$. The Pythagoreans called these pairs *side* and *diagonal* numbers: a square of side y has a diagonal close to x.

Archimedes approximated $\sqrt{3}$ by $1351/780$. He did not explain where this came from, but $1351^2 - 3.780^2 = 1$. The famous 'cattle problem' attributed to Archimedes also involves Pell's Equation, see Lenstra [80]. Diophantus studied a problem equivalent to Pell's Equation. Around AD 628 the Indian mathematician Brahmagupta studied Pell's Equation, in particular solving it when $d = 92$. Here the smallest solution is $x = 1151, y = 120$. The smallest values of x, y vary wildly with d, and there are no obvious patterns. For example, when $d = 60$ we find $x = 31, y = 4$; when $d = 61$ the smallest solution is $x = 1766319049, y = 226153980$; and when $d = 62$

we find $x = 63, y = 8$. Brahmagupta discovered a key feature, which in its simplest form can be expressed as:

Theorem 4.1. (Brahmagupta's Theorem) *If* (x_1, y_1) *and* (x_2, y_2) *satisfy* (4.1), *then so does* $(x_1x_2 + dy_1y_2, y_1x_2 + x_1y_2)$.

Proof: Direct computation shows that

$$(x_1^2 + dy_1)^2(x_2^2 + dy_2)^2 = (x_1x_2 + dy_1y_2)^2 + d(x_1y_2 + y_1x_2)^2 \tag{4.4}$$

For solutions, both terms on the left-hand side equal 1, so the right-hand side also equals 1. □

The main point of this result is not the formula as such: it is that solutions to the equation can be combined to give new solutions.

In AD 1150 Bhāskara II developed Brahmagupta's methods to give an algorithm for solving Pell's Equation, the *chakravala* (cyclic) method, and used it to find the above solution for the difficult case $d = 61$. (He did not prove that the algorithm always terminates, but actually it does.)

By the 17th century, some European mathematicians knew how to solve Pell's Equation. Fermat solved it for $d \leq 150$ and challenged John Wallis and Brouncker to solve it for $d = 151$ and 313. Both of them did so, although Wallis may have got his solution from Brouncker. It was at this point that Pell inadvertently managed to get his name attached to the equation.

In 1766–1769 Joseph-Louis Lagrange put the entire topic on firm foundations by relating it to the continued fractions of quadratic irrationals, and proved that the algorithm of Brouncker and Wallis always stops with a solution.

4.2 Relation with Algebraic Numbers

A modern treatment of Pell's Equation sets it in the context of the quadratic number field $\mathbb{Q}(\sqrt{d})$. The key point is that the norm of this field is

$$\mathrm{N}(x + y\sqrt{d}) = x^2 - dy^2$$

We can therefore write (4.1) as $\mathrm{N}(x+y\sqrt{d}) = 1$ and (4.2) as $\mathrm{N}(x+y\sqrt{d}) = -1$. Since $\mathbb{Q}(\sqrt{d}) \subseteq \mathbb{R}$, these conditions specify the *units* of $\mathbb{Q}(\sqrt{d})$. Dirichlet's Units Theorem, which we prove in Chapter 10, then comes into play. We postpone discussion of this viewpoint to that chapter, and instead we give the classical proof using continued fractions. The existence of a minimal solution can be proved without using continued fractions or Dirichlet's Units Theorem; see Conrad [21, 22].

The connection with the norm 'explains' Brahmagupta's method for combining solutions. The norm is multiplicative, so if $\mathrm{N}(x_1 + y_1\sqrt{d}) = \pm 1$ and $\mathrm{N}(x_2 + y_2\sqrt{d}) = \pm 1$ then $\mathrm{N}(x_1 + y_1\sqrt{d})(x_2 + y_2\sqrt{d})) = \pm 1$, with appropriate signs. Moreover,

$$(x_1 + y_1\sqrt{d})(x_2 + y_2\sqrt{d}) = (x_1x_2 + dy_1y_2) + (x_1y_2 + x_2y_1)\sqrt{d}$$

(It is arguably more accurate to say that Brahmagupta's Theorem explains why this particular norm is multiplicative, but multiplicativity is a more general property.)

This method works even when the two solutions are the same. For example, again with $d = 2$, there is a simpler solution $(x, y) = (3, 2)$. Now

$$(3 + 2\sqrt{2})^2 = 17 + 12\sqrt{2}$$

gives the solution $(17, 12)$ mentioned above. Continuing in this manner,

$$(3 + 2\sqrt{2})^3 = 99 + 70\sqrt{2}$$

gives a new solution $(99, 70)$, and so on. So a single solution generates infinitely many. Buy one, get infinitely many free.

We will prove:

Theorem 4.2. *For any nonsquare d, Pell's Equation* (4.1) *has infinitely many integer solutions. They can all be deduced from a specific positive solution (x, y) by forming powers of $(x + y\sqrt{d})$.*

We call this unique positive solution the *minimal* solution (another common term is *fundamental* solution). We clarify the meaning of 'minimal' in Section 4.3.

Proving Theorem 4.2 requires two steps:

(1) Prove that all solutions can be derived from the minimal solution if one exists.

(2) Prove that there is at least one nontrivial solution. This implies the existence and uniqueness of the minimal solution.

The first step is easier than the second. We deal with the first step in Section 4.3 and the second in Section 4.6.

Since $(\pm 1)(\pm 1) = \pm 1$, similar results can be obtained for the modified Equation (4.2). Indeed, two solutions of (4.2) can be combined to give a solution of (4.1), and a solution of (4.1) can be combined with one of (4.2) to give a solution of (4.1). Also solutions of (4.3) for k_1 and k_2 can be combined to give solutions for k_1k_2. For simplicity we concentrate on (4.1) and relegate the related equations to the Exercises.

4.3 Geometry of Pell's Equation

From now on, 'solution' of Pell's Equation means 'solution in integers'. We assume there exists a nontrivial positive solution to Pell's Equation, and prove that the smallest positive solution generates all of them. 'Smallest' here refers either to x or to y, because it is easy to see that if (x_1, y_1) and (x_2, y_2) are positive solutions then $x_1 < x_2 \Leftrightarrow y_1 < y_2$.

Theorem 4.3. *Assume that* (4.1) *has at least one positive solution. Let* (x_1, y_1) *be a positive solution with* x_1 *minimal. Then every positive solution* (x_n, y_n) *is generated by taking powers of* $x_1 + y_1\sqrt{d}$:

$$x_n + y_n\sqrt{d} = (x_1 + y_1\sqrt{d})^n$$

Below we reinterpret this theorem in matrix terms as Theorem 4.4. A conceptually simple proof is then obtained by considering the geometry of Pell's Equation.

Figure 4.1 shows the main features, not drawn to scale because the result of doing that would be unintelligible. Allowing x and y to be real numbers, the curve $x^2 - dy^2 = 1$ in the plane is a hyperbola H. Its asymptotes satisfy $x^2 - dy^2 = 0$; that is, they are the two lines $y = \pm x/\sqrt{d}$. The hyperbola is symmetric under reflection in either axis, and we draw the positive quadrant. The dashed line A is the asymptote that intersects this quadrant.

A positive solution (x, y) of Pell's Equation lies on H and has positive integer coordinates. We also consider the non-negative trivial solution $(1, 0)$ because it plays a key role.

Let x_1, y_1 be the unique positive solution for which x_1 is minimal. Then $x_1 > 1$. Define the matrix

$$M = \begin{bmatrix} x_1 & dy_1 \\ y_1 & x_1 \end{bmatrix}$$

When using M we consider points in the plane as column vectors:

$$(x, y) = \begin{bmatrix} x \\ y \end{bmatrix}$$

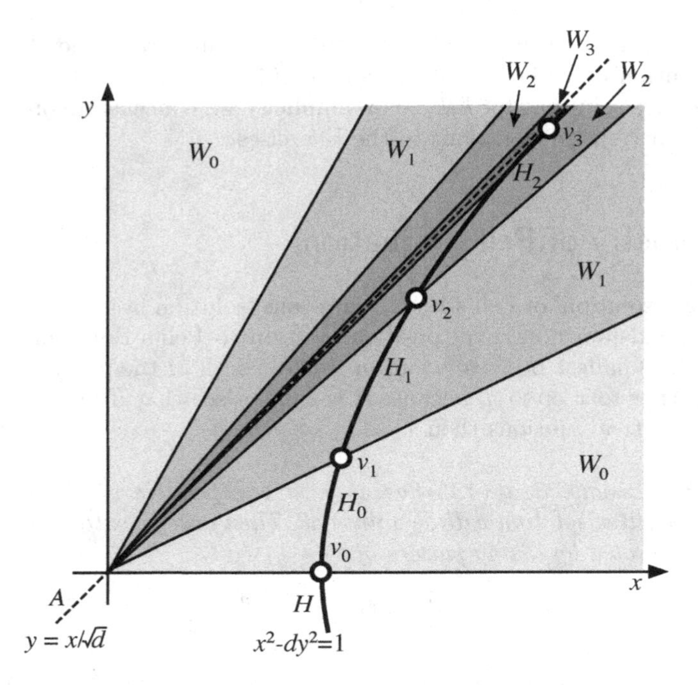

Figure 4.1. Geometry of Pell's Equation. The curve H is part of the hyperbola $x^2 - dy^2 = 1$ where $x, y \in \mathbb{R}$ and the dashed line A is one asymptote. The wedges W_j are shaded. The minimal solution is v_1 and the remaining v_j are the other solutions.

For typographical reasons, the usual (x, y) notation is more convenient except when performing matrix operations.

We introduce the matrix M because it arises naturally from Brahmagupta's formula (4.4), which tells us that if (x_2, y_2) is any solution then $(x_1x_2 + dy_1y_2, y_2x_1 + x_2y_1)$ is also a solution. In matrix notation,

$$\begin{bmatrix} x_1x_2 + dy_1y_2 \\ y_1x_2 + x_1y_2 \end{bmatrix} = \begin{bmatrix} x_1 & dy_1 \\ y_1 & x_1 \end{bmatrix} \begin{bmatrix} x_2 \\ y_2 \end{bmatrix}$$

In other words, if $(x_2, y_2) = v \in \mathbb{Z}^2$ is a solution then so is Mv. Obviously if v is a positive solution then so is Mv. Theorem 4.3 can now be reinterpreted as:

Theorem 4.4. *Assume that* (4.1) *has at least one positive solution and that* $v_1 = (x_1, y_1)$ *is the unique positive solution with* x_1 *minimal. Then the positive solutions are precisely the vectors* $(x_n, y_n) = v_n$ *where* $v_n = M^{n-1}v_1$, $n \geq 1$.

Proof: The routine calculation that proves (4.4), applied to real numbers x_1, x_2, y_1, y_2, shows that M leaves H invariant; that is:

$$\text{if } v \in \mathbb{R}^2 \text{ and } v \in H \text{ then } Mv \in H \tag{4.5}$$

For any subset $S \subseteq \mathbb{R}^2$ we define

$$MS = \{Ms : s \in S\}$$

Then we can express (4.5) as: $MH \subseteq H$.

Clearly $\det M = x_1^2 - dy_1^2 = 1$, so M is an area-preserving linear transformation of $\mathbb{R}^2$. The inverse matrix is

$$M^{-1} = \begin{bmatrix} x_1 & -dy_1 \\ -y_1 & x_1 \end{bmatrix}$$

and, like M, this matrix has integer entries. If v is a solution of Pell's Equation then so is $M^{-1}v$; however, M^{-1} can map positive solutions to nonpositive ones. For example

$$M^{-1}v = \begin{bmatrix} x_1 & -dy_1 \\ -y_1 & x_1 \end{bmatrix} \begin{bmatrix} x_1 \\ y_1 \end{bmatrix} = \begin{bmatrix} x_1^2 - dy_1^2 \\ x_1y_1 - x_1y_1 \end{bmatrix} = \begin{bmatrix} 1 \\ 0 \end{bmatrix} \tag{4.6}$$

Some geometric intuition (treated more precisely in the Exercises) now helps us to understand the effect of the linear map M.

Consider the unit square S with vertices at $(0,0)$, $(1,0)$, $(0,1)$, $(1,1)$. Then MS is a parallelogram with vertices at $(0,0)$, (x_1, y_1), (dy_1, x_1), $(x_1 + dy_1, x_1 + y_1)$. This parallelogram has area 1, is longer and thinner than S, and contains a segment of A (because S contains such a segment and A is M-invariant). Iterating M leads to increasingly longer and thinner parallelograms that converge towards the asymptote A.

Parametrise the positive quadrant of H by arc-length s, measured from v_0. Imagine a point moving along H starting from v_0, with s increasing monotonically from 0. Then its x-coordinate increases monotonically from 1 and its y-coordinate increases monotonically from 0. We can therefore use any of s, x, y as a measure of the size of a solution, in the sense that 'greater than' and 'less than' have the same meaning for any of them. It is convenient to use x, but s helps when visualising the proof because it is intrinsic to H.

We use M and the presumed minimal solution $v_1 = (x_1, y_1)$ to define an infinite sequence $v_0, v_1, v_2, \ldots$ of points on H. Set $v_0 = (0,1)$, the trivial solution, and define

$$v_n = M^n v_0$$

Then $v_1 = (x_1, y_1)$ so the notation is consistent with the statement of Theorem 4.4.

By (4.5) the matrix M leaves H invariant; that is, $MH \subseteq H$. The asymptote A is also invariant under M so $MA \subseteq A$. This can be proved in several ways: by direct calculation, because H determines A uniquely, or by observing that the asymptotes of H are the eigenspaces of M.

Denote the (closed) positive quadrant by W_0. Then $W_1 = MW_0 \subseteq W_0$. Since M is linear, W_1 is a closed wedge bounded by the images under M of the two axes. Define $W_n = M^n W_0$. Then $M^k W_n \subseteq W_{n+k}$ for $k \geq 0$, so in particular

$$W_0 \supsetneq W_1 \supsetneq W_2 \supsetneq \cdots$$

The earlier discussion of the effect of M on parallelograms shows that as n increases, these wedges become thinner and thinner, and their intersection over all $n \geq 0$ is the non-negative segment of A. (This can also be checked algebraically.)

The hyperbola H splits into arcs $H_n = H \cap W_n$. The endpoints of H_n are v_n and v_{n+1}. Moreover, $MH_n = M(W_n \cap H) = MW_n \cap MH = W_{n+1} \cap H = H_{n+1}$. So M slides points along H, moving everything one step along. Therefore, M^{-1} slides points in $H \setminus H_0$ the other way. See the remark below.

Define $v_n = M^n v_0$. By (4.4) each v_n gives a positive solution of Pell's Equation. We claim that the v_n are the *only* positive solutions.

Suppose, for a contradiction, that there is another positive solution v. Then v lies on the interior of some arc H_n, and $n > 0$ since v_0 is the smallest positive solution. We now use powers of M^{-1} to 'slide' v to the left along the curve H until it reaches H_0. Acting by M^{-1} preserves solutions, so the same goes for M^{-k} for any $k > 0$. Since $M^k W_n \subseteq W_{n+k}$ for $k \geq 0$, we also have $M^{-k} W_n \subseteq W_{n-k}$ for $n \geq k$. Therefore $u = M^{-n} v \in W_0$ and $u \in H_0$. The point u has integer coordinates. Since v is in the interior of H_n, the point u is in the interior of H_0, so its coordinates are positive. But this implies that u is a positive solution of Pell's Equation that is smaller than v. This is a contradiction, so no such v exists. Therefore the v_n are all the positive integer solutions. □

Remark. We can continue the 'sliding' process using negative powers of M, that is, powers of M^{-1}. This does not generate positive solutions; instead, it generates solutions along the lower half (negative y) of the same branch of the hyperbola H. See Exercise 4.12.

It is possible to rewrite the above proof without reference to the geometry, using estimates and inequalities. A similar (but different) proof along these lines is in Conrad [22].

4.4 Continued Fractions: Basic Theory

It remains to prove that some nontrivial solution of Pell's Equation exists. Lagrange proved this result using continued fractions. These are expressions of the form

$$\alpha = a_0 + \cfrac{1}{a_1 + \cfrac{1}{a_2 + \cfrac{1}{a_3 + \cdots}}} \tag{4.7}$$

where the a_j are real numbers. (Technically these are *simple* continued fractions, for which all numerators equal 1. We do not consider other kinds of continued fraction, and omit 'simple'.)

In this section we discuss some basic properties of continued fractions, omitting proofs where these are routine. We consider only properties needed to solve Pell's Equation. For details and further information on continued fractions, see Hardy and Wright [58], Khinchin [71], Olds [102], and Unger [138].

We are mainly interested in cases where $a_0 \geq 0$ and $a_j > 0$ are integers, but some parts of the theory require real numbers. The expression (4.7) can either stop at some stage, or continue indefinitely, as we explain below. For typographical convenience we replace the right-hand side of (4.7) by the notation

$$\begin{array}{ll} [a_0, a_1, a_2, a_3, \ldots, a_n] & \text{in the finite case} \\ [a_0, a_1, a_2, a_3, \ldots] & \text{in the infinite case} \end{array}$$

Finite Continued Fractions

A finite continued fraction is just a complicated arithmetical expression for some real number (rational number if the a_i are integers). For example

$$\begin{aligned} [1,2,3,4] &= 1 + \frac{1}{2 + \frac{1}{3+\frac{1}{4}}} = 1 + \frac{1}{2 + \frac{1}{13/4}} = 1 + \frac{1}{2 + \frac{4}{13}} \\ &= 1 + \frac{1}{30/13} = 1 + \frac{13}{30} = \frac{43}{30} \end{aligned}$$

Definition 4.5. The finite continued fraction $[a_0, a_1, a_2, a_3, \ldots, a_n]$ is defined recursively by

$$\begin{aligned} [a_0] &= a_0 \\ [a_0, a_1, \ldots, a_{k+1}] &= [a_0, a_1, \ldots, a_k + \frac{1}{a_{k+1}}] \qquad k < n \end{aligned}$$

We usually assume two conditions:

(A) a_0 is an integer.

(B) The a_j are positive integers for $j \geq 1$.

Clearly if all a_j are integers, and nonzero except perhaps for $j = 0$, then $[a_0, a_1, a_2, a_3, \ldots, a_n]$ evaluates to a specific rational number α, so we can write

$$\alpha = [a_0, a_1, a_2, a_3, \ldots, a_n]$$

Moreover, if conditions (A) and (B) hold then $\alpha > 0$. Conversely, every positive rational α has a continued fraction expansion satisfying (A) and (B). Indeed, the continued fraction for p/q with p, q positive integers closely parallels the Euclidean algorithm for $\gcd(p, q)$, see Exercise 4.5.

Infinite Continued Fractions

An infinite continued fraction is defined as the limit of a sequence of finite ones:

Definition 4.6.

$$[a_0, a_1, \ldots] = \lim_{n \to \infty} [a_0, a_1, \ldots, a_n] \tag{4.8}$$

provided the sequence converges.

In general the sequence need not converge, but later we prove that it does converge if $a_0 \in \mathbb{Z}$ and condition (B) holds. As motivation, consider the following calculation. This uses the *floor function* (or *integer part*) $\lfloor x \rfloor$, which is the largest integer $n \leq x$.

Example 4.7. Let $\alpha = \sqrt{2}$. Then $\lfloor \alpha \rfloor = 1$, so $a_0 = 1$. Thus

$$\sqrt{2} = 1 + (\sqrt{2} - 1)$$

Now $0 < \sqrt{2} - 1 < 1$, so $r_1 = (\sqrt{2} - 1)^{-1} > 1$ and

$$\sqrt{2} = 1 + 1/r_1$$

In fact, $r_1 = \sqrt{2} + 1$ with integer part 2. Thus $r_1 = 2 + r_2$ where $r_2 = r_1 - 2 = \sqrt{2} - 1$. In this case, $r_2 = r_1$, so repeating the process indefinitely we obtain, formally:

$$\sqrt{2} = [1, 2, 2, 2, \ldots]$$

This example motivates an algorithm to find the continued fraction expansion of any $\alpha > 0$ to any finite number of terms. The a_j can be computed recursively using the floor function. The process continues forever if and only if α is irrational. It stops when some $r_k = 0$.

Algorithm for Continued Fraction

$$\begin{array}{lll} a_0 = \lfloor \alpha \rfloor & t_0 = \alpha - a_0 & r_1 = 1/t_0 \\ a_1 = \lfloor s_0 \rfloor & t_1 = r_0 - a_1 & r_2 = 1/t_1 \\ a_2 = \lfloor s_1 \rfloor & t_2 = r_1 - a_2 & r_3 = 1/t_2 \\ \vdots & \vdots & \vdots \\ a_n = \lfloor s_{n-1} \rfloor & t_n = r_{n-1} - a_n & r_{n+1} = 1/t_n \\ \vdots & \vdots & \vdots \end{array} \tag{4.9}$$

By construction, the a_j satisfy conditions (A) and (B).

Definition 4.8. The number r_k is the kth *residue* of α.

In more compact form,

$$a_0 = \lfloor \alpha \rfloor \qquad r_1 = \frac{1}{\alpha - a_0} \qquad a_k = \lfloor r_k \rfloor \qquad r_{k+1} = \frac{1}{\alpha - a_k}$$

By construction, or arguing inductively,

$$\alpha = [a_0, a_1, \ldots, a_{n-1}, r_n] \qquad (n \geq 1) \tag{4.10}$$

Example 4.9. Let $\alpha = \sqrt{3}$. Then the above algorithm gives:

$$\begin{array}{lll} a_0 = 1 & t_0 = \sqrt{3} - 1 & r_0 = (1 + \sqrt{3})/2 \\ a_1 = 1 & t_1 = (-1 + \sqrt{3})/2 & r_1 = 1 + \sqrt{3} \\ a_2 = 2 & t_2 = \sqrt{3} - 1 & r_2 = r_0 \end{array} \tag{4.11}$$

after which the final two lines repeat. So

$$\sqrt{3} = [1, 1, 2, 1, 2, 1, \ldots]$$

repeating $1, 2$ indefinitely.

Definition 4.10. The *continued fraction (expansion)* of α is the expression $[a_0, a_1, \ldots]$ generated by the above algorithm. The a_j satisfy (A) and (B). In Theorem 4.17 we prove that the corresponding sequence of finite continued fractions converges, so we can write

$$\alpha = [a_0, a_1, \ldots]$$

Since α has a finite continued fraction satisfying (A) and (B) if and only if it is rational, a real number α has an infinite continued fraction if and only if it is irrational.

Convergents

Next we introduce some basic terminology and prove standard properties of continued fractions. These apply to finite or infinite continued fractions, with the same proofs, because they are defined in terms of finite continued fractions.

To avoid repeating clumsy alternatives, the notation $[a_0, a_1, \ldots]$ now refers either to a finite continued fraction or to an infinite one. In the finite case, where this denotes some expression $[a_0, a_1, \ldots a_n]$, any reference to a_k must have $0 \leq k \leq n$, in order for the definition to make sense. We do not make this explicit.

Definition 4.11. *The nth* convergent *of* $[a_0, a_1, a_2, a_3, \ldots]$ *is the number* $[a_0, a_1, a_2, a_3, \ldots, a_n]$.

To compute the nth convergent, we define two sequences $(p_n), (q_n)$ of real numbers recursively by

$$\begin{array}{llll} p_{-1} = 1 & p_0 = a_0 & p_n = a_n p_{n-1} + p_{n-2} & n \geq 1 \\ q_{-1} = 0 & q_0 = 1 & q_n = a_n q_{n-1} + q_{n-2} & n \geq 1 \end{array} \tag{4.12}$$

Proposition 4.12. *The nth* convergent *of* $[a_0, a_1, a_2, a_3, \ldots]$ *is the rational number* p_n/q_n.

Proof: By Definition 4.5,

$$[a_0, a_1, \ldots, a_{n+1}] = [a_0, a_1, a_2, a_3, \ldots, a_n + \frac{1}{a_{n+1}}]$$

The required result then follows by induction on n. We omit details, which are routine. □

The omitted details work for all real a_n, so by the same reasoning, (4.10) leads to:

Proposition 4.13. *If* α *is irrational then, for all* $n \geq 1$,

$$\alpha = \frac{r_n p_{n-1} + p_{n-2}}{r_n q_{n-1} + q_{n-2}}$$

□

Proposition 4.14. *For all* $n \geq 1$:

(a) $p_n q_{n-1} - p_{n-1} q_n = (-1)^{n-1}$.

(b) $p_n q_{n-2} - p_{n-2} q_n = (-1)^n a_n$.

Proof: This is an easy induction using (4.12). □

Proposition 4.15. *With assumptions* (A) *and* (B), p_k *and* q_k *are coprime integers for any* $k \geq 0$. *Thus* p_n/q_n *is rational and in lowest terms.*

Proof: By an easy induction, p_n and q_n are non-negative integers.

By Proposition 4.14 (a), any common divisor of p_n and q_n divides ± 1. Thus they are coprime. □

We can now unambiguously refer to *the* nth convergent as the fraction p_n/q_n, since this is unique when specified in lowest terms.

Example 4.16. The 5th convergent for $\sqrt{3}$ is p_5/q_5, which equals

$$[1, 1, 2, 1, 2, 1] = 26/15$$

So $p_5 = 26, q_5 = 15$.

Convergence

We now prove that (assuming conditions (A) and (B)) the infinite continued fraction expansion of any irrational real number converges, so (4.8) makes sense.

Theorem 4.17. *For any irrational real* α *the sequence* (p_n/q_n) *converges to* α.

Proof: Using (4.14) routine calculations show that

$$\alpha - \frac{p_n}{q_n} = \frac{(-1)^n}{q_n(\alpha_{n+1}q_n + q_{n-1})} \tag{4.13}$$

Therefore

$$|\alpha - \frac{p_n}{q_n}| = \frac{1}{q_n(\alpha_{n+1}q_n + q_{n-1})} < \frac{1}{q_n q_{n+1}} \tag{4.14}$$

By (4.14) (b), the sequence q_n is strictly increasing, so $q_n \to \infty$ as $n \to \infty$. Therefore $|\alpha - \frac{p_n}{q_n}| \to 0$ as $n \to \infty$. □

4.5 Continued Fractions of Quadratic Irrationals

This section and the next are based in part on Unger [138], which in turn is based in part on Olds [102]. We begin to make contact between continued fractions and Pell's Equation, by considering the continued fractions of 'quadratic irrationals'.

Assume as usual that d is a non-square positive integer. Any element of the number field $\mathbb{Q}(\sqrt{d})$ that is not in $\mathbb{Q}$ is a *quadratic irrational.* Thus it has the form $\alpha = a + b\sqrt{d}$ with $a, b \in \mathbb{Q}$ and $b \neq 0$. Its conjugate is $\alpha' = a - b\sqrt{d}$.

Definition 4.18. The quadratic irrational α is *reduced* if $\alpha > 1$ and $-1 < \alpha' < 0$.

Theorem 4.19. *If α_n is reduced in $\mathbb{Q}(\sqrt{d})$ and $\alpha_n = a_n + 1/\alpha_{n+1}$ where $a_n = \lfloor \alpha_n \rfloor$, then α_{n+1} is reduced in $\mathbb{Q}(\sqrt{d})$.*

Proof: Clearly $\alpha_{n+1} \in \mathbb{Q}(\sqrt{d})$.

We have

$$0 < \alpha_n - a_n < 1 \qquad 0 < 1/\alpha_{n+1} < 1$$

so in particular $\alpha_{n+1} > 1$. Since conjugation is an automorphism,

$$(\alpha_n - a_n)' = (1/\alpha_{n+1})'$$

so

$$\alpha_n' - a_n = 1/\alpha_{n+1}'$$

Now $a_n > 1$ and $-1 < \alpha_n' < 0$, so

$$1 < \alpha_n' - a_n = -1/\alpha_{n+1}'$$

whence $-1 < \alpha_{n+1}' < 0$. □

Theorem 4.20. *For each d, the number of reduced quadratic irrationals in $\mathbb{Q}(\sqrt{d})$ is finite.*

Proof: Any quadratic irrational can be written in the form $(A \pm \sqrt{d})/B$ where $A, B \in \mathbb{Z}$ and $B > 0$. If $\alpha = (A + \sqrt{d})/B$ then $\alpha' = (A - \sqrt{d})/B$. Suppose that α is reduced. Then $1 < (A + \sqrt{d})/B$ so $B < A + \sqrt{d}$, and $-1 < (A - \sqrt{d})/B < 0$ so $-B < A - \sqrt{d}$ and $\sqrt{d} - A < B$. Also $(A - \sqrt{d})/B < 0$ so $A < \sqrt{d}$.

Next, observe that $0 = -1 + 1 < \alpha' + \alpha$, so $0 < 2A/B$ and $A > 0$. Now

$$0 < \sqrt{d} - A < B < \sqrt{d} + A < 2\sqrt{d}$$

Therefore A lies between 0 and $\sqrt{d}$, and B lies between 0 and $2\sqrt{d}$. Thus the number of such A, B is finite.

Lemma 4.21. *If $\alpha = a + x = b + y$ where $a, b \in \mathbb{Z}$ and $0 < x, y < 1$ then $a = b$ and $x = y$.*

Proof: We have $a = \lfloor \alpha \rfloor = b$. Now $x = y$. □

We say that the continued fraction $[a_0, a_1, a_2 \ldots]$ is *purely periodic* if there exists $m > 1$ such that $a_i = a_{m+i}$ for all $i \geq 0$. It then repeats the sequence $a_0, a_1, \ldots, a_m$ indefinitely.

Theorem 4.22. *If α is a reduced quadratic irrational, then its continued fraction expansion is purely periodic.*

Proof: By Theorem 4.19, every residue r_j of α is reduced, and Theorem 4.20 tells us that the number of such residues is finite. Since α is irrational its continued fraction never terminates, so some value of r_j must occur twice: $r_j = r_k$ for $j \neq k$. Now $a_j = \lfloor r_j \rfloor$ and $a_k = \lfloor r_k \rfloor$, so $a_j = a_k$. Now

$$a_j + \frac{1}{r_{j+1}} = a_k + \frac{1}{r_{k+1}}$$

so $r_{j+1} = r_{k+1}$. Inductively we find that $r_{j+l} = r_{k+l}$ for all $l > 0$.

This shows that the continued fraction is eventually periodic, but not yet that it is purely periodic. To complete the proof we show that $r_{j-1} = r_{k-1}$. Continuing to use $'$ to indicate the conjugate, we have

$$r_{j-1} = a_{j-1} + \frac{1}{r_j} \qquad r_{k-1} = a_{k-1} + \frac{1}{r_k}$$

and

$$r'_{j-1} = a_{j-1} + \frac{1}{r'_j} \qquad r'_{k-1} = a_{k-1} + \frac{1}{r'_k}$$

Since $r_j = r_k$ we also have $r'_j = r'_k$, and

$$r'_{j-1} - a_{j-1} = r'_{k-1} - a_{k-1}$$

so

$$a_{j-1} - r'_{j-1} = a_{k-1} - r'_{k-1}$$

Since r_{j-1} and r_{k-1} are reduced, Lemma 4.21 implies that $a_{j-1} = a_{k-1}$. Inductively, $a_{j-l} = a_{k-l}$ for all $l > 0$ such that these numbers are defined. Thus the continued fraction is purely periodic.

Theorem 4.23. *If d is a non-square positive integer, then the continued fraction for $\sqrt{d}$ has the form*

$$\sqrt{d} = [a_0, \overline{a_1, a_2, \ldots, a_{m-1}, 2a_0}] \tag{4.15}$$

where the overline indicates periodic repetition.

Proof: Since $1 < \sqrt{d}$ its conjugate $-\sqrt{d} < -1$ so $\sqrt{d}$ is not reduced. However, if $a_0 = \lfloor \sqrt{d} \rfloor$ then $\sqrt{d} - a_0$ is reduced. By Theorem 4.22 its continued fraction expansion is purely periodic. Thus

$$a_0 + \sqrt{d} = [\overline{2a_0, a_2, \ldots, a_{m-1}}]$$

which continues as

$$a_0 + \sqrt{d} = [2a_0, a_2, \ldots, a_{m-1}, 2a_0, a_2, \ldots, a_{m-1}, \ldots]$$

Subtract a_0 to get

$$\sqrt{d} = [a_0, \overline{a_1, a_2, \ldots, a_{m-1}, 2a_0}]$$

as claimed. □

4.6 Minimal Solution of Pell's Equation by Continued Fractions

Theorem 4.24. *Let d be a non-square positive integer, so that $\sqrt{d}$ has the continued fraction*

$$\sqrt{d} = [a_0, \overline{a_1, a_2, \ldots, a_{m-1}, 2a_0}]$$

with period length m.

If m is even then $x = p_{m-1}, y = q_{m-1}$ is a nontrivial positive solution of Pell's Equation $x^2 - dy^2 = 1$.

If m is odd then $x = p_{2m-1}, y = q_{2m-1}$ is a nontrivial positive solution of Pell's Equation $x^2 - dy^2 = 1$.

Proof: Equation (4.15) implies that

$$\sqrt{d} = [a_0, a_2, \ldots, a_{m-1} + \frac{1}{a_0 + \sqrt{d}}]$$

By Proposition 4.13,

$$\begin{aligned}\sqrt{d} &= \frac{(a_0+\sqrt{d})p_{m-1}+p_{m-2}}{(a_0+\sqrt{d})q_{m-1}+q_{m-2}}\\ (a_0+\sqrt{d})q_{m-1}\sqrt{d}+q_{m-2}\sqrt{d} &= (a_0+\sqrt{d})p_{m-1}+p_{m-2}\\ q_{m-1}d+(a_0q_{m-1}+q_{m-2})\sqrt{d} &= a_0p_{m-1}+p_{m-2}+p_{m-1}\sqrt{d}\end{aligned}$$

Since 1 and $\sqrt{d}$ are linearly independent over $\mathbb{Q}$,

$$q_{m-1}d = a_0p_{m-1}+p_{m-2} \qquad a_0q_{m-1}+q_{m-2}=p_{m-1}$$

so

$$p_{m-2}=q_{m-1}d-a_0p_{m-1} \qquad q_{m-2}=p_{m-1}-a_0q_{m-1}$$

By Proposition 4.14,

$$\begin{aligned}(-1)^{m-2} &= p_{m-1}q_{m-2}-p_{m-2}q_{m-1}\\ &= p_{m-1}(p_{m-1}-a_0q_{m-1})-(q_{m-1}d-a_0p_{m-1})q_{m-1}\\ &= p_{m-1}^2-dq_{m-1}^2\end{aligned}$$

When m is even this proves that $x=p_{m-1}, y=q_{m-1}$ is a nontrivial positive solution of Pell's Equation, as required.

When m is odd we argue as above for *two* periods of the continued fraction. Now the period is $2m$, which is even. Therefore, by the first part, $x=p_{2m-1}, y=q_{2m-1}$ is a nontrivial positive solution of Pell's Equation.

Alternatively, $x=p_{m-1}, y=q_{m-1}$ is a nontrivial positive solution of the negative Pell equation. Combining this with itself gives a nontrivial positive solution of Pell's Equation. In fact this is the solution $x=p_{2m-1}, y=q_{2m-1}$.

□

Example 4.25. We solve $x^2-7y^2=1$ by the continued fraction method. We have $\sqrt{7}=[2,\overline{1,1,1,4}]$. Here $m=4$ so we can find the solution from the convergent $p_3/q_3=[2,1,1,1]$. This equals 8/3, so $x=8, y=3$. As a check, $8^2-7.3^2=64-63=1$.

Example 4.26. As a more complicated example, we solve $x^2-114y^2=1$. It can be shown that $\sqrt{114}=[10,\overline{1,2,10,2,1,20}]$. Here $m=6$, so we can find the solution from the convergent $p_5/q_5=[10,1,2,10,2,1]$. This equals 1025/96, so $x=1025, y=96$. As a check, $1025^2-114.96^2=1050625-1050624=1$.

4.7 Exercises

4.1 If (x, y) solve Pell's equation $x^2 - 2y^2 = 1$, and x and y are large, show that $\frac{x}{y}$ is a good approximation to $\sqrt{2}$. Prove that the approximation gets closer as x, y become larger.

4.2 Using computer algebra if necessary, find minimal solutions for Pell's Equation for non-square d between 2 and 20. *Hint*: The hardest cases are $d = 19, d = 13$.

4.3 Use Brahmagupta's formula to show that if $x_1^2 - dy_1^2 = k_1$ and $x_2^2 - dy_2^2 = k_2$ then $(x_1x_2 + dy_1y_2)^2 - d(x_1y_2 + x_2y_1)^2 = k_1k_2$. Explain how to derive a nontrivial solution of Pell's Equation from one of the negative Pell's Equation. Show that if $x^2 - dy^2 = k$ has a nontrivial solution then it has infinitely many.

4.4 Show that there are no integer solutions to $x^2 - 3y^2 = 7$. (*Hint*: Consider the equation (mod 4).)

4.5 Compute the first five positive solutions (x, y) of Pell's Equation for $d = 2$. Calculate the fractions x/y to five decimal places and compare them with $\sqrt{2}$ to five decimal places.

4.6 Relate the Euclidean algorithm to find the $\gcd(m, n)$ of two positive integers $m \geq n$ to the continued fraction expansion of m/n. Deduce that every positive rational has a finite continued fraction expansion.

4.7 Find an integer solution of $x^2 - 5y^2 = 11$.

Show how to generate infinitely many solutions.

4.8 Using computer algebra if you wish, show that $x = 158070671986249$, $y = 15140424455100$ is a solution of $x^2 - 109y^2 = 1$. (It is the minimal solution!)

4.9 Show that except for trivial exceptions, every positive rational has exactly *two* distinct finite continued fraction expansions. Which numbers are exceptions and how many continued fraction expansions do they have? (*Hint*: For example, $3/2 = [1, 2] = [1, 1, 1]$.)

4.10 Let α be irrational. Prove that the convergents p_n/q_n are alternately greater than and less than α.

4.11 Calculate the continued fractions of $\sqrt{5}$, $\sqrt{6}$, $\sqrt{7}$, $\sqrt{11}$, and $(1+\sqrt{5})/2$. (Continue until the continued fraction becomes periodic.)

4.12 Calculate the first five terms $[a_0, a_1, a_2, a_3, a_4]$ of the continued fraction expansion of π, and find the convergents to this order. Compare their numerical values to that of π.

(Use computer algebra or a high-precision calculator if you wish.)

4.13 Do the same for e.

4.14 Let $d \in \mathbb{Z}$. Show that the set of all matrices

$$\begin{bmatrix} x & dy \\ y & x \end{bmatrix}$$

with $x^2 - dy^2 = 1$ is a group under multiplication.

Show that the subset for which $x, y \in \mathbb{Z}$ is a subgroup and consists of integer matrices.

4.15 Let (x, y) be a minimal solution of Pell's Equation. Prove that the first columns of the matrices $\pm M^n$ for $n \in \mathbb{Z}$ give all solutions.

4.16 How are these solutions arranged along the right-hand branch of the hyperbola H in Figure 4.1? How do (positive or negative) powers of M act on these solutions?

4.17 The group $\mathbf{SL}_2(\mathbb{R})$ comprises all 2×2 matrices over $\mathbb{R}$ with determinant 1, under matrix multiplication.

Show that the subset $\mathbf{SL}_2(\mathbb{Z})$ of matrices with integer entries is a subgroup.

When (x, y) is a minimal solution of Pell's Equation, show that $M \in \mathbf{SL}_2(\mathbb{Z})$.

4.18 Find the eigenvalues and eigenvectors of M. Verify that the eigenspaces are the asymptotes of H.

4.19 Read the beautiful article by Lenstra [80] on Archimedes's cattle problem, which reduces the problem to Pell's Equation with $d = 410286423278424$ and provides a compact formula for the solution(s).

5

Factorisation into Irreducibles

After the optional diversion into Pell's Equation, we resume pursuit of Kummer's special case of Fermat's Last Theorem, turning to the vexed but vital question of uniqueness of factorisation in the ring of integers of an algebraic number field. Historically, early experience with unique factorisation of integers and polynomials over a field led to a general intuition that factorisation of algebraic integers should also be unique. In the early days of algebraic number theory many experts, including Euler, simply assumed uniqueness without perceiving any need for a proof, and used it implicitly to 'prove' results that were later found to be based on a false or unproved assumption.

The reason for this misconception is subtle and has its origins in the definition of a prime number. Two distinct properties can serve as a definition. The most familiar is that a prime number cannot be factorised into the product of two positive integers other than itself and 1. For a prime p, this property may be written more generally as:

(a) If $p = ab$ then one of a or b must be a unit.

In $\mathbb{Z}$ the only units are ± 1, so this reduces to the usual definition. However, there is a second property of prime numbers that is also of interest, namely that if a prime number p divides a product of two numbers then it must divide one or the other:

(b) If $p|ab$ then $p|a$ or $p|b$.

Although these properties are equivalent in the ring of integers—and even in some algebraic number rings—they are not equivalent in all cases. Of deeper psychological significance is that the more familiar property (a) turns out to be less powerful than property (b). Property (b) can be shown in general to imply (a); moreover, it is property (b) that guarantees uniqueness of factorisation, not the more comfortable property (a). In contrast, property (a) does not imply (b), and (a) turns out to be inadequate to give uniqueness of factorisation.

The way out of this dilemma is to use the less familiar (b) to define a prime. An element p satisfying the weaker assumption (a) is no longer called a prime: it is said to be irreducible. (We have used that word in the same sense earlier in this text. Sometimes we say that an element is *an* irreducible, to avoid the clumsy phrase 'irreducible element', so the word can be used both as an adjective and as a noun. 'Product of irreducibles' is an example.) It can then be proved that if factorisation into primes (defined in the new sense) is possible, then it is unique. In contrast, factorisation into irreducibles may not be unique even when it is possible.

Example 5.1. In $\mathbb{Z}[\sqrt{-6}]$ there are two factorisations $6 = 2 \cdot 3$ and $6 = (+\sqrt{-6}) \cdot (-\sqrt{-6})$. The elements $2, 3, \pm\sqrt{-6}$ are all irreducible (because they cannot be written as a product of nontrivial factors in $\mathbb{Z}[\sqrt{-6}]$); however, they are not prime. For instance, $\sqrt{-6}$ is a factor of $6 = 2 \cdot 3$, but it is not a factor of either 2 or 3 in the ring $\mathbb{Z}[\sqrt{-6}]$.

We must therefore proceed with care. For instance, when trying to factorise an element x in a domain D, it is natural to seek proper factors $x = ab$ ('proper' meaning that neither a nor b is a unit). If either of these factors is further reducible, we factorise that, and so on, seeking a factorisation

$$x = a_1 a_2 \ldots a_n$$

into factors that cannot be reduced any further. Reflecting on what we are doing, we see that if this search for a factorisation terminates, then it naturally leads to elements that are irreducible, definition (a), rather than what we now call primes, definition (b). So initially we concern ourselves with factorisation into irreducibles.

Naively, it might seem that if we break a number into smaller pieces (factors) and keep going until no more pieces can be broken, the result ought to be the same, no matter how we break it apart—rather like breaking up an already assembled jigsaw puzzle. Example 5.1 shows that this intuition is misleading. An algebraic integer is more like a jigsaw puzzle, some of whose pieces start to glue together when you try to break it up, with the curious feature that which ones glue together depends on how you set about breaking it.

Another potential difficulty is that in general, factorisation into irreducibles may not be possible, because the 'break-up' procedure may continue indefinitely. However, it can be proved that this obstacle does not apply to the ring $\mathfrak{O}$ of integers of any number field. To prove this we introduce the notion of a noetherian ring, named for Amalie Emmy Noether, one of the founders of abstract algebra. We show that factorisation is always possible if the domain D is noetherian; we then demonstrate that $\mathfrak{O}$ is noetherian.

Even though factorisation into irreducibles is always possible in $\mathfrak{O}$, we give an extensive list of examples where such a factorisation is not unique. In other cases, however, the existence of a generalised version of the division algorithm (which we term a 'Euclidean function') implies that every irreducible is prime. We see that factorisation into primes is unique, when it is possible, so some rings $\mathfrak{O}$ possess unique factorisation. In particular we characterise such $\mathfrak{O}$ for the fields $\mathbb{Q}(\sqrt{d})$ with d a negative rational integer: there are exactly five of them, corresponding to $d = -1, -2, -3, -7, -11$. We also prove the existence of a Euclidean function for some fields $\mathbb{Q}(\sqrt{d})$ with d positive. In Chapter 12 we see that $\mathfrak{O}$ may have unique factorisation without possessing a Euclidean function.

To begin the chapter we consider a little history, and look at an example of the intuitive use of unique factorisation, to motivate the ideas.

5.1 Historical Background

In the 18th century and the first part of the 19th there were varying standards of rigour in number theory. Euler, the most prolific mathematician of the 18th century, was primarily interested in obtaining *results* rather than setting up general theories, though he did that too. He sometimes used intuitive methods of proof which the hindsight of history has shown to be incorrect. For instance, in his famous textbook on algebra, he made several elegant applications of unique factorisation to 'prove' number theoretic results in cases where unique factorisation is false. Gauss, on the other hand, was fully aware of the dangers, and he found it necessary to demonstrate rigorously that the Gaussian integers $\mathbb{Z}[i]$ factorise uniquely.

In the Preface we mentioned that in 1847, Gabriel Lamé announced to a meeting of the Paris Academy that he had proved Fermat's Last Theorem, but his proof was seen to depend on uniqueness of factorisation for cyclotomic integers, and was shown to be inadequate. Kummer had, in fact, published a paper three years earlier that demonstrated the failure of unique factorisation for cyclotomic integers, thus destroying Lamé's proof. This result, which formed part of his habilitation thesis and was published as a pamphlet, went unnoticed at the time. (It is sometimes stated that at one stage Kummer thought he had proved Fermat's Last Theorem using this erroneous method, until Peter Lejeune Dirichlet pointed out the mistake, but this tale seems to be based on a misunderstanding.)

Gotthold Eisenstein put his finger on the property that characterises unique factorisation in a letter of 1844, which translates as:

> If one had the theorem which states that the product of two complex numbers can be divisible by a prime number only when one of the factors is—which seems completely obvious—then one would have the whole theory at a single blow; but this theorem is totally false.

By 'the whole theory' he was referring to consequences of unique factorisation, which in particular is relevant to Fermat's Last Theorem.

In Eisenstein's letter, 'prime' meant definition (a) of this chapter, and his comment translated into the terminology of this book is 'if every irreducible is prime, then unique factorisation holds'. It is also clear from his comment that he knew instances of irreducibles that are not prime, leading to non-unique factorisation. All this must have seemed very confusing to average 19th century mathematicians, who were accustomed to using intuitive ideas about factorisation without question.

To give an idea of what this style of mathematical 'proof' was like, before we develop a rigorous theory of unique factorisation, we exhibit a concocted, fallacious, but plausible proof of a statement of Fermat using this intuitive approach. Fermat's proof has not survived, and we are not suggesting that it resembled our faulty but instructive attempt below. Indeed it is not hard to reconstruct a *rigorous* proof using ideas that were known to Fermat; see Weil [143].

A Statement of Fermat. *The only integer solutions of the equation* $y^2+2=x^3$ *are* $y=\pm 5$, $x=3$.

Intuitive 'Proof'. Clearly y cannot be even, for then the right-hand side would be divisible by 8, but the left-hand side only by 2. Factorise in the ring $\mathbb{Z}[\sqrt{-2}]$, consisting of all $a+b\sqrt{-2}$ for $a,b\in\mathbb{Z}$, to obtain

$$(y+\sqrt{-2})(y-\sqrt{-2})=x^3$$

A common factor $c+d\sqrt{-2}$ of $y+\sqrt{-2}$ and $y-\sqrt{-2}$ would also divide their sum $2y$ and their difference $2\sqrt{-2}$. Taking norms,

$$c^2+2d^2|4y^2 \qquad c^2+2d^2|8,$$

hence $c^2+2d^2|4$. The only solutions of this relation are $c=\pm 1$, $d=0$, or $c=0$, $d=\pm 1$, or $c=\pm 2$, $d=0$. None of these give proper factors of $y+\sqrt{-2}$, so $y+\sqrt{-2}$ and $y-\sqrt{-2}$ are coprime. Now the product of two coprime numbers is a cube only when each is a cube (both units ± 1 are cubes), so

$$y+\sqrt{-2}=(a+b\sqrt{-2})^3$$

and comparing coefficients of $\sqrt{-2}$,

$$1 = b(3a^2 - 2b^2)$$

for which the only solutions are $b = 1$, $a = \pm 1$. Then $x = 3$, $y = \pm 5$. □

The flaw is that we are carrying over the language of factorisation of integers to factorisation in $\mathbb{Z}[\sqrt{-2}]$, without checking that the usual properties actually hold in $\mathbb{Z}[\sqrt{-2}]$. (As it happens, they do, see Theorem 5.28, but this needs to be proved.) In this chapter we develop the appropriate theory and investigate when it generalises to a ring of integers in a number field.

5.2 Trivial Factorisations

If u is a unit in a ring R, then any element $x \in R$ can be trivially factorised as $x = uy$, where $y = u^{-1}x$. Such considerations lead to:

Definition 5.2. An element y is an *associate* of x if $x = uy$ for a unit u.

Recall that a factorisation of $x \in R$, $x = yz$ is proper if neither y nor z is a unit. We call a factorisation that is not proper *trivial.* Then one factor is a unit and the other is an associate of x. Before going on to proper factorisations, we therefore look at elementary properties of units and associates.

Definition 5.3. The set of units in a ring R is denoted by $U(R)$.

It is routine to prove:

Proposition 5.4. *The units $U(R)$ of a ring R form a group under multiplication.* □

Example 5.5.

(a) $R = \mathbb{Q}$. The units are $U(\mathbb{Q}) = \mathbb{Q}\backslash\{0\}$, which is an infinite group.

(b) $R = \mathbb{Z}$. The units are ± 1, so $U(\mathbb{Z})$ is cyclic of order 2.

(c) $R = \mathbb{Z}[\mathrm{i}]$, the Gaussian integers $a + \mathrm{i}b$ ($a, b \in \mathbb{Z}$). The element $a + \mathrm{i}b$ is a unit if and only if there exists $c + \mathrm{i}d$ ($c, d \in \mathbb{Z}$) such that

$$(a + \mathrm{i}b)(c + \mathrm{i}d) = 1$$

This implies $ac - bd = 1$, $ad + bc = 0$, whence $c = a/(a^2 + b^2)$, $d = -b/(a^2+b^2)$. These have integer solutions only when $a^2+b^2 = 1$, so $a = \pm 1$, $b = 0$, or $a = 0$, $b = \pm 1$. Hence the units are $\{1, -1, \mathrm{i}, -\mathrm{i}\}$ and $U(R)$ is cyclic of order 4.

By using norms, we can extend the results of Example 5.5 to the more general case of the units in the ring of integers of $\mathbb{Q}(\sqrt{d})$ for d negative and squarefree:

Proposition 5.6. *The group of units U of the integers in $\mathbb{Q}(\sqrt{d})$ where d is negative and squarefree is as follows:*

(a) For $d = -1$, $U = \{\pm 1, \pm \mathrm{i}\}$.
(b) For $d = -3$, $U = \{\pm 1, \pm\omega, \pm\omega^2\}$ where $\omega = e^{2\pi \mathrm{i}/3}$.
(c) For all other $d < 0$, $U = \{\pm 1\}$.

Proof: Suppose α is a unit in the ring of integers of $\mathbb{Q}(\sqrt{d})$ with inverse β. Then $\alpha\beta = 1$, so taking norms

$$\mathrm{N}(\alpha)\mathrm{N}(\beta) = 1$$

But $\mathrm{N}(\alpha)$, $\mathrm{N}(\beta)$ are rational integers, so $\mathrm{N}(\alpha) = \pm 1$. Writing $\alpha = a + b\sqrt{d}$ $(a, b \in \mathbb{Q})$, it is clear that $\mathrm{N}(\alpha) = a^2 - db^2$ is positive (for negative d), so $\mathrm{N}(\alpha) = +1$. Hence finding units reduces to solving Pell's Equation

$$a^2 - db^2 = 1$$

If $a, b \in \mathbb{Z}$, then for $d = -1$ this is

$$a^2 + b^2 = 1$$

with solutions $a = \pm 1$, $b = 0$, or $a = 0$, $b = \pm 1$, already found in Example 5.5(c). This gives (a).

For $d < -3$ we immediately conclude that $b = 0$ (otherwise $a^2 - db^2$ would exceed 1), so the only rational integer solutions are $a = \pm 1$, $b = 0$. If $d \not\equiv 1 \pmod 4$, then $a, b \in \mathbb{Z}$, so the only solutions are those discovered. For $d \equiv 1 \pmod 4$, however, we must also consider the additional possibility $a = A/2$, $b = B/2$ where both A and B are odd rational integers. In this case

$$A^2 - dB^2 = 4$$

Since $d < -3$ we deduce $B = 0$ and there are no additional solutions. This completes (c).

For $d = -3$, we find additional solutions $A = \pm 1$, $B = \pm 1$. The case $A = 1$, $B = 1$ gives

$$\alpha = \tfrac{1}{2}(-1 + \sqrt{-3}) = e^{2\pi \mathrm{i}/3}$$

which we usually denote by ω. The other three cases give $-\omega, \omega^2, -\omega^2$. These allied with the solutions already found give (b). □

We postpone the general case of units in a ring of integers in a number field until Chapter 10. We now return to simple properties of units and associates.

Proposition 5.4 easily implies that 'being associates' is an equivalence relation on R. The only associate of 0 is 0 itself. Recall that a non-unit $x \in R$ is irreducible if it has no proper factors. The zero element $0 = 0.0$ has factors, neither of which is a unit, so in particular an irreducible is nonzero. We now list a few elementary properties of units, associates, and irreducibles. To prove some of these we require the cancellation law, so the ring must be a domain.

Proposition 5.7. *For a domain D,*

(a) *An element x is a unit if and only if $x|1$.*

(b) *Any two units are associates and any associate of a unit is a unit.*

(c) *Elements x, y are associates if and only if $x|y$ and $y|x$.*

(d) *An element x is irreducible if and only if every divisor of x is an associate of x or a unit.*

(e) *An associate of an irreducible is irreducible.*

Proof: Most of these follow straight from the definitions. We prove (c) which requires the cancellation law. Suppose $x|y$ and $y|x$; then there exist $a, b \in D$ such that $y = ax$, $x = by$. Substituting, $x = bax$. Now either $x = 0$, in which case $y = 0$ also and they are associates, or $x \neq 0$ and we cancel x to find $1 = ba$, so a and b are units. Hence x, y are associates. The converse is trivial. □

Some of these concepts may usefully be expressed in terms of ideals:

Proposition 5.8. *If D is a domain and x, y are nonzero elements of D then*

(a) *$x|y$ if and only if $\langle x \rangle \supseteq \langle y \rangle$.*

(b) *x and y are associates if and only if $\langle x \rangle = \langle y \rangle$.*

(c) *x is a unit if and only if $\langle x \rangle = D$.*

(d) *x is irreducible if and only if $\langle x \rangle$ is maximal among the proper principal ideals of D.*

Proof: (a) If $x|y$ then $y = zx \in \langle x \rangle$ for some $z \in D$, hence $\langle y \rangle \subseteq \langle x \rangle$. Conversely, if $\langle y \rangle \subseteq \langle x \rangle$ then $y \in \langle x \rangle$, so $y = zx$ for some $z \in D$.

(b) is immediate from (a).

(c) If x is a unit then $xv = 1$ for some $v \in D$, hence for any $y \in D$ we have $y = xvy \in \langle x \rangle$ and $D = \langle x \rangle$. If $D = \langle x \rangle$ then since $1 \in D$, $1 = zx$ for some $z \in D$ and x is a unit.

(d) Suppose x is irreducible, with $\langle x \rangle \subsetneqq \langle y \rangle \subsetneqq D$. Then $y|x$ but is neither a unit, nor an associate of x, contradicting Proposition 5.7(d). Conversely, if no such y exists, then every divisor of x is either a unit or an associate, so x is irreducible. □

5.3 Factorisation in Noetherian Rings

In a domain D, if a non-unit x is reducible then $x = ab$ for $a, b \in D$ where neither is a unit. If either of a or b is reducible, we can express it as a product of proper factors; then carry on the process, seeking to write

$$x = p_1 p_2 \dots p_m$$

where each p_i is irreducible. We say that *factorisation into irreducibles is possible* in D if every $x \in D$, not a unit nor zero, is a product of a finite number of irreducibles.

In general, such a factorisation may not be possible. An example is ready to hand, namely the ring $\mathbb{B}$ of all algebraic integers. For if α is not zero or a unit, neither is $\sqrt{\alpha}$. Now $\alpha = \sqrt{\alpha} \cdot \sqrt{\alpha}$ and $\sqrt{\alpha}$ is an integer, so α is reducible. Thus $\mathbb{B}$ has no irreducibles at all, but it does have nonzero non-units, so factorisation into irreducibles is not possible.

This trouble does not arise in the ring $\mathfrak{O}$ of integers of a number field—another reason why we concentrate on such rings instead of the whole of $\mathbb{B}$. We prove the possibility of factorisation in $\mathfrak{O}$ by introducing a more general notion that makes the arguments involved more transparent.

Definition 5.9. A ring R is *noetherian* if every ideal in R is finitely generated.

As already mentioned, the adjective commemorates Emmy Noether, who introduced the concept, along with many of the standard 'group, ring, field' ideas in abstract algebra.

Below, we demonstrate the possibility of factorisation in any noetherian ring. Then we show that $\mathfrak{O}$ is noetherian, so factorisation is possible here also. First, we introduce two useful properties. In Proposition 5.10 we prove that each is equivalent to the noetherian condition.

The Ascending Chain Condition. Given an ascending chain of ideals:

$$I_0 \subseteq I_1 \subseteq \ldots \subseteq I_n \subseteq \ldots \tag{5.1}$$

there exists some N for which $I_n = I_N$ for all $n \geq N$. That is, every ascending chain *stops* (or *terminates*).

The Maximal Condition. Every non-empty set of ideals has a maximal element, that is an element which is not properly contained in every other element.

This maximal element need not contain *all* the other ideals in the given set: we require only that there is no other element in the set that properly contains *it*.

Proposition 5.10. *The following conditions are equivalent for a ring R:*
(a) *R is noetherian.*
(b) *R satisfies the ascending chain condition.*
(c) *R satisfies the maximal condition.*

Proof: We show that (a) $\Rightarrow$ (b) $\Rightarrow$ (c) $\Rightarrow$ (a).

Assume (a). Consider an ascending chain as in (5.1). Let $I = \cup_{n=1}^{\infty} I_n$. Then I is an ideal, so it is finitely generated: say $I = \langle x_1, \ldots, x_m \rangle$. Each x_i belongs to some $I_{n(i)}$. If we let $N = \max_i n(i)$, then we have $I = I_N$ and $I_n = I_N$ for all $n \geq N$, proving (b).

Now suppose (b) and consider a non-empty set S of ideals. Suppose for a contradiction that S does not have a maximal element. Pick $I_0 \in S$. Since I_0 is not maximal we can pick $I_1 \in S$ with $I_0 \subsetneqq I_1$. Inductively, having found I_n, since this is not maximal, we can pick $I_{n+1} \in S$ with $I_n \subsetneqq I_{n+1}$. But now we have an ascending chain that does not stop, which is a contradiction. So (b) implies (c). (You may wish to ponder the use of the axiom of choice in this proof.)

Finally, suppose (c). Let I be any ideal, and let S be the set of all finitely generated ideals contained in I. Then $\{0\} \in S$, so S is non-empty and thus has a maximal element J. If $J \neq I$, pick $x \in I \setminus J$. Then $\langle J, x \rangle$ is finitely generated and strictly larger than J, a contradiction. Hence $J = I$ and I is finitely generated. □

Example 5.11. Figure 5.1 shows some of the ideals of $\mathbb{Z}$, namely those that contain $\langle 60 \rangle$.

Any ascending chain of ideals that contains $\langle 60 \rangle$ must terminate after at most four more steps, since that is the length of the longest upward chain from $\langle 60 \rangle$.

More generally, $\mathbb{Z}$ is a principal ideal domain. Let I be an ideal, so that $I = \langle n \rangle$ for some (non-negative) integer n. The ideals that contain n are

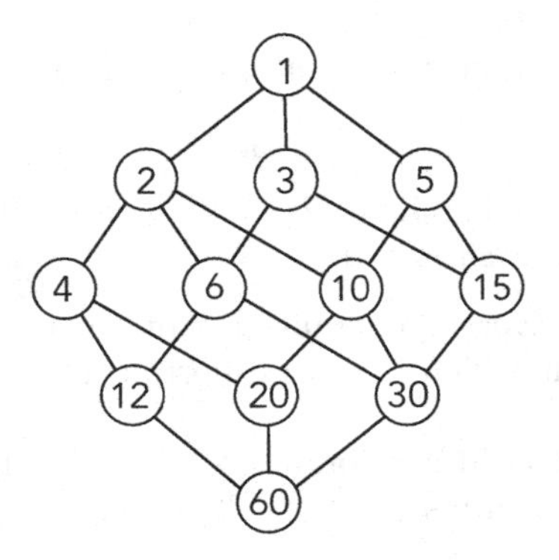

Figure 5.1. All ideals of $\mathbb{Z}$ that contain $\langle 60\rangle$. Numbers indicate the generator of the ideal. Sequences of upward sloping lines indicate containment.

$\langle d\rangle$ where d divides n. Since n has only a finite number of divisors, any ascending chain containing I must terminate.

A simpler proof is to observe that in a principal ideal domain, every ideal is finitely generated (by *one* generator). Now use Definition 5.9.

Theorem 5.12. *If a domain D is noetherian, factorisation into irreducibles is possible in D.*

Proof: Suppose that D is noetherian, but there exists a non-unit $x \neq 0$ in D that cannot be expressed as a product of a finite number of irreducibles. Choose x so that $\langle x\rangle$ is maximal subject to these conditions on x, which is possible by the maximal condition. By definition, x is not irreducible, so $x = yz$ where y and z are not units. Then $\langle y\rangle \supseteq \langle x\rangle$ by Proposition 5.7(a). If $\langle y\rangle = \langle x\rangle$ then x and y are associates by Proposition 5.7(b), but this is not the case because it implies that z is a unit. So $\langle y\rangle \supsetneqq \langle x\rangle$, and similarly $\langle z\rangle \supsetneqq \langle x\rangle$. By maximality of $\langle x\rangle$,

$$y = p_1 \dots p_r \qquad z = q_1 \dots q_s$$

where each p_i and q_j is irreducible. Multiply these together to express x as a product of irreducibles, a contradiction. Hence the assumption that there exists a non-unit $\neq 0$ that is not a finite product of irreducibles is false, and factorisation into irreducibles is always possible. □

We are now in business:

Theorem 5.13. *The ring of integers $\mathfrak{O}$ in a number field K is noetherian.*

Proof: We prove that every ideal I of $\mathfrak{O}$ is finitely generated. Now $(\mathfrak{O}, +)$ is free abelian of rank n equal to the degree of K by Theorem 2.28. Hence $(I, +)$ is free abelian of rank $s \leq n$ by Theorem 1.29. If $\{x_1, \dots, x_s\}$ is a $\mathbb{Z}$-basis for $(I, +)$, then clearly $\langle x_1, \dots, x_s \rangle = I$, so I is finitely generated and $\mathfrak{O}$ is noetherian. □

Corollary 5.14. *Factorisation into irreducibles is possible in $\mathfrak{O}$.* □

To get very far in the theory, we need an easy way to detect units and irreducibles in $\mathfrak{O}$. The norm proves to be a convenient tool:

Proposition 5.15. *Let $\mathfrak{O}$ be the ring of integers in a number field K, and let $x, y \in \mathfrak{O}$. Then*

(a) *x is a unit if and only if $\mathrm{N}(x) = \pm 1$.*
(b) *If x and y are associates, then $\mathrm{N}(x) = \pm\mathrm{N}(y)$.*
(c) *If $\mathrm{N}(x)$ is a rational prime, then x is irreducible in $\mathfrak{O}$.*

Proof: (a) If $xu = 1$, then $\mathrm{N}(x)\mathrm{N}(u) = 1$. Since $\mathrm{N}(x), \mathrm{N}(u) \in \mathbb{Z}$, we have $\mathrm{N}(x) = \pm 1$. Conversely, if $\mathrm{N}(x) = \pm 1$, then

$$\sigma_1(x)\sigma_2(x)\dots\sigma_n(x) = \pm 1$$

where the σ_i are the monomorphisms $K \to \mathbb{C}$. One factor, without loss in generality $\sigma_1(x)$, is equal to x; all the other $\sigma_i(x)$ are integers. Put

$$y = \pm\sigma_2(x)\dots\sigma_n(x)$$

Then $xy = 1$, so $y = x^{-1} \in K$. Hence $y \in K \cap \mathbb{B} = \mathfrak{O}$, and x is a unit.

(b) If x, y are associates, then $x = uy$ for a unit u, so $\mathrm{N}(x) = \mathrm{N}(uy) = \mathrm{N}(u)\mathrm{N}(y) = \pm\mathrm{N}(y)$ by (a).

(c) Let $x = yz$. Then $\mathrm{N}(y)\mathrm{N}(z) = \mathrm{N}(yz) = \mathrm{N}(x) = p$, a rational prime; so one of $\mathrm{N}(y)$ and $\mathrm{N}(z)$ is $\pm p$ and the other is ± 1. By (a), one of y and z is a unit, so x is irreducible. □

We have not asserted converses to parts (b) and (c) because these are generally false, as examples in the next section reveal.

Units are important because they determine which elements are associates. Proposition 5.15 is useful when seeking units, but in general it can be difficult to solve the equation $\mathrm{N}(x) = \pm 1$ explicitly. Dirichlet's Units Theorem 10.9, which is much deeper, helps.

5.4 Examples of Non-Unique Factorisation into Irreducibles

Definition 5.16. Factorisation in a domain D is *unique* if, whenever

$$p_1 \dots p_r = q_1 \dots q_s$$

where every p_i and q_j is irreducible in D, then

(a) $r = s$.

(b) There is a permutation π of $\{1, \dots, r\}$ such that p_i and $q_{\pi(i)}$ are associates for all $i = 1, \dots, r$.

In view of our earlier remarks about trivial factorisations, this is the best we can hope for. It says that a factorisation into irreducibles (if it exists) is unique except for the order of the factors and the possible presence of units. Variation to this extent is necessary, since even in $\mathbb{Z}$ we have, for instance,

$$3 \cdot 5 = 5 \cdot 3 = 1 \cdot 3 \cdot 5 = (-3)(-5) = 5(-3)(-1) = 3 \cdot 5 \cdot (-1)^{1066}$$

and so on. Unfortunately, factorisation into irreducibles need not be unique in a ring of integers of an algebraic number field. Examples are easy to come by, and to drive the point home we give quite a lot of them. They are drawn from quadratic fields, and we state them as positive theorems. The easiest come from imaginary quadratic fields:

Theorem 5.17. *Factorisation into irreducibles is not unique in the ring of integers of* $\mathbb{Q}(\sqrt{d})$ *for (at least) the following values of* d: -5, -6, -10, -13, -14, -15, -17, -21, -22, -23, -26, -29, -30.

Proof: In $\mathbb{Q}(\sqrt{-5})$ we have the factorisations

$$6 = 2 \cdot 3 = (1 + \sqrt{-5})(1 - \sqrt{-5})$$

We claim that $2, 3, 1 + \sqrt{-5}$ and $1 - \sqrt{-5}$ are irreducible in the ring $\mathfrak{O}$ of integers of $\mathbb{Q}(\sqrt{-5})$. Since the norm is

$$\mathrm{N}(a + b\sqrt{-5}) = a^2 + 5b^2$$

their norms are 4, 9, 6, 6, respectively. If $2 = xy$ where $x, y \in \mathfrak{O}$ are non-units, then $4 = \mathrm{N}(2) = \mathrm{N}(x)\mathrm{N}(y)$ so that $\mathrm{N}(x) = \pm 2$, $\mathrm{N}(y) = \pm 2$. Similarly nontrivial divisors of 3 must, if they exist, have norm ± 3, and nontrivial divisors of $1 \pm \sqrt{-5}$ have norm ± 2 or ± 3. Since $-5 \not\equiv 1 \pmod 4$, the integers $\mathfrak{O}$ are of the form $a + b\sqrt{-5}$ for $a, b \in \mathbb{Z}$ (Theorem 3.3) so

$$a^2 + 5b^2 = \pm 2 \text{ or } \pm 3 \qquad (a, b \in \mathbb{Z})$$

Now $|b| \geq 1$ implies $\left|a^2 + 5b^2\right| \geq 5$, so the only possibility is $|b| = 0$; but then we have $a^2 = \pm 2$ or ± 3, which is impossible in integers. Thus the putative divisors do not exist, and the four factors are all irreducible. Since $\mathrm{N}(2) = 4$, $\mathrm{N}(1 \pm \sqrt{-5}) = 6$, by Proposition 5.15(b), 2 is not an associate of $1 + \sqrt{-5}$ or $1 - \sqrt{-5}$, so factorisation is not unique.

The other stated values of d are dealt with in exactly the same way (with a few slight subtleties noted at the end of the proof) starting from the following factorisations:

$$\begin{array}{llll}
\mathbb{Q}(\sqrt{-6}): & 6 & = 2 \cdot 3 & = (\sqrt{-6})(-\sqrt{-6}) \\
\mathbb{Q}(\sqrt{-10}): & 14 & = 2 \cdot 7 & = (2 + \sqrt{-10})(2 - \sqrt{-10}) \\
\mathbb{Q}(\sqrt{-13}): & 14 & = 2 \cdot 7 & = (1 + \sqrt{-13})(1 - \sqrt{-13}) \\
\mathbb{Q}(\sqrt{-14}): & 15 & = 3 \cdot 5 & = (1 + \sqrt{-14})(1 - \sqrt{-14}) \\
\mathbb{Q}(\sqrt{-15}): & 4 & = 2 \cdot 2 & = \left(\frac{1+\sqrt{-15}}{2}\right)\left(\frac{1-\sqrt{-15}}{2}\right) \\
\mathbb{Q}(\sqrt{-17}): & 18 & = 2 \cdot 3 \cdot 3 & = (1 + \sqrt{-17})(1 - \sqrt{-17}) \\
\mathbb{Q}(\sqrt{-21}): & 22 & = 2 \cdot 11 & = (1 + \sqrt{-21})(1 - \sqrt{-21}) \\
\mathbb{Q}(\sqrt{-22}): & 26 & = 2 \cdot 13 & = (2 + \sqrt{-22})(2 - \sqrt{-22}) \\
\mathbb{Q}(\sqrt{-23}): & 6 & = 2 \cdot 3 & = \left(\frac{1+\sqrt{-23}}{2}\right)\left(\frac{1-\sqrt{-23}}{2}\right) \\
\mathbb{Q}(\sqrt{-26}): & 27 & = 3 \cdot 3 \cdot 3 & = (1 + \sqrt{-26})(1 - \sqrt{-26}) \\
\mathbb{Q}(\sqrt{-29}): & 30 & = 2 \cdot 3 \cdot 5 & = (1 + \sqrt{-29})(1 - \sqrt{-29}) \\
\mathbb{Q}(\sqrt{-30}): & 34 & = 2 \cdot 17 & = (2 + \sqrt{-30})(2 - \sqrt{-30}).
\end{array}$$

In cases -15 and -23, note that $d \equiv 1 \pmod 4$ and be careful. For -26 it is easy to prove 3 as irreducible. For $1 - \sqrt{-26}$ we are led to the equation $\mathrm{N}(x)\mathrm{N}(y) = 27$, so $\mathrm{N}(x) = \pm 9$, $\mathrm{N}(y) = \pm 3$, or the other way round. This leads to $a^2 + 26b^2 = \pm 9$ or ± 3. There *is* a solution for ± 9, but not for ± 3, and the latter is sufficient to show $1 + \sqrt{-26}$ is irreducible. □

Examining this list, we see that in the ring of integers of $\mathbb{Q}(\sqrt{-17})$ there is an example to show that even the *number* of irreducible factors may differ; the case $\mathbb{Q}(\sqrt{-26})$ shows that the number of distinct factors may differ and that even a (rational) prime power may factorise non-uniquely.

For real quadratic fields there are similar results, but these are harder to find. Also, since the norm is $a^2 - db^2$, it is harder to prove given numbers irreducible. With the same range of values as in Theorem 5.17 we find:

Theorem 5.18. *Factorisation into irreducibles is not unique in the ring of integers of $\mathbb{Q}(\sqrt{d})$ for (at least) the following values of d:*

$$10, 15, 26, 30$$

Proof: In the integers of $\mathbb{Q}(\sqrt{10})$ we have factorisations:

$$6 = 2 \cdot 3 = (4 + \sqrt{10})(4 - \sqrt{10})$$

We prove $2, 3, 4 \pm \sqrt{10}$ irreducible. Looking at norms, this amounts to proving that the equations

$$a^2 - 10b^2 = \pm 2 \text{ or } \pm 3$$

have no solutions in integers a, b. It is no longer helpful to look at the size of $|b|$, because of the minus sign. However, the equation implies

$$a^2 \equiv \pm 2 \text{ or } \pm 3 \qquad (\text{mod } 10)$$

or equivalently

$$a^2 = 2, 3, 7 \text{ or } 8 \qquad (\text{mod } 10)$$

The squares (mod 10) are, in order, 0, 1, 4, 9, 6, 5, 6, 9, 4, 1; by a seemingly remarkable coincidence, the numbers we are looking for are precisely those that do not occur. Hence no solutions exist and the four factors are irreducible. Now 2 and $4 \pm \sqrt{10}$ are not associates, since their norms are 4, 6, respectively.

Similarly:

$$\begin{array}{lcl} \mathbb{Q}(\sqrt{15}) & : & 10 = 2 \cdot 5 = (5 + \sqrt{15})(5 - \sqrt{15}) \\ \mathbb{Q}(\sqrt{26}) & : & 10 = 2 \cdot 5 = (6 + \sqrt{26})(6 - \sqrt{26}) \\ \mathbb{Q}(\sqrt{30}) & : & 6 = 2 \cdot 3 = (6 + \sqrt{30})(6 - \sqrt{30}) \end{array}$$

We omit details. □

The values of d considered in Theorems 5.17 and 5.18 have not, despite appearances, been chosen at random. If we try similar tricks with other d in the range -30 to 30, nothing seems to work. Thus in $\mathbb{Q}(\sqrt{-19})$ we get

$$\left(\frac{1 + \sqrt{-19}}{2}\right)\left(\frac{1 - \sqrt{-19}}{2}\right) = 5$$

but all this shows is that 5 is reducible.

Trying another obvious product in the integers of $\mathbb{Q}(\sqrt{-19})$, we find

$$(2 + \sqrt{-19})(2 - \sqrt{-19}) = 23$$

which just tells us that 23 is also reducible. The case

$$\left(\frac{3+\sqrt{-19}}{2}\right)\left(\frac{3-\sqrt{-19}}{2}\right)=7$$

shows 7 is reducible. After more of these calculations we may alight on

$$35 = 5 \cdot 7 = (4+\sqrt{-19})(4-\sqrt{-19}) \tag{5.2}$$

Does this prove non-uniqueness? No, because neither 5 nor 7 is irreducible, as we have seen; and neither is $4 \pm \sqrt{-19}$.

The complete factorisation of 35 is

$$\left(\frac{1+\sqrt{-19}}{2}\right)\left(\frac{1-\sqrt{-19}}{2}\right)\left(\frac{3+\sqrt{-19}}{2}\right)\left(\frac{3-\sqrt{-19}}{2}\right)$$

and the two apparently distinct factorisations in (5.2) come from different groupings of these in pairs. Eventually we are led to conjecture that the integers of $\mathbb{Q}(\sqrt{-19})$ have unique factorisation. This is indeed true, but we shall not be able to prove it until Chapter 10. In fact the ring of integers of $\mathbb{Q}(\sqrt{d})$ for negative squarefree d has unique factorisation into irreducibles if and only if d takes one of the values:

$$-1 \quad -2 \quad -3 \quad -7 \quad -11 \quad -19 \quad -43 \quad -67 \quad -163$$

Numerical evidence available in the time of Gauss pointed to this result. In 1934 Heilbronn and Linfoot [61] showed that at most one further value of d can occur, and that if it does then $|d|$ is very large. In 1952 Heegner [60] offered a proof, but it was thought to contain a gap. In 1967 Stark [126] found a proof, as did Baker [3] soon after. Finally Birch [8], Deuring [33] and Siegel [122] filled in the gap in Heegner's proof. The methods of this book are not appropriate to give any of these proofs, but we prove in Chapter 10 that for these nine values factorisation is unique.

The situation for positive d is not at all well understood. Factorisation is unique in many more cases, for instance 2, 3, 5, 6, 7, 11, 13, 14, 17, 19, 21, 22, 23, 29, 31, 33, 37, 38, 41, 43, 46, 47, 53, 57, 59, 61, 62, 67, 69, 71, 73, 77, 83, 86, 89, 93, 94, 97, ..., these being all for d less than 100. Gauss conjectured (in the context of his work on quadratic forms) a result equivalent to there being infinitely many real quadratic fields with unique factorisation, but this has neither been proved nor disproved. Henri Cohen and Hendrik Lenstra have given a heuristic (that is, plausible but non-rigorous) argument suggesting that about 75.446% of real quadratic fields $\mathbb{Q}(\sqrt{p})$ for prime p have unique factorisation. Computational experiments agree with this prediction, see Cohen [20] and te Riele and Williams [135].

5.5 Prime Factorisation

So far we have not proved uniqueness of factorisation for the ring of integers in any number fields (apart from $\mathbb{Z}$). We now introduce a criterion for factorisation to be unique, stated in terms of a special property of the irreducibles. We have already noted that an irreducible p in $\mathbb{Z}$ satisfies the additional property

$$p|mn \quad \text{implies} \quad p|m \quad \text{or} \quad p|n$$

In this section we show that this property characterises uniqueness of factorisation.

Definition 5.19. An element x of a domain D is *prime* if it is not zero or a unit and

$$x|ab \quad \text{implies} \quad x|a \quad \text{or} \quad x|b$$

The zero element satisfies the given property in a domain, but we exclude it to correspond with the definition of prime in $\mathbb{Z}$, where 0 is not usually considered a prime. This convention allows us to state:

Proposition 5.20. *A prime in a domain D is always irreducible.*

Proof: Suppose that D is a domain, $x \in D$ is prime, and $x = ab$. Then $x|ab$, so $x|a$ or $x|b$.

If $x|a$, then $a = xc$ $(c \in D)$, so $x = xcb$. Cancelling x (which is nonzero) we see that $1 = cb$ and b is a unit. In the same way, $x|b$ implies a is a unit. □

The converse of this result is false, as Eisenstein lamented in 1844; in many domains there exist irreducibles that are not primes. For example in $\mathbb{Z}[\sqrt{-5}]$

$$6 = 2 \cdot 3 = (1 + \sqrt{-5})(1 - \sqrt{-5})$$

but 2 does not divide either of $1 + \sqrt{-5}$ or $1 - \sqrt{-5}$, as shown in the proof of Theorem 5.17. So 2 is an irreducible in $\mathbb{Z}[\sqrt{-5}]$, but not prime. The factorisations in the proofs of Theorems 5.17 and 5.18 readily yield other examples. The next theorem tells us that such examples are entirely typical—every domain with non-unique factorisation contains irreducibles that are not prime:

Theorem 5.21. *A domain in which factorisation into irreducibles is possible, factorisation is unique if and only if every irreducible is prime.*

Proof: Let D be the domain. It is convenient to rephrase the possibility of factorisation for all nonzero $x \in D$ as

$$x = up_1 \dots p_r$$

where u is a unit and $p_1, \dots, p_r$ are irreducibles. When $r = 0$ this can then be interpreted as $x = u$ is a unit and when $r \geq 1$, then up_1 is an irreducible, so x is a product of the irreducibles $up_1, p_2, \dots, p_r$.

Now for the proof. Suppose first that factorisation is unique and p is an irreducible. We must show p is prime. That is, if $p|ab$ then $p|a$ or $p|b$. Now $p|ab$ if and only if $pc = ab$ for some $c \in D$. We need to consider only the nontrivial case $a \neq 0$, $b \neq 0$ which implies $c \neq 0$ also. Factorise, a, b, c into irreducibles:

$$a = u_1 p_1 \dots p_n \qquad b = u_2 q_1 \dots q_m \qquad c = u_3 r_1 \dots r_s$$

where each u_i is a unit and p_i, q_i and r_i are irreducibles. Then

$$p(u_3 r_1 \dots r_s) = (u_1 p_1 \dots p_n)(u_2 q_1 \dots q_m)$$

and unique factorisation implies p is an associate (hence divides) one of the p_i or q_j, so divides a or b. Hence p is prime.

Conversely, suppose that every irreducible is prime. We demonstrate that if

$$u_1 p_1 \dots p_m = u_2 q_1 \dots q_n \tag{5.3}$$

where u_1, u_2 are units and the p_i, q_j, are irreducibles, then $m = n$ and there is a permutation π of $\{1, \dots, m\}$ such that p_i and $q_{\pi(i)}$ are associates $(1 \leq i \leq m)$.

This is trivially true for $m = 0$.

For $m \geq 1$, if (5.3) holds, then $p_m \, |u_2 q_1 \dots q_n$. But p_m is prime, so by induction on n we find that $p_m | u_2$ or $p_m | q_j$ for some j. The first of these possibilities implies that p_m is a unit by Proposition 5.7(a), so $p_m | q_j$. Renumber so that $j = n$. Then $p_m | q_n$ and $q_n = p_m u$ where u is a unit, so

$$u_1 p_1 \dots p_m = u_2 q_1 \dots q_{n-1} u p_m$$

Cancel p_m:

$$u_1 p_1 \dots p_{m-1} = (u_2 u) q_1 \dots q_{n-1}$$

By induction we may suppose that $m - 1 = n - 1$ and there is a permutation of $1, \dots, m-1$ such that $p_i, q_{\pi(i)}$ are associates $(1 \leq i \leq m-1)$. We can then extend π to $\{1, \dots, m\}$ by defining $\pi(m) = m$. □

Definition 5.22. A domain D is a *unique factorisation domain* if factorisation into irreducibles is possible and unique.

In a unique factorisation domain every irreducible is prime, so we may speak of a factorisation into irreducibles as a 'prime factorisation'. Theorem 5.21 tells us that a prime factorisation is unique in the usual sense.

We can immediately generalise many ideas on factorisation to any unique factorisation domain. For instance:

Definition 5.23. If a, b belong to a unique factorisation domain D, the *highest common factor* (hcf) or *greatest common divisor* (gcd) h of a, b is an element that satisfies

(1) $h|a$, $h|b$,

(2) If $h'|a$, $h'|b$, then $h'|h$.

The *lowest common multiple* or *least common multiple* (lcm) l of a, b is an element that satisfies

(3) $a|l$, $b|l$,

(4) If $a|l'$, $b|l'$, then $l|l'$.

If a is zero, the highest common factor of a, b is b. For $a, b \neq 0$, in a unique factorisation domain we can write

$$\begin{aligned} a &= u_1 p_1^{e_1} \dots p_n^{e_n} \\ b &= u_2 p_1^{f_1} \dots p_n^{f_n} \end{aligned}$$

where u_1, u_2 are units and the p_i are distinct (that is, non-associate) primes with non-negative integer exponents e_i, f_i. Then it is easy to show that

$$h = u p_1^{m_1} \dots p_n^{m_n}$$

where u is any unit and m_i is the smaller of e_i, f_i $(1 \leq i \leq n)$. The highest common factor is unique up to multiplication by a unit. We can say that a, b are *coprime* if their highest common factor is 1 (or any other unit).

In the same way, for nonzero a, b the lowest common multiple is:

$$l = u p_1^{k_1} \dots p_n^{k_n}$$

where k_i is the larger of e_i, f_i in the factorisations noted above.

Without uniqueness of factorisation we can no longer guarantee the existence of highest common factors and lowest common multiples; see Exercise 5.9. The language of factorisation of integers can be carried over sensibly only to a unique factorisation domain.

In the next section we see that if a domain has a property analogous to the division algorithm of Section 1.2, then every irreducible is prime and factorisation is unique. In later chapters we develop more advanced techniques, which prove unique factorisation for a wider class of domains that do not have this property.

5.6 Euclidean Domains

The crucial property in the usual proofs of unique factorisation in $\mathbb{Z}$ or $K[t]$ for a field K is the existence of a division algorithm, an idea that can be found in Euclid's *Elements.* A reasonable generalisation of this is:

Definition 5.24. Let D be a domain. A *Euclidean function* for D is a function $\phi : D \setminus \{0\} \to \mathbb{N}$ such that

(1) If $a, b \in D\backslash\{0\}$ and $a|b$ then $\phi(a) \leq \phi(b)$.

(2) If $a, b \in D\backslash\{0\}$ then there exist $q, r \in D$ such that $a = bq + r$ where either $r = 0$ or $\phi(r) < \phi(b)$.

A domain that has a Euclidean function is a *Euclidean domain.*

A *principal ideal domain* is one in which every ideal is principal.

Thus for $\mathbb{Z}$ the function $\phi(n) = |n|$ and for $K[t]$, $\phi(p) = \partial p$ (the degree of the polynomial p) are Euclidean functions.

We prove that a Euclidean domain has unique factorisation by showing that in such a domain, every irreducible is prime. The route is this: first we show that any Euclidean domain is a principal ideal domain. Then we show that the latter property implies that all irreducibles are primes.

Theorem 5.25. *Every Euclidean domain is a principal ideal domain.*

Proof: Let D be Euclidean, I an ideal of D. If $I = 0$ it is principal, so we may assume there exists a nonzero element x of I. Choose x to make $\phi(x)$ as small as possible. If $y \in I$ then by (2) $y = qx + r$ where either $r = 0$ or $\phi(r) < \phi(x)$. Now $r \in I$ so we cannot have $\phi(r) < \phi(x)$ because $\phi(x)$ is minimal. This means that $r = 0$, so y is a multiple of x. Therefore $I = \langle x \rangle$ is principal. □

Theorem 5.26. *Every principal ideal domain is a unique factorisation domain.*

Proof: Let D be a principal ideal domain. Then every ideal has *one* generator, so D is noetherian; hence factorisation into irreducibles is possible by Theorem 5.12. To prove uniqueness we show that every irreducible is prime.

Suppose p is irreducible, then $\langle p \rangle$ is maximal among all principal ideals of D by Proposition 5.8(d), but since every ideal is principal, this means that $\langle p \rangle$ is maximal among all ideals.

Suppose $p|ab$ but $p \nmid a$. The fact that $p \nmid a$ implies $\langle p, a\rangle \supsetneqq \langle p\rangle$, so by maximality, $\langle p, a\rangle = D$. Then $1 \in \langle p, a\rangle$, so

$$1 = cp + da \quad (c, d \in D)$$

Multiply by b:

$$b = cpb + dab$$

Since $p|ab$, we find that $p|(cpb + dab)$, so $p|b$. This proves p is prime. □

Theorem 5.27. *A Euclidean domain is a unique factorisation domain.* □

5.7 Euclidean Quadratic Fields

In order to apply Theorem 5.27 it is necessary to exhibit some number fields whose ring of integers is Euclidean. We restrict ourselves to the simplest case of quadratic fields $\mathbb{Q}(\sqrt{d})$ for squarefree d, beginning with the easier situation when d is negative.

Theorem 5.28. *The ring of integers $\mathfrak{O}$ of $\mathbb{Q}(\sqrt{d})$ is Euclidean for $d = -1, -2, -3, -7, -11$, with Euclidean function*

$$\phi(\alpha) = |\mathrm{N}(\alpha)|$$

Proof: To begin with, we consider the suitability of the function ϕ defined in the theorem. For this to be a Euclidean function, the following two conditions must be satisfied for all $\alpha, \beta \in \mathfrak{O} \setminus 0$:

(a) If $\alpha|\beta$ then $|\mathrm{N}(a)| \leq |\mathrm{N}(\beta)|$.

(b) There exist $\gamma, \delta \in \mathfrak{O}$ such that $\alpha = \beta\gamma + \delta$ where either $\delta = 0$ or $|\mathrm{N}(\delta)| < |\mathrm{N}(\beta)|$.

It is clear that (a) holds, for if $\alpha|\beta$ then $\beta = \lambda\alpha$ for $\lambda \in \mathfrak{O}$ and then

$$|\mathrm{N}(\beta)| = |\mathrm{N}(a\lambda)| = |\mathrm{N}(\alpha)\mathrm{N}(\lambda)| = |\mathrm{N}(\alpha)||\mathrm{N}(\lambda)|$$

with rational integer values for the various norms.

To prove (b), we consider the alternative statement:

(c) For any $\varepsilon \in \mathbb{Q}(\sqrt{d})$ there exists $\kappa \in \mathfrak{O}$ such that

$$|\mathrm{N}(\varepsilon - \kappa)| < 1 \tag{5.4}$$

We prove that (c) is equivalent to (b). First, suppose (b) holds. By Lemma 2.22, $c\varepsilon \in \mathfrak{O}$ for some $c \in \mathbb{Z}$. Applying (b) with $\alpha = c\varepsilon$, $\beta = c$ we get two possibilities:

(1) $\delta = 0$ and $c\varepsilon = c\gamma$ for $\gamma \in \mathfrak{O}$. Then $\varepsilon = \gamma \in \mathfrak{O}$ and we may take $\kappa = \varepsilon$.

(2) $c\varepsilon = c\gamma + \delta$ where $|\mathrm{N}(\delta)| < |\mathrm{N}(c)|$. Now $c \neq 0$, so this implies

$$|\mathrm{N}(\delta/c)| < 1$$

which is the same as

$$|\mathrm{N}(\varepsilon - \gamma)| < 1$$

so we may take $\kappa = \gamma$. Hence (b) implies (c). To prove that (c) implies (b) we put $\varepsilon = \alpha/\beta$ and argue similarly.

This allows us to concentrate on condition (c), which is relatively easy to handle: in spirit it says that everything in $\mathbb{Q}(\sqrt{d})$ is 'near to' an integer.

Suppose $\varepsilon = r + s\sqrt{d}$ $(r, s \in \mathbb{Q})$. If $d \not\equiv 1 \pmod 4$ we have to find $\kappa = x + y\sqrt{d}$ $(x, y \in \mathbb{Z})$ with

$$|(r-x)^2 - d(s-y)^2| < 1$$

For $d = -1, -2$ we may do this by taking x and y to be the rational integers nearest to r and s, respectively, for then

$$\left|(r-x)^2 - d(s-y)^2\right| \leq \left|\left(\tfrac{1}{2}\right)^2 + 2\left(\tfrac{1}{2}\right)^2\right| = \tfrac{3}{4} < 1$$

The remaining three values of d to be considered have $d \equiv 1 \pmod 4$. In this case we must find

$$\kappa = x + y\left(\frac{1+\sqrt{d}}{2}\right) \qquad (x, y \in \mathbb{Z})$$

such that

$$|(r - x - \tfrac{1}{2}y)^2 - d(s - \tfrac{1}{2}y)^2| < 1$$

Certainly we can take y to be the rational integer nearest to $2s$, so that $|2s - y| \leq \frac{1}{2}$; and then we may find $x \in \mathbb{Z}$ so that $|r - x - \frac{1}{2}y| \leq \frac{1}{2}$. For $d = -3, -7$, or -11 this means that

$$|(r - x - \tfrac{1}{2}y)^2 - d(s - \tfrac{1}{2}y)^2| \leq \left|\tfrac{1}{4} + \tfrac{11}{16}\right| = \tfrac{15}{16} < 1$$

as required. □

To complete the picture for negative d we have:

Theorem 5.29. *For squarefree $d < -11$ the ring of integers of $\mathbb{Q}(\sqrt{d})$ is not Euclidean.*

Proof: Let $\mathfrak{O}$ be the ring of integers of $\mathbb{Q}(\sqrt{d})$ and suppose for a contradiction that there exists a Euclidean function ϕ. (We do *not* assume $\phi = |\mathrm{N}|$.) Choose $\alpha \in \mathfrak{O}$ such that $\alpha \neq 0$, α is not a unit, and $\phi(\alpha)$ is minimal subject to this. Let β be any element of $\mathfrak{O}$. Now there exist γ, δ such that $\beta = \alpha\gamma + \delta$ with $\delta = 0$ or $\phi(\delta) < \phi(\alpha)$. By choice of α the latter condition implies that either $\delta = 0$ or δ is a unit.

For $d < -11$, Proposition 5.6 shows that the only units of $\mathbb{Q}(\sqrt{d})$ are ± 1. Hence for every $\beta \in \mathfrak{O}$ we have $\beta \equiv -1, 0$, or $1 \pmod{\langle\alpha\rangle}$, so $|\mathfrak{O}/\langle\alpha\rangle| \leq 3$.

Now we compute $|\mathfrak{O}/\langle\alpha\rangle|$ using Theorem 1.30. By Theorem 2.28 $(\mathfrak{O}, +)$ is free abelian of rank 2. If $d \not\equiv 1 \pmod 4$ a $\mathbb{Z}$-basis for $\langle\alpha\rangle$ is $\{\alpha, \alpha\sqrt{d}\}$ since a $\mathbb{Z}$-basis for $\mathfrak{O}$ is $\{1, \sqrt{d}\}$. If $\alpha = a + b\sqrt{d}$ $(a, b \in \mathbb{Z})$ the $\mathbb{Z}$-basis for $\langle\alpha\rangle$ is

$$\{a + b\sqrt{d}, db + a\sqrt{d}\}$$

Hence by Theorem 1.30

$$|\mathfrak{O}/\langle\alpha\rangle| = \left|\det\begin{bmatrix} a & b \\ db & a \end{bmatrix}\right| = \left|a^2 - db^2\right| = |\mathrm{N}(\alpha)|$$

Similar calculations apply for $d \equiv 1 \pmod 4$ with the same end result. (These calculations are a special case of Corollary 6.18 below.) It follows that $|\mathrm{N}(\alpha)| \leq 3$. Thus if $d \not\equiv 1 \pmod 4$ we have $|a^2 - db^2| \leq 3$ with $a, b \in \mathbb{Z}$. If $d \equiv 1 \pmod 4$ then $a = A/2$, $b = B/2$, for $A, B \in \mathbb{Z}$; and then $|A^2 - dB^2| \leq 12$. Since $d < -11$ the only solutions are $a = \pm 1$, $b = 0$; so $|\mathrm{N}(\alpha)| = 1$ and hence α is a unit. This contradicts the choice of α. □

These two theorems together show that for negative d the ring of integers of $\mathbb{Q}(\sqrt{d})$ is Euclidean if and only if $d = -1, -2, -3, -7, -11$. Further, when it is Euclidean it has as Euclidean function the absolute value of the norm. For brevity, call such fields *norm-Euclidean.*

The determination of the norm-Euclidean quadratic fields with d positive has been a long process involving many mathematicians. Leonard Dickson proved that $\mathbb{Q}(\sqrt{d})$ is norm-Euclidean for $d = 2, 3, 5, 13$, mistakenly asserting there are no others. Oskar Perron added 6, 7, 11, 17, 21, 29 to the list. Oppenheimer, Robert Remak, and László Rédei added 19, 33, 37, 41, 57, 73. Rédei claimed 97 as well, but this was disproved by Eric Barnes and Peter Swinnerton-Dyer. Hans Heilbronn proved the list finite in 1934, and the problem was finished off by Harold Chatland and Harold Davenport [18] in 1950 (and independently Kustaa Inkeri [67] in 1949) who proved:

Theorem 5.30. *The ring of integers of $\mathbb{Q}(\sqrt{d})$, for positive d, is norm-Euclidean if and only if d =* 2, 3, 5, 6, 7, 11, 13, 17, 19, 21, 29, 33, 37, 41, 57, 73. □

We cannot prove this theorem here. However, we can prove:

Theorem 5.31. *The ring of integers of* $\mathbb{Q}(\sqrt{d})$ *is norm-Euclidean if* $d = 2, 3, 5$.

Proof: Let $K = \mathbb{Q}(\sqrt{d})$ with ring of integers $\mathfrak{O}$. Using (5.4), we have to prove that if $\varepsilon = r + s\sqrt{d}$ $(r, s \in \mathbb{Q})$ then there exists $\kappa = x + y\sqrt{d} \in \mathfrak{O}$ such that

$$|(r-x)^2 - d(s-y)^2| < 1$$

If $d = 2$ we have $\mathfrak{O} = \mathbb{Z}[\sqrt{2}]$. Let x be the integer nearest to r, and y the integer nearest to s. Then

$$0 \leq |r - x| \leq \tfrac{1}{2} \qquad 0 \leq |s - y| \leq \tfrac{1}{2}$$

Now

$$-2(s-y)^2 \leq (r-x)^2 - 2(s-y)^2 \leq (r-x)^2$$

so

$$-\tfrac{1}{2} \leq (r-x)^2 - 2(s-y)^2 \leq \tfrac{1}{4}$$

This implies that

$$|(r-x)^2 - 2(s-y)^2| \leq \tfrac{1}{2} < 1$$

and we are done.

The case $d = 3$ is similar, but now we have

$$-3(s-y)^2 \leq (r-x)^2 - 3(s-y)^2 \leq (r-x)^2$$

so

$$-\tfrac{3}{4} \leq (r-x)^2 - 3(s-y)^2 \leq \tfrac{1}{4}$$

This implies that

$$|(r-x)^2 - 3(s-y)^2| \leq \tfrac{3}{4} < 1$$

When $d = 5$ we use the same idea, but observe that $\mathfrak{O} = \mathbb{Z}[\frac{1+\sqrt{5}}{2}]$. Now

$$\mathrm{N}\left(a + b\frac{1+\sqrt{5}}{2}\right) = \left(a + \frac{b}{2}\right)^2 - 5\left(\frac{b^2}{4}\right)$$

Then

$$\mathrm{N}\left((r-x) + (s-y)\frac{1+\sqrt{5}}{2}\right) = \left((r-x) + \frac{s-y}{2}\right)^2 - 5\left(\frac{(s-y)^2}{4}\right)$$

and

$$-5\frac{(s-y)^2}{4} \le \mathrm{N}\left((r-x)+(s-y)\frac{1+\sqrt{5}}{2}\right) \le \left((r-x)+\frac{s-y}{2}\right)^2$$

so

$$-\frac{5}{16} \le \mathrm{N}\left((r-x)+(s-y)\frac{1+\sqrt{5}}{2}\right) \le \frac{3}{4}$$

This implies that

$$\left|\mathrm{N}\left((r-x)+(s-y)\frac{1+\sqrt{5}}{2}\right)\right| \le \frac{3}{4} < 1$$

□

Narkiewicz [97] surveys what was known at the time about real norm-Euclidean quadratic fields and related questions. The cases $d = 2, 3, 5, 6, 7, 13, 17, 21$, and 29 are proved in Hardy and Wright [58, Theorem 248].

Unlike the case $d < 0$, a real quadratic field $\mathbb{Q}(\sqrt{d})$ can be Euclidean but not norm-Euclidean, as Clark [19] proved in 1994 for $d = 69$. Samuel [114] suggested that $\mathbb{Z}(\sqrt{14})$ might also have such properties. Malcolm Harper proved this in his PhD thesis in 2000 and published a proof [59] in 2004. Using similar methods he also proved that for positive d, whenever the discriminant of $\mathbb{Q}(\sqrt{d})$ is less than or equal to 500, the ring of integers is Euclidean if and only if it is a principal ideal domain.

5.8 Geometric Approach

A geometric approach to norm-Euclidean rings is illuminating. We begin with imaginary quadratic fields $\mathbb{Q}(\sqrt{d})$ with $d < 0$, the simplest case.

The starting point is (5.4), which is valid if and only if every number $\varepsilon = p + q\sqrt{d} \in \mathbb{Q}(\sqrt{d})$ with $0 \le p, q < 1$ lies inside the union of the sets

$$N_{ab} = \{s + t\sqrt{d} : |\mathrm{N}((s-a)+(t-b)\sqrt{d})| < 1\}$$

These sets are translates of N_{00} by the lattice $\mathcal{L}$.

We draw the lattice $\mathcal{L} = \{a + b\sqrt{d} : a, b \in \mathbb{Z}\}$ by identifying it with the integer lattice $\mathbb{Z}^2 = \{(a,b) : a, b \in \mathbb{Z}\}$ considered as a subset of the plane $\mathbb{R}^2$. When $d < 0$ the norm is non-negative, and N_{00} is the interior of the curve $E = \{(x,y) : x^2 - dy^2 = 1\}$. This is an *ellipse*, with centre the origin and passing through $(\pm 1, 0)$ and $\left(0, \pm\sqrt{\frac{-1}{d}}\right)$.

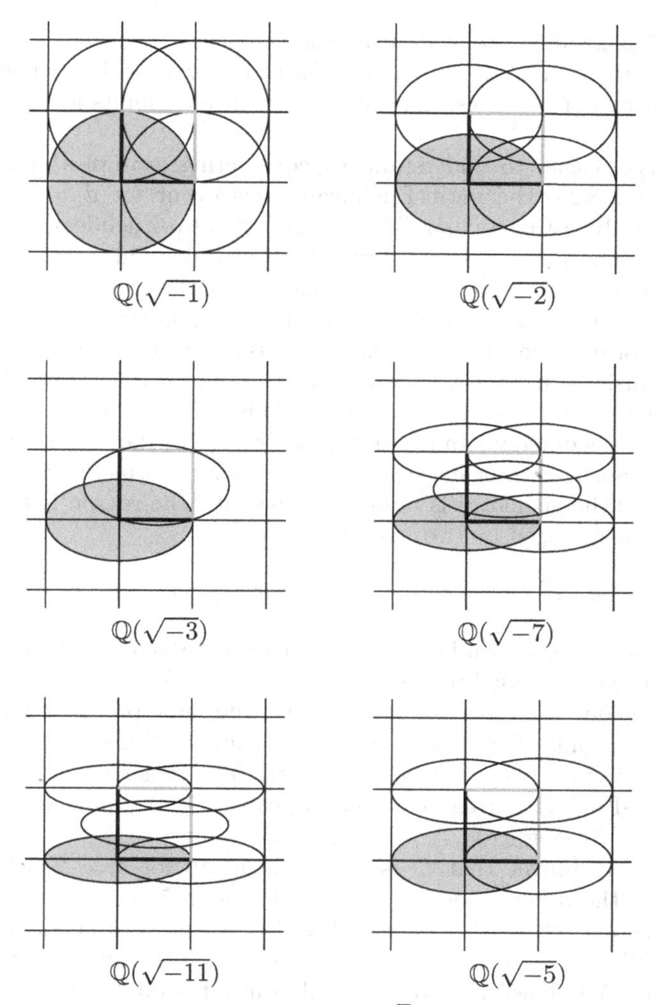

Figure 5.2. Geometric proofs that $\mathbb{Q}(\sqrt{d})$ is norm-Euclidean for $d = -1, -2, -3, -7, -11$, but not -5. Shaded ellipse is N_{00}, other ellipses are translates by algebraic integers. Thin lines are the integer grid. The unit square is marked with thicker black and grey lines, where grey indicates omitted edges.

Condition (5.4) states that the subset $S = [0,1) \times [0,1) \subseteq \mathbb{R}^2$ is contained in the union of all translates of this ellipse by integers $\mathfrak{O}$ of $\mathbb{Q}(\sqrt{d})$. Equivalently, some finite subset of these translates completely covers S; that is, S is contained in the union of this subset.

The set $\mathfrak{O}$ is a sublattice of $\mathcal{L}$. It equals $\mathcal{L}$ unless $d \equiv 1 \pmod 4$, when it is a proper sublattice of $\mathcal{L}$. The subset S is the interior of the unit square, together with two of its edges, not including their end points at $(0,1)$ and $(1,0)$.

All of this is easier to understand by considering examples. By Theorems 5.28 and 5.29, the norm-Euclidean rings occur for $d = -1, -2, -3, -7, -11$ only. In particular $\mathbb{Q}(\sqrt{-5})$ is not norm-Euclidean. In fact, Theorem 5.17 shows that it is not a principal ideal domain.

Figure 5.2 shows N_{00} and several translates for the cases $d = -1, -2, -3, -7, -11$, and -5. The set S is outlined in thick black and grey lines, at the centre of each figure. The ellipse N_{00} is shaded; its *interior* is the set whose translates must cover S if $\mathfrak{O}$ is to be norm-Euclidean. In the first five figures, translates of N_{00} cover S. (This can be verified rigorously using coordinate geometry.) In the sixth, the translates shown do not cover S. No other translates intersect S, so $\mathbb{Q}(\sqrt{-5})$ is not norm-Euclidean.

When $d > 0$ the geometry is more complicated. The reason is that the ellipse E is now replaced by two hyperbolas

$$H_1 = \{x + y\sqrt{d} : x^2 - dy^2 = 1\} \qquad H_2 = \{x + y\sqrt{d} : x^2 - dy^2 = -1\}$$

and the set N_{00} is the region between (but not on) these curves. In particular N_{00} is neither bounded nor convex.

The same geometric criterion is valid, with the same proof: $\mathfrak{O}$ is norm-Euclidean if and only if S is contained in the union of translates of N_{00} by $\mathfrak{O}$. But verifying this condition is trickier. The richer geometry leads to more norm-Euclidean rings than in the case $d < 0$, but the rigorous calculations are messier.

Theorem 5.31 proves that $\mathfrak{O}$ is norm-Euclidean for $d = 2, 3, 5$. We discuss $d = 2$; the other cases are similar. Theorem 5.30, which we have not proved, asserts that other values of d also give norm-Euclidean rings of integers. We illustrate the geometric approach with the first of these, $d = 6$. Figure 5.3 shows N_{00} and several translates for these two cases. When $d = 2$, two translates of N_{00} cover S. Six translates deal with the case $d = 6$.

5.9 Consequences of Unique Factorisation

When the integers in a number field have unique factorisation, we can carry over many arguments of the type used in the factorisation of rational integers (taking a little care at first). For example, the proof of the statement of Fermat in Section 5.1 makes it clear that, since $\mathbb{Z}[\sqrt{-2}]$ (the ring of

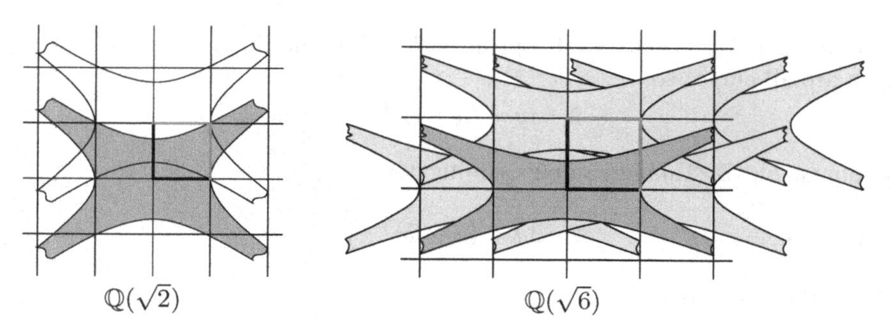

Figure 5.3. Geometric proofs that $\mathbb{Q}(\sqrt{d})$ is norm-Euclidean for $d = 2, 6$. Darker shaded area is the central part of N_{00}, other shaded areas are translates by algebraic integers. Thin lines are the integer grid. The unit square is marked with thicker black and grey lines, where grey indicates omitted edges.

integers in $\mathbb{Q}(\sqrt{-2})$) has unique factorisation, the intuitive 'proof' given there is, in fact, valid. We now prove another example of the same sort of thing, again a statement of Fermat:

Theorem 5.32. *The only integer solutions of the equation*

$$y^2 + 4 = z^3 \tag{5.5}$$

are $y = \pm 11$, $z = 5$ *and* $y = \pm 2$, $z = 2$.

Proof: We work in the ring $\mathbb{Z}[\mathrm{i}]$, which is a unique factorisation domain by Theorem 5.28. First, suppose that y is odd. Then (5.5) factorises as

$$(2 + \mathrm{i}y)(2 - \mathrm{i}y) = z^3$$

A common factor $a + \mathrm{i}b$ of $2 + \mathrm{i}y$, $2 - \mathrm{i}y$ is also a factor of their sum, 4, and difference, $2y$, so taking norms

$$a^2 + b^2 | 16 \qquad a^2 + b^2 | 4y^2$$

implying

$$a^2 + b^2 | 4$$

The only solutions of this relation are $a = \pm 1$, $b = 0$, or $a = 0$, $b = \pm 1$, or $a = \pm 1$, $b = \pm 1$, none of which turns out to give a proper factor $a + \mathrm{i}b$ of $2 + \mathrm{i}y$. Hence $2 + \mathrm{i}y$, $2 - \mathrm{i}y$ are coprime. By unique factorisation in $\mathbb{Z}[\mathrm{i}]$, if their product is a cube then one is $\varepsilon\alpha^3$ and the other is $\varepsilon^{-1}\beta^3$ where ε

is a unit, and $\alpha, \beta \in \mathbb{Z}[\mathrm{i}]$. By Proposition 5.6 the units in $\mathbb{Z}[\mathrm{i}]$ are $\pm\mathrm{i}$, ± 1, which are all cubes, so

$$2 + \mathrm{i}y = (a + \mathrm{i}b)^3$$

for some $a, b \in \mathbb{Z}$. Taking complex conjugates,

$$2 - \mathrm{i}y = (a - \mathrm{i}b)^3$$

Adding the two equations,

$$4 = 2a(a^2 - 3b^2)$$

so

$$a(a^2 - 3b^2) = 2$$

Now a divides 2, so $a = \pm 1$ or ± 2, and the choice of a determines b. It is easy to see that the only solutions are $a = -1$, $b = \pm 1$, or $a = 2$, $b = \pm 1$. Then

$$z^3 = ((a + \mathrm{i}b)(a - \mathrm{i}b))^3 = (a^2 + b^2)^3$$

so $z = a^2 + b^2 = 2, 5$, respectively. Then $y^2 + 4 = 8, 125$, so $y = \pm 2, \pm 11$. This gives the solutions with $y = \pm 11$ as the only ones for y are odd.

Now suppose that y is even, so $y = 2Y$. Then z is even as well, say $z = 2Z$, and

$$Y^2 + 1 = 2Z^3$$

Then Y must be odd, say $Y = 2k + 1$. The highest common factor of $Y + \mathrm{i}$ and $Y - \mathrm{i}$ divides the difference $2\mathrm{i} = (1 + \mathrm{i})^2$. Now $1 + \mathrm{i}$ divides $Y + \mathrm{i}$ and $Y - \mathrm{i}$ but $(1 + \mathrm{i})^2$ does not, so the highest common factor of $Y + \mathrm{i}$ and $Y - \mathrm{i}$ is $1 + \mathrm{i}$. But

$$(1 + \mathrm{i}Y)(1 - \mathrm{i}Y) = 2Z^3$$

and the common factor $1 + \mathrm{i}$ occurs twice on the left (bearing in mind that $1 + \mathrm{i}Y = \mathrm{i}(Y - \mathrm{i})$, $1 - \mathrm{i}Y = -\mathrm{i}(Y + \mathrm{i})$). Hence there must be a factorisation

$$1 + \mathrm{i}Y = (1 + \mathrm{i})(a + \mathrm{i}b)^3$$

whence as before

$$1 = (a + b)(a^2 - 4ab + b^2)$$

so $a = \pm 1$, $b = 0$, or $a = 0$, $b = \pm 1$. These imply that $y = \pm 2$, yielding the other two solutions stated. □

5.10 Primes in the Gaussian Integers

The Gaussian numbers $\mathbb{Q}(\mathrm{i})$ have ring of integers $\mathbb{Z}[\mathrm{i}]$. We examine whether primes in $\mathbb{Z}$ remain prime in $\mathbb{Z}[\mathrm{i}]$. By Theorem 5.28 the ring $\mathbb{Z}[\mathrm{i}]$ is Euclidean, hence a principal ideal domain, so each prime element p generates a prime ideal $\langle p\rangle$, and every prime ideal is of this form.

As a warm-up, we consider the ideals $\langle 2\rangle$, $\langle 3\rangle$, and $\langle 5\rangle$ in $\mathbb{Z}[\mathrm{i}]$, which correspond to the first three primes in $\mathbb{Z}$.

Example 5.33.

In $\mathbb{Z}[\mathrm{i}]$ the ideal $\langle 2\rangle$ is not prime. It factorises as

$$\langle 2\rangle = \langle 1+\mathrm{i}\rangle^2$$

because $(1+\mathrm{i})^2 = 2\mathrm{i}$ and i is a unit. Each factor is a prime ideal in $\mathbb{Z}[\mathrm{i}]$ since it has norm 2, a rational prime. See Theorem 6.23(a).

The ideal $\langle 3\rangle$ remains prime in $\mathbb{Z}[\mathrm{i}]$. Its norm is $3^2 = 9$. If it is not prime then any prime divisor $a+ib$ has norm ± 3. So $a^2+b^2 = \pm 3$, which is impossible.

The ideal $\langle 5\rangle$ is not prime in $\mathbb{Z}[\mathrm{i}]$. Since $5 = (2+\mathrm{i})(2-\mathrm{i})$, we have

$$\langle 5\rangle = \langle 2+\mathrm{i}\rangle\langle 2-\mathrm{i}\rangle$$

Each factor has norm 5, which is a rational prime, so by Proposition 5.15(c) the factors are prime.

These calculations motivate:

Theorem 5.34. *The prime ideals in the Gaussian integers are:*

(a) *Ideals $\langle p\rangle$ where p is a rational prime of the form $4k+3$, with norm p^2.*

(b) *Ideals $\langle a+\mathrm{i}b\rangle$ where $a^2+b^2 = p$, a rational prime of the form $4k+1$, with norm p.*

(c) *The ideal $\langle 1+\mathrm{i}\rangle$, with norm 2.*

Proof: We appeal to two results that are proved later.

Let $\mathfrak{p}$ be a prime ideal. Since $\mathbb{Z}[\mathrm{i}]$ is a unique factorisation domain, $\mathfrak{p}$ is principal, say $\mathfrak{p} = \langle a+ib\rangle$. By Theorem 6.23(c) below, $a^2+b^2 = \mathrm{N}(a+ib) = p^k$ for some rational prime p, where $k \leq [\mathbb{Q}(\mathrm{i}) : \mathbb{Q}] = 2$. Therefore either $a^2+b^2 = \mathrm{N}(a+ib) = p$ or $a^2+b^2 = \mathrm{N}(a+ib) = p^2$.

If $p = 2$ then $a^2+b^2 = 2$ or 4. So $\mathfrak{p}$ is a prime divisor of $\langle 2\rangle$. We saw in Example 5.33 that $p = \langle 1+\mathrm{i}\rangle$.

If p is of the form $p = 4k+3$ then $a^2+b^2 = p$ is impossible (consider this equation (mod 4)). So $\langle p\rangle$ is a prime ideal.

If p is of the form $4k+1$ we invoke the Two-Squares Theorem, proved later as Theorem 8.4. This states that $p = a^2 + b^2$ for integers a, b. Then $p = (a + \mathrm{i}b)(a - \mathrm{i}b)$ so

$$\langle p \rangle = \langle a + \mathrm{i}b \rangle \langle a - \mathrm{i}b \rangle$$

Each factor has norm p so is a prime ideal. □

5.11 Exercises

5.1 Which of the following elements of $\mathbb{Z}[\mathrm{i}]$ are irreducible: $1+\mathrm{i}$, $3-7\mathrm{i}$, 5, 7, $12\mathrm{i}$, $-4+5\mathrm{i}$?

5.2 Write down the group of units of the ring of integers of: $\mathbb{Q}(\mathrm{i})$, $\mathbb{Q}(\sqrt{-2})$, $\mathbb{Q}(\sqrt{-3})$, $\mathbb{Q}(\sqrt{-5})$, $\mathbb{Q}(\sqrt{-6})$.

5.3 Is the group of units of the integers in $\mathbb{Q}(\sqrt{3})$ finite?

5.4 Show that a homomorphic image of a noetherian ring is noetherian.

5.5 Find all ideals of $\mathbb{Z}$ that contain $\langle 120 \rangle$.
Show that every ascending chain of ideals of $\mathbb{Z}$ starting with $\langle 120 \rangle$ stops, by direct examination of the possibilities.

5.6 Find a ring that is not noetherian.

5.7 Check the calculations required to complete Theorems 5.17 and 5.18.

5.8 Is $10 = (3+\mathrm{i})(3-\mathrm{i}) = 2 \cdot 5$ an example of non-unique factorisation in $\mathbb{Z}[\mathrm{i}]$? Give reasons for your answer.

5.9 Show that 6 and $2(1+\sqrt{-5})$ both have 2 and $1+\sqrt{-5}$ as factors, but do not have a highest common factor in $\mathbb{Z}[\sqrt{-5}]$. Do they have a least common multiple? (*Hint:* Consider norms.)

5.10 Let D be a domain. Suppose that $x \in D$ has a factorisation $x = up_1 \dots p_n$, where u is a unit and $p_1, \dots, p_n$ are *primes*. Show that given any factorisation $x = vq_1 \dots q_m$, where v is a unit and $q_1, \dots, q_m$ are irreducibles, then $m = n$ and there exists a permutation π of $\{1, \dots, n\}$ such that p_i, $q_{\pi(i)}$ are associates $(1 \leq i \leq n)$.

5.11 Show in $\mathbb{Z}[\sqrt{-5}]$ that $\sqrt{-5} | (a + b\sqrt{-5})$ if and only if $5|a$. Deduce that $\sqrt{-5}$ is prime in $\mathbb{Z}[\sqrt{-5}]$. Hence conclude that the element 5 factorises uniquely into irreducibles in $\mathbb{Z}[\sqrt{-5}]$, although $\mathbb{Z}[\sqrt{-5}]$ does not have unique factorisation.

5.12 Suppose D is a unique factorisation domain, and a, b are coprime non-units. Deduce that if

$$ab = c^n$$

for $c \in D$, there exists a unit $e \in D$ such that ea and $e^{-1}b$ are nth powers in D.

5.13 Let p be an odd rational prime and $\zeta = e^{2\pi i/p}$. If α is a prime element in $\mathbb{Z}[\zeta]$, prove that the rational integers that are divisible by α are precisely the rational integer multiples of some prime rational integer q. (*Hint:* $\alpha|\mathrm{N}(\alpha)$, so α divides some rational prime factor q of $\mathrm{N}(\alpha)$. Now show α is not a factor of any $m \in \mathbb{Z}$ prime to q.)

5.14 Let $\mathbb{Q}_2$ be the set of all rational numbers a/b, where $a, b \in \mathbb{Z}$ and b is odd. Prove that $\mathbb{Q}_2$ is a domain, and that the only irreducibles in $\mathbb{Q}_2$ are 2 and its associates.

5.15 Generalise Exercise 5.16 to the ring $\mathbb{Q}_\pi$, where π is a finite set of primes in $\mathbb{Z}$, this ring being defined as the set of all rationals a/b with b prime to the elements of π.

5.16 The following purports to be a proof that in any number field K the ring of integers $\mathfrak{O}$ contains infinitely many non-associate irreducibles. Find the error.

Since not all elements are units, $\mathfrak{O}$ contains at least one irreducible. Assume $\mathfrak{O}$ has only finitely many non-associate irreducibles $p_1, \ldots, p_n$. The number $1 + p_1 \ldots p_n$ must be divisible by some irreducible q, and this cannot be an associate of any of $p_1, \ldots, p_n$. This is a contradiction. (*Hint*: The 'proof' does not use any properties of $\mathfrak{O}$ beyond the existence of irreducible factorisation. Now $\mathbb{Q}_2$ in Exercise 5.14 has these properties ...)

5.17 Give a correct proof of the statement in Exercise 5.18.

5.18 Is every subring of a noetherian ring noetherian?

5.19 If a ring R has an ideal I such that I and R/I are noetherian, must R be noetherian?

5.20 David Hilbert liked to explain why unique factorisation is not obvious by considering the semigroup of positive integers of the form $4k + 1$ under multiplication. Show that every element is a product of finitely many irreducibles, but this factorisation is not unique.

6

Ideals

After the somewhat traumatic realisation that factorisation into irreducibles is unique in some rings of integers but not in others, we seek some way to minimise the damage. Kummer, and then Dedekind, took steps to develop more insightful theories. Kummer had the bright idea that if a number does not factorise uniquely in a given ring of integers, then perhaps the ring extends to a bigger one in which further factorisation might be not only possible, but unique. For example, we pointed out in Chapter 4 that there are two factorisations $6 = 2 \cdot 3 = \sqrt{-6} \cdot \sqrt{-6}$ in $\mathbb{Z}[\sqrt{-6}]$, but $\sqrt{-6}$ does not divide 2 or 3 in this ring. In fact, $2/\sqrt{-6} = \sqrt{-2/3}$, and $3/\sqrt{-6} = \sqrt{-3/2}$, neither of which belongs to $\mathbb{Z}[\sqrt{-6}]$. Kummer's idea: throw extra elements into the pot to create a larger ring. We cannot use $\sqrt{-2/3}$ or $\sqrt{-3/2}$, since these are not integers. But if $\sqrt{2}$ is adjoined then so is $\sqrt{-3}$, and now $6 = \sqrt{2} \cdot \sqrt{2} \cdot \sqrt{-3} \cdot \sqrt{-3} \cdot (-1)$, which can be rearranged to give the two factorisations above.

Kummer called the new elements introduced in this way 'ideal numbers'. Dedekind formulated similar ideas from a different, and conceptually simpler direction, introducing the notion of an 'ideal' in ring theory. An ideal is not a new kind of number, but a subset of the ring of integers with special properties. He showed that although unique factorisation may fail for numbers, a simple and elegant theory of unique factorisation can be developed for ideals. In this theory, the essential building blocks are 'prime ideals', which are defined by adapting the concept of a prime element, Definition 5.19.

Just as it is often easier to work with both a ring of integers and its corresponding field of quotients, we generalise the concept of an ideal to a 'fractional ideal'. This generalisation has the advantage that the nonzero fractional ideals form a group under multiplication, whereas the ideals form only a semigroup. Unique factorisation of ideals into prime ideals can then be proved, and several standard consequences of unique factorisation are easy consequences.

We define the norm of an ideal as a generalisation of the norm of an element and prove that the new norm has the corresponding multiplicative property. We use this to show that every ideal can be generated by at most two elements. In the previous chapter we saw that factorisation of elements is unique in a principal ideal domain (where every ideal is generated by a single element). We refine this result to show that factorisation of elements into irreducibles is unique in a ring of integers if and only if every ideal is principal.

6.1 Historical Background

To motivate Kummer's introduction of ideal numbers and Dedekind's reformulation in terms of ideals, we look more closely at some examples where unique factorisation fails, in the hope that some pattern might emerge. Many previous examples exhibit no obvious pattern, but others seem to have significant features. For instance, consider:

$$\begin{aligned}
\mathbb{Q}\left(\sqrt{15}\right): &\quad 2\cdot 5=(5+\sqrt{15})(5-\sqrt{15})\\
\mathbb{Q}\left(\sqrt{30}\right): &\quad 2\cdot 3=(6+\sqrt{30})(6-\sqrt{30})\\
\mathbb{Q}\left(\sqrt{-10}\right): &\quad 2\cdot 7=(2+\sqrt{-10})(2-\sqrt{-10})
\end{aligned}$$

In these we see a curious phenomenon: there is a prime p occurring on the left, and on the right a factor $a+b\sqrt{d}$ where a and d are multiples of p. It looks as though $\sqrt{p}$ is somehow a common factor of both sides—but $\sqrt{p}$ does not lie in the given number field. As a specific case, consider the first example; here $\sqrt{5}$ looks a likely candidate for a common factor but $\sqrt{5}$ is not an element of $\mathbb{Q}(\sqrt{15})$. Leaving aside the niceties for the moment, introduce $\sqrt{5}$ into the factorisation to get

$$\begin{aligned}
5+\sqrt{15} &= \sqrt{5}(\sqrt{5}+\sqrt{3})\\
5-\sqrt{15} &= \sqrt{5}(\sqrt{5}-\sqrt{3})
\end{aligned}$$

Multiply and cancel the 5 to get

$$2=(\sqrt{5}+\sqrt{3})(\sqrt{5}-\sqrt{3})$$

It is then apparent that the two different factorisations of 10 are obtained by grouping the factors in

$$(\sqrt{5})(\sqrt{5})(\sqrt{5}+\sqrt{3})(\sqrt{5}-\sqrt{3})$$

in two different ways.

Perhaps introducing new numbers such as $\sqrt{5}$ restores unique factorisation. Can the problem be that we are not factorising in the right context? In other words, if factorisation of some element in the ring of integers of the given number field K is not unique, can we extend K to a field L where it *is*? In the example we factorised the element 10 by extending $\mathbb{Q}(\sqrt{15})$ to $\mathbb{Q}(\sqrt{3}, \sqrt{5})$.

What about the other examples of non-unique factorisation? The factorisations of 14 in $\mathbb{Q}(\sqrt{-10})$ can be found by extending to $\mathbb{Q}(\sqrt{2}, \sqrt{-5})$ to get two possible groupings of the factors in

$$14 = (\sqrt{2})(\sqrt{2})(\sqrt{2}+\sqrt{-5})(\sqrt{2}-\sqrt{-5})$$

The case of 6 in $\mathbb{Q}(\sqrt{30})$ is even more interesting: we have

$$6 = (\sqrt{2})(\sqrt{2})(\sqrt{3})(\sqrt{3})(\sqrt{6}+\sqrt{5})(\sqrt{6}-\sqrt{5})$$

and the last two factors are *units* because $(\sqrt{6}+\sqrt{5})(\sqrt{6}-\sqrt{5}) = 6-5 = 1$.

This is one way to view Kummer's theory. Start with a number field K and extend to a field L. Then $\mathfrak{O}_K \subseteq \mathfrak{O}_L$. Neither ring of integers need have unique factorisation, but an element in $\mathfrak{O}_K$ might factorise uniquely into elements in $\mathfrak{O}_L$.

At the outset, Kummer did not describe the theory in this way. His method involved detailed computations, described in Edwards [37, 38], and a radically new notion of 'ideal' prime factors for elements that have no prime factors in $\mathfrak{O}_K$. These extra 'ideal numbers' can be interpreted as elements introduced from $\mathfrak{O}_L$ for factorisation purposes, but Kummer's description was more mysterious: things that behave like numbers but are not.

Dedekind's simpler and more natural formulation of the theory in terms of ideals clarified matters. To motivate this approach, consider a factorisation $x = ab$ in a ring R. Recall from Chapter 1 that the product of ideals IJ is just the set of finite sums $\sum x_i y_i$ $(x_i \in I, y_i \in J)$. Therefore the ideal generated by x is the product of the ideals generated by a and by b:

$$\langle x \rangle = \langle a \rangle \langle b \rangle$$

More generally a product $x = p_1 \dots p_n$ of elements in R corresponds to a product of principal ideals

$$\langle x \rangle = \langle p_1 \rangle \dots \langle p_n \rangle$$

When considering unique factorisation, the formulation in terms of ideals is marginally better, for if we replace p_1 by up_1 where u is a unit, the ideals $\langle p_1 \rangle$ and $\langle up_1 \rangle$ are the same, see Proposition 5.7(b). Thus, when the factors are unique up to multiplication by units and order, the ideals $\langle p_1 \rangle, \dots, \langle p_n \rangle$ are *unique* (up to order). Passing to ideals eliminates the complications related to units—or, at least, sweeps them under the carpet.

How does this tie in with the earlier discussion? First consider the example

$$10 = (\sqrt{5})(\sqrt{5})(\sqrt{5}+\sqrt{3})(\sqrt{5}-\sqrt{3})$$

in the integers of $\mathbb{Q}(\sqrt{3}, \sqrt{5})$. Let $K = \mathbb{Q}(\sqrt{15})$, $L = \mathbb{Q}(\sqrt{3}, \sqrt{5})$; then this factorisation holds in the ring of integers $\mathfrak{O}_L$. In this ring we also have the corresponding factorisation of principal ideals:

$$\langle 10 \rangle = \langle \sqrt{5} \rangle \langle \sqrt{5} \rangle \langle \sqrt{5}+\sqrt{3} \rangle \langle \sqrt{5}-\sqrt{3} \rangle$$

We may intersect the ideals in this factorisation with $\mathfrak{O}_K$, and once more we get ideals in $\mathfrak{O}_K$, but these ideals *need not be principal*, and in this case they are not. For instance, let $I = \langle \sqrt{5}+\sqrt{3} \rangle \cap \mathfrak{O}_K$. Then $\sqrt{3}(\sqrt{5}+\sqrt{3}) = 3+\sqrt{15} \in I$ and $\sqrt{5}(\sqrt{5}+\sqrt{3}) = 5+\sqrt{15} \in I$.

Suppose, for a contradiction, that I is principal, say $I = \langle k \rangle$. Now $\mathrm{N}(5+\sqrt{15}) = 10$ and $\mathrm{N}(3+\sqrt{15}) = -6$, so $\mathrm{N}(k)|2$. We know that $\mathrm{N}(k) \neq \pm 1$ since I is proper. If $\mathrm{N}(k) = \pm 2$ then there exist $a, b \in \mathbb{Z}$ with $a^2 - 15b^2 = \pm 2$. But, taken modulo 5, this leads to a contradiction. So I is not principal.

The moral is now clear. If we wish to factorise the principal ideal $\langle x \rangle$ in a ring of integers $\mathfrak{O}_K$, we might get a unique factorisation of ideals

$$\langle x \rangle = I_1 \dots I_n$$

but the ideals $I_1, \dots, I_n$ might not be principal.

Factorisation into ideals is extremely useful, in part because the ideals in $\mathfrak{O}_K$ are not far off being principal, having at most two generators; see Theorem 6.29.

6.2 Prime Factorisation of Ideals

Throughout this chapter $\mathfrak{O}$ is the ring of integers of a number field K of degree n. We use small Gothic/Fraktur letters $\mathfrak{a}, \mathfrak{b}, \mathfrak{c}, \dots$ to denote ideals (and later 'fractional ideals') of $\mathfrak{O}$. We are interested in two special types of ideal, which we define in a general context as follows.

Definition 6.1. An ideal $\mathfrak{a}$ of a ring R is *maximal* if $\mathfrak{a}$ is a proper ideal of R and there are no ideals of R strictly between $\mathfrak{a}$ and R.

The ideal $\mathfrak{a} \neq R$ of R is *prime* if, whenever $\mathfrak{b}$ and $\mathfrak{c}$ are ideals of R with $\mathfrak{bc} \subseteq \mathfrak{a}$, then either $\mathfrak{b} \subseteq \mathfrak{a}$ or $\mathfrak{c} \subseteq \mathfrak{a}$.

We can see where the latter definition comes from by considering the special case where all three ideals concerned are principal, say $\mathfrak{a} = \langle a \rangle$, $\mathfrak{b} = \langle b \rangle$, $\mathfrak{c} = \langle c \rangle$. Since $x|y$ is equivalent to $\langle y \rangle \subseteq \langle x \rangle$ by Proposition 5.8(a), the statement

$$\mathfrak{bc} \subseteq \mathfrak{a} \text{ implies either } \mathfrak{b} \subseteq \mathfrak{a} \text{ or } \mathfrak{c} \subseteq \mathfrak{a}$$

translates into

$$a|bc \text{ implies either } a|b \text{ or } a|c$$

If R is an integral domain, then the zero ideal is prime, and here we find $\langle p \rangle$ is prime if and only if p is a prime or zero. (See Exercise 6.1.) Excluding 0 from the list of prime elements but including $\langle 0 \rangle$ as a prime ideal is a quirk of historical development. Elements came first and 0 was excluded from the list of primes of $\mathbb{Z}$. On the other hand, the definition we have given for a prime ideal implies the following simple characterisations:

Lemma 6.2. *Let R be a ring and $\mathfrak{a}$ an ideal of R. Then*
(a) *$\mathfrak{a}$ is maximal if and only if $R/\mathfrak{a}$ is a field.*
(b) *$\mathfrak{a}$ is prime if and only if $R/\mathfrak{a}$ is a domain.*

Proof: The ideals of $R/\mathfrak{a}$ are in bijective correspondence with the ideals of R lying between $\mathfrak{a}$ and R. Hence $\mathfrak{a}$ is maximal if and only if $R/\mathfrak{a}$ has no nonzero proper ideals. Now it is easy to show that a ring S has no nonzero proper ideals if and only if S is a field. (Recall that we assume throughout that rings are commutative with 1; this result is false for more general rings.) Taking $S = R/\mathfrak{a}$ proves (a).

To prove (b), first suppose that $\mathfrak{a}$ is prime. If $x, y \in R$ are such that in $R/\mathfrak{a}$ we have

$$(\mathfrak{a} + x)(\mathfrak{a} + y) = 0$$

then $xy \in \mathfrak{a}$, so $\langle x \rangle \langle y \rangle \subseteq \mathfrak{a}$. Hence either $\langle x \rangle \subseteq \mathfrak{a}$ or $\langle y \rangle \subseteq \mathfrak{a}$, so either $x \in \mathfrak{a}$ or $y \in \mathfrak{a}$. Hence one of $(\mathfrak{a} + x)$ or $(\mathfrak{a} + y)$ is zero in $R/\mathfrak{a}$, and therefore $R/\mathfrak{a}$ has no zero-divisors so it is a domain.

Conversely suppose that $R/\mathfrak{a}$ is a domain. Then $|R/\mathfrak{a}| \neq 1$ so $\mathfrak{a} \neq R$. Suppose if possible that $\mathfrak{bc} \subseteq \mathfrak{a}$ but $\mathfrak{b} \nsubseteq \mathfrak{a}$, $\mathfrak{c} \nsubseteq \mathfrak{a}$. Then we can find elements $b \in \mathfrak{b}$, $c \in \mathfrak{c}$, with $b, c \notin \mathfrak{a}$ but $bc \in \mathfrak{a}$. This means that $(\mathfrak{a} + b)$ and $(\mathfrak{a} + c)$ are zero-divisors in $R/\mathfrak{a}$, which is a contradiction. □

Corollary 6.3. *Every maximal ideal is prime.* □

Next we list some important properties of the ring of integers of a number field:

Theorem 6.4. *The ring of integers $\mathfrak{O}$ of a number field K has the following properties:*

(a) *$\mathfrak{O}$ is a domain, with field of fractions K.*

(b) *$\mathfrak{O}$ is noetherian.*

(c) *If $\alpha \in K$ satisfies a monic polynomial equation with coefficients in $\mathfrak{O}$ then $\alpha \in \mathfrak{O}$.*

(d) *Every nonzero prime ideal of $\mathfrak{O}$ is maximal.*

Proof: Part (a) is obvious. For part (b): by Theorem 2.28 the additive group $(\mathfrak{O}, +)$ is free abelian of rank n. Theorem 1.29 implies that if $\mathfrak{a}$ is an ideal of $\mathfrak{O}$ then $(\mathfrak{a}, +)$ is free abelian of rank $\leq n$. Now any $\mathbb{Z}$-basis for $(\mathfrak{a}, +)$ generates $\mathfrak{a}$ as an ideal, so every ideal of $\mathfrak{O}$ is finitely generated and $\mathfrak{O}$ is noetherian. Part (c) is immediate from Theorem 2.20. To prove part (d) let $\mathfrak{p}$ be a prime ideal of $\mathfrak{O}$. Let $0 \neq \alpha \in \mathfrak{p}$. Then

$$N = \mathrm{N}(\alpha) = \alpha_1, \dots \alpha_n \in \mathfrak{p}$$

(the α_i being the conjugates of α) since $\alpha_1 = \alpha$. Therefore $\langle N \rangle \subseteq \mathfrak{p}$, so $\mathfrak{O}/\mathfrak{p}$ is a quotient ring of $\mathfrak{O}/N\mathfrak{O}$ which, being a finitely generated abelian group with every element of finite order, is finite. Since $\mathfrak{O}/\mathfrak{p}$ is a domain by Lemma 6.2(b) and is finite, it is a field by Theorem 1.6. Hence $\mathfrak{p}$ is a maximal ideal by Lemma 6.2(a). □

From this proof we extract another useful result:

Corollary 6.5. *Let $\mathfrak{O}$ be the ring of integers of a number field K and let $\mathfrak{a} \neq 0$ be an ideal of $\mathfrak{O}$. Then the quotient ring $\mathfrak{O}/\mathfrak{a}$ is finite.*

Proof: The additive groups $(\mathfrak{O}, +)$ and $(\mathfrak{a}, +)$ are free abelian of rank n. By Theorem 1.29, there is a basis $u_1, \dots, u_n$ for $(\mathfrak{O}, +)$ such that $\alpha_1 u_1, \dots, \alpha_n u_n$ is a basis for $(\mathfrak{a}, +)$ where the α_j are positive rational integers. Therefore every element of $(\mathfrak{O}/\mathfrak{a}, +)$ has finite order, and since this group is finitely generated abelian, it is finite. □

This motivates:

Definition 6.6. Let K have ring of integers $\mathfrak{O}$ and let $\mathfrak{a}$ be an ideal of $\mathfrak{O}$. Then the *residue ring* of $\mathfrak{a}$ is the ring $\mathfrak{O}/\mathfrak{a}$.

Corollary 6.5 now states that the residue ring of $\mathfrak{a}$ is finite.

Part (d) of Theorem 6.4 is by no means typical of general rings. For example if $R = \mathbb{R}[x, y]$, the ring of polynomials in indeterminates x, y with real coefficients, the ideal $\langle x \rangle$ is prime but not maximal because $R/\langle x \rangle \cong \mathbb{R}[y]$ is a domain but not a field.

Conditions (a)–(d) of Theorem 6.4 motivate a general definition:

Definition 6.7. A *Dedekind ring* or *Dedekind domain* is a ring D satisfying the following conditions:

(a) D is a domain.

(b) D is noetherian.

(c) Let K be the field of fractions of D. If $\alpha \in K$ satisfies a monic polynomial equation with coefficients in D then $\alpha \in D$. (That is, D is *integrally closed.*)

(d) Every prime ideal of D is maximal.

Theorem 6.4 states that the ring of integers $\mathfrak{O}$ of a number field K is always a Dedekind domain. The proof of unique factorisation of ideals, which we give shortly, uses only these four conditions (a)–(d), so it is valid for all Dedekind rings. In this text we mainly require the special case when the ring is a ring $\mathfrak{O}$ of integers in a number field, so we state it for that case.

To prove uniqueness we need to study the 'arithmetic' of nonzero ideals of $\mathfrak{O}$, especially their behaviour under multiplication. Clearly this multiplication is commutative and associative, with $\mathfrak{O}$ as an identity. However, inverses need not exist, so we do not have a group structure. It turns out that we can capture a group if we spread our net wider. An ideal is an $\mathfrak{O}$-submodule of $\mathfrak{O}$, so it is natural to look at $\mathfrak{O}$-submodules of the field K. The submodules that give the required group structure turn out to be characterised by the following property:

Definition 6.8. An $\mathfrak{O}$-submodule $\mathfrak{a}$ of K is a *fractional ideal* of $\mathfrak{O}$ if there exists some nonzero $c \in \mathfrak{O}$ such that $c\mathfrak{a} \subseteq \mathfrak{O}$.

In other words, the set $\mathfrak{b} = c\mathfrak{a}$ is an ideal of $\mathfrak{O}$, and $\mathfrak{a} = c^{-1}\mathfrak{b}$; thus the fractional ideals of $\mathfrak{O}$ are subsets of K of the form $c^{-1}\mathfrak{b}$ where $\mathfrak{b}$ is an ideal of $\mathfrak{O}$ and c is a nonzero element of $\mathfrak{O}$. This explains the name.

Example 6.9. The fractional ideals of $\mathbb{Z}$ are the subsets $r\mathbb{Z}$ where $r \in \mathbb{Q}$.

Of course if every ideal of $\mathfrak{O}$ is principal, then the fractional ideals are of the form $c^{-1}\langle d \rangle = c^{-1}d\mathfrak{O}$ where d is a generator. By Theorem 6.4(a) the fractional ideals in a principal ideal domain $\mathfrak{O}$ are just the subsets $\alpha\mathfrak{O}$ where $\alpha \in K$. The interest in fractional ideals is greater because $\mathfrak{O}$ need not be a principal ideal domain.

In general, an ideal is clearly a fractional ideal and, conversely, a fractional ideal $\mathfrak{a}$ is an ideal if and only if $\mathfrak{a} \subseteq \mathfrak{O}$. The product of fractional ideals is once more a fractional ideal. In fact, if $\mathfrak{a}_1 = c_1^{-1}\mathfrak{b}_1$, $\mathfrak{a}_2 = c_2^{-1}\mathfrak{b}_2$ where $\mathfrak{b}_1$, $\mathfrak{b}_2$, are ideals and c_1, c_2 are nonzero elements of $\mathfrak{O}$, then $\mathfrak{a}_1\mathfrak{a}_2 = (c_1c_2)^{-1}\mathfrak{b}_1\mathfrak{b}_2$. Multiplication of fractional ideals is commutative and associative, with $\mathfrak{O}$ acting as an identity.

It is convenient to prove the next two theorems together. First we state them, then give the proofs.

Theorem 6.10. *The set $\mathbb{F}$ of nonzero fractional ideals of $\mathfrak{O}$ is an abelian group under multiplication.*

Theorem 6.11. *Every nonzero ideal of $\mathfrak{O}$ is a product of prime ideals, uniquely up to the order of the factors.*

Proof: We prove Theorems 6.10 and 6.11 together in a series of nine steps.

(1) *Let $\mathfrak{a} \neq 0$ be an ideal of $\mathfrak{O}$. Then there exist prime ideals $\mathfrak{p}_1, \ldots, \mathfrak{p}_r$ such that $\mathfrak{p}_1 \ldots \mathfrak{p}_r \subseteq \mathfrak{a}$.*

For a contradiction, suppose not. By Theorem 6.4(b) $\mathfrak{O}$ is noetherian, so we may choose $\mathfrak{a}$ to be maximal subject to the non-existence of such $\mathfrak{p}$'s. Then $\mathfrak{a}$ is not prime (since we could then take $\mathfrak{p}_1 = \mathfrak{a}$), so there exist ideals $\mathfrak{b}$, $\mathfrak{c}$ of $\mathfrak{O}$ with $\mathfrak{b}\mathfrak{c} \subseteq \mathfrak{a}$, $\mathfrak{b} \not\subseteq \mathfrak{a}$, $\mathfrak{c} \not\subseteq \mathfrak{a}$. Let

$$\mathfrak{a}_1 = \mathfrak{a} + \mathfrak{b} \qquad \mathfrak{a}_2 = \mathfrak{a} + \mathfrak{c}$$

Then $\mathfrak{a}_1\mathfrak{a}_2 \subseteq \mathfrak{a}$, $\mathfrak{a}_1 \supsetneq \mathfrak{a}$, $\mathfrak{a}_2 \supsetneq \mathfrak{a}$. By maximality of $\mathfrak{a}$ there exist prime ideals $\mathfrak{p}_1, \ldots, \mathfrak{p}_s, \mathfrak{p}_{s+1}, \ldots, \mathfrak{p}_r$ such that

$$\begin{aligned} \mathfrak{p}_1 \ldots \mathfrak{p}_s &\subseteq \mathfrak{a}_1 \\ \mathfrak{p}_{s+1} \ldots \mathfrak{p}_r &\subseteq \mathfrak{a}_2 \end{aligned}$$

Hence

$$\mathfrak{p}_1 \ldots \mathfrak{p}_r \subseteq \mathfrak{a}_1\mathfrak{a}_2 \subseteq \mathfrak{a}$$

contrary to the choice of $\mathfrak{a}$.

(2) *Definition of what will turn out to be the inverse of an ideal:*

For each ideal $\mathfrak{a}$ of $\mathfrak{O}$, define

$$\mathfrak{a}^{-1} = \{x \in K | x\mathfrak{a} \subseteq \mathfrak{O}\}$$

It is clear that $\mathfrak{a}^{-1}$ is an $\mathfrak{O}$-submodule. If $\mathfrak{a} \neq 0$ then for any $c \in \mathfrak{a}$, $c \neq 0$, we have $c\mathfrak{a}^{-1} \subseteq \mathfrak{O}$, so $\mathfrak{a}^{-1}$ is a fractional ideal. Clearly $\mathfrak{O} \subseteq \mathfrak{a}^{-1}$, so $\mathfrak{a} = \mathfrak{a}\mathfrak{O} \subseteq \mathfrak{a}\mathfrak{a}^{-1}$. From the definition,

$$\mathfrak{a}\mathfrak{a}^{-1} = \mathfrak{a}^{-1}\mathfrak{a} \subseteq \mathfrak{O}$$

Therefore the fractional ideal $\mathfrak{a}\mathfrak{a}^{-1}$ is actually an *ideal.* Our aim is to prove that $\mathfrak{a}\mathfrak{a}^{-1} = \mathfrak{O}$.

A further useful fact for ideals $\mathfrak{p}$, $\mathfrak{a}$ is that $\mathfrak{a} \subseteq \mathfrak{p}$ implies $\mathfrak{O} \subseteq \mathfrak{p}^{-1} \subseteq \mathfrak{a}^{-1}$.

(3) *If $\mathfrak{a}$ is a proper ideal, then $\mathfrak{a}^{-1} \supsetneq \mathfrak{O}$.*

Since $\mathfrak{a} \subseteq \mathfrak{p}$ for some maximal ideal $\mathfrak{p}$, whence $\mathfrak{p}^{-1} \subseteq \mathfrak{a}^{-1}$, it is sufficient to prove that $\mathfrak{p}^{-1} \neq \mathfrak{O}$ for $\mathfrak{p}$ maximal. We must therefore find a non-integer in $\mathfrak{p}^{-1}$. Start with any $a \in \mathfrak{p}$, $a \neq 0$. Using step (1), choose the smallest r such that

$$\mathfrak{p}_1 \dots \mathfrak{p}_r \subseteq \langle a \rangle$$

for $\mathfrak{p}_1, \dots, \mathfrak{p}_r$ prime. Since $\langle a \rangle \subseteq \mathfrak{p}$ and $\mathfrak{p}$ is prime (remember maximal implies prime), some $\mathfrak{p}_i \subseteq \mathfrak{p}$. Without loss of generality $\mathfrak{p}_1 \subseteq \mathfrak{p}$. Hence $\mathfrak{p}_1 = \mathfrak{p}$ since prime ideals in $\mathfrak{O}$ are maximal by Theorem 6.4(d). Moreover,

$$\mathfrak{p}_2 \dots \mathfrak{p}_r \not\subseteq \langle a \rangle$$

by minimality of r. Hence we can find $b \in \mathfrak{p}_2 \dots \mathfrak{p}_r \setminus \langle a \rangle$. Now $b\,\mathfrak{p} \subseteq \langle a \rangle$ so $ba^{-1}\mathfrak{p} \subseteq \mathfrak{O}$ and $ba^{-1} \in \mathfrak{p}^{-1}$. But $b \notin a\mathfrak{O}$ and so $ba^{-1} \notin \mathfrak{O}$, whence $\mathfrak{p}^{-1} \neq \mathfrak{O}$.

(4) *If $\mathfrak{a}$ is a nonzero ideal and $\mathfrak{a}S \subseteq \mathfrak{a}$ for any subset $S \subseteq K$, then $S \subseteq \mathfrak{O}$.*

We must show that if $\mathfrak{a}\theta \subseteq \mathfrak{a}$ for $\theta \in S$, then $\theta \in \mathfrak{O}$. Because $\mathfrak{O}$ is noetherian, $\mathfrak{a} = \langle a_1, \dots, a_m \rangle$ where not all a_i are zero. Then $\mathfrak{a}\theta \subseteq \mathfrak{a}$ implies

$$\begin{aligned} a_1\theta &= b_{11}a_1 + \dots + b_{1m}a_m \\ &\vdots \qquad\qquad\qquad\qquad\qquad (b_{ij} \in \mathfrak{O}) \\ a_m\theta &= b_{m1}a_1 + \dots + b_{mm}a_m \end{aligned}$$

The equations

$$\begin{aligned} (b_{11} - \theta)\, x_1 + \dots + b_{1m}x_m &= 0 \\ &\vdots \\ b_{m1}x_1 + \dots + (b_{mm} - \theta)\, x_m &= 0 \end{aligned}$$

have a nonzero solution $x_1 = a_1, \dots, x_m = a_m$, so, as in Lemma 2.18, the determinant of the array of coefficients is zero. This gives a monic polynomial equation in θ with coefficients in $\mathfrak{O}$, so $\theta \in \mathfrak{O}$ by Theorem 6.4(c). (We could shortcut part of this proof by noting, as in the proof of Lemma 2.8, that the b_{ij} may be taken to be *rational* integers, which gives $\theta \in \mathfrak{O}$ directly.)

We are now in a position to take an important step in the proof of Theorem 6.10:

(5) *If* $\mathfrak{p}$ *is a maximal ideal then* $\mathfrak{p}\mathfrak{p}^{-1} = \mathfrak{O}$.

From (2), $\mathfrak{p}\mathfrak{p}^{-1}$ is an ideal where $\mathfrak{p} \subseteq \mathfrak{p}\mathfrak{p}^{-1} \subseteq \mathfrak{O}$. Since $\mathfrak{p}$ is maximal, $\mathfrak{p}\mathfrak{p}^{-1}$ is equal to $\mathfrak{p}$ or $\mathfrak{O}$. But if $\mathfrak{p}\mathfrak{p}^{-1} = \mathfrak{p}$ then (4) implies that $\mathfrak{p}^{-1} \subseteq \mathfrak{O}$, contradicting (3). So $\mathfrak{p}\mathfrak{p}^{-1} = \mathfrak{O}$.

We can now extend (5) to any ideal $\mathfrak{a}$:

(6) *For every ideal* $\mathfrak{a} \neq 0$, $\mathfrak{a}\mathfrak{a}^{-1} = \mathfrak{O}$.

If not, choose $\mathfrak{a}$ maximal subject to $\mathfrak{a}\mathfrak{a}^{-1} \neq \mathfrak{O}$. Then $\mathfrak{a} \subseteq \mathfrak{p}$ where $\mathfrak{p}$ is maximal. From (2), $\mathfrak{O} \subseteq \mathfrak{p}^{-1} \subseteq \mathfrak{a}^{-1}$, so

$$\mathfrak{a} \subseteq \mathfrak{a}\mathfrak{p}^{-1} \subseteq \mathfrak{a}\mathfrak{a}^{-1} \subseteq \mathfrak{O}$$

In particular, $\mathfrak{a}\mathfrak{p}^{-1} \subseteq \mathfrak{O}$ implies that $\mathfrak{a}\mathfrak{p}^{-1}$ is an ideal. We cannot have $\mathfrak{a} = \mathfrak{a}\mathfrak{p}^{-1}$ since this implies that $\mathfrak{p}^{-1} \subseteq \mathfrak{O}$ by (4), contradicting (3) once more. So $\mathfrak{a} \subsetneqq \mathfrak{a}\mathfrak{p}^{-1}$ and the maximality condition on $\mathfrak{a}$ implies that $\mathfrak{a}\mathfrak{p}^{-1}$ satisfies

$$\mathfrak{a}\mathfrak{p}^{-1}\left(\mathfrak{a}\mathfrak{p}^{-1}\right)^{-1} = \mathfrak{O}$$

By the definition of $\mathfrak{a}^{-1}$ this implies:

$$\mathfrak{p}^{-1}\left(\mathfrak{a}\mathfrak{p}^{-1}\right)^{-1} \subseteq \mathfrak{a}^{-1}$$

Thus

$$\mathfrak{O} = \mathfrak{a}\mathfrak{p}^{-1}\left(\mathfrak{a}\mathfrak{p}^{-1}\right)^{-1} \subseteq \mathfrak{a}\mathfrak{a}^{-1} \subseteq \mathfrak{O}$$

so $\mathfrak{a}\mathfrak{a}^{-1} = \mathfrak{O}$.

(7) *Every fractional ideal* $\mathfrak{a}$ *has an inverse* $\mathfrak{a}^{-1}$ *such that* $\mathfrak{a}\mathfrak{a}^{-1} = \mathfrak{O}$.

The set $\mathbb{F}$ of fractional ideals is a commutative semigroup, so given a fractional ideal $\mathfrak{a}$, we need only find another fractional ideal $\mathfrak{a}'$ such that $\mathfrak{a}\mathfrak{a}' = \mathfrak{O}$. Then $\mathfrak{a}'$ is the required inverse. But there exists an ideal $\mathfrak{b}$ and a nonzero element $c \in \mathfrak{O}$ such that $\mathfrak{a} = c^{-1}\mathfrak{b}$. Let $\mathfrak{a}' = c\mathfrak{b}^{-1}$; then $\mathfrak{a}\mathfrak{a}' = \mathfrak{O}$ as required.

This, of course, proves Theorem 6.10.

(8) *Every nonzero ideal* $\mathfrak{a}$ *is a product of prime ideals.*

If not, let $\mathfrak{a}$ be maximal subject to not being a product of prime ideals. Then $\mathfrak{a}$ is not prime, but $\mathfrak{a} \subseteq \mathfrak{p}$ for some maximal (hence prime) ideal, and as in (6),

$$\mathfrak{a} \subsetneqq \mathfrak{a}\mathfrak{p}^{-1} \subseteq \mathfrak{O}$$

By the maximality condition on $\mathfrak{a}$,

$$\mathfrak{a}\mathfrak{p}^{-1} = \mathfrak{p}_2 \dots \mathfrak{p}_r$$

for prime ideals $\mathfrak{p}_2, \ldots, \mathfrak{p}_r$, whence

$$\mathfrak{a} = \mathfrak{p}\mathfrak{p}_2 \ldots \mathfrak{p}_r$$

(9) *Prime factorisation is unique.*

By analogy with factorisation of elements, for ideals $\mathfrak{a}$, $\mathfrak{b}$ we say that $\mathfrak{a}$ divides $\mathfrak{b}$ (written $\mathfrak{a}|\mathfrak{b}$) if there is an ideal $\mathfrak{c}$ such that $\mathfrak{b} = \mathfrak{a}\mathfrak{c}$. This condition is equivalent to $\mathfrak{a} \supseteq \mathfrak{b}$ since we may then take $\mathfrak{c} = \mathfrak{a}^{-1}\mathfrak{b}$. The definition of a prime ideal $\mathfrak{p}$ shows that if $\mathfrak{p}|\mathfrak{a}\mathfrak{b}$ then either $\mathfrak{p}|\mathfrak{a}$ or $\mathfrak{p}|\mathfrak{b}$. If we now have prime ideals $\mathfrak{p}_1, \ldots, \mathfrak{p}_r, \mathfrak{q}_1, \ldots, \mathfrak{q}_s$ with

$$\mathfrak{p}_1 \ldots \mathfrak{p}_r = \mathfrak{q}_1 \ldots \mathfrak{q}_s$$

then $\mathfrak{p}_1$ divides some $\mathfrak{q}_i$, so by maximality $\mathfrak{p}_1 = \mathfrak{q}_i$. Multiplying by $\mathfrak{p}_1^{-1}$ and using induction, we obtain uniqueness of prime factorisation up to the order of the factors. This proves Theorem 6.11. □

The fractional ideals also factorise uniquely if we allow negative powers of prime ideals, which make sense in the group $\mathbb{F}$ of fractional ideals:

Theorem 6.12. *Every fractional ideal $\mathfrak{a} \in \mathbb{F}$ can be expressed as a product*

$$\mathfrak{a} = \prod_{\mathfrak{p}} \mathfrak{p}^{a(\mathfrak{p})} \tag{6.1}$$

where $\mathfrak{p}$ runs through all prime ideals of $\mathfrak{O}$, the $a(\mathfrak{p}) \in \mathbb{Z}$, and all but finitely many $a(\mathfrak{p})$ are zero.

The identity element $\mathfrak{O} \in \mathbb{F}$ corresponds to $a(\mathfrak{p}) = 0$ for all $\mathfrak{p}$.

This expression is unique up to the order of the terms in the product.

Proof: Every element of $\mathfrak{O}$ is such a product, with $a(\mathfrak{p}) \geq 0$. By Definition 6.8, every fractional ideal $\mathfrak{a}$ is equal to $c^{-1}\mathfrak{b}$ where $0 \neq c \in \mathfrak{O}$ and $\mathfrak{b}$ is an ideal of $\mathfrak{O}$. So $\mathfrak{a} = \langle c \rangle^{-1}\mathfrak{b}$. Write $\mathfrak{b} = \prod_{\mathfrak{p}} \mathfrak{p}^{b(\mathfrak{p})}$, where $b(\mathfrak{p}) \geq 0$, and $\langle c \rangle = \prod_{\mathfrak{p}} \mathfrak{p}^{c(\mathfrak{p})}$, where $c(\mathfrak{p}) \geq 0$. Then

$$\mathfrak{a} = c^{-1}\mathfrak{b} = \prod_{\mathfrak{p}} \mathfrak{p}^{-c(\mathfrak{p})} \prod_{\mathfrak{p}} \mathfrak{p}^{b(\mathfrak{p})} = \prod_{\mathfrak{p}} \mathfrak{p}^{b(\mathfrak{p})-c(\mathfrak{p})}$$

which has the required form.

For uniqueness, suppose that

$$\prod_{\mathfrak{p}} \mathfrak{p}^{a(\mathfrak{p})} = \prod_{\mathfrak{p}} \mathfrak{p}^{b(\mathfrak{p})}$$

For any $\mathfrak{p}$ such that $a(\mathfrak{p}) \neq 0$ or $b(\mathfrak{p}) \neq 0$ choose $c(\mathfrak{p})$ so that $a(\mathfrak{p})+c(\mathfrak{p}) \geq 0$ and $b(\mathfrak{p}) + c(\mathfrak{p}) \geq 0$. Multiply by $\prod_{\mathfrak{p}} \mathfrak{p}^{c(\mathfrak{p})}$ to get

$$\prod_{\mathfrak{p}} \mathfrak{p}^{a(\mathfrak{p})+c(\mathfrak{p})} = \prod_{\mathfrak{p}} \mathfrak{p}^{b(\mathfrak{p})+c(\mathfrak{p})}$$

By uniqueness of factorisation into prime ideals, $a(\mathfrak{p}) + c(\mathfrak{p}) = b(\mathfrak{p}) + c(\mathfrak{p})$ for all $\mathfrak{p}$. Therefore $a(\mathfrak{p}) = b(\mathfrak{p})$ for all $\mathfrak{p}$. □

One result in the proofs of Theorems 6.10 and 6.11, which is worth isolating, occurs in step (9):

Proposition 6.13. *For ideals* $\mathfrak{a}$, $\mathfrak{b}$ *of* $\mathfrak{O}$,

$$\mathfrak{a}|\mathfrak{b} \quad \text{if and only if} \quad \mathfrak{a} \supseteq \mathfrak{b}$$

□

This tells us that in $\mathfrak{O}$ the factors of an ideal $\mathfrak{b}$ are precisely the ideals containing $\mathfrak{b}$. The definition of a prime ideal $\mathfrak{p}$ also translates into a notation directly analogous to that of a prime element:

$$\mathfrak{p}|\mathfrak{ab} \quad \text{implies} \quad \mathfrak{p}|\mathfrak{a} \quad \text{or} \quad \mathfrak{p}|\mathfrak{b}$$

Greatest Common Divisor and Least Common Multiple

Once unique factorisation is proved, several useful consequences follow in the usual way. In particular, by analogy with Definition 5.23, any two nonzero ideals $\mathfrak{a}$ and $\mathfrak{b}$ have a unique *greatest common divisor* $\mathfrak{g}$ and *least common multiple* $\mathfrak{l}$, with the following properties:

$$\mathfrak{g}|\mathfrak{a},\ \mathfrak{g}|\mathfrak{b} \text{ and if } \mathfrak{g}' \text{ has the same properties then } \mathfrak{g}'|\mathfrak{g}$$
$$\mathfrak{a}|\mathfrak{l},\ \mathfrak{b}|\mathfrak{l} \text{ and if } \mathfrak{l}' \text{ has the same properties then } \mathfrak{l}|\mathfrak{l}'$$

Theorem 6.14. *If* $\mathfrak{a}$ *and* $\mathfrak{b}$ *factorise into primes as:*

$$\mathfrak{a} = \prod \mathfrak{p}_i^{a_i} \qquad \mathfrak{b} = \prod \mathfrak{p}_i^{b_i}$$

with distinct prime ideals $\mathfrak{p}_i$, *then*

$$\mathfrak{g} = \prod \mathfrak{p}_i^{\min(a_i,b_i)} \qquad \mathfrak{l} = \prod \mathfrak{p}_i^{\max(a_i,b_i)}$$

Proof: This is straightforward from the definitions using unique factorisation for prime ideals.

There are useful alternative expressions:

Lemma 6.15. *If $\mathfrak{a}$ and $\mathfrak{b}$ are ideals of $\mathfrak{O}$ and $\mathfrak{g}$, $\mathfrak{l}$ are their greatest common divisor and least common multiple, respectively, then*

$$\mathfrak{g} = \mathfrak{a} + \mathfrak{b} \qquad \mathfrak{l} = \mathfrak{a} \cap \mathfrak{b}.$$

Proof: By Proposition 6.13 $\mathfrak{x}|\mathfrak{a}$ if and only if $\mathfrak{x} \supseteq \mathfrak{a}$. Therefore $\mathfrak{g}$ is the smallest ideal containing $\mathfrak{a}$ and $\mathfrak{b}$, and $\mathfrak{c}$ is the largest ideal contained in $\mathfrak{a}$ and $\mathfrak{b}$. The rest is obvious. □

6.3 Norm of an Ideal

We now define the norm of an ideal and prove that it is multiplicative. Later we develop other properties of the norm that help to streamline the calculations for the extended example in Section 6.4.

Corollary 6.5 shows that if $\mathfrak{a}$ is a nonzero ideal of $\mathfrak{O}$ then the residue ring $\mathfrak{O}/\mathfrak{a}$ is finite. This motivates:

Definition 6.16. The *norm* of $\mathfrak{a}$ is

$$\mathrm{N}(\mathfrak{a}) = |\mathfrak{O}/\mathfrak{a}|$$

Clearly $\mathrm{N}(\mathfrak{a})$ is a positive integer.

There is no reason to confuse this norm with the old norm of an element $\mathrm{N}(a)$ since it applies only to ideals. But in fact there is a close connection between the two norms, as we see in Corollary 6.18 below.

Theorem 6.17. (a) *Every ideal $\mathfrak{a}$ of $\mathfrak{O}$ with $\mathfrak{a} \neq 0$ has a $\mathbb{Z}$-basis $\{\alpha_1, \ldots, \alpha_n\}$ where n is the degree of K.*

(b) *We have*

$$\mathrm{N}(\mathfrak{a}) = \left| \frac{\Delta[\alpha_1, \ldots, \alpha_n]}{\Delta} \right|^{1/2}$$

where Δ is the discriminant of K.

Proof: By Theorem 2.28 $(\mathfrak{O}, +)$ is free abelian of rank n. Since $\mathfrak{O}/\mathfrak{a}$ is finite, Theorem 1.30 implies that $(\mathfrak{a}, +)$ is also free abelian of rank n, hence has a $\mathbb{Z}$-basis $\{\alpha_1, \ldots, \alpha_n\}$. This proves (a).

For part (b) let $\{\omega_1, \ldots, \omega_n\}$ be a $\mathbb{Z}$-basis for $\mathfrak{O}$, and suppose that $\alpha_i = \sum c_{ij}\omega_j$. By Theorem 1.30,

$$\mathrm{N}(a) = |\mathfrak{O}/\mathfrak{a}| = |\det c_{ij}|$$

By Equation (2.4),

$$\Delta[\alpha_1,\ldots,\alpha_n] = (\det c_{ij})^2 \Delta[\omega_1,\ldots,\omega_n] = (\mathrm{N}(\mathfrak{a}))^2 \Delta$$

Now take square roots and remember that $\mathrm{N}(\mathfrak{a})$ is positive. □

Corollary 6.18. *If* $\mathfrak{a} = \langle a \rangle$ *is a principal ideal then* $\mathrm{N}(\mathfrak{a}) = |\mathrm{N}(a)|$.

Proof: A $\mathbb{Z}$-basis for $\mathfrak{a}$ is $\{a\omega_1,\ldots,a\omega_n\}$. Now use the definition of $\Delta[\alpha_1,\ldots,\alpha_n]$ and Theorem 6.17. □

This corollary helps us to perform a straightforward calculation of the norm of a principal ideal.

Example 6.19. If $\mathfrak{O}$ is the ring of integers of $\mathbb{Q}(\sqrt{d})$ for a squarefree rational integer d, then

$$\mathrm{N}(\langle a + b\sqrt{d} \rangle) = |a^2 - bd^2|$$

In particular in $\mathfrak{O} = \mathbb{Z}[\sqrt{-17}]$

$$\mathrm{N}(\langle 18 \rangle) = 18^2$$

Next we prove, in Theorem 6.22 below, that the new norm of Definition 6.16, like the old norm, is *multiplicative*. First, we need a lemma and a simple corollary.

Lemma 6.20. *Let* $\mathfrak{p} \neq 0$ *be a prime ideal of* $\mathfrak{O}$. *Then, as additive groups,*

$$\mathfrak{O}/\mathfrak{p} \cong \mathfrak{p}^k/\mathfrak{p}^{k+1} \qquad (k \geq 1) \tag{6.2}$$

Proof: By unique prime ideal factorisation, $\mathfrak{p}^{k+1} \subsetneq \mathfrak{p}^k$, so we may take $a \in \mathfrak{p} \setminus \mathfrak{p}^{k+1}$, which in particular implies that $a \neq 0$. Therefore the map $x \mapsto ax$ for $x \in \mathfrak{O}$ induces isomorphisms $\mathfrak{O} \cong a\mathfrak{O}$ and $\mathfrak{p} \cong a\mathfrak{p}$, which in turn induce an isomorphism

$$\mathfrak{O}/\mathfrak{p} \cong a\mathfrak{O}/a\mathfrak{p} \cong \mathfrak{a}\mathfrak{p}^k/\mathfrak{a}\mathfrak{p}^{k+1}$$

where $\langle a \rangle = \mathfrak{a}\mathfrak{p}^k$ with $\mathfrak{a}$ prime to $\mathfrak{p}$. Now

$$\mathfrak{a}\mathfrak{p}^k + \mathfrak{p}^{k+1} = \mathfrak{p}^k \qquad \mathfrak{a}\mathfrak{p}^k \cap \mathfrak{p}^{k+1} = \mathfrak{a}\mathfrak{p}^{k+1}$$

Therefore the map $x \mapsto x + \mathfrak{p}^{k+1}$, for $x \in \mathfrak{a}\mathfrak{p}^k$, induces an isomorphism

$$\mathfrak{a}\mathfrak{p}^k/\mathfrak{a}\mathfrak{p}^{k+1} \cong \mathfrak{p}^k/\mathfrak{p}^{k+1}$$

□

Corollary 6.21. *As additive groups,*

$$|\mathfrak{O}/\mathfrak{p}^k| = |\mathfrak{O}/\mathfrak{p}|^k \tag{6.3}$$

Proof: Consider the series of additive subgroups $\mathfrak{O} \supseteq \mathfrak{p} \supseteq \mathfrak{p}^2 \supseteq \cdots \supseteq \mathfrak{p}^k$. Each successive quotient has order equal to $|\mathfrak{O}/\mathfrak{p}|$. □

We can now prove multiplicativity:

Theorem 6.22. *If $\mathfrak{a}$ and $\mathfrak{b}$ are nonzero ideals of $\mathfrak{O}$, then*

$$\mathrm{N}(\mathfrak{a}\mathfrak{b}) = \mathrm{N}(\mathfrak{a})\mathrm{N}(\mathfrak{b}) \tag{6.4}$$

Proof: Suppose first that $\mathfrak{a}$ and $\mathfrak{b}$ are coprime. Then $\mathfrak{a}\mathfrak{b} = \mathfrak{a} \cap \mathfrak{b}$, so Corollary 1.2 implies that $\mathfrak{O}/(\mathfrak{a}\mathfrak{b}) \cong \mathfrak{O}/\mathfrak{a} \times \mathfrak{O}/\mathfrak{b}$. Therefore (6.4) holds in this case.

By uniqueness of factorisation into prime ideals, and induction on the number of factors, it remains to prove that $\mathrm{N}(\mathfrak{p}^k) = \mathrm{N}(\mathfrak{p})^k$. But this is (6.3). □

6.4 Extended Example

We discuss a more complicated example: the factorisation of the ideal $\langle 18 \rangle$ in $\mathbb{Z}[\sqrt{-17}]$.

Theorem 5.17 displays the factorisation

$$18 = 2 \cdot 3 \cdot 3 = (1 + \sqrt{-17})(1 - \sqrt{-17}) \tag{6.5}$$

Consider the ideal $\mathfrak{p}_1 = \langle 2, 1 + \sqrt{-17} \rangle$ whose generators are both factors of 18. Clearly $18 \in \mathfrak{p}_1$, so $\langle 18 \rangle \subseteq \mathfrak{p}_1$, which means that $\mathfrak{p}_1$ is a factor of $\langle 18 \rangle$. Now

$$1 - \sqrt{-17} = 2 - (1 + \sqrt{-17}) \in \mathfrak{p}_1$$

so

$$18 = (1 + \sqrt{-17})(1 - \sqrt{-17}) \in \mathfrak{p}_1^2$$

Therefore $\langle 18 \rangle \subseteq \mathfrak{p}_1^2$, and $\mathfrak{p}_1^2$ is a factor of $\langle 18 \rangle$. The elements of $\mathfrak{p}_1$ have the form

$$2(a + b\sqrt{-17}) + (1 + \sqrt{-17})(c + d\sqrt{-17})$$

$$= (2a + c - 17d) + (2b + c + d)\sqrt{-17}$$
$$= r + s\sqrt{-17}$$

where

$$r - s = 2a - 2b - 18d$$

which is always even. Clearly r may be taken to be any integer and then s may be any integer of the same parity (odd or even). This implies that $\mathfrak{p}_1$ is not the whole ring $\mathbb{Z}[\sqrt{-17}]$. On the other hand, $\mathfrak{p}_1$ is maximal, for if $m + n\sqrt{-17}$ is any element not in $\mathfrak{p}_1$ then one of m, n is even and the other odd, so

$$\langle \mathfrak{p}_1, m + n\sqrt{-17} \rangle = \mathbb{Z}\left[\sqrt{-17}\right]$$

Similarly, considering

$$\mathfrak{p}_2 = \langle 3, 1 + \sqrt{-17} \rangle$$

an element of $\mathfrak{p}_2$ has the form

$$r + s\sqrt{-17} = (3a + c - 17d) + (3b + c + d)\sqrt{-17}$$

where $r - s = 3(a + b - 6d)$. Thus r, s can be any integers subject to the constraint

$$r \equiv s \pmod 3$$

Once more we find $\mathfrak{p}_2$ maximal and $18 = 2 \cdot 3 \cdot 3 \in \mathfrak{p}_2^2$, so $\mathfrak{p}_2^2$ is a factor of $\langle 18 \rangle$.

Finally, considering

$$\mathfrak{p}_3 = \langle 3, 1 - \sqrt{-17} \rangle$$

we get another prime ideal such that $\mathfrak{p}_3^2$ is a factor of $\langle 18 \rangle$, and a calculation similar to the previous ones shows that $r + s\sqrt{-17} \in \mathfrak{p}_3$ if and only if

$$r + s \equiv 0 \pmod 3$$

Using the factorisation theory of Theorem 6.11,

$$\mathfrak{p}_1^2 \mathfrak{p}_2^2 \mathfrak{p}_3^2 \supseteq \langle 18 \rangle$$

The final step, to show that $\langle 18 \rangle = \mathfrak{p}_1^2 \mathfrak{p}_2^2 \mathfrak{p}_3^2$, is best performed using a counting argument. Since every element in $\mathbb{Z}[\sqrt{-17}]$ is either in $\mathfrak{p}_1$ or of the form $1 + x$ for $x \in \mathfrak{p}_1$, the number of elements in the quotient ring $\mathbb{Z}[-17]/\mathfrak{p}_1$ is

$$|\mathbb{Z}[\sqrt{-17}]/\mathfrak{p}_1| = 2$$

Similarly

$$|\mathbb{Z}[\sqrt{-17}]/\mathfrak{p}_r| = 3 \quad (r = 2, 3)$$

Recall from Definition 6.16 that the norm of an ideal $\mathfrak{a}$ is $\mathrm{N}(\mathfrak{a}) = |\mathfrak{O}/\mathfrak{p}|$. By (6.4), $\mathrm{N}(\mathfrak{ab}) = \mathrm{N}(\mathfrak{a})\mathrm{N}(\mathfrak{b})$. Therefore

$$\mathrm{N}(\mathfrak{p}_1^2\mathfrak{p}_2^2\mathfrak{p}_3^2) = 2^2 \cdot 3^2 \cdot 3^2 = 18^2$$

Now the norm of the ideal $\langle 18 \rangle$ is

$$\mathrm{N}(\langle 18 \rangle) = |\mathbb{Z}[\sqrt{-17}]/\langle 18 \rangle|$$

and since every element of $\mathbb{Z}[\sqrt{-17}]$ is uniquely of the form

$$a + b\sqrt{-17} + x$$

where a, b are integers in the range 0 to 17 and $x \in \langle 18 \rangle$, we find 18 choices each for a, b so

$$\mathrm{N}(\langle 18 \rangle) = 18^2$$

Suppose $\langle 18 \rangle$ factorises as $\langle 18 \rangle = \mathfrak{p}_1^2\mathfrak{p}_2^2\mathfrak{p}_3^2\mathfrak{a}$ for some ideal $\mathfrak{a}$. Then taking norms and using the multiplicative property, we find $\mathrm{N}(\mathfrak{a}) = 1$, so $\mathfrak{a}$ is the whole ring and

$$\langle 18 \rangle = \mathfrak{p}_1^2\mathfrak{p}_2^2\mathfrak{p}_3^2 \tag{6.6}$$

We can now explain how the two factorisations in (6.5) relate to the prime ideal factors in (6.6).

If we consider the factorisation of elements $18 = 2 \cdot 3 \cdot 3$, we obtain

$$\langle 2 \rangle \langle 3 \rangle^2 = \mathfrak{p}_1^2\mathfrak{p}_2^2\mathfrak{p}_3^2 \tag{6.7}$$

By unique factorisation for ideals, both $\langle 2 \rangle$, $\langle 3 \rangle$ are products of prime ideals from the set $\{\mathfrak{p}_1, \mathfrak{p}_2, \mathfrak{p}_3\}$. Now $2 \in \mathfrak{p}_1$ but $2 \notin \mathfrak{p}_2$, $2 \notin \mathfrak{p}_3$, so $\mathfrak{p}_1 | \langle 2 \rangle$, $\mathfrak{p}_2 \nmid \langle 2 \rangle$, $\mathfrak{p}_3 \nmid \langle 2 \rangle$, thus $\langle 2 \rangle = \mathfrak{p}_1^q$. Similarly $3 \notin \mathfrak{p}_1$, $3 \in \mathfrak{p}_2$, $3 \in \mathfrak{p}_3$ implies that $\langle 3 \rangle = \mathfrak{p}_2^r\mathfrak{p}_3^s$. Substitute in Equation (6.7) to get $\mathfrak{p}_1^q\mathfrak{p}_2^{2r}\mathfrak{p}_3^{2s} = \mathfrak{p}_1^2\mathfrak{p}_2^2\mathfrak{p}_3^2$. Unique factorisation of ideals implies that $q = 2$, $r = s = 1$. Therefore

$$\langle 2 \rangle = \mathfrak{p}_1^2 \qquad \langle 3 \rangle = \mathfrak{p}_2\mathfrak{p}_3 \tag{6.8}$$

(It is instructive to check these by direct calculation.)

A similar argument using

$$\langle 18 \rangle = \langle 1 + \sqrt{-17} \rangle \langle 1 - \sqrt{-17} \rangle = \mathfrak{p}_1^2\mathfrak{p}_2^2\mathfrak{p}_3^2 \tag{6.9}$$

where $1+\sqrt{-17} \in \mathfrak{p}_1, \mathfrak{p}_2$; $1+\sqrt{-17} \notin \mathfrak{p}_3$; $1-\sqrt{-17} \in \mathfrak{p}_1, \mathfrak{p}_3$; $1-\sqrt{-17} \notin \mathfrak{p}_2$ gives

$$\langle 1 + \sqrt{-17} \rangle = \mathfrak{p}_1^m\mathfrak{p}_2^n \qquad \langle 1 - \sqrt{-17} \rangle = \mathfrak{p}_1^r\mathfrak{p}_3^s$$

Substitute in (6.9) to get $m = r = 1$, $n = s = 2$, so

$$\langle 1+\sqrt{-17}\rangle = \mathfrak{p}_1\mathfrak{p}_2^2 \qquad \langle 1-\sqrt{-17}\rangle = \mathfrak{p}_1\mathfrak{p}_3^2 \tag{6.10}$$

By (6.8) and (6.10) the two alternative factorisations of the element 18 come from alternative groupings of the ideals:

$$\begin{aligned}\langle 18\rangle &= \left(\mathfrak{p}_1^2\right)\left(\mathfrak{p}_2\mathfrak{p}_3\right)^2 = \langle 2\rangle\,\langle 3\rangle^2 \\ &= \left(\mathfrak{p}_1\mathfrak{p}_2^2\right)\left(\mathfrak{p}_1\mathfrak{p}_3^2\right) = \langle 1+\sqrt{-17}\rangle\,\langle 1-\sqrt{-17}\rangle\end{aligned}$$

6.5 Further Properties of Norms

It is convenient to introduce yet another usage for the word 'divides'. If $\mathfrak{a}$ is an ideal of $\mathfrak{O}$ and b an element of $\mathfrak{O}$ such that $\mathfrak{a}|\langle b\rangle$, then we also write $\mathfrak{a}|b$ and say that $\mathfrak{a}$ divides b. It is clear that $\mathfrak{a}|b$ if and only if $b \in \mathfrak{a}$; however, the new notation has certain distinct advantages. For example, if $\mathfrak{p}$ is a prime ideal and $\mathfrak{p}|\langle a\rangle\langle b\rangle$, then we must have $\mathfrak{p}|\langle a\rangle$ or $\mathfrak{p}|\langle b\rangle$. Thus for $\mathfrak{p}$ prime,

$$\mathfrak{p}|ab \quad \text{implies} \quad \mathfrak{p}|a \text{ or } \mathfrak{p}|b \quad .$$

This new notation allows us to emphasise the correspondence between factorisation of elements and principal ideals which would otherwise be less evident.

Theorem 6.23. *Let $\mathfrak{a}$ be an ideal of $\mathfrak{O}$, $\mathfrak{a} \neq 0$.*

(a) *If $\mathrm{N}(\mathfrak{a})$ is prime, then so is $\mathfrak{a}$.*
(b) *$\mathrm{N}(\mathfrak{a})$ is an element of $\mathfrak{a}$, or equivalently $\mathfrak{a}|\mathrm{N}(\mathfrak{a})$.*
(c) *If $\mathfrak{a}$ is prime it divides exactly one rational prime p, and then*

$$\mathrm{N}(\mathfrak{a}) = p^m$$

where $m \leq n$, the degree of K.

Proof: For part (a) write $\mathfrak{a}$ as a product of prime ideals and equate norms. For part (b), the definition $\mathrm{N}(\mathfrak{a}) = |\mathfrak{O}/\mathfrak{a}|$ implies that for any $x \in \mathfrak{O}$ we have $\mathrm{N}(\mathfrak{a})x \in \mathfrak{a}$. Now put $x = 1$. To prove part (c), by part (b)

$$\mathfrak{a}|\mathrm{N}(\mathfrak{a}) = p_1^{m_1} \ldots p_r^{m_r}$$

Replacing the p_i by principal ideals $\langle p_i\rangle$, we see that $\mathfrak{a}|p_i$ for some rational prime p_i. If p and q are distinct rational primes, both divisible by $\mathfrak{a}$, we can find integers u, v such that $up + vq = 1$, so $\mathfrak{a}|1$ and $\mathfrak{a} = \mathfrak{O}$, a contradiction. Then

$$\mathrm{N}(\mathfrak{a})|\mathrm{N}(\langle p\rangle) = p^n$$

so $\mathrm{N}(\mathfrak{a}) = p^m$ for some $m \leq n$. □

Example 6.24. Let $\mathfrak{O} = \mathbb{Z}[\sqrt{-17}]$, $\mathfrak{p}_1 = \langle 2, 1 + \sqrt{-17}\rangle$. Because $\mathrm{N}(\mathfrak{p}_1) = 2$ we immediately deduce that $\mathfrak{p}_1$ is prime. Moreover, $\mathrm{N}(\mathfrak{p}_1) = 2 \in \mathfrak{p}_1$ as asserted in Theorem 6.23(b).

Example 6.25. A prime ideal $\mathfrak{a}$ can satisfy $\mathrm{N}(\mathfrak{a}) = p^m$ where $m > 1$, so the norm of a prime ideal need not be (a rational) prime. For instance $\mathfrak{O} = \mathbb{Z}[\mathrm{i}]$, $\mathfrak{a} = \langle 3\rangle$. Here 3 is irreducible in $\mathbb{Z}[\mathrm{i}]$, hence prime because $\mathbb{Z}[\mathrm{i}]$ has unique factorisation. It is an easy deduction (Exercise 6.1) that if an element is prime, so is the ideal it generates. Hence $\langle 3\rangle$ is prime in $\mathbb{Z}[\mathrm{i}]$, but $\mathrm{N}(\langle 3\rangle) = 3^2$.

The next theorem collects together several useful finiteness assertions:

Theorem 6.26. (a) *Every nonzero ideal of* $\mathfrak{O}$ *has a finite number of divisors.*

(b) *A nonzero rational integer belongs to only a finite number of ideals of* $\mathfrak{O}$.

(c) *Only finitely many ideals of* $\mathfrak{O}$ *have given norm.*

Proof: (a) is an immediate consequence of prime factorisation, (b) is a special case of (a), and (c) follows from (b) using Theorem 6.23(b). □

Example 6.27. Consider the calculation in Section 6.4:

$$\langle 18\rangle = \mathfrak{p}_1^2\mathfrak{p}_2^2\mathfrak{p}_3^2$$

in $\mathbb{Z}[\sqrt{-17}]$ where $\mathfrak{p}_1 = \langle 2, 1 + \sqrt{-17}\rangle$, $\mathfrak{p}_2 = \langle 3, 1 + \sqrt{-17}\rangle$, and $\mathfrak{p}_3 = \langle 3, 1 - \sqrt{-17}\rangle$. The only prime divisors of $\langle 18\rangle$ are $\mathfrak{p}_1, \mathfrak{p}_2, \mathfrak{p}_3$. If 18 belongs to some ideal $\mathfrak{a}$, then $\langle 18\rangle \subseteq \mathfrak{a}$, whence $\mathfrak{a}|\langle 18\rangle$, so $\mathfrak{a}|\mathfrak{p}_1^2\mathfrak{p}_2^2\mathfrak{p}_3^2$ and $\mathfrak{a} = \mathfrak{p}_1^q\mathfrak{p}_2^r\mathfrak{p}_3^s$ where q, r, s are 0, 1, or 2. Thus 18 belongs to only a finite number of ideals.

How many ideals $\mathfrak{a}$ have norm 18? By Theorem 6.23(b) this can happen only when $\mathfrak{a}|18$, so

$$\mathfrak{a} = \mathfrak{p}_1^q\mathfrak{p}_2^r\mathfrak{p}_3^s$$

which implies

$$\mathrm{N}(\mathfrak{a}) = 2^q \cdot 3^r \cdot 3^s$$

This norm is 18 only when $q = 1$ and $r + s = 2$, which means that $\mathfrak{a}$ is $\mathfrak{p}_1\mathfrak{p}_2^2$, $\mathfrak{p}_1\mathfrak{p}_2\mathfrak{p}_3$ or $\mathfrak{p}_1\mathfrak{p}_3^2$.

6.6 Two-Generator Theorem

We know that every ideal of $\mathfrak{O}$ is finitely generated. In fact, two generators suffice. To show this, we first prove:

Lemma 6.28. *If $\mathfrak{a}$, $\mathfrak{b}$ are nonzero ideals of $\mathfrak{O}$ then there exists $\alpha \in \mathfrak{a}$ such that*

$$\alpha\mathfrak{a}^{-1} + \mathfrak{b} = \mathfrak{O}$$

Proof: If $\alpha \in \mathfrak{a}$ then $\mathfrak{a}|\alpha$ so that $\alpha\mathfrak{a}^{-1}$ is an ideal and not just a fractional ideal. Now $\alpha\mathfrak{a}^{-1} + \mathfrak{b}$ is the greatest common divisor of $\alpha\mathfrak{a}^{-1}$ and $\mathfrak{b}$, so it is sufficient to choose $\alpha \in \mathfrak{a}$ so that

$$\alpha\mathfrak{a}^{-1} + \mathfrak{p}_i = \mathfrak{O} \qquad (i = 1, \ldots, r)$$

where $\mathfrak{p}_1, \ldots, \mathfrak{p}_r$ are the distinct prime ideals dividing $\mathfrak{b}$. We can find such an α if

$$\mathfrak{p}_i \nmid \alpha\mathfrak{a}^{-1}$$

since $\mathfrak{p}_i$ is a maximal ideal. So it is sufficient to choose $\alpha \in \mathfrak{a} \setminus \mathfrak{a}\mathfrak{p}_i$ for all $i = 1, \ldots, r$.

If $r = 1$ this is easy, for unique factorisation of ideals implies $\mathfrak{a} \neq \mathfrak{a}\mathfrak{p}_i$. For $r > 1$ let

$$\mathfrak{a}_i = \mathfrak{a}\mathfrak{p}_1 \ldots \mathfrak{p}_{i-1}\mathfrak{p}_{i+1} \ldots \mathfrak{p}_r$$

By the case $r = 1$ we can choose

$$\alpha_i \in \mathfrak{a}_i \setminus \mathfrak{a}_i\mathfrak{p}_i$$

Define

$$\alpha = \alpha_1 + \ldots + \alpha_r$$

Then each $\alpha_i \in \mathfrak{a}_i \subseteq \mathfrak{a}$, so $\alpha \in \mathfrak{a}$. Suppose if possible that $\alpha \in \mathfrak{a}\mathfrak{p}_i$. If $j \neq i$ then $\alpha_j \in \mathfrak{a}_j \subseteq \mathfrak{a}\mathfrak{p}_i$, so

$$\alpha_i = \alpha - \alpha_1 - \ldots - \alpha_{i-1} - \alpha_{i+1} - \ldots - \alpha_r \in \mathfrak{a}\mathfrak{p}_i$$

Hence $\mathfrak{a}\mathfrak{p}_i | \langle\alpha_i\rangle$. On the other hand $\mathfrak{a}_i | \langle\alpha_i\rangle$. We have $\mathfrak{a}_i\mathfrak{p}_i | \langle\alpha_i\rangle$. This contradicts the choice of α_i. □

Theorem 6.29. (Two-Generator Theorem) *Let $\mathfrak{a} \neq 0$ be an ideal of $\mathfrak{O}$, and $0 \neq \beta \in \mathfrak{a}$. Then there exists $\alpha \in \mathfrak{a}$ such that $\mathfrak{a} = \langle\alpha, \beta\rangle$.*

Proof: Let $\mathfrak{b} = \beta\mathfrak{a}^{-1}$. By Lemma 6.28 there exists $\alpha \in \mathfrak{a}$ such that

$$\alpha\mathfrak{a}^{-1} + \mathfrak{b} = \alpha\mathfrak{a}^{-1} + \beta\mathfrak{a}^{-1} = \mathfrak{O}$$

hence

$$(\langle\alpha\rangle + \langle\beta\rangle)\,\mathfrak{a}^{-1} = \mathfrak{O}$$

so that

$$\mathfrak{a} = \langle\alpha\rangle + \langle\beta\rangle = \langle\alpha, \beta\rangle$$ □

This theorem demonstrates that the extended example in Section 6.4, where each ideal considered has at most two generators, is typical.

We are now in a position to characterise those $\mathfrak{O}$ for which factorisation of elements into irreducibles is unique:

Theorem 6.30. *Factorisation of elements of $\mathfrak{O}$ into irreducibles is unique if and only if every ideal of $\mathfrak{O}$ is principal.*

Proof: If every ideal is principal, then unique factorisation of elements follows by Theorem 5.26. To prove the converse, if factorisation of elements is unique, it is enough to prove that every *prime* ideal is principal, since every other ideal, being a product of prime ideals, is then principal.

Let $\mathfrak{p} \neq 0$ be a prime ideal of $\mathfrak{O}$. By Theorem 6.23(b) there exists a rational integer $N = \mathrm{N}(\mathfrak{p})$ such that $\mathfrak{p}|N$. We can factorise N as a product of irreducible elements in $\mathfrak{O}$, say

$$N = \pi_1 \ldots \pi_s$$

Since $\mathfrak{p}|N$ and $\mathfrak{p}$ is a prime ideal, $\mathfrak{p}|\pi_i$, or equivalently, $\mathfrak{p}|\,\langle\pi_i\rangle$. Since factorisation is unique in $\mathfrak{O}$, the irreducible π_i is actually *prime* by Theorem 5.21, so the principal ideal $\langle\pi_i\rangle$ is prime (Exercise 6.1). Thus $\mathfrak{p}|\,\langle\pi_i\rangle$, where both $\mathfrak{p}, \langle\pi_i\rangle$ are prime, and by uniqueness of factorisation, $\mathfrak{p} = \langle\pi_i\rangle$, so $\mathfrak{p}$ is principal. □

Using this theorem we can nicely round off the relationship between factorisation of elements and ideals. To do this, consider an element π that is irreducible but not prime. Then the ideal $\langle\pi\rangle$ is not prime, so it has a proper factorisation into prime ideals:

$$\langle\pi\rangle = \mathfrak{p}_1 \ldots \mathfrak{p}_r$$

None of these $\mathfrak{p}_i$ can be principal, for if $\mathfrak{p}_i = \langle a\rangle$ then $\langle a\rangle \mid \langle\pi\rangle$, implying $a|\pi$. Since π is irreducible, a is either a unit, contradicting $\langle a\rangle$ being prime, or an associate of π, whence $\langle\pi\rangle = \mathfrak{p}_i$, contradicting $\langle\pi\rangle$ having a proper factorisation.

Tying up the loose ends, we see that if $\mathfrak{O}$ has unique factorisation of elements into irreducibles then these irreducibles are all primes, and factorisation of elements corresponds precisely to factorisation of the corresponding principal ideals. On the other hand, if $\mathfrak{O}$ does not have unique factorisation of elements, then not all irreducibles are prime, and any nonprime irreducible generates a principal ideal which has a proper factorisation into nonprincipal ideals. We may add in the latter case that such nonprincipal ideals have precisely two generators.

Example 6.31. In $\mathbb{Z}[\sqrt{-17}]$, the elements 2, 3 are irreducible (proved by considering norms) and not prime, with

$$\begin{aligned}\langle 2\rangle &= \left\langle 2, 1+\sqrt{-17}\right\rangle^2 \\ \langle 3\rangle &= \left\langle 3, 1+\sqrt{-17}\right\rangle\left\langle 3, 1-\sqrt{-17}\right\rangle\end{aligned}$$

6.7 Non-Unique Factorisation in Cyclotomic Fields

We mentioned in the Preface that unique prime factorisation *fails* in the cyclotomic field of 23rd roots of unity. The failure, rather than the precise value $n = 23$, is the crucial point; however, it can be proved that 23 is the smallest prime for which uniqueness fails. In this section we use the tools developed in this chapter to demonstrate this failure of unique factorisation. The calculations are somewhat tedious and have been abbreviated where feasible: it is a challenging exercise to check the details. A few tricks, inspired by the structure of the group of units of the ring $\mathbb{Z}_{23}$, are used, but we lack the space to motivate them. For further details see the admirable book by Edwards [38].

Let $\zeta = e^{2\pi i/23}$, and let $K = \mathbb{Q}(\zeta)$. By Theorem 3.9 the ring of integers $\mathfrak{O}_K$ is $\mathbb{Z}[\zeta]$. The group of units of $\mathbb{Z}_{23}$ is generated by -2, whose powers in order are

$$1, -2, 4, -8, -7, 14, -5, \ldots \tag{6.11}$$

For reasons that will emerge later, we introduce two elements

$$\begin{aligned}\theta_0 &= \zeta + \zeta^4 + \zeta^{-7} + \zeta^{-5} + \ldots \\ \theta_1 &= \zeta^{-2} + \zeta^{-8} + \zeta^{14} + \ldots\end{aligned}$$

The powers that occur are alternate elements in the sequence (6.11), and they ensure that $\mathbb{Q}(\theta_0)$ is a quadratic extension of $\mathbb{Q}$. Specifically:

$$\theta_0 + \theta_1 = \zeta + \zeta^2 + \ldots + \zeta^{22} = -1 \qquad \theta_0\theta_1 = 6$$

so θ_0 and θ_1 are the zeros of the quadratic polynomial $t^2 + t + 6 = 0$, whose zeros are $\frac{1}{2}(-1 \pm \sqrt{-23})$. The norm of a general element $f(\zeta)$, with f a polynomial over $\mathbb{Z}$ of degree ≤ 22, can be broken up as

$$\begin{aligned} \mathrm{N}(f(\zeta)) &= \prod_{j=1}^{22} \mathrm{N}f(\zeta^j) \\ &= \prod_{j \text{ even}} \mathrm{N}f(\zeta^j) \cdot \prod_{j \text{ odd}} \mathrm{N}f(\zeta^j) \\ &= G(\zeta^2)G(\zeta^{-2}) \end{aligned}$$

where

$$G(\zeta) = f(\zeta)f(\zeta^4)f(\zeta^{-7})f(\zeta^{-5})f(\zeta^3)f(\zeta^{-11})f(\zeta^2)f(\zeta^8)f(\zeta^9)f(\zeta^{-10})f(\zeta^6)$$

By definition, $G(\zeta)$ is invariant under the linear mapping α sending ζ^i to ζ^{4i}. But it is easy to check that an element fixed by α must be of the form $a + b\theta_0$ for a, $b \in \mathbb{Z}$. (Use a direct argument based on the linear independence over $\mathbb{Q}$ of $\{1, \zeta, \ldots, \zeta^{21}\}$.)

We pull out of a hat the element

$$\mu = 1 - \zeta + \zeta^{21} = 1 - \zeta + \zeta^{-2}$$

which Kummer found by a great deal of (fairly systematic) experimentation. Using (6.11) and a lot of paper and ink we eventually find that

$$\mathrm{N}(\mu) = (-31 + 28\theta_0)(-31 + 28\theta_1) = 6533 = 47 \cdot 139$$

By Theorem 6.23 the principal ideal $\mathfrak{m} = \langle \mu \rangle$ cannot be prime, hence it must be a nontrivial product of prime ideals, say

$$\mathfrak{m} = \mathfrak{p}\mathfrak{q}$$

Taking norms we must (without loss of generality) have $\mathrm{N}(\mathfrak{p}) = 47$, $\mathrm{N}(\mathfrak{q}) = 139$. If K has unique factorisation then every ideal, in particular $\mathfrak{p}$, is principal by Theorem 6.30. Hence $\mathfrak{p} = \langle v \rangle$ for some $v \in \mathbb{Z}[\zeta]$. Clearly $\mathrm{N}(v) = \pm\, 47$ by Corollary 6.18.

We claim this is impossible. We have already observed that $G(\zeta)$ can always be expressed in the form $a + b\theta_0$ $(a, b \in \mathbb{Z})$; and then $G(\zeta^{-2})$ must be equal to $a + b\theta_1$. Set $f(\zeta) = v$:

$$\pm 47 = (a + b\theta_0)(a + b\theta_1) = a^2 - ab + 6b^2$$

Multiplying by 4 and regrouping,

$$(2a - b)^2 + 23b^2 = \pm\, 188$$

The sign must be positive. A simple trial-and-error analysis (involving only two cases) shows that 188 cannot be written in the form $P^2 + 23Q^2$. This contradiction establishes that prime factorisation of elements cannot be unique in $\mathfrak{O}_K$.

The complete list of cyclotomic fields with unique prime factorisation is known:

Theorem 6.32. *The ring of integers* $\mathbb{Z}[\zeta_n]$ *of the cyclotomic field* $\mathbb{Q}(\zeta_n)$ *has unique prime factorisation, or equivalently is a principal ideal domain, if and only if* n *is one of the* 30 *following integers:*

1	3	4	5	7	8	9	11	12	13	15	16	17	19	20
21	24	25	27	28	32	33	35	36	40	44	45	48	60	84

Proof: See Masley and Montgomery [87]. (Cases $2n$ where n is odd give the same field as n, and are omitted.)

6.8 Exercises

6.1 In a domain D, show that a principal ideal $\langle p\rangle$ is prime if and only if p is prime or zero.

6.2 In $\mathbb{Z}[\sqrt{-5}]$, define the ideals

$$\begin{aligned}\mathfrak{p} &= \langle 2, 1+\sqrt{-5}\rangle \\ \mathfrak{q} &= \langle 3, 1+\sqrt{-5}\rangle \\ \mathfrak{r} &= \langle 3, 1-\sqrt{-5}\rangle\end{aligned}$$

Prove that these are maximal ideals, hence prime. Show that

$$\mathfrak{p}^2 = \langle 2\rangle \qquad \mathfrak{q}\mathfrak{r} = \langle 3\rangle$$
$$\mathfrak{p}\mathfrak{q} = \langle 1+\sqrt{-5}\rangle \quad \mathfrak{p}\mathfrak{r} = \langle 1-\sqrt{-5}\rangle$$

Show that the factorisations of 6 given in the proof of Theorem 5.17 come from two different groupings of the factorisation into prime ideals $\langle 6\rangle = \mathfrak{p}^2\mathfrak{q}\mathfrak{r}$.

6.3 Calculate the norms of the ideals mentioned in Exercise 6.2 and check multiplicativity.

6.4 Prove that the ideals $\mathfrak{p}$, $\mathfrak{q}$, $\mathfrak{r}$ of Exercise 6.2 cannot be principal.

6.5 Show the principal ideals $\langle 2\rangle$, $\langle 3\rangle$ in Exercise 6.2 are generated by irreducible elements but the ideals are not prime.

6.6 In $\mathbb{Z}[\sqrt{-6}]$ we have

$$6 = 2 \cdot 3 = (\sqrt{-6})(-\sqrt{-6})$$

Factorise these elements further in the extension ring $\mathbb{Z}[\sqrt{2}, \sqrt{-3}]$ as

$$6 = (-1)\sqrt{2}\sqrt{2}\sqrt{-3}\sqrt{-3}$$

Show that if $\mathfrak{q}$ is the principal ideal in $\mathbb{Z}[\sqrt{2}, \sqrt{-3}]$ generated by $\sqrt{2}$, then

$$\mathfrak{p}_1 = \mathfrak{q} \cap \mathbb{Z}[\sqrt{-6}] = \langle 2, \sqrt{-6}\rangle$$

Demonstrate that $\mathfrak{p}_1$ is maximal in $\mathbb{Z}[\sqrt{-6}]$, hence prime, and find another prime ideal $\mathfrak{p}_2$ in $\mathbb{Z}[\sqrt{-6}]$ such that $\langle 6\rangle = \mathfrak{p}_1^2\mathfrak{p}_2^2$.

6.7 Factorise $14 = 2 \cdot 7 = (2 + \sqrt{-10})(2 - \sqrt{-10})$ further in $\mathbb{Z}[\sqrt{2}, \sqrt{-5}]$ and by intersecting appropriate ideals with $\mathbb{Z}[\sqrt{-10}]$, factorise the ideal $\langle 14\rangle$ into prime (maximal) ideals in $\mathbb{Z}[\sqrt{-10}]$.

6.8 Suppose that $\mathfrak{p}$, $\mathfrak{q}$ are distinct prime ideals in $\mathfrak{O}$. Show that $\mathfrak{p}+\mathfrak{q} = \mathfrak{O}$ and $\mathfrak{p} \cap \mathfrak{q} = \mathfrak{p}\mathfrak{q}$.

6.9 Find all fractional ideals of $\mathbb{Z}$ and of $\mathbb{Z}[\mathrm{i}]$.

6.10 In $\mathbb{Z}[\sqrt{-5}]$, find a $\mathbb{Z}$-basis $\{\alpha_1, \alpha_2\}$ for the ideal $\langle 2, 1 + \sqrt{-5}\rangle$. Check the formula

$$\mathrm{N}\left(\langle 2, 1 + \sqrt{-5}\rangle\right) = \left|\frac{\Delta[\alpha_1, \alpha_2]}{d}\right|^{1/2}$$

of Theorem 6.17.

6.11 Find all the ideals in $\mathbb{Z}[\sqrt{-5}]$ that contain the element 6.

6.12 If K is a number field of degree n with ring of integers $\mathfrak{O}$, show that if $m \in \mathbb{Z}$ and $\langle m\rangle$ is the ideal in $\mathfrak{O}$ generated by m, then

$$\mathrm{N}\left(\langle m\rangle\right) = |m|^n$$

6.13 In $\mathbb{Z}[\sqrt{-29}]$

$$30 = 2 \cdot 3 \cdot 5 = (1 + \sqrt{-29})(1 - \sqrt{-29})$$

Show that $\langle 30\rangle \subseteq \langle 2, 1 + \sqrt{-29}\rangle$ and verify that $\mathfrak{p}_1 = \langle 2, 1 + \sqrt{-29}\rangle$ has norm 2 and is thus prime. Check that $1 - \sqrt{-29} \in \mathfrak{p}_1$ and deduce

$\langle 30\rangle \subseteq \mathfrak{p}_1^2$. Find prime ideals $\mathfrak{p}_2, \mathfrak{p}_2', \mathfrak{p}_3, \mathfrak{p}_3'$ with norms 3 or 5 such that $\langle 30\rangle \subseteq \mathfrak{p}_i\mathfrak{p}_i'$ $(i = 2, 3)$. Deduce that $\mathfrak{p}_1^2\mathfrak{p}_2\mathfrak{p}_2'\mathfrak{p}_3\mathfrak{p}_3' | \langle 30\rangle$ and by calculating norms, or otherwise, show that

$$\langle 30\rangle = \mathfrak{p}_1^2\mathfrak{p}_2\mathfrak{p}_2'\mathfrak{p}_3\mathfrak{p}_3'$$

Comment on how this relates to the two factorisations:

$$\begin{aligned}\langle 30\rangle &= \langle 2\rangle \langle 3\rangle \langle 5\rangle \\ \langle 30\rangle &= \langle 1+\sqrt{-29}\rangle \langle 1-\sqrt{-29}\rangle\end{aligned}$$

6.14 Find all ideals in $\mathbb{Z}[\sqrt{-29}]$ containing the element 30.

6.15 Find an integral domain A with a prime ideal $\mathfrak{p} \neq 0$ and a subring R such that $R \cap \mathfrak{p} = \{0\}$.

6.16 Find a noetherian ring R with a prime ideal that is not maximal.

6.17 Let $K = \mathbb{Q}(\sqrt{5})$, with ring of integers $\mathfrak{O} = \mathbb{Z}[\frac{1}{2}(1+\sqrt{5})]$. Show that $\mathbb{Z}[\sqrt{5}]$ is a subring of $\mathfrak{O}$, but is not an ideal of $\mathfrak{O}$. Show that factorisation into irreducibles is always possible in $\mathbb{Z}[\sqrt{5}]$.

Do the factorisations $4 = 2.2 = (-1)(1+\sqrt{5})(1-\sqrt{5})$ show that factorisation into irreducibles is not unique in $\mathbb{Z}[\sqrt{5}]$?

Is $\mathbb{Z}[\sqrt{5}]$ a Dedekind domain? If not, which conditions (a)–(d) of the definition of a Dedekind domain fail? Justify your answer.

6.18 Let K be a field, and let $R = K[t^2, t^3] \subseteq K[t]$. Prove that R is noetherian. Prove that R is not a Dedekind domain.

6.19 Let $\mathfrak{a}$ be a nonzero ideal of the ring of integers $\mathfrak{O}$ of a number field K. Let $R = \mathfrak{O}/\mathfrak{a}$ be the quotient ring of $\mathfrak{O}$ by $\mathfrak{a}$. By Corollary 6.5 R is finite. Prove:

(a) If $\mathfrak{a} = \langle p\rangle$ for a rational prime p, then R is a finite-dimensional associative algebra over the field $\mathbb{Z}_p$.

(b) If the dimension of R as a vector space over $\mathbb{Z}_p$ is m, then $\mathrm{N}(\mathfrak{a}) = p^m$.

(c) For any $\mathfrak{a}$, prime or not, there is a natural bijection between the set of ideals $\mathfrak{b}$ of $\mathfrak{O}$ such that $\mathfrak{b} \supseteq \mathfrak{a}$ and the set of ideals of R. This bijection preserves containment.

(d) Every ideal of R has a unique factorisation into prime ideals of R.

(e) Every maximal ideal of R is a prime ideal.

(f) The ring R is not a Dedekind domain.

II

Geometric Methods

7

Lattices

At this stage we take a radical new view of the theory, turning from purely algebraic methods to techniques inspired by geometry. A few geometric ideas have already been described, but now we apply the geometry systematically to obtain deeper insights. This approach requires a different attitude of mind, in which formal ideas are built on a visual foundation.

We begin with basic properties of lattices: subsets of $\mathbb{R}^n$ that generalise how $\mathbb{Z}$ is embedded in $\mathbb{R}$. We characterise lattices topologically as the discrete subgroups of $\mathbb{R}^n$. We introduce the fundamental domain and quotient torus corresponding to a lattice and relate the two concepts. Finally we define a concept of volume for subsets of the quotient torus.

7.1 Lattices

Lattices are certain additive subgroups of $\mathbb{R}^n$, which can be characterised algebraically or topologically. Many structures in algebraic number theory turn out to be lattices. (Do not confuse this term with any other uses of the word 'lattice' in algebra.) We begin with an algebraic definition:

Definition 7.1. Let $e_1, \ldots, e_m$ be a linearly independent set of vectors in $\mathbb{R}^n$ (so that $m \leq n$). The additive subgroup of $(\mathbb{R}^n, +)$ generated by $e_1, \ldots, e_m$ is called a *lattice* of *dimension* m, *generated* by $e_1, \ldots, e_m$. It consists of all elements

$$a_1e_1 + \cdots + a_me_m \qquad a_i \in \mathbb{Z},\ 1 \leq i \leq m$$

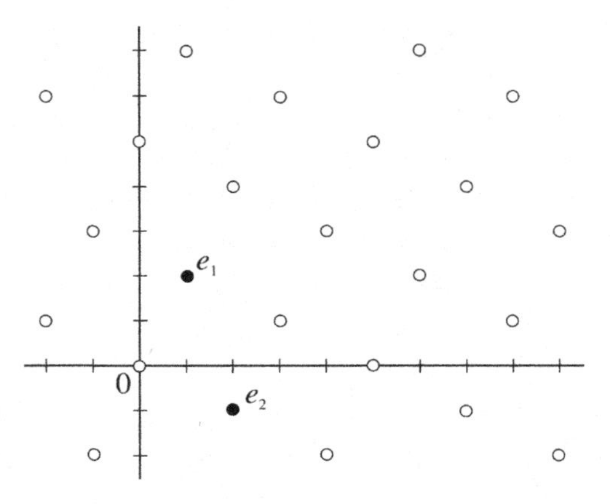

Figure 7.1. The lattice in $\mathbb{R}^2$ generated by $e_1 = (1, 2)$ and $e_2 = (2, -1)$.

As a typical example, Figure 7.1 shows a lattice of dimension 2 in $\mathbb{R}^2$ generated by $e_1 = (1, 2)$ and $e_2 = (2, -1)$. Obviously, as regards the group-theoretic structure, a lattice of dimension m is a free abelian group of rank m, so we can apply the terminology and theory of free abelian groups to lattices.

Next, we give a topological characterisation of lattices. Equip $\mathbb{R}^n$ with the usual metric (à la Pythagoras), where $\|x - y\|$ denotes the distance between x and y. Denote the closed ball with centre x and radius r by $B_r[x]$. Recall that a subset $X \subseteq \mathbb{R}^n$ is *bounded* if $X \subseteq B_r[0]$ for some r. A key concept is:

Definition 7.2. A subset of $\mathbb{R}^n$ is *discrete* if and only if it intersects every ball $B_r[0]$ in a finite set.

The topological characterisation of lattices is:

Theorem 7.3. *An additive subgroup of $\mathbb{R}^n$ is a lattice if and only if it is discrete.*

Proof: Suppose L is a lattice in $\mathbb{R}^n$. By passing to the subspace spanned by L we may assume L has dimension n. Let L be generated by $e_1, \ldots, e_n$; then these vectors form a basis for the space $\mathbb{R}^n$. Every $v \in \mathbb{R}^n$ has a unique representation

$$v = \lambda_1 e_1 + \cdots + \lambda_n e_n \quad (\lambda_i \in \mathbb{R})$$

Define $f : \mathbb{R}^n \to \mathbb{R}^n$ by

$$f(\lambda_1 e_1 + \cdots + \lambda_n e_n) = (\lambda_1, \ldots, \lambda_n)$$

Then $f(B_r[0])$ is bounded, say

$$\|f(v)\| \le k \text{ for } v \in B_r[0]$$

If $\sum a_i e_i \in B_r[0]$ $(a_i \in \mathbb{Z})$, then certainly $\|(a_1, \ldots, a_n)\| \le k$. This implies

$$|a_i| \le \| (a_1, \ldots, a_n) \| \le k \tag{7.1}$$

The number of integer solutions of (7.1) is finite and so $L \cap B_r[0]$, being a subset of the solutions of (7.1), is also finite, and L is discrete.

Conversely, let G be a discrete subgroup of $\mathbb{R}^n$. We prove by induction on n that G is a lattice. Let $\{g_1, \ldots, g_m\}$ be a maximal linearly independent subset of G, let V be the subspace spanned by $\{g_1, \ldots, g_{m-1}\}$, and let $G_0 = G \cap V$. Then G_0 is discrete so, by induction, G_0 is a lattice. Hence there exist linearly independent elements $h_1, \ldots, h_{m'}$ generating G_0. Since the elements $g_1, \ldots, g_{m-1} \in G_0$ we have $m' = m - 1$, and we can replace $\{g_1, \ldots, g_{m-1}\}$ by $\{h_1, \ldots, h_{m-1}\}$, or equivalently assume that every element of G_0 is a $\mathbb{Z}$-linear combination of $g_1, \ldots, g_{m-1}$. Let T be the subset of all $x \in G$ of the form

$$x = a_1 g_1 + \cdots + a_m g_m$$

with $a_i \in \mathbb{R}$, such that

$$\begin{aligned} &0 \le a_i < 1 \quad (i = 1, \ldots, m-1) \\ &0 \le a_m \le 1 \end{aligned}$$

Then T is bounded, hence finite since G is discrete, and we may therefore choose $x' \in T$ with smallest nonzero coefficient a_m, say

$$x' = b_1 g_1 + \cdots + b_m g_m$$

Certainly $\{g_1, \ldots, g_{m-1}, x'\}$ is linearly independent. Now starting with any vector $g \in G$ we can select integer coefficients c_i so that

$$g' = g - c_m x' - c_1 g_1 - \ldots - c_{m-1} g_{m-1}$$

lies in T, and the coefficient of g_m in g' is less than b_m, but non-negative. By choice of x' this coefficient must be zero, so $g' \in G_0$. Hence $\{x', g_1, \ldots, g_{m-1}\}$ generates G, and G is a lattice. □

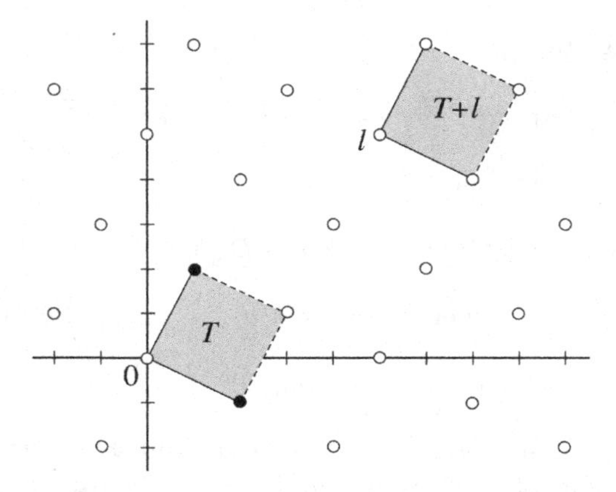

Figure 7.2. A fundamental domain T for the lattice of Figure 7.1, and a translate $T+l$. Dotted lines indicate omission of boundaries.

Definition 7.4. If L is a lattice generated by $\{e_1, \ldots, e_n\}$, then the corresponding *fundamental domain* T consists of all elements $\sum a_i e_i$ $(a_i \in \mathbb{R})$ such that $0 \leq a_i < 1$.

The fundamental domain depends on the choice of generators.

Lemma 7.5. *Each element of* $\mathbb{R}^n$ *lies in exactly one of the sets* $T + l$ *for* $l \in L$.

Proof: Chop off the integer parts of the coefficients. □

Figure 7.2 illustrates the concept of a fundamental domain, and the result of Lemma 7.5, for the lattice of Figure 7.1.

7.2 Quotient Torus

Let L be a lattice in $\mathbb{R}^n$, and assume initially that L has dimension n. We study the quotient group $\mathbb{R}^n/L$.

Let $\mathbb{S}^1$ denote the set of all complex numbers of modulus 1. Under multiplication $\mathbb{S}^1$ is a group, called for obvious reasons the *circle group*.

Lemma 7.6. *The quotient group* $\mathbb{R}/\mathbb{Z}$ *is isomorphic to the circle group* $\mathbb{S}^1$.

Proof: Define a map $\phi : \mathbb{R} \to \mathbb{S}^1$ by $\phi(x) = e^{2\pi i x}$. Then ϕ is a surjective homomorphism with kernel $\mathbb{Z}$. □

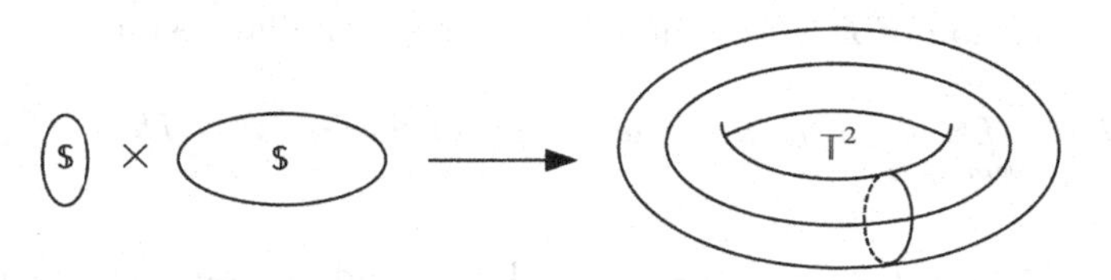

Figure 7.3. The Cartesian product of two circles is a torus.

Definition 7.7. The *n-dimensional torus* is the direct product of n copies of the circle group $\mathbb{S}^1$. We denote it by $\mathbb{T}^n$.

For instance, $\mathbb{T}^2 = \mathbb{S}^1 \times \mathbb{S}^1$ is the usual torus (with a group structure) as sketched in Figure 7.3. Tori are important because they arise as quotients of $\mathbb{R}^n$ by a lattice:

Theorem 7.8. *If L is an n-dimensional lattice in $\mathbb{R}^n$ then $\mathbb{R}^n/L$ is isomorphic to the n-dimensional torus $\mathbb{T}^n$.*

Proof: Let $\{e_1, \ldots, e_n\}$ be generators for L. Then $\{e_1, \ldots, e_n\}$ is a basis for $\mathbb{R}^n$. Define $\phi : \mathbb{R}^n \to \mathbb{T}^n$ by

$$\phi(a_1 e_1 + \cdots + a_n e_n) = (\mathrm{e}^{2\pi \mathrm{i} a_1}, \ldots, \mathrm{e}^{2\pi \mathrm{i} a_n})$$

Then ϕ is a surjective homorphism, and the kernel of ϕ is L. □

Lemma 7.9. *The map ϕ defined above, when restricted to the fundamental domain T, yields a bijection $T \to \mathbb{T}^n$.* □

Geometrically, $\mathbb{T}^n$ is obtained by 'gluing' (that is, identifying) opposite faces of the closure of the fundamental domain, as in Figure 7.4.

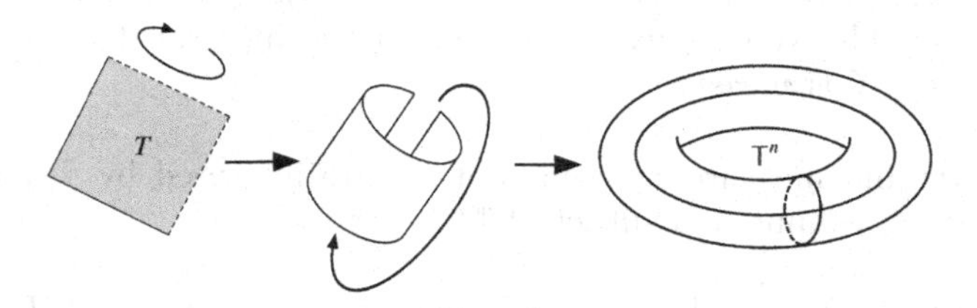

Figure 7.4. The quotient of Euclidean space by a lattice of the same dimension is a torus, obtained by identifying opposite edges of a fundamental domain.

If the dimension of L is less than n, we have a similar result:

Theorem 7.10. *Let L be an m-dimensional lattice in $\mathbb{R}^n$. Then $\mathbb{R}^n/L$ is isomorphic to $\mathbb{T}^m \times \mathbb{R}^{n-m}$.*

Proof: Let V be the subspace spanned by L, and choose a complement W so that $\mathbb{R}^n = V \oplus W$. Then $L \subseteq V$, $V/L \cong \mathbb{T}^m$ by Theorem 7.8, and $W \cong \mathbb{R}^{n-m}$. □

For example, $\mathbb{R}^2/\mathbb{Z} \cong \mathbb{T}^1 \times \mathbb{R}$, which geometrically is a cylinder as in Figure 7.5.

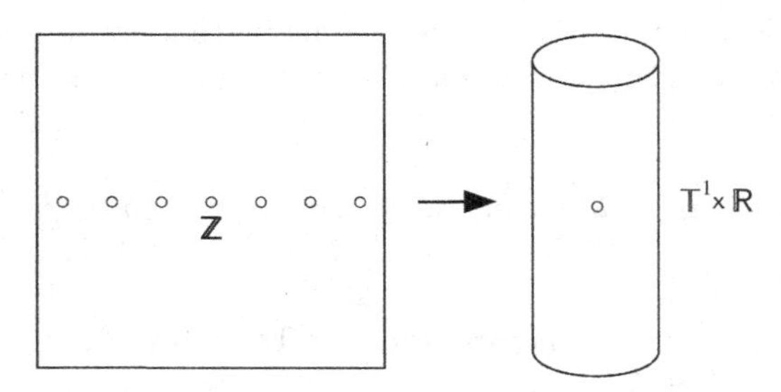

Figure 7.5. The quotient of Euclidean space by a lattice of smaller dimension is a cylinder.

There is a standard notion of 'volume' for subsets of $\mathbb{R}^n$:

Definition 7.11. The *volume* $v(X)$ of a subset $X \subseteq \mathbb{R}^n$ is defined in the usual way. For precision, we take it to be the value of the multiple integral

$$\int_X \mathrm{d}x_1 \ldots \mathrm{d}x_n$$

where $(x_1, \ldots, x_n)$ are coordinates.

More generally, we could use Lebesgue measure μ on $\mathbb{R}^n$, in which case $v(X) = \mu(X)$. The volume exists only when the integral does; that is, when the subset X is *measurable.*

The metric and measure structures of $\mathbb{R}^n$ are inherited by $\mathbb{T}^n$, which lets us define the volume of a subset of $\mathbb{T}^n$:

Definition 7.12. Let $L \subseteq \mathbb{R}^n$ be a lattice of dimension n, so that $\mathbb{R}^n/L \cong \mathbb{T}^n$. Let T be a fundamental domain of L. Consider the natural bijection

$$\phi : T \to \mathbb{T}^n$$

For any subset X of $\mathbb{T}^n$, the *volume* of X is

$$v(X) = v(\phi^{-1}(X))$$

The volume exists if and only if $\phi^{-1}(X)$ has a well-defined volume in $\mathbb{R}^n$.

Remark. In one dimension, $\mathbb{R}$ and $\mathbb{T}$, we usually use 'length', not volume. In two dimensions, $\mathbb{R}^2$ and $\mathbb{T}^2$, we usually use 'area', not volume.

Let $\nu : \mathbb{R}^n \to \mathbb{T}^n$ be the natural homomorphism with kernel L. It is intuitively clear that ν is 'locally volume-preserving', that is, for each $x \in \mathbb{R}^n$ there exists a ball $B_\varepsilon[x]$ such that for all subsets $X \subseteq B_\varepsilon[x]$ for which $v(X)$ exists,

$$v(X) = v(\nu(X))$$

It is also intuitively clear that if an injective map is locally volume-preserving, then it is volume-preserving. We prove a result that combines these two intuitive ideas:

Theorem 7.13. *If X is a bounded subset of $\mathbb{R}^n$ and $v(X)$ exists, and if $v(\nu(X)) \neq v(X)$, then $\nu|_X$ is not injective.*

Proof: Assume $\nu|_X$ is injective. Now X, being bounded, intersects only a finite number of the sets $T + l$, for T a fundamental domain and $l \in L$. Put

$$X_l = X \cap (T + l)$$

Then

$$X = X_{l_1} \cup \ldots \cup X_{l_n}$$

For each l_i define

$$Y_{l_i} = X_{l_i} - l_i$$

so that $Y_{l_i} \subseteq T$. We claim that the Y_{l_i} are disjoint. Since $\nu(x - l_i) = \nu(x)$ for all $x \in \mathbb{R}^n$ this follows from the assumed injectivity of ν. Now

$$v(X_{l_i}) = v(Y_{l_i})$$

for all i. Also

$$\nu(X_{l_i}) = \phi(Y_{l_i})$$

where ϕ is the bijection $T \to \mathbb{T}^n$. Now we compute:

$$\begin{aligned} v(\nu(X)) &= v\left(\nu\left(\bigcup X_{l_i}\right)\right) = v\left(\bigcup Y_{l_i}\right) \\ &= \sum v(Y_{l_i}) \text{ by disjointness} \\ &= \sum v(X_{l_i}) = v(X) \end{aligned}$$

a contradiction. □

The idea of the proof is illustrated in Figure 7.6.

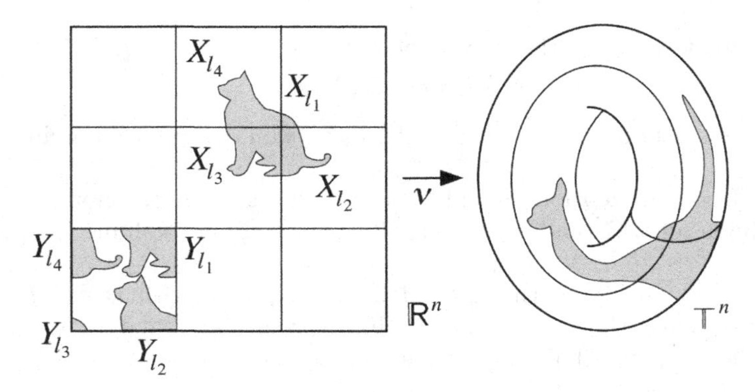

Figure 7.6. Proof of Theorem 7.13: if a locally volume-preserving map does not preserve volume globally, then it cannot be injective. The volume-preserving case is illustrated here; the parts of the cat do not overlap. A cat with a larger volume than the torus would have to overlap itself.

7.3 Exercises

7.1 Let L be a lattice in $\mathbb{R}^2$ with $L \subseteq \mathbb{Z}^2$. Prove that the volume of a fundamental domain T is equal to the number of points of $\mathbb{Z}^2$ lying in T.

7.2 Generalise the previous exercise to $\mathbb{R}^n$ and link this to Lemma 11.4 in Chapter 11 by using Theorem 1.30.

7.3 Sketch the lattices in $\mathbb{R}^2$ generated by:

(a) $(0, 1)$ and $(1, 0)$

(b) $(-1, 2)$ and $(2, 2)$

(c) $(1, 1)$ and $(2, 3)$

(d) $(-2, -7)$ and $(4, -3)$

(e) $(1, 6)$ and $(1, -6)$

(f) $(1, \pi)$ and $(1, -\pi)$

7.4 Sketch fundamental domains for these lattices.

7.5 Hence show that the fundamental domain of a lattice is not uniquely determined until we specify a set of generators.

7.6 Verify that nonetheless the volume of a fundamental domain of a given lattice is independent of the set of generators chosen.

7.7 Find the volume of a fundamental domain for the lattice in $\mathbb{R}^3$ generated by $(0, 0, a)$, $(0, b, c)$, (d, e, f), for integers a, b, c, d, e, f.

7.8 A sublattice of $\mathbb{Z}^3$ has $\mathbb{Z}$-basis $(3, 4, 5), (5, 12, 13), (a, b, c)$. Can a, b, c be chosen so that a fundamental domain has volume 1? If not, which values of the volume are possible by choice of a, b, c?

7.9 Draw a version of Figure 7.6 for the case when the area of the cat is larger than that of the fundamental domain.

8

Minkowski's Theorem

The aim of this chapter is to prove a marvellous theorem due to Hermann Minkowski in 1896. This asserts the existence, within a suitable set X, of a nonzero point of a lattice L, provided the volume of X is sufficiently large relative to that of a fundamental domain of L. The idea behind the proof is deceptive in its simplicity: X cannot be squashed into a space whose volume is less than that of X, unless X is allowed to overlap itself. See Figure 7.6. Minkowski discovered that this essentially trivial observation has many nontrivial and important consequences, and used it as a foundation for an extensive theory of the 'geometry of numbers'. As immediate and accessible instances of its application, we prove the Two- and Four-Squares Theorems of classical number theory.

8.1 Minkowski's Theorem

We begin by defining two important properties of subsets of $\mathbb{R}^n$: convexity and central symmetry.

Definition 8.1. A subset $X \subseteq \mathbb{R}^n$ is *convex* if whenever $x, y \in X$ then all points on the straight line segment joining x to y also lie in X. In algebraic terms, X is convex if, whenever $x, y \in X$ and $0 \leq \lambda \leq 1$, the point

$$\lambda x + (1 - \lambda)y$$

belongs to X.

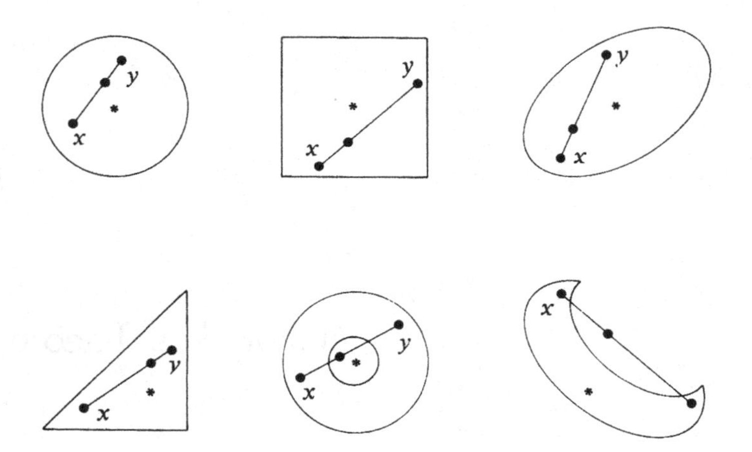

Figure 8.1. Convex and nonconvex sets. The circular disc, square, ellipse, and triangle are convex; the annulus and crescent are not. The circle, square, ellipse, and annulus are centrally symmetric about *; the triangle and crescent are not.

For example a circle, a square, an ellipse, or a triangle is convex in $\mathbb{R}^2$, but an annulus or crescent is not; see Figure 8.1.

Definition 8.2. A subset $X \subseteq \mathbb{R}^n$ is (*centrally*) *symmetric* if $x \in X$ implies $-x \in X$. Geometrically this means that X is invariant under reflection through the origin.

Among the sets in Figure 8.1, assuming the origin to be at the positions marked with an asterisk, the circle, square, ellipse, and annulus are symmetric, but the triangle and crescent are not.

We may now state Minkowski's Theorem.

Theorem 8.3. (Minkowski's Theorem) *Let L be an n-dimensional lattice in $\mathbb{R}^n$ with fundamental domain T, and let X be a bounded symmetric convex subset of $\mathbb{R}^n$. If*

$$v(X) > 2^n v(T)$$

then X contains a nonzero point of L.

Proof: Double the size of L to obtain a lattice $2L$ with fundamental domain $2T$ of volume $2^n v(T)$. Consider the torus

$$\mathbb{T}^n = \mathbb{R}^n / 2L$$

By definition,

$$v(\mathbb{T}^n) = v(2T) = 2^n v(T)$$

The natural map $\nu : \mathbb{R}^n \to \mathbb{T}^n$ cannot preserve the volume of X, since this is strictly larger than $v(\mathbb{T}^n)$. Since $\nu(X) \subseteq \mathbb{T}^n$ we have

$$v(\nu(X)) \leq v(\mathbb{T}^n) = 2^n v(T) < v(X)$$

By Theorem 7.13, $\nu|_X$ is not injective, so there exist $x_1, x_2 \in X$, with $x_1 \neq x_2$, such that $\nu(x_1) = \nu(x_2)$, or equivalently

$$x_1 - x_2 \in 2L \tag{8.1}$$

But $x_2 \in X$, so $-x_2 \in X$ by symmetry; and now by convexity

$$\tfrac{1}{2}(x_1) + \tfrac{1}{2}(-x_2) \in X$$

that is,

$$\tfrac{1}{2}(x_1 - x_2) \in X$$

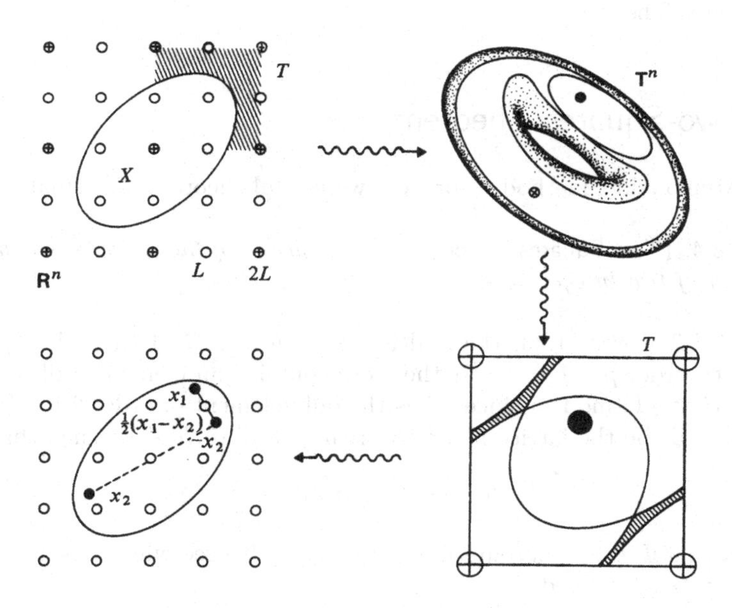

Figure 8.2. Proof of Minkowski's Theorem. Expand the original lattice ($\circ$) to double the size ($\oplus$) and form the quotient torus. By computing volumes, the natural quotient map is not injective when restricted to the given convex set. From point x_1 and x_2 with the same image we may construct a nonzero lattice point $\frac{1}{2}(x_1 - x_2)$.

But by (8.1)

$$\tfrac{1}{2}(x_1 - x_2) \in L$$

so

$$0 \neq \tfrac{1}{2}(x_1 - x_2) \in X \cap L$$

as required. □

The geometrical reasoning is illustrated in Figure 8.2. The decisive step in the proof is that since $\mathbb{T}^n$ has smaller volume than X it is impossible to squash X into $\mathbb{T}^n$ without overlap: the ancient platitude of quarts and pint pots (which we might update to large cats and small tori). That such olde-worlde wisdom became, in the hands of Minkowski, a weapon of devastating power, was the wonder of the 19th century and a lesson for the 21st. We unleash this power at several crucial stages in the forthcoming battle. (Our original thespian metaphor has been abandoned in favour of a military one, reinforcing the change of viewpoint from that of the algebraic *voyeur* to that of the geometric participant.) As a more immediate affirmation, we now give two traditional applications to number theory: the Two-Squares and Four-Squares Theorems.

8.2 Two-Squares Theorem

We use Minkowski's method to prove a wonderful theorem of Fermat:

Theorem 8.4. (Two-Squares Theorem) *If p is prime of the form $4k+1$ then p is a sum of two integer squares.*

Proof: By Theorem 1.33, the multiplicative group G of the field $\mathbb{Z}_p$ is cyclic, with order $p - 1 = 4k$. It therefore contains an element u of order 4. Then $u^2 \equiv -1 \pmod{p}$ since -1 is the only element of order 2 in G.

Let $L \subseteq \mathbb{Z}^2$ be the lattice in $\mathbb{R}^2$ consisting of all $(a, b) \in \mathbb{Z}^2$ such that

$$b \equiv ua \pmod{p}$$

We claim that L is a subgroup of $\mathbb{Z}^2$ of index p. To see why, consider the map $\phi : \mathbb{Z}^2 \to \mathbb{Z}_p$ defined by

$$\phi(a, b) = b - ua \pmod{p}$$

This is clearly a surjective homomorphism, and its kernel is L, proving the claim.

Therefore the volume (here 'area') of a fundamental domain for L is p. By Minkowski's Theorem any circle, centre the origin, of radius r, which has area $\pi r^2 > 4p$, contains a nonzero point of L. This is the case for $r^2 = 3p/2$. So there exists a point $(a, b) \in L$, not the origin, for which

$$0 \neq a^2 + b^2 \leq r^2 = 3p/2 < 2p$$

But modulo p,

$$a^2 + b^2 \equiv a^2 + u^2 a^2 \equiv 0$$

Therefore $a^2 + b^2$ is a multiple of p lying strictly between 0 and $2p$, so it must equal p. □

It is instructive to draw the lattice L and the appropriate circle in a few cases ($p = 5, 13, 17$) and check that the relevant lattice point exists and provides suitable a, b.

Theorem 8.4 goes back to Fermat, who stated it in a letter to Marin Mersenne in 1640. He sent a sketch proof, quite different from the geometric one given here, to Pierre de Carcavi in 1659. Euler gave a complete proof in 1754. Recently a cunning, and very simple, alternative proof of the Two-Squares Theorem has been found by Dolan [35].

8.3 Four-Squares Theorem

Refining this argument leads to another famous theorem, first proved by Lagrange:

Theorem 8.5. (Four-Squares Theorem) *Every positive integer is a sum of four integer squares.*

Proof: We prove the theorem for primes p, and then extend the result to all integers. Now

$$2 = 1^2 + 1^2 + 0^2 + 0^2$$

so we may suppose p is odd. We claim that the congruence

$$u^2 + v^2 + 1 \equiv 0 \pmod{p}$$

has a solution $u, v \in \mathbb{Z}$. This is because u^2 takes exactly $(p+1)/2$ distinct values as u runs through $0, \ldots, p-1$; and $-1 - v^2$ also takes on $(p+1)/2$ values. For the congruence to have no solution, all these values, $p + 1$ in total, are distinct, but then $p + 1 \leq p$ which is absurd.

For such a choice of u and v, consider the lattice $L \subseteq \mathbb{Z}^4$ consisting of all $(a, b, c, d) \in \mathbb{Z}^4$ such that

$$c \equiv ua + vb, \quad d \equiv ub - va \pmod{p}$$

Then L has index p^2 in $\mathbb{Z}^4$ by a similar argument to that in the proof of Theorem 8.4, so the volume of a fundamental domain is p^2. Now a 4-dimensional sphere, centre the origin, radius r, has volume $\pi^2 r^4/2$; see for example Coxeter [26] section 7.3.

Choose r to make this greater than $16p^2$; say $r^2 = 1.9p$. By Minkowski's Theorem, there exists a lattice point $0 \neq (a, b, c, d)$ in this 4-sphere, so

$$0 \neq a^2 + b^2 + c^2 + d^2 \leq r^2 = 1.9p < 2p$$

Modulo p, it is easy to verify that $a^2 + b^2 + c^2 + d^2 \equiv 0$, hence as before it must equal p.

To deal with an arbitrary integer n, it suffices to factorise n into primes and then use the remarkable identity

$$\begin{aligned}(a^2 + b^2 + c^2 + d^2)&(A^2 + B^2 + C^2 + D^2) \\ &= (aA - bB - cC - dD)^2 + (aB + bA + cD - dC)^2 \\ &\quad +(aC - bD + cA + dB)^2 + (aD + bC - cB + dA)^2\end{aligned}$$

□

Theorem 8.5 also goes back to Fermat. Euler spent forty years trying to prove it, and Lagrange succeeded in 1770.

8.4 Exercises

8.1 Which of the following solids (interiors included) are convex? Sphere, pyramid, icosahedron, cube, torus, ellipsoid, parallelepiped.

8.2 How many different convex solids can be made by joining n unit cubes face to face, so that their vertices coincide, for $n = 1, 2, 3, 4, 5, 6$; counting two solids as different if and only if they cannot be mapped to each other by rigid motions? What is the result for general n?

8.3 Prove that the intersection of any collection of convex sets is convex.

8.4 Is the union of any collection of convex sets convex? Justify your answer.

8.5 Prove that the intersection of any collection of centrally symmetric sets is centrally symmetric.

8.6 Is the union of any collection of centrally symmetric sets centrally symmetric? Justify your answer.

8.7 Deduce from Exercise 8.3 that every set $X \subseteq \mathbb{R}^n$ has a *convex hull*; that is, a set $X' \subseteq \mathbb{R}^n$ such that:

$X' \supseteq X$

X' is convex

If Y has the above two properties then $X' \subseteq Y$

Prove that X' is unique.

8.8 Let $X \subseteq \mathbb{R}^n$. Use Exercise 8.5 to prove that there exists a *symmetrisation* $\tilde{X} \subseteq \mathbb{R}^n$ such that

$\tilde{X} \supseteq X$

$\tilde{X}$ is centrally symmetric

If Y has the above two properties then $X \subseteq \tilde{Y}$

Prove that $\tilde{X}$ is unique.

8.9 Are $\widetilde{(X')}$ and $(\tilde{X})'$ always equal?

8.10 Verify the Two-Squares Theorem for all primes less than 200.

8.11 Verify the Four-Squares Theorem for all integers less than 100.

8.12 Prove that not every integer is a sum of three squares.

8.13 More generally, prove that no number of the form $4^a(8b+7)$ is a sum of three squares.

8.14 Prove that the number $\mu(n)$ of pairs of integers (x, y) with $x^2+y^2 < n$ satisfies $\mu(n)/n \to \pi$ as $n \to \infty$.

8.15 Prove that not every integer is a sum of eight positive cubes.

8.16 Prove that if a Gaussian integer $a + bi$ is a sum of any finite number of squares of Gaussian integers, then b must be even.

Prove that if b is even then $a+bi$ is a sum of three squares of Gaussian integers.

8.17 The Three-Squares Theorem states that evernumber *not* of the form $4^a(8b+7)$ is a sum of three squares (see Theorem 23.41 below). Show that the Three-Squares Theorem implies a stronger version of the Four-Squares Theorem: Every $n \in \mathbb{N}$ has the form $n = a^2+b^2+c^2+d^2$ where $a = 0$ or $a = 2^k$ where $k \geq 1$.

9

Geometric Representation of Algebraic Numbers

The purpose of this chapter is to develop a method for embedding a number field K in a real vector space of dimension equal to the degree of K, in such a way that ideals in K map to lattices in this vector space. This clever idea opens the way to applications of Minkowski's Theorem. The embedding is defined in terms of the monomorphisms $K \to \mathbb{C}$, so in some sense it encodes some of the field structure of K. We must distinguish between those monomorphisms that map K into $\mathbb{R}$ and those that do not.

9.1 The Space $\mathbb{L}^{st}$

Let $K = \mathbb{Q}(\theta)$ be a number field of degree n, where θ is an algebraic integer. Let $\sigma_1, \ldots, \sigma_n$ be the set of all monomorphisms $K \to \mathbb{C}$ (see Theorem 2.7). We now distinguish two types of monomorphism:

Definition 9.1. If $\sigma_i(K) \subseteq \mathbb{R}$, which happens if and only if $\sigma_i(\theta) \in \mathbb{R}$, we say that σ_i, is *real*; otherwise σ_i is *complex.*

As usual denote complex conjugation by a bar and define

$$\bar{\sigma}_i(\alpha) = \overline{\sigma_i(\alpha)}$$

Since complex conjugation is an automorphism of $\mathbb{C}$, the map $\bar{\sigma}_i$ is a monomorphism $K \to \mathbb{C}$, so $\bar{\sigma}_i = \sigma_j$ for some j. Now $\sigma_i = \bar{\sigma}_i$ if and only if σ_i is real, and $\overline{\overline{\sigma}}_i = \sigma_i$, so the complex monomorphisms come in conjugate pairs. Hence

$$n = s + 2t \tag{9.1}$$

where s is the number of real monomorphisms and $2t$ is the number of complex ones. We standardise the numeration in such a way that the system of all monomorphisms $K \to \mathbb{C}$ is

$$\sigma_1, \ldots, \sigma_s; \sigma_{s+1}, \bar{\sigma}_{s+1}, \ldots, \sigma_{s+t}, \bar{\sigma}_{s+t}$$

where $\sigma_1, \ldots, \sigma_s$ are real and the rest complex.

Further define

$$\mathbb{L}^{st} = \mathbb{R}^s \times \mathbb{C}^t$$

This is the set of all $(s+t)$-tuples

$$x = (x_1, \ldots, x_s; x_{s+1}, \ldots, x_{s+t})$$

where $x_1, \ldots, x_s \in \mathbb{R}$ and $x_{s+1}, \ldots, x_{s+t} \in \mathbb{C}$. Then $\mathbb{L}^{st}$ is a vector space over $\mathbb{R}$, and a ring (with coordinatewise operations): in fact it is an $\mathbb{R}$-algebra. As vector space over $\mathbb{R}$ it has dimension $s + 2t = n$.

Definition 9.2. For $x \in \mathbb{L}^{st}$, define the *norm*

$$\begin{aligned} \mathrm{N}(x) &= x_1 \ldots x_s . x_{s+1}\overline{x}_{s+1} \ldots x_{s+t}\overline{x}_{s+t} \\ &= x_1 \ldots x_s |x_{s+1}|^2 \ldots |x_{s+t}|^2 \end{aligned} \tag{9.2}$$

There should be no confusion with other uses of the word 'norm', and we will see why it is desirable to use this apparently overworked word in a moment. The norm has two obvious properties:

(a) $\mathrm{N}(x)$ is real for all x.

(b) $\mathrm{N}(xy) = \mathrm{N}(x)\mathrm{N}(y)$.

Define a map $\sigma : K \to \mathbb{L}^{st}$ by:

$$\sigma(\alpha) = (\sigma_1(\alpha), \ldots, \sigma_s(\alpha); \sigma_{s+1}(\alpha), \ldots, \sigma_{s+t}(\alpha))$$

for $\alpha \in K$. Clearly

$$\begin{aligned} \sigma(\alpha + \beta) &= \sigma(\alpha) + \sigma(\beta) \\ \sigma(\alpha\beta) &= \sigma(\alpha)\sigma(\beta) \end{aligned} \tag{9.3}$$

for all $\alpha, \beta \in K$, so σ is a ring homomorphism. If r is a rational number then

$$\sigma(r\alpha) = r\sigma(\alpha)$$

so σ is a $\mathbb{Q}$-algebra homomorphism. Furthermore,

$$N(\sigma(\alpha)) = N(\alpha) \tag{9.4}$$

since $N(\alpha)$ is defined to be

$$\sigma_1(\alpha)\dots\sigma_s(\alpha)\sigma_{s+1}(\alpha)\bar{\sigma}_{s+1}(\alpha)\dots\sigma_{s+t}(\alpha)\bar{\sigma}_{s+t}(\alpha)$$

which equals $N(\sigma(\alpha))$ by (9.2).

Example 9.3. Let $K = \mathbb{Q}(\theta)$ where $\theta = \sqrt[3]{2} \in \mathbb{R}$, which satisfies the polynomial equation $\theta^3 - 2 = 0$. The conjugates of θ are $\theta, \omega\theta, \omega^2\theta$ where ω is a complex cube root of unity. The monomorphisms $K \to \mathbb{C}$ are defined by:

$$\sigma_1(\theta) = \theta \quad \sigma_2(\theta) = \omega\theta \quad \bar{\sigma}_2(\theta) = \omega^2\theta$$

Clearly σ_1 is real, while σ_2 and $\bar{\sigma}_2$ are a conjugate complex pair. Hence $s = t = 1$.

An element of K, say

$$x = q + r\theta + s\theta^2 \quad (q, r, s \in \mathbb{Q})$$

maps into $\mathbb{L}^{1,1}$ according to

$$\sigma(x) = (q + r\theta + s\theta^2, q + r\omega\theta + s\omega^2\theta^2)$$

The kernel of σ is an ideal of K since σ is a ring homomorphism. Since K is a field, σ is either identically zero or injective. But $\sigma(1) = (1, 1, \dots, 1) \neq 0$, so σ is injective.

The next result proves a much stronger property than injectivity:

Theorem 9.4. *If $\alpha_1, \dots, \alpha_n$ is a basis for K over $\mathbb{Q}$ then $\sigma(\alpha_1), \dots, \sigma(\alpha_n)$ are linearly independent over $\mathbb{R}$.*

Proof: Linear independence over $\mathbb{Q}$ is immediate since σ is injective, but we need more than this. Let

$$\begin{aligned} \sigma_k(\alpha_l) &= x_k^{(l)} & (k = 1, \dots, s) \\ \sigma_{s+j}(\alpha_l) &= y_j^{(l)} + \mathrm{i}z_j^{(l)} & (j = 1, \dots, t) \end{aligned}$$

where $x_k^{(l)}, y_k^{(l)}, z_k^{(l)}$ are real. Then

$$\sigma(\alpha_l) = (x_1^{(l)}, \dots, x_s^{(l)}; y_1^{(l)} + \mathrm{i}z_1^{(l)}, \dots, y_t^{(l)} + \mathrm{i}z_t^{(l)})$$

and it is sufficient to prove that the determinant

$$D = \begin{vmatrix} x_1^{(1)} & \dots & x_s^{(1)} & y_1^{(1)} & z_1^{(1)} & \dots & y_t^{(1)} & z_t^{(1)} \\ \vdots & & \vdots & \vdots & \vdots & & \vdots & \vdots \\ x_1^{(n)} & \dots & x_s^{(n)} & y_1^{(n)} & z_1^{(n)} & \dots & y_t^{(n)} & z_t^{(n)} \end{vmatrix}$$

is nonzero. Put

$$\begin{aligned} E &= \begin{vmatrix} x_1^{(1)} & \dots & x_s^{(1)} & y_1^{(1)} + \mathrm{i}z_1^{(1)} & y_1^{(1)} - \mathrm{i}z_1^{(1)} & \dots \\ \vdots & & \vdots & \vdots & \vdots & \vdots \\ x_1^{(n)} & \dots & x_s^{(n)} & y_1^{(n)} + \mathrm{i}z_1^{(n)} & y_1^{(n)} - \mathrm{i}z_1^{(n)} & \dots \end{vmatrix} \\ &= \begin{vmatrix} \sigma_1(\alpha_1) & \dots & \sigma_s(\alpha_1) & \sigma_{s+1}(\alpha_1)\bar{\sigma}_{s+1}(\alpha_1) & \dots \\ \vdots & & \vdots & \vdots & \vdots \\ \sigma_1(\alpha_n) & \dots & \sigma_s(\alpha_n) & \sigma_{s+1}(\alpha_n)\bar{\sigma}_{s+1}(\alpha_n) & \dots \end{vmatrix} \end{aligned}$$

Then

$$E^2 = \Delta[\alpha_1, \dots, \alpha_n]$$

by definition of the discriminant, Definition 2.12. Also $E^2 \neq 0$ by Theorem 2.15. Now elementary properties of determinants (column operations) yield $E = (-2\mathrm{i})^t D$, so $D \neq 0$ as required. □

Corollary 9.5. *$\mathbb{Q}$-linearly independent elements of the number field K map under σ to $\mathbb{R}$-linearly independent elements of $\mathbb{L}^{st}$.* □

Corollary 9.6. *Suppose that G is a finitely generated subgroup of $(K, +)$ with $\mathbb{Z}$-basis $\{\alpha_1, \dots, \alpha_m\}$. Then the image of G in $\mathbb{L}^{st}$ is a lattice with generators $\sigma(\alpha_1), \dots, \sigma(\alpha_m)$.* □

The 'geometric representation' of K in $\mathbb{L}^{st}$ defined by σ, in combination with Minkowski's Theorem, provides the key to several of the deeper parts of the theory in Chapters 10–12. For these applications we need a notion of 'distance' on $\mathbb{L}^{st}$. Since $\mathbb{L}^{st}$ is isomorphic to $\mathbb{R}^{s+2t}$ as a real vector space, the natural idea is to transfer the usual Euclidean metric from $\mathbb{R}^{s+2t}$ to $\mathbb{L}^{st}$. This amounts to choosing a basis in $\mathbb{L}^{st}$ and defining an inner product with respect to which this basis is orthonormal. A natural basis is:

$$\begin{cases} (1,0,\ldots,0;0,\ldots,0) \\ (0,1,\ldots,0;0,\ldots,0) \\ \vdots \\ (0,0,\ldots,1;0,\ldots,0) \\ (0,0,\ldots,0;1,0,\ldots,0) \\ (0,0,\ldots,0;\mathrm{i},0,\ldots,0) \\ \vdots \\ (0,0,\ldots,0;0,0,\ldots,1) \\ (0,0,\ldots,0;0,0,\ldots,\mathrm{i}) \end{cases} \tag{9.5}$$

With respect to this basis, the element

$$(x_1,\ldots,x_s;y_1+\mathrm{i}z_1,\ldots,y_t+\mathrm{i}z_t)$$

of $\mathbb{L}^{st}$ has coordinates

$$(x_1,\ldots,x_s,y_1,z_1,\ldots,y_t,z_t)$$

Changing notation slightly, if we take

$$x=(x_1,\ldots,x_{s+2t}) \qquad x'=(x'_1,\ldots,x'_{s+2t})$$

with respect to the new coordinates (9.5), then the inner product becomes

$$(x,x')=x_1x'_1+\ldots+x_{s+2t}x'_{s+2t}$$

The *length* of a vector x is then $\|x\|=\sqrt{(x,x)}$, and the *distance* between x and x' is $\|x-x'\|$.

Referred to our original mixture of real and complex coordinates, any element

$$x=(x_1,\ldots,x_s;y_1+\mathrm{i}z_1,\ldots,y_t+\mathrm{i}z_t)$$

has length

$$\|x\|=\sqrt{x_1^2+\ldots+x_s^2+y_1^2+z_1^2+\ldots+y_t^2+z_t^2}$$

For later use, we record:

Theorem 9.7. *The sign of the discriminant d_K of K is $(-1)^t$, where $2t$ is the number of complex embeddings of K in $\mathbb{C}$.*

Proof: Let $\{\omega_i\}$ be an integral basis for K, and let $\Delta=\det(\sigma_j(\omega_i))$. By definition $d_K=\Delta^2$. Let τ be complex conjugation. Now $\tau(\Delta)=\Delta$ if $d_K>0$, while $\tau(\Delta)=-\Delta$ if $d_K<0$. Since τ fixes the first s columns of Δ and interchanges the remaining $2t$ columns in pairs, $\tau(\Delta)=(-1)^t\Delta$. □

9.2 Exercises

9.1 Find the monomorphisms $\sigma_i : K \to \mathbb{C}$ for the following fields. Determine the number s of the σ_i satisfying $\sigma_i(K) \subseteq \mathbb{R}$, and the number t of distinct conjugate pairs σ_i, σ_j such that $\bar{\sigma}_i = \sigma_j$.

(i) $\mathbb{Q}(\sqrt{5})$

(ii) $\mathbb{Q}(\sqrt{-5})$

(iii) $\mathbb{Q}(\sqrt[4]{5})$

(iv) $\mathbb{Q}(\zeta)$ where $\zeta = e^{2\pi i/7}$

(v) $\mathbb{Q}(\zeta)$ where $\zeta = e^{2\pi i/p}$ for a rational prime p

9.2 For $K = \mathbb{Q}(\sqrt{d})$ where d is a squarefree integer, calculate $\sigma : K \to \mathbb{L}^{st}$, distinguishing the cases $d < 0$, $d > 0$. Compute $\mathrm{N}(x)$ for $x \in \mathbb{L}^{st}$ and by direct calculation verify that

$$\mathrm{N}(\alpha) = \mathrm{N}(\sigma(\alpha)) \quad (\alpha \in K)$$

9.3 Let $K = \mathbb{Q}(\theta)$ where the algebraic integer θ has minimum polynomial f. If f factorises over $\mathbb{R}$ into irreducibles as

$$f(t) = g_1(t) \ldots g_q(t) h_1(t) \ldots h_r(t)$$

where g_i is linear and h_j quadratic, prove that $q = s$ and $r = t$ in the notation of (9.1).

9.4 Let $K = \mathbb{Q}(\theta)$ where $\theta \in \mathbb{R}$ and $\theta^3 = 3$. What is the map σ in this case? Pick a basis for K and verify Theorem 9.4 for it.

9.5 Find a map from $\mathbb{R}^2$ to itself under which $\mathbb{Q}$-linearly independent sets map to $\mathbb{Q}$-linearly independent sets, but some $\mathbb{R}$-linearly independent set does not map to an $\mathbb{R}$-linearly independent set.

9.6 If $K = \mathbb{Q}(\theta)$ where $\theta \in \mathbb{R}$ and $\theta^3 = 3$, verify Corollary 9.6 for the additive subgroup of K generated by $1 + \theta$ and $\theta^2 - 2$.

10

Dirichlet's Units Theorem

We now look a little more deeply at properties of the units in the integers of a number field. These properties are significant for the general theory, but are not essential to our development of Fermat's Last Theorem. Units are important because, while ideals are best suited to technicalities, there may come a point at which it is necessary to return to elements. (This is why we used the phrase 'sweeps them under the carpet' in Section 6.1.) But the generator of a principal ideal is ambiguous up to multiples by a unit. To translate results about ideals to their corresponding generators, we therefore need to know about units in the ring of integers.

The most fundamental and far-reaching theorem on units is that of Dirichlet, which gives an almost complete description, in abstract terms, of the group of units of the ring of integers of any number field. In particular it implies that this group is finitely generated. We prove Dirichlet's Units Theorem using geometric methods, notably Minkowski's Theorem and a 'logarithmic' variant of the space $\mathbb{L}^{st}$.

10.1 Introduction

We have already described the group of units U in the integers of $\mathbb{Q}(\sqrt{d})$ for negative squarefree d in Proposition 5.6. For $d = -1$ the units are $\{\pm 1, \pm \mathrm{i}\}$, for $d = -3$ they are $\{\pm 1, \pm\omega, \pm\omega^2\}$ where $\omega = e^{2\pi \mathrm{i}/3}$, and for all other $d < 0$, the units are just $\{\pm 1\}$.

In all cases U is a finite cyclic group of even order (2, 4, or 6) whose elements are roots of unity. It is in any case obvious that every unit of finite order is a root of unity in any number field.

For other number fields, the structure of the group of units U is more complicated. For example in $\mathbb{Q}(\sqrt{2})$

$$(1+\sqrt{2})(-1+\sqrt{2}) = 1$$

so $\varepsilon = 1+\sqrt{2}$ is a unit. Now ε is not a root of unity since $|\varepsilon| = 1+\sqrt{2} \neq 1$. Therefore ε has infinite order, all the elements $\pm\ \varepsilon^n$ $(n \in \mathbb{Z})$ are distinct units, and U is an infinite group. In fact, though we do not prove it here, the $\pm\varepsilon^n$ are all the units of $\mathbb{Q}(\sqrt{2})$, so U is isomorphic to $\mathbb{Z}_2 \times \mathbb{Z}$.

Dirichlet's Units Theorem shows that this more complicated structure of U is in some sense typical.

10.2 Logarithmic Space

Let K be a number field of degree $n = s + 2t$, as in Chapter 9, and let $\mathbb{L}^{st}$ be as described there. We use the notation of Chapter 9. Define a map

$$l : \mathbb{L}^{st} \to \mathbb{R}^{s+t}$$

as follows. For $x = (x_1, \ldots, x_s; x_{s+1}, \ldots, x_{s+t}) \in \mathbb{L}^{st}$ put

$$l_k(x) = \begin{cases} \log|x_k| & \text{for } k = 1, \ldots, s \\ \log|x_k|^2 & \text{for } k = s+1, \ldots s+t \end{cases}$$

Then set

$$l(x) = (l_1(x), \ldots, l_{s+t}(x))$$

The additive property of the logarithm leads at once to the property

$$l(xy) = l(x) + l(y) \tag{10.1}$$

for $x, y \in \mathbb{L}^{st}$. The set of elements of $\mathbb{L}^{st}$ with all coordinates nonzero is a group under multiplication, and l is a homomorphism from this group into $\mathbb{R}^{s+t}$. By (9.2),

$$\sum_{h=1}^{s+t} l_k(x) = \log|\mathrm{N}(x)| \tag{10.2}$$

Definition 10.1. For $\alpha \in K$ define

$$l(\alpha) = l(\sigma(\alpha))$$

where $\sigma : K \to \mathbb{L}^{st}$ is the standard map. This ambiguity in the use of l causes no confusion, and is tantamount to an identification of α with $\sigma(\alpha)$. Explicitly,

$$l(\alpha) = (\log|\sigma_1(\alpha)|, \ldots, \log|\sigma_s(\alpha)|, \log|\sigma_{s+1}(\alpha)|^2, \ldots, \log|\sigma_{s+t}(\alpha)|^2)$$

The map $l : K \to \mathbb{R}^{s+t}$ is the *logarithmic representation* of K, and $\mathbb{R}^{s+t}$ is the *logarithmic space*.

By (9.3) and (10.1),

$$l(\alpha\beta) = l(\alpha) + l(\beta) \quad (\alpha, \beta \in K)$$

so l is a homomorphism from the multiplicative group $K^* = K \setminus \{0\}$ of K to the additive group of $\mathbb{R}^{s+t}$. Further, setting $l_k(\alpha) = l_k(\sigma(\alpha))$, we have

$$\sum_{k=1}^{s+t} l_k(\alpha) = \log|\mathrm{N}(\alpha)|$$

using (9.4) and (10.2).

10.3 Embedding the Unit Group in Logarithmic Space

Why all these logarithms? Because the group of units is multiplicative, whereas Minkowski's Theorem applies to lattices, which are additive. We must pass from one context to the other, and it is just for this purpose that logarithms were created.

Let U be the group of units of $\mathfrak{O}$, the ring of integers of K. By restriction we obtain a homomorphism

$$l : U \to \mathbb{R}^{s+t}$$

It is not injective, but the kernel is easily described, see Lemma 10.3 below. To prove this lemma we need a curious result about polynomials and roots of unity:

Lemma 10.2. *If $p(t) \in \mathbb{Z}[t]$ is a monic polynomial, all of whose zeros in $\mathbb{C}$ have absolute value* 1, *then every zero is a root of unity.*

Proof: Let $\alpha_1, \ldots, \alpha_k$ be the zeros of $p(t)$. For each integer $l > 0$ the polynomial

$$p_l(t) = (t - \alpha_1^l) \ldots (t - \alpha_k^l)$$

lies in $\mathbb{Z}[t]$ by the usual argument on symmetric polynomials. If

$$p_l(t) = t^k + a_{k-1}t^{k-1} + \cdots + a_0$$

then

$$|a_j| \le \binom{k}{j} \quad (j = 0, \ldots, k-1)$$

by estimating the size of elementary symmetric polynomials in the α_j and using $|\alpha_j| = 1$. But only finitely many distinct polynomials over $\mathbb{Z}$ can satisfy this system of inequalities, so for some $m \neq l$, $p_l(t) = p_m(t)$. Hence there exists a permutation π of $\{1, \ldots, k\}$ such that

$$\alpha_j^l = \alpha_{\pi(j)}^m$$

for $j = 1, \ldots, k$. Inductively,

$$\alpha_j^{l^r} = \alpha_{\pi^r(j)}^{m^r}$$

Since $\pi^{k!}(j) = j$, we have $\alpha_j^{l^{k!}} = \alpha_j^{m^{k!}}$ so

$$\alpha_j^{(l^{k!} - m^{k!})} = 1$$

Since $l^{k!} \neq m^{k!}$ it follows that α_j is a root of unity. □

Applying the above result, we get:

Lemma 10.3. *The kernel W of the map $l : U \to \mathbb{R}^{s+t}$ is the set of all roots of unity belonging to U. This is a finite cyclic group of even order.*

Proof: We have $l(\alpha) = 0$ if and only if $|\sigma_i(\alpha)| = 1$ for all i. The field polynomial

$$\prod_i (t - \sigma_i(\alpha))$$

lies in $\mathbb{Z}[t]$ by Theorem 2.11(a) combined with Lemma 2.24. We can therefore appeal to Lemma 10.2 to conclude that all the $\sigma_i(\alpha)$ are roots of unity, in particular α itself.

The image $\sigma(\mathfrak{O})$ in $\mathbb{L}^{st}$ is a lattice by Corollary 9.6, so it is discrete by Theorem 7.3. The unit circle in $\mathbb{C}$ maps to a bounded subset in $\mathbb{L}^{st}$, so $\mathfrak{O}$ contains only finitely many roots of unity and W is finite. By Theorem 1.37, any finite subgroup of K^* is cyclic. Finally, W contains -1, which has order 2, so W has even order. □

Obviously the next thing to find out is the image E of U in $\mathbb{R}^{s+t}$:

Lemma 10.4. *The image E of U in $\mathbb{R}^{s+t}$ is a lattice of dimension $\leq s+t-1$.*

Proof: The norm of any unit is ± 1, so for any unit ε

$$\sum_{k=1}^{s+t} l_k(\varepsilon) = \log |\mathrm{N}(\varepsilon)| = \log 1 = 0$$

Hence all points of E lie in the subspace V of $\mathbb{R}^{s+t}$ consisting of those elements $(x_1, \ldots, x_{s+t})$ such that

$$x_1 + \cdots + x_{s+t} = 0$$

This has dimension $s + t - 1$.

To prove E is a lattice it is sufficient to prove it discrete by Theorem 7.3. Let $\|\cdot\|$ be the usual length function on $\mathbb{R}^{s+t}$. Suppose that $0 < r \in \mathbb{R}$, and $\|l(\varepsilon)\| < r$. Now $|l_k(\varepsilon)| \leq \|l(\varepsilon)\| < r$, so

$$|\sigma_k(\varepsilon)| < \mathrm{e}^r \quad (k = 1, \ldots, s)$$
$$|\sigma_{s+j}(\varepsilon)|^2 < \mathrm{e}^r \quad (j = 1, \ldots, t)$$

Hence the set of points $\sigma(\varepsilon)$ in $\mathbb{L}^{st}$ corresponding to units with $\|l(\varepsilon)\| < r$ is bounded, so finite by Corollary 9.6. Hence E intersects each closed ball in $\mathbb{R}^{s+t}$ in a finite set, so E is discrete. Therefore E is a lattice. Since $E \subseteq V$ it has dimension $\leq s + t - 1$. □

Already we know quite a lot about U. In particular, U is finitely generated, because W is finite and $U/W \cong E$ is a lattice, so free abelian, with rank $\leq s + t - 1$. All that remains is to find the exact dimension of the lattice E. In fact it is $s + t - 1$, as we prove in the next section.

10.4 Dirichlet's Units Theorem

The main thing we lack is a topological criterion to decide whether a lattice L in a vector space V has the same dimension as V. We remedy this lack with:

Lemma 10.5. *Let L be a lattice in $\mathbb{R}^m$. Then L has dimension m if and only if there exists a bounded subset B of $\mathbb{R}^m$ such that*

$$\mathbb{R}^m = \bigcup_{x \in L} (x + B)$$

Proof: If L has dimension m, then we may take B to be a fundamental domain of L and appeal to Lemma 7.5.

Suppose conversely that B exists but, for a contradiction, that L has dimension $d < m$. An intuitive argument goes like this: the quotient $\mathbb{R}^m/L$ is, by Theorem 7.10, the direct product of a torus and $\mathbb{R}^{m-d}$. The condition on B says that the image of B under the natural map $\nu : \mathbb{R}^m \to \mathbb{R}^m/L$ is the whole of $\mathbb{R}^m/L$. But because B is bounded this contradicts the presence of a direct factor $\mathbb{R}^{m-d}$ which is unbounded. By taking more account of the topology than we have done hitherto, this argument can easily be made rigorous. Alternatively, we operate in $\mathbb{R}^m$ instead of $\mathbb{R}^m/L$ as follows.

Let V be the subspace of $\mathbb{R}^m$ spanned by L. If L has dimension less than m then $\dim V < \dim \mathbb{R}^m$. Hence we can find an orthogonal complement V' to V in $\mathbb{R}^m$. The condition on B implies that $\mathbb{R}^m = \cup_{v \in V}(v + B)$, so V' is the image of B under the projection $\pi : \mathbb{R}^m \to V'$. But π is distance-preserving, so V' is bounded, contradiction. □

In fact, what we are saying topologically is that L has dimension m if and only if the quotient topological group $\mathbb{R}^m/L$ is *compact.*

Before proving that E has dimension $s+t-1$ it is convenient to extract one computation from the proof:

Lemma 10.6. *Let $y \in \mathbb{L}^{st}$ and let $\lambda_y : \mathbb{L}^{st} \to \mathbb{L}^{st}$ be defined by $\lambda_y(x) = yx$. Then λ_y is a linear map and*

$$\det \lambda_y = \mathrm{N}(y)$$

Proof: It is obvious that λ_y is linear. To compute $\det \lambda_y$ we use the basis (9.5). If

$$y = (x_1, \ldots, x_s; y_1 + \mathrm{i}z_1, \ldots, y_t + \mathrm{i}z_t)$$

then we obtain for $\det \lambda_y$ the expression

$$\det \lambda_y = \begin{vmatrix} x_1 & & & & & & & \\ & \ddots & & & & & \mathbf{0} & \\ & & x_s & & & & & \\ & & & y_1 & -z_1 & & & \\ & & & z_1 & y_1 & & & \\ & \mathbf{0} & & & & \ddots & & \\ & & & & & & y_t & -z_t \\ & & & & & & z_t & y_t \end{vmatrix}$$

which is

$$x_1 \ldots x_s(y_1^2 + z_1^2) \ldots (y_t^2 + z_t^2) = \mathrm{N}(y)$$

□

Lemma 10.7. *If M is a lattice in $\mathbb{L}^{st}$ of dimension $s+2t$ having fundamental domain of volume V, and if $c_1, \ldots, c_{s+t}$ are positive real numbers whose product satisfies*

$$c_1 \ldots c_{s+t} > \left(\frac{4}{\pi}\right)^t V$$

then there exists a nonzero $x = (x_1, \ldots, x_{s+t}) \in M$ such that

$$|x_1| < c_1, \ldots, |x_s| < c_s$$
$$|x_{s+1}|^2 < c_{s+1}, \ldots, |x_{s+t}|^2 < c_{s+t}$$

Proof: Let X be the set of all points $x \in \mathbb{L}^{st}$ for which the conclusion holds. Compute

$$\begin{aligned} v(X) &= \int_{-c_1}^{c_1} dx_1 \cdots \int_{-c_s}^{c_s} dx_s \times \iint_{y_1^2+z_1^2<c_{s+1}} dy_1 dz_1 \cdots \\ &\quad \times \iint_{y_t^2+z_t^2<c_{s+t}} dy_t dz_t \\ &= 2c_1 \cdot 2c_2 \cdots 2c_s \cdot \pi c_{s+1} \cdots \pi c_{s+t} \\ &= 2^s \pi^t c_1 \ldots c_{s+t} \end{aligned}$$

Now X is a cartesian product of line segments and circular discs, so X is bounded, symmetric, and convex. Minkowski's Theorem yields the required result provided

$$2^s \pi^t c_1 \ldots c_{s+t} > 2^{s+2t} V$$

that is

$$c_1 \ldots c_{s+t} > \left(\frac{4}{\pi}\right)^t V \qquad \square$$

The way is now clear for the proof of:

Theorem 10.8. *The image E of U in $\mathbb{R}^{s+t}$ is a lattice of dimension $s+t-1$.*

Proof: As before let V be the subspace of $\mathbb{R}^{s+t}$ whose elements satisfy

$$x_1 + \cdots + x_{s+t} = 0$$

Then $E \subseteq V$, and dim $V = s + t - 1$. To prove the theorem we appeal to Lemma 10.5; it is sufficient to find in V some bounded subset B such that

$$V = \bigcup_{e \in E} (e + B)$$

This additive property translates into a multiplicative property in $\mathbb{L}^{st}$. Every point in $\mathbb{R}^{s+t}$ is the image under l of some point in $\mathbb{L}^{st}$, so every point in V is the image of some point in $\mathbb{L}^{st}$. In fact, for $x \in \mathbb{L}^{st}$, we have $l(x) \in V$ if and only if $|\mathrm{N}(x)| = 1$. So if we let

$$S = \{x \in \mathbb{L}^{st} : |\mathrm{N}(x)| = 1\}$$

then $l(S) = V$. If $X_0 \subseteq S$ is bounded, then so is $l(X_0)$, as may be verified easily. If $x \in S$ then multiplicativity of the norm implies that $xX_0 \subseteq S$ if $X_0 \subseteq S$. In particular if ε is a unit then $\sigma(\varepsilon)X_0 \subseteq S$. So if we can find a bounded subset X_0 of S such that

$$S = \bigcup_{\varepsilon \in U} \sigma(\varepsilon)X_0 \tag{10.3}$$

then $B = l(X_0)$ will do what is required in V.

Now we find a suitable X_0. Let M be the lattice in $\mathbb{L}^{st}$ corresponding to $\mathfrak{O}$ under σ. Consider the linear transformation $\lambda_y : \mathbb{L}^{st} \to \mathbb{L}^{st}$ $(y \in \mathbb{L}^{st})$ of Lemma 10.6. If $y \in S$ then the determinant of λ_y is $\mathrm{N}(y)$, which is ± 1. Therefore λ_y is unimodular. This easily implies that any fundamental domain for the lattice yM $(= \lambda_y(M))$ has the same volume as a fundamental domain for M. Call this volume v.

Choose real numbers $c_i > 0$ with

$$Q = c_1 \dots c_{s+t} > \left(\frac{4}{\pi}\right)^t v$$

Let X be the set of $x \in \mathbb{L}^{st}$ for which

$$|x_k| < c_k \quad (k = 1, \dots, s)$$
$$|x_{s+j}|^2 < c_{s+j} \quad (j = 1, \dots, t)$$

By Lemma 10.7 there exists in yM a nonzero point $x \in X$. We have

$$x = y\sigma(\alpha) \quad (0 \neq \alpha \in \mathfrak{O})$$

Now

$$\mathrm{N}(x) = \mathrm{N}(y)\mathrm{N}(\alpha) = \pm\mathrm{N}(\alpha)$$

so

$$|\mathrm{N}(\alpha)| < Q$$

By Theorem 6.26 (c) only finitely many ideals of $\mathfrak{O}$ have norm $< Q$. Consider principal ideals and recall that their generators are ambiguous up to unit multiples. Then there exist in $\mathfrak{O}$ only finitely many pairwise non-associated numbers

$$\alpha_1, \ldots, \alpha_N$$

whose norms are $< Q$ in absolute value. Thus for some $i = 1, \ldots, N$ we have $\alpha\varepsilon = \alpha_i$ for a unit ε. Therefore

$$y = x\sigma(\alpha_i^{-1})\sigma(\varepsilon) \tag{10.4}$$

Now define

$$X_0 = S \cap \left(\bigcup_{i=1}^{N} \sigma(\alpha_i^{-1})X \right) \tag{10.5}$$

Since X is bounded so are the sets $\sigma(\alpha_i^{-1})X$, and since N is finite X_0 is bounded. Obviously X_0 does not depend on the choice of $y \in S$.

But now, since y and $\sigma(\varepsilon) \in S$, we have $x\sigma(\alpha_i^{-1}) \in S$, hence $x\sigma(\alpha_i^{-1}) \in X_0$. Then (10.4) shows that

$$y \in \sigma(\varepsilon)X_0$$

Hence (10.3) holds for an arbitrary element $y \in S$. □

We put this result into a more explicit form, obtaining *Dirichlet's Units Theorem*:

Theorem 10.9. (Dirichlet's Units Theorem) *The group of units of $\mathfrak{O}$ is isomorphic to*

$$W \times \mathbb{Z} \times \ldots \times \mathbb{Z}$$

where W is as described in Lemma 10.3 and there are $s + t - 1$ direct factors $\mathbb{Z}$.

Proof: By Theorem 10.8, $U/W \cong \mathbb{Z} \times \ldots \times \mathbb{Z} = \mathbb{Z}^{s+t-1}$. Since W is finite, U is a finitely generated abelian group, hence a direct product of cyclic groups, see Fraleigh [41] theorem 9.3, p. 90. Since W is finite and U/W is torsion-free, W is the set of elements of U of finite order, which is the product of all the finite cyclic factors in the direct decomposition. The other factors are all infinite cyclic; looking at U/W tells us there are exactly $s + t - 1$ of them. □

In more classical terms, Dirichlet's Units Theorem asserts the existence of a system of $s + t - 1$ *fundamental units*

$$\eta_1, \ldots, \eta_{s+t-1}$$

such that every unit of $\mathfrak{O}$ is representable *uniquely* in the form

$$\zeta \cdot \eta_1^{r_1} \ldots \eta_{s+t-1}^{r_{s+t-1}}$$

for a root of unity ζ and rational integers r_i.

We return briefly to $\mathbb{Q}(\sqrt{2})$, which we looked at in Section 10.1. For this field, $s = 2$, $t = 0$, so $s + t - 1 = 1$. Hence U is of the form $W \times \mathbb{Z}$ where W consists of the roots of unity in $\mathbb{Q}(\sqrt{2})$. These are just ± 1, so we get $U \cong \mathbb{Z}_2 \times \mathbb{Z}$ as asserted in Section 10.1. However, we still have not proved that $1 + \sqrt{2}$ is a fundamental unit. In fact, this is true in general of Dirichlet's Units Theorem: it does not determine any *specific* system of fundamental units. Other methods can be developed to solve this problem, and the theorem is still needed to tell us when we have found sufficiently many units.

10.5 Pell's Equation Revisited

In Chapter 4 we mentioned that the Dirichlet Units Theorem can be used to prove the existence of a fundamental solution to Pell's Equation. We now explain this connection. Recall that Pell's Equation is the Diophantine equation

$$x^2 - dy^2 = 1 \tag{10.6}$$

where d is a nonsquare positive integer. A fundamental solution is a positive solution (x_1, y_1) with x_1 minimal. Theorem 4.3 proves that every positive solution (x_n, y_n) is generated by the powers $x_n + y_n\sqrt{d} = (x_1 + y_1\sqrt{d})^n$ for $n \geq 0$.

To apply Theorem 10.9, consider the field $K = \mathbb{Q}(\sqrt{d})$. Equation 10.6 can be stated as

$$\mathrm{N}(x + y\sqrt{d}) = 1$$

so each solution (x, y) corresponds to a unit. However, units do not quite correspond to solutions, because the norm of a unit is ± 1. Therefore some units solve the related equation $x^2 - dy^2 = -1$.

First, suppose that the unit has norm $+1$. There are two monomorphisms $K \to \mathbb{C}$:

$$\begin{aligned} \sigma_1(x + y\sqrt{d}) &= x + y\sqrt{d} \\ \sigma_2(x + y\sqrt{d}) &= x - y\sqrt{d} \end{aligned}$$

Each σ_i is real; that is, its image is contained in $\mathbb{R}$. Thus $s = 2, t = 0$ in the notation of Section 9.1. By Theorem 10.9, the units of K form a group isomorphic to $W \times \mathbb{Z}$, where $W = \pm 1$ by Lemma 10.3. There are two possible generators for the direct factor $\mathbb{Z}$, namely $x \pm y\sqrt{d}$. Choose the sign to make x, y have the same sign; if necessary then multiply by the unit -1 to make $x, y > 0$. Now $x + y\sqrt{d}$ is a fundamental solution to Pell's equation.

Finally, suppose that the unit under consideration has norm -1. Then its square has norm 1.

Unlike the classical proof obtained by combining Theorem 4.4 and Theorem 4.24, this approach is an existence proof, rather than an explicit construction.

10.6 Exercises

10.1 Find units, not equal to 1, in the rings of integers of the fields $\mathbb{Q}(\sqrt{d})$ for $d = 2, 3, 5, 6, 7, 10$.

10.2 Prove that $1 + \sqrt{2}$ is a fundamental unit for $\mathbb{Q}(\sqrt{2})$.

10.3 Let $\eta_1, \ldots, \eta_{s+t-1}$ be a system of fundamental units for a number field K. Show that the *regulator*

$$R = |\det(\log|\sigma_i(\eta_j)|)|$$

is independent of the choice of $\eta_1, \ldots, \eta_{s+t-1}$.

10.4 Show that the group of units of a number field K is finite if and only if $K = \mathbb{Q}$ or K is an imaginary quadratic field.

10.5 Show that a number field of odd degree contains only two roots of unity.

10.6 Prove that the group of units of $\mathbb{Q}(\sqrt{3})$ is $\{\pm(2 + \sqrt{3})^n : n \in \mathbb{Z}\}$.

11

Class-Group and Class-Number

We now use the geometric ideas developed in Chapters 9 and 10 to build further insight into the property of unique factorisation. We already know that in the ring of integers of any number field, factorisation into primes is unique if and only if every ideal is principal. We refine this statement to provide a quantitative measure of *how non-unique* factorisation can be. This is done using fractional ideals, introduced in Chapter 6. Theorem 6.10 states that the fractional ideals form an abelian group $\mathbb{F}$. The class-group of a number field is defined to be the quotient of $\mathbb{F}$ by the subgroup $\mathbb{P}$ of *principal* fractional ideals, which is normal since $\mathbb{F}$ is abelian. The class-number is the order of this group. This gives the required measure: factorisation in a ring of integers is unique if and only if every fractional ideal is principal; that is, the class-number is 1. If the class-number is greater than 1 then factorisation is non-unique. Intuitively, the larger the class-number, the more complicated the possibilities for non-uniqueness. In a sense, prime factorisation becomes 'less unique' as the class-number increases.

The class group originated in the work of Gauss on quadratic forms over $\mathbb{Z}$ in two variables. He defined a way to compose two such forms—actually, equivalence classes of forms under unimodular linear changes of variables. This operation generalises Brahmagupta's Theorem 4.1, and Gauss proved that it determines a group structure on the classes.

It turns out that the class-number is always finite, an important fact that we prove using Minkowski's Theorem. Simple group-theoretic considerations then yield useful conditions for an ideal to be principal. These conditions lead to a proof that every ideal becomes principal in a suitable extension field, which is one formulation of the basic idea of Kummer's 'ideal numbers' within Dedekind's ideal-theoretic framework.

The importance of the class-number can only be hinted at here. It is crucial in the proof of Kummer's special case of Fermat's Last Theorem in Chapter 13. Many deep and delicate results in the theory of numbers are related to arithmetic properties of the class-number, or to algebraic properties of the class-group.

11.1 Class-Group

As usual let $\mathfrak{O}$ be the ring of integers of a number field K of degree n. Theorem 6.30 tells us that prime factorisation in $\mathfrak{O}$ is unique if and only if every ideal of $\mathfrak{O}$ is principal. Our aim here is to find a way of measuring how far prime factorisation fails to be unique in the case where $\mathfrak{O}$ contains nonprincipal ideals, or equivalently how far away the ideals of $\mathfrak{O}$ are from being principal.

To this end we use the group of fractional ideals defined in Chapter 6. Say that a fractional ideal of $\mathfrak{O}$ is *principal* if it is of the form $c^{-1}\mathfrak{a}$ where $0 \neq c \in \mathfrak{O}$ and $\mathfrak{a}$ is a principal ideal of $\mathfrak{O}$. Let $\mathbb{F}$ be the group of fractional ideals under multiplication. It is easy to check that the set $\mathbb{P}$ of principal fractional ideals is a subgroup of $\mathbb{F}$.

Definition 11.1. The *class-group* of $\mathfrak{O}$ is the quotient group

$$\mathbb{H} = \mathbb{F}/\mathbb{P}$$

The *class-number* $h = h(\mathfrak{O})$ is the order of $\mathbb{H}$.

Since each of $\mathbb{F}, \mathbb{P}$ is an infinite group we have no immediate way of deciding whether h is finite. In fact it is, and we develop a proof of this deep and important fact. We begin by reformulating the definition of the class-group in terms of ideals rather than fractional ideals.

Definition 11.2. Two fractional ideals are *equivalent* if they belong to the same coset of $\mathbb{P}$ in $\mathbb{F}$, or in other words if they map to the same element of $\mathbb{F}/\mathbb{P}$. If $\mathfrak{a}$ and $\mathfrak{b}$ are fractional ideals we write

$$\mathfrak{a} \sim \mathfrak{b}$$

if $\mathfrak{a}$ and $\mathfrak{b}$ are equivalent. We use

$$[\mathfrak{a}]$$

to denote the equivalence class of $\mathfrak{a}$.

The class-group $\mathbb{H}$ is the set of these equivalence classes. The equivalence class $[\mathfrak{O}]$ consists of all principal fractional ideals and is the identity element of $\mathbb{H}$.

If $\mathfrak{a}$ is a fractional ideal then $\mathfrak{a} = c^{-1}\mathfrak{b}$ where $c \in \mathfrak{O}$ and $\mathfrak{b}$ is an ideal. Therefore

$$\mathfrak{b} = c\mathfrak{a} = \langle c \rangle \mathfrak{a}$$

and since $\langle c \rangle \in \mathbb{P}$ this means that $\mathfrak{a} \sim \mathfrak{b}$. In other words, *every equivalence class contains an ideal.*

Now let $\mathfrak{x}$ and $\mathfrak{y}$ be equivalent ideals. (These symbols are Gothic/Fraktur x and y, despite appearances.) Then $\mathfrak{x} = \mathfrak{c}\mathfrak{y}$ where $\mathfrak{c}$ is a principal fractional ideal, say $\mathfrak{c} = d^{-1}\mathfrak{e}$ for $d \in \mathfrak{O}$, $\mathfrak{e}$ a principal ideal. Therefore

$$\mathfrak{x} \langle d \rangle = \mathfrak{y}\mathfrak{e}$$

Conversely if $\mathfrak{x}\mathfrak{b} = \mathfrak{y}\mathfrak{e}$ for $\mathfrak{b}$, $\mathfrak{e}$ principal ideals then $\mathfrak{x} \sim \mathfrak{y}$.

This leads to an alternative description of $\mathbb{H}$: on the set $\mathbb{F}$ of all fractional ideals, define a relation $\sim$ by $\mathfrak{x} \sim \mathfrak{y}$ if and only if there exist *principal* ideals $\mathfrak{b}$, $\mathfrak{e}$ with $\mathfrak{x}\mathfrak{b} = \mathfrak{y}\mathfrak{e}$. This is an equivalence relation, and $\mathbb{H}$ is the set of equivalence classes $[\mathfrak{x}]$ with group operation

$$[\mathfrak{x}][\mathfrak{y}] = [\mathfrak{x}\mathfrak{y}]$$

The significance of the class-group is that it captures the extent to which factorisation is not unique. In particular:

Theorem 11.3. *Factorisation in $\mathfrak{O}$ is unique if and only if the class-group $\mathbb{H}$ has order* 1*, or equivalently the class-number $h = 1$.*

Proof: By Theorem 6.30, factorisation is unique if and only if every ideal of $\mathfrak{O}$ is principal, which is true if and only if every fractional ideal is principal, which is equivalent to $\mathbb{F} = \mathbb{P}$, which is equivalent to $|\mathbb{H}| = h = 1$. □

The rest of this chapter proves that h is finite and deduces a few useful consequences. In the next chapter we develop some methods whereby h, and the structure of $\mathbb{H}$, may be computed: such methods are an obvious necessity for applications of the class-group in particular cases.

11.2 Existence Theorem

The finiteness of h rests on an application of Minkowski's Theorem to the space $\mathbb{L}^{st}$. It is possible to give a more elementary proof that h is finite, see Lang [75], but Minkowski's Theorem gives a better bound, and is in any case needed elsewhere. In this section we state and prove the relevant result, leaving the finiteness theorem to the next section.

Let K be a number field of degree $n = s + 2t$ as in (9.1), with ring of integers $\mathfrak{O}$; and let $\mathfrak{a}$ be an ideal of $\mathfrak{O}$. Then $(\mathfrak{a}, +)$ is a free abelian group of rank n by Theorem 2.28, so by Corollary 9.6 its image $\sigma(\mathfrak{a})$ in $\mathbb{L}^{st}$ is a lattice of dimension n. To apply Lemma 10.7 we must know the volume of a fundamental domain for $\sigma(\mathfrak{a})$. A useful general result is:

Lemma 11.4. *Let L be an n-dimensional lattice in $\mathbb{R}^n$ with basis $\{e_1, \ldots, e_n\}$. Suppose that*

$$e_i = (a_{1i}, \ldots, a_{ni}) \qquad \text{where } 1 \leq i \leq n$$

Then the volume of the fundamental domain T of L defined by this basis is

$$v(T) = |\det a_{ij}|$$

Proof: The volume is

$$v(T) = \int_T dx_1 \ldots dx_n$$

Define new variables by

$$x_i = \sum_j a_{ij} y_j$$

The Jacobian of this transformation is $\det a_{ij}$, and T is the set of points $\sum a_{ij} y_i$ with $0 \leq y_i < 1$. By the transformation formula for multiple integrals, Apostol [2] p. 271,

$$\begin{aligned} v(T) &= \int_T |\det a_{ij}|\, dy_1 \ldots dy_n \\ &= |\det a_{ij}| \int_0^1 dy_1 \ldots \int_0^1 dy_n \\ &= |\det a_{ij}| \end{aligned}$$

□

Given a lattice L there exist many different $\mathbb{Z}$-bases for L, hence many distinct fundamental domains. However, since distinct $\mathbb{Z}$-bases are related by a unimodular matrix, Lemma 11.4 implies that the volumes of these distinct fundamental domains are all equal.

Theorem 11.5. *Let K be a number field of degree $n = s + 2t$ as in (9.1), with ring of integers $\mathfrak{O}$, and let $0 \neq \mathfrak{a}$ be an ideal of $\mathfrak{O}$. Then the volume of a fundamental domain for $\sigma(\mathfrak{a})$ in $\mathbb{L}^{st}$ is*

$$2^{-t}\mathrm{N}(\mathfrak{a})\sqrt{|d_K|}$$

where d_K is the discriminant of K.

Proof: Let $\{\alpha_1, \ldots, \alpha_n\}$ be a $\mathbb{Z}$-basis for $\mathfrak{a}$. Then, in the notation of Theorem 9.4, a $\mathbb{Z}$-basis for $\sigma(\mathfrak{a})$ in $\mathbb{L}^{st}$ is

$$\begin{gathered}(x_1^{(1)}, \ldots, x_s^{(1)}, y_1^{(1)}, z_1^{(1)}, \ldots, y_t^{(1)}, z_t^{(1)}) \\ \vdots \\ (x_1^{(n)}, \ldots, x_s^{(n)}, y_1^{(n)}, z_1^{(n)}, \ldots, y_t^{(n)}, z_t^{(n)})\end{gathered}$$

By Lemma 11.4, if T is a fundamental domain for $\sigma(\mathfrak{a})$ then $v(T) = |D|$, where D is as in Theorem 9.4. In the notation of that theorem, $D = (-2\mathrm{i})^{-t}E$, so that $|D| = 2^{-t}|E|$. Now $E^2 = \Delta[\alpha_1, \ldots, \alpha_n]$, and

$$\mathrm{N}(\mathfrak{a}) = \left|\frac{\Delta[\alpha_1, \ldots, \alpha_n]}{d_K}\right|^{1/2}$$

by Theorem 6.17. □

Lemma 10.7 and Theorem 11.5 now yield the important:

Theorem 11.6. *If $\mathfrak{a} \neq 0$ is an ideal of $\mathfrak{O}$ then $\mathfrak{a}$ contains an integer α with*

$$|\mathrm{N}(\alpha)| \leq \left(\frac{2}{\pi}\right)^t \mathrm{N}(\mathfrak{a})\sqrt{|d_K|}$$

Proof: For fixed but arbitrary $\varepsilon > 0$ choose positive real numbers $c_1, \ldots, c_{s+t}$ with

$$c_1 \ldots c_{s+t} = \left(\frac{2}{\pi}\right)^t \mathrm{N}(\mathfrak{a})\sqrt{|d_K|} + \varepsilon$$

By Lemma 10.7 and Theorem 11.5 there exists $0 \neq \alpha \in \mathfrak{a}$ such that

$$\begin{aligned}|\sigma_1(\alpha)| &< c_1, \ldots, |\sigma_s(\alpha)| < c_s \\ |\sigma_{s+1}(\alpha)|^2 &< c_{s+1}, \ldots, |\sigma_{s+t}(\alpha)|^2 < c_{s+t}\end{aligned}$$

Multiply all these inequalities together:

$$|\mathrm{N}(\alpha)| < c_1 \ldots c_s c_{s+1} \ldots c_{s+t} = \left(\frac{2}{\pi}\right)^t \mathrm{N}(\mathfrak{a})\sqrt{|d_K|} + \varepsilon$$

Since a lattice is discrete, the set A_ε of such α is finite. Also $A_\varepsilon \neq \emptyset$, so $A = \cap_\varepsilon A_\varepsilon \neq \emptyset$. It we pick $\alpha \in A$ then

$$|\mathrm{N}(\alpha)| \leq \left(\frac{2}{\pi}\right)^t \mathrm{N}(\mathfrak{a})\sqrt{|d_K|}$$

□

Corollary 11.7. *Every nonzero ideal $\mathfrak{a}$ of $\mathfrak{O}$ is equivalent to an ideal whose norm is $\leq (2/\pi)^t\sqrt{|d_K|}$.*

Proof: The class of fractional ideals equivalent to $\mathfrak{a}^{-1}$ contains an ideal $\mathfrak{c}$, so $\mathfrak{a}\mathfrak{c} \sim \mathfrak{O}$. Use Theorem 11.6 to find an integer $\gamma \in \mathfrak{c}$ such that

$$|\mathrm{N}(\gamma)| \leq \left(\frac{2}{\pi}\right)^t \mathrm{N}(\mathfrak{c})\sqrt{|d_K|}$$

Since $\mathfrak{c}|\gamma$ we have $\langle\gamma\rangle = \mathfrak{c}\mathfrak{b}$ for some ideal $\mathfrak{b}$. Since $\mathrm{N}(\mathfrak{b})\mathrm{N}(\mathfrak{c}) = \mathrm{N}(\mathfrak{b}\mathfrak{c}) = \mathrm{N}(\langle\gamma\rangle) = |\mathrm{N}(\gamma)|$,

$$\mathrm{N}(\mathfrak{b}) \leq \left(\frac{2}{\pi}\right)^t \sqrt{|d_K|}$$

We claim that $\mathfrak{b} \sim \mathfrak{a}$. This is clear since $\mathfrak{c} \sim \mathfrak{a}^{-1}$ and $\mathfrak{b} \sim \mathfrak{c}^{-1}$. □

Example 11.8. We now give an explicit computation using the above ideas.

Let $K = \mathbb{Q}(\sqrt{-5})$. Then $\mathfrak{O} = \mathbb{Z}[\sqrt{-5}]$ does not have unique factorisation, so $h > 1$. Because the monomorphisms $\sigma_i : K \to \mathbb{C}$ are σ_1, σ_2 where $\sigma_1 \neq \sigma_2$ and $\bar{\sigma}_1 = \sigma_2$, we have $s = 0, t = 1$. The discriminant d_K of K is $d_K = -20$, so

$$\left(\frac{2}{\pi}\right)^t \sqrt{|d_K|} = \frac{2\sqrt{20}}{\pi} < 2.85$$

By Corollary 11.7, every ideal of $\mathfrak{O}$ is equivalent to an ideal of norm less than 2.85, which means a norm of 1 or 2. An ideal of norm 1 is the whole ring $\mathfrak{O}$, hence principal. An ideal $\mathfrak{a}$ of norm 2 satisfies $\mathfrak{a}|2$ by Theorem 6.23 (b), so $\mathfrak{a}$ is a factor of $\langle 2\rangle$. But

$$\langle 2\rangle = \langle 2, 1+\sqrt{-5}\rangle^2$$

where $\langle 2, 1+\sqrt{-5}\rangle$ is prime and has norm 2. So $\langle 2, 1+\sqrt{-5}\rangle$ is the *only* ideal of norm 2. Hence every ideal of $\mathfrak{O}$ is equivalent to $\mathfrak{O}$ or $\langle 2, 1+\sqrt{-5}\rangle$ which are themselves inequivalent (since $\langle 2, 1+\sqrt{-5}\rangle$ is not principal), proving that $h = 2$.

11.3 Finiteness of the Class-Group

Theorem 11.9. *The class-group of a number field is a finite abelian group. The class-number h is finite.*

Proof: Let K be a number field of discriminant d_K and degree $n = s + 2t$ as usual. The class-group $\mathbb{H} = \mathbb{F}/\mathbb{P}$ is abelian, so it remains to prove $\mathbb{H}$ finite. This is true if and only if the number of distinct equivalence classes of fractional ideals is finite. Let $[\mathfrak{c}]$ be such an equivalence class. Then $[\mathfrak{c}]$ contains an ideal $\mathfrak{a}$, and by Corollary 11.7 $\mathfrak{a}$ is equivalent to an ideal $\mathfrak{b}$ with $\mathrm{N}(\mathfrak{b}) \leq (2/\pi)^t\sqrt{|d_K|}$. By Theorem 6.26(c), only finitely many ideals have a given norm, so there are only finitely many choices for $\mathfrak{b}$. Since $[\mathfrak{c}] = [\mathfrak{a}] = [\mathfrak{b}]$ there are only finitely many equivalence classes $[\mathfrak{c}]$. Thus $\mathbb{H}$ is a finite group and $h = |\mathbb{H}|$ is finite. □

From simple group-theoretic facts we obtain the useful:

Proposition 11.10. *Let K be a number field of class-number h, and $\mathfrak{a}$ an ideal of the ring of integers $\mathfrak{O}$. Then*

(a) *$\mathfrak{a}^h$ is principal.*

(b) *If q is prime to h and $\mathfrak{a}^q$ is principal, then $\mathfrak{a}$ is principal.*

Proof: Since $h = |\mathbb{H}|$ we have $[\mathfrak{a}]^h = [\mathfrak{O}]$ for all $[\mathfrak{a}] \in \mathbb{H}$, because $[\mathfrak{O}]$ is the identity element of $\mathbb{H}$. Hence $[\mathfrak{a}^h] = [\mathfrak{a}]^h = [\mathfrak{O}]$, so $\mathfrak{a}^h \sim \mathfrak{O}$, so $\mathfrak{a}^h$ is principal. This proves (a). For (b) choose u and $v \in \mathbb{Z}$ such that $uh + vq = 1$. Then $[\mathfrak{a}]^q = [\mathfrak{O}]$, so

$$[\mathfrak{a}] = [\mathfrak{a}]^{uh+vq} = \left([\mathfrak{a}]^h\right)^u \left([\mathfrak{a}]^q\right)^v = [\mathfrak{O}]^u[\mathfrak{O}]^v = [\mathfrak{O}]$$

and again $\mathfrak{a}$ is principal. □

11.4 How to Make an Ideal Principal

Given an ideal $\mathfrak{a}$ in the ring $\mathfrak{O}$ of integers of a number field K, we already know that $\mathfrak{a}$ has at most two generators

$$\mathfrak{a} = \langle \alpha, \beta \rangle \quad (\alpha, \beta \in \mathfrak{O})$$

In this section we demonstrate that there exists an extension number field $L \supseteq K$ with integers $\mathfrak{O}'$, such that the extended ideal $\mathfrak{O}'\mathfrak{a}$ in $\mathfrak{O}'$ is principal. As standard notation we retain the symbols $\langle \alpha \rangle$, $\langle \alpha, \beta \rangle$ to denote the ideals in $\mathfrak{O}$ generated by α and by $\alpha, \beta \in \mathfrak{O}$. We write the ideal in $\mathfrak{O}'$ generated by $S \subseteq \mathfrak{O}'$ as $\mathfrak{O}'S$. For example $\mathfrak{O}'\kappa$ denotes the principal ideal in $\mathfrak{O}'$ generated by $\kappa \in \mathfrak{O}'$.

Lemma 11.11. *If S_1, S_2 are subsets of $\mathfrak{O}'$, then*

$$\mathfrak{O}'(S_1S_2) = (\mathfrak{O}'S_1)(\mathfrak{O}'S_2)$$

Proof: Trivial (remembering $1 \in \mathfrak{O}'$). □

The central result is:

Theorem 11.12. *Let K be a number field, $\mathfrak{a}$ an ideal in the ring of integers $\mathfrak{O}$ of K. Then there exists an algebraic integer κ such that if $\mathfrak{O}'$ is the ring of integers of $K(\kappa)$, then*

(a) $\mathfrak{O}'\kappa = \mathfrak{O}'\mathfrak{a}$.
(b) $(\mathfrak{O}'\kappa) \cap \mathfrak{O} = \mathfrak{a}$.
(c) *If $\mathbb{B}$ is the ring of all algebraic integers, then* $(\mathbb{B}\kappa) \cap K = \mathfrak{a}$.
(d) *If $\mathfrak{O}''\gamma = \mathfrak{O}''\mathfrak{a}$ for any $\gamma \in \mathbb{B}$, and any ring $\mathfrak{O}''$ of integers, then $\gamma = u\kappa$ where u is a unit of $\mathbb{B}$.*

Proof: By Proposition 11.10, $\mathfrak{a}^h$ is principal, say $\mathfrak{a}^h = \langle\theta\rangle$. Let $\kappa = \theta^{1/h} \in \mathbb{B}$, and consider $L = K(\kappa)$. Let $\mathfrak{O}' = \mathbb{B} \cap L$ be the ring of integers in L; clearly $\kappa \in \mathfrak{O}'$. Since $\mathfrak{a}^h = \langle\theta\rangle$ Lemma 11.11 implies that

$$(\mathfrak{O}'\mathfrak{a})^h = \mathfrak{O}'(\mathfrak{a}^h) = \mathfrak{O}'\theta = \mathfrak{O}'\kappa^h = (\mathfrak{O}'\kappa)^h$$

Uniqueness of factorisation of ideals in $\mathfrak{O}'$ easily yields $\mathfrak{O}'a = \mathfrak{O}'\kappa$, proving (a).

Since (c) implies (b), we now consider (c). The inclusion $\mathfrak{a} \subseteq \mathbb{B}\kappa \cap K$ is straightforward. Conversely, suppose $\gamma \subseteq \mathbb{B}\kappa \cap K$. Then $\gamma = \lambda\kappa \quad (\lambda \in \mathbb{B})$, and we must show that $\gamma \in \mathfrak{a}$.

Since $\gamma \in K$, $\kappa \in L$, we have $\lambda = \gamma\kappa^{-1} \in L$, so $\lambda \in L \cap \mathbb{B} = \mathfrak{O}'$. This gives

$$\gamma^h = \lambda^h\kappa^h = \lambda^h\theta \quad (\gamma \in K, \lambda \in \mathfrak{O}', \theta \in \mathfrak{O})$$

so $\gamma^h \in \mathbb{B}$, and by Theorem 2.20, $\gamma \in \mathbb{B}$. Thus $\gamma \in \mathbb{B} \cap K = \mathfrak{O}$. Considering the equation $\gamma^h = \lambda^h\theta$ again, we find

$$\lambda^h = \gamma^h\theta^{-1} \in K$$

so $\lambda^h \in K \cap \mathbb{B} = \mathfrak{O}$. Thus we finish up with

$$\gamma^h = \lambda^h\theta \quad (\gamma, \lambda^h, \theta \in \mathfrak{O})$$

Taking ideals in $\mathfrak{O}$,

$$\langle\gamma\rangle^h = \langle\lambda^h\rangle\langle\theta\rangle = \langle\lambda^h\rangle\mathfrak{a}^h$$

Unique factorisation in $\mathfrak{O}$ implies that $\langle\lambda^h\rangle = \mathfrak{b}^h$ for some ideal $\mathfrak{b}$, so $\langle\gamma\rangle^h = \mathfrak{b}^h\mathfrak{a}^h$. Unique factorisation once more implies that $\langle\gamma\rangle = \mathfrak{b}\mathfrak{a}$, so $\gamma \in \mathfrak{a}$ as required.

To prove (d), observe that by Theorem 6.29, $\mathfrak{a} = \langle\alpha, \beta\rangle$ for $\alpha, \beta \in \mathfrak{O}$. Substituting in (d) gives $\mathfrak{O}''\gamma = \mathfrak{O}''\langle\alpha, \beta\rangle$. Thus $\gamma = \lambda\alpha + \mu\beta$ where $\lambda, \mu \in \mathfrak{O}''$, so certainly $\lambda, \mu \in \mathbb{B}$. From (a), $\alpha, \beta \in \mathfrak{O}'\kappa$, so

$$\alpha = \eta\kappa, \quad \beta = \zeta\kappa \quad (\eta, \zeta \in \mathfrak{O}' \subseteq \mathbb{B})$$

Hence $\gamma = \lambda\eta\kappa + \mu\zeta\kappa$ and $\kappa|\gamma$ in $\mathbb{B}$. Finally, interchange the roles of γ, κ to prove (d). □

Theorem 11.12 can be improved, for as it stands the extension ring $\mathfrak{O}'$ in which $\mathfrak{O}'\mathfrak{a}$ is principal depends on $\mathfrak{a}$. We can actually find a single extension ring in which the extension of every ideal is principal. This depends on the following lemma and the finiteness of the class-number:

Lemma 11.13. *If $\mathfrak{a}, \mathfrak{b}$ are equivalent ideals in the ring $\mathfrak{O}$ of integers of a number field and $\mathfrak{O}'\mathfrak{a}$ is principal, then so is $\mathfrak{O}'\mathfrak{b}$.*

Proof: By the definition of equivalence, there exist principal ideals $\mathfrak{d}, \mathfrak{e}$ of $\mathfrak{O}$ such that $\mathfrak{a}\mathfrak{d} = \mathfrak{b}\mathfrak{e}$. Hence

$$(\mathfrak{O}'\mathfrak{a})(\mathfrak{O}'\mathfrak{d}) = (\mathfrak{O}'\mathfrak{b})(\mathcal{D}'\mathfrak{e})$$

where now $\mathfrak{O}'\mathfrak{a}$, $\mathfrak{O}'\mathfrak{d}$, $\mathfrak{O}'\mathfrak{e}$ are all principal. Since the set $\mathbb{P}$ of principal fractional ideals of $\mathfrak{O}'$ is a group, $\mathfrak{O}'\mathfrak{b}$ is a principal fractional ideal which is also an ideal, so $\mathfrak{O}'\mathfrak{b}$ is a principal ideal. □

Theorem 11.14. *Let K be a number field with integers $\mathfrak{O}_K$. Then there exists a number field $L \supseteq K$ with ring $\mathfrak{O}_L$ of integers such that for every ideal $\mathfrak{a}$ in $\mathfrak{O}_K$:*

(a) $\mathfrak{O}_L\mathfrak{a}$ is a principal ideal.

(b) $(\mathfrak{O}_L\mathfrak{a}) \cap \mathfrak{O}_K = \mathfrak{a}$.

Proof: Since h is finite, select a representative set of ideals $\mathfrak{a}_1, \ldots, \mathfrak{a}_h$, one from each class. Choose algebraic integers $\kappa_1, \ldots, \kappa_h$ such that $\mathfrak{O}_i\mathfrak{a}_i$ is principal where $\mathfrak{O}_i$ is the ring of integers of $K(\kappa_i)$. Let $L = K(\kappa_1, \ldots, \kappa_h)$; its ring of integers $\mathfrak{O}_L$ contains all the $\mathfrak{O}_i$. Hence each ideal $\mathfrak{O}_L\mathfrak{a}_i$ is principal in $\mathfrak{O}_L$. Since every ideal $\mathfrak{a}$ in $\mathfrak{O}$ is equivalent to some $\mathfrak{a}_i$, the ideal

$\mathfrak{O}_L\mathfrak{a}$ is principal by Lemma 11.13. That is, for some $\alpha \in \mathbb{B}$

$$\mathfrak{O}_L\mathfrak{a} = \mathfrak{O}_L\alpha$$

This proves (a).

Now we prove (b). Clearly $\mathfrak{a} \subseteq (\mathfrak{O}_L\mathfrak{a}) \cap \mathfrak{O}_K$. For the converse inclusion, Theorem 11.12(d) implies that $\alpha = u\kappa$ where u is a unit in $\mathbb{B}$. Now

$$(\mathfrak{O}_L\mathfrak{a}) \cap \mathfrak{O}_K = (\mathfrak{O}_L\alpha) \cap \mathfrak{O}_K \subseteq (\mathbb{B}\alpha) \cap K = (\mathbb{B}\kappa) \cap K = \mathfrak{a}$$

by Theorem 11.12(c). □

Remark. For many years it was an open question, going back to David Hilbert, whether every number field can be embedded in one with unique factorisation. However, in 1964 Golod and Šafarevič [53] showed that this is not always possible, citing the explicit example

$$\mathbb{Q}(\sqrt{(-3 \cdot 5 \cdot 7 \cdot 11 \cdot 13 \cdot 17 \cdot 19)})$$

The proof is ingenious rather than hard, but it uses ideas we have not developed.

11.5 Unique Factorisation of Elements in an Extension Ring

The results of the last section can be translated from principal ideals back to elements to give the version of Kummer's theory alluded to in the introduction to Chapter 5. There we considered examples of non-unique factorisation such as

$$10 = 2 \cdot 5 = (5 + \sqrt{15})(5 - \sqrt{15})$$

in the ring of integers of $\mathbb{Q}(\sqrt{15})$. Viewing this as an equation in $\mathbb{Q}(\sqrt{3}, \sqrt{5})$, we saw that the factors can be further reduced as

$$\begin{aligned} 2 &= (\sqrt{5} + \sqrt{3})(\sqrt{5} - \sqrt{3}) \\ 5 &= \sqrt{5}\sqrt{5} \\ 5 + \sqrt{15} &= \sqrt{5}(\sqrt{5} + \sqrt{3}) \\ 5 - \sqrt{15} &= \sqrt{5}(\sqrt{5} - \sqrt{3}) \end{aligned}$$

and the two factorisations of 10 found above are just regroupings of the factors

$$10 = \sqrt{5}\sqrt{5}(\sqrt{5} + \sqrt{3})(\sqrt{5} - \sqrt{3})$$

We now show that a similar phenomenon occurs for all non-unique prime factorisations in all rings of integers.

Theorem 11.15. *Suppose K is a number field with integers $\mathfrak{O}_K$. Then there exists an extension field $L \supseteq K$ with integers $\mathfrak{O}_L$ such that every nonzero, non-unit $a \in \mathfrak{O}_K$ has a factorisation*

$$a = p_1 \dots p_r \quad (p_i \in \mathfrak{O}_L)$$

where the p_i are non-units in $\mathfrak{O}_L$, and the following property is satisfied. Given any factorisation $a = a_1 \dots a_s$ in $\mathfrak{O}_K$, where the a_i are non-units in $\mathfrak{O}_K$, there exist integers $1 \leq n_1 < \dots < n_s = r$ and a permutation π of $\{1, \dots, r\}$ such that the following elements are associates in $\mathfrak{O}_L$:

$$a_1, p_{\pi(1)} \cdots p_{\pi(n_1)}$$

$$\vdots$$

$$a_s, p_{\pi(n_{s-1}+1)} \cdots p_{\pi(n_s)}$$

Remark. What this theorem says in plain language is that the factorisations of elements into irreducibles in $\mathfrak{O}_K$ may not be unique, but all factorisations of an element in $\mathfrak{O}_K$ come from different groupings of associates of a single factorisation in $\mathfrak{O}_L$. In this sense elements in $\mathfrak{O}_K$ have unique factorisation into elements in $\mathfrak{O}_L$.

Proof: There is a unique factorisation of $\langle a \rangle$ into prime ideals in $\mathfrak{O}_K$, say

$$\langle a \rangle = \mathfrak{p}_1 \dots \mathfrak{p}_r$$

Since a is a non-unit, $r \geq 1$. Let $\mathfrak{O}_L$ be a ring of integers as in Theorem 11.14 such that every ideal of $\mathfrak{O}_K$ extends to a principal ideal of $\mathfrak{O}_L$. Suppose that

$$\mathfrak{O}_L \mathfrak{p}_i = \mathfrak{O}_L p_i \qquad (p_i \in \mathfrak{O}_L)$$

Then $a = up_1 \dots p_r$ where u is a unit in $\mathfrak{O}_L$, and since $r \geq 1$, we may replace p_1 by $up_1 \in \mathfrak{O}_L p_1$ to get a factorisation of the form $a = p_1 \dots p_r$. Given any factorisation of elements $a = a_1 \dots a_s$, where the a_i are non-units in $\mathfrak{O}_K$, we obtain $\langle a \rangle = \langle a_1 \rangle \dots \langle a_s \rangle$, where all the $\langle a_i \rangle$ are proper ideals of $\mathfrak{O}_K$. Unique factorisation in $\mathfrak{O}_K$ provides integers n_i and a permutation π such that

$$\begin{aligned} \langle a_1 \rangle &= \mathfrak{p}_{\pi(1)} \cdots \mathfrak{p}_{\pi(n_1)} \\ &\vdots \\ \langle a_s \rangle &= \mathfrak{p}_{\pi(n_{s-1}+1)} \cdots \mathfrak{p}_{\pi(n_s)} \end{aligned}$$

Now take ideals in $\mathfrak{O}_L$ generated by these ideals. □

Example 11.16. From the explicit computation of Example 11.8, if $K = \mathbb{Q}(\sqrt{-5})$, then $h = 2$ and a representative set of ideals consists of $\mathfrak{O}$, and $\langle 2, 1+\sqrt{-5}\rangle$ where $\langle 2, 1+\sqrt{-5}\rangle^2 = \langle 2\rangle$. Hence we may take $L = K(\sqrt{2}) = \mathbb{Q}(\sqrt{-5}, \sqrt{2})$. Theorem 11.15 tells us that *every* element of $\mathbb{Z}[\sqrt{-5}]$ factorises uniquely in the integers of $\mathbb{Q}(\sqrt{-5}, \sqrt{2})$. The case of the factorisation of the element 6 is dealt with in Exercise 11.7, where

$$6 = \sqrt{2}\sqrt{2}\left(\frac{\sqrt{2}+\sqrt{-10}}{2}\right)\left(\frac{\sqrt{2}-\sqrt{-10}}{2}\right)$$

That $\frac{\sqrt{2}\pm\sqrt{-10}}{2}$ really are integers may be dealt with by computing their minimal polynomials over $\mathbb{Q}$. Granted this, it is an easy matter to check that the two alternative factorisations

$$6 = 2 \cdot 3 = (1+\sqrt{-5})(1-\sqrt{-5})$$

in $\mathbb{Z}[\sqrt{-5}]$ are just different groupings of the factors in the integers of $\mathbb{Q}(\sqrt{-5}, \sqrt{2})$.

The above example underlines a basic problem when factorising elements in an extension ring. We have not given a general method for computing the integers of a number field. To date, we have explicitly calculated only the integers in quadratic and cyclotomic fields—and even those calculations were not trivial. There is also another weakness when factorising elements in an extension ring. The elements p_i occurring in the factorisation of a in Theorem 11.15 need not be irreducible. (For instance we might work in a slightly larger ring $\mathfrak{O}_{L'}$ containing $\sqrt{p_i}$; the method of adjoining $\kappa = \theta^{1/h}$ may very well add such roots.) However, the proof of Theorem 11.15 tells us that the factorisation of the element $\mathfrak{a}$ in $\mathfrak{O}_L$ that gives the unique factorisation properties is given by the factorisation of the ideal $\langle a\rangle$ in the ring $\mathfrak{O}_K$. For this reason we may just as well stick to ideals in the original ring rather than embellish the situation by factorising elements outside. Our computations in future will be concerned mainly with ideals—unique factorisation of ideals proves so much easier to handle.

11.6 Exercises

11.1 Let $K = \mathbb{Q}(\sqrt{-5})$, and let $\mathfrak{p}$, $\mathfrak{q}$, $\mathfrak{r}$ be the ideals defined in Exercise 6.2. Let $\mathbb{H}$ be the class group. Show that in $\mathbb{H}$

$$[\mathfrak{p}]^2 = [\mathfrak{O}] \qquad [\mathfrak{p}][\mathfrak{q}] = [\mathfrak{O}] \qquad [\mathfrak{p}][\mathfrak{r}] = [\mathfrak{O}]$$

and hence show that $\mathfrak{p}$, $\mathfrak{q}$, $\mathfrak{r}$ are equivalent.

11.2 Verify by explicit computation that $\mathfrak{p}$, $\mathfrak{q}$, $\mathfrak{r}$ in Exercise 11.1 are equivalent.

11.3 Using Corollary 11.7, show that for $K = \mathbb{Q}(\sqrt{-6})$ every ideal is equivalent to one of norm at most 3. Verify that

$$\langle 2 \rangle = \langle 2, \sqrt{-6} \rangle^2 \qquad \langle 3 \rangle = \langle 3, \sqrt{-6} \rangle^2$$

and conclude that the only ideals of norm 2, 3 are $\langle 2, \sqrt{-6} \rangle$, $\langle 3, \sqrt{-6} \rangle$. Deduce that $h \leq 3$. Using $\langle 2, \sqrt{-6} \rangle^2 = \langle 2 \rangle$, or otherwise, show that $h = 2$.

11.4 Find principal ideals $\mathfrak{a}$, $\mathfrak{b}$ in $\mathbb{Z}[\sqrt{-6}]$ such that

$$\mathfrak{a} \langle 2, \sqrt{-6} \rangle = \mathfrak{b} \langle 3, \sqrt{-6} \rangle$$

11.5 Find all squarefree integers d in $-10 < d < 10$ such that the class-number of $\mathbb{Q}(\sqrt{d})$ is 1. (*Hint:* Look up a few theorems.)

11.6 In $\mathbb{Z}[\sqrt{-10}]$ we have the factorisations into irreducibles

$$14 = 2 \cdot 7 = (2 + \sqrt{-10})(2 - \sqrt{-10})$$

Find an extension ring $\mathfrak{O}_L$ of $\mathbb{Z}[\sqrt{-10}]$ and a factorisation of 14 in $\mathfrak{O}_L$ such that the given factorisations are found by different groupings of the factors.

11.7 Factorise $6 = 2 \cdot 3 = (4 + \sqrt{10})(4 - \sqrt{10}) \in \mathbb{Z}[\sqrt{10}]$ in an extension ring to exhibit the given factors as different groupings of the new ones.

11.8 Relate the factorisation

$$10 = \sqrt{5}\sqrt{5}(\sqrt{5} + \sqrt{3})(\sqrt{5} - \sqrt{3})$$

in the integers of $\mathbb{Q}(\sqrt{3}, \sqrt{5})$ to the factorisation of $\langle 10 \rangle$ into prime ideals in the integers of $\mathbb{Q}(\sqrt{15})$. Explain how this gives rise to the different factorisations

$$10 = 2 \cdot 5 = (5 + \sqrt{15})(5 - \sqrt{15})$$

into irreducibles in the integers of $\mathbb{Q}(\sqrt{15})$.

11.9 Let $d \equiv 1 \pmod 4$. Show that K contains a principal ideal of norm 2 if and only if $a^2 - db^2 = \pm 8$ has a solution in integers a, b. Deduce that $\mathfrak{O}$ is a principal ideal domain for $d = 17, 33$, and that its class-number is 2 when $d = -15$.

11.10 Show that the class-number of $\mathbb{Q}(\sqrt{-6})$ is 2. (*Hint*: Every ideal class contains an ideal of norm 1, 2, or 3.)

11.11 Given that the class-number of $\mathbb{Q}\sqrt{-5})$ is 2, show that the only solution in integers of $y^2 = x^3 - 20$ is $(x, y) = (6, \pm 14)$.

III

Number-Theoretic Applications

12

Computational Methods

The results of this chapter, although apparently diverse, all have a strong bearing on the question of practical computation of the class-number, within the limits of the techniques now at our command. We focus on computations performed by hand. For complicated calculations, mathematicians and computer scientists have developed many software packages for algebraic computations. The website numbertheory.org provides a lengthy list of specialised packages, along with links to other useful information such as tables.

In the first section we study a special case of how a rational prime breaks up into prime ideals in a number field. The second section supplements this by showing that the distinct classes of fractional ideals may be found from the prime ideals dividing a finite set of rational primes, this set being in some sense small provided the degree of a number field K and its discriminant are not too large. Several specific cases are studied, especially quadratic fields: in particular we complete the list of fields $\mathbb{Q}(\sqrt{d})$ with negative d and with class-number 1. We do not prove this list is complete, however, because the methods required are beyond the scope of this book.

12.1 Factorisation of a Rational Prime

If p is a prime number in $\mathbb{Z}$, it is not generally true that $\langle p \rangle$ is a prime ideal in the ring of integers $\mathfrak{O}$ of a number field K. For instance, in $\mathbb{Q}(\mathrm{i})$ we have the factorisation

$$\langle 2 \rangle = \langle 1 + \mathrm{i} \rangle^2 \tag{12.1}$$

It is obviously useful to compute the prime factors of $\langle p \rangle$. In the case where the ring of integers is generated by a single element, which includes quadratic and cyclotomic fields, the following theorem of Dedekind ([31] section 2) is decisive.

Theorem 12.1. (Dedekind's Criterion) *Let K be a number field of degree n with ring of integers $\mathfrak{O} = \mathbb{Z}[\theta]$ generated by $\theta \in \mathfrak{O}$. Given a rational prime p, suppose the minimal polynomial f of θ over $\mathbb{Q}$ gives rise to the factorisation into irreducibles over $\mathbb{Z}_p$:*

$$\bar{f} = \bar{f}_1^{e_1} \dots \bar{f}_r^{e_r}$$

where the bar denotes the natural map $\mathbb{Z}[t] \to \mathbb{Z}_p[t]$. Then if $f_i \in \mathbb{Z}[t]$ is any polynomial mapping onto $\bar{f}_i$, the ideal

$$\mathfrak{p}_i = \langle p \rangle + \langle f_i(\theta) \rangle$$

is prime and the prime factorisation of $\langle p \rangle$ in $\mathfrak{O}$ is

$$\langle p \rangle = \mathfrak{p}_1^{e_1} \dots \mathfrak{p}_r^{e_r}$$

Proof: Let θ_i be a zero of $\bar{f}_i$ in $\mathbb{Z}_p[\theta_i] \cong \mathbb{Z}_p[t]/\langle \bar{f}_i \rangle$. There is a natural map $\nu_i : \mathbb{Z}[\theta] \to \mathbb{Z}_p[\theta_i]$ given by

$$\nu_i(p(\theta)) = \bar{p}(\theta_i)$$

The image of ν_i is $\mathbb{Z}_p[\theta_i]$, which is a field, so $\ker \nu_i$ is a prime ideal of $\mathbb{Z}[\theta] = \mathfrak{O}$. Clearly

$$\langle p \rangle + \langle f_i(\theta) \rangle \subseteq \ker \nu_i$$

But if $g(\theta) \in \ker \nu_i$, then $\bar{g}(\theta_i) = 0$, so $\bar{g} = \bar{f}_i \bar{h}$ for some $\bar{h} \in \mathbb{Z}_p[t]$; this means that $g - f_i h \in \mathbb{Z}[t]$ has coefficients divisible by p. Thus

$$g(\theta) = (g(\theta) - f_i(\theta)h(\theta)) + f_i(\theta)h(\theta) \in \langle p \rangle + \langle f_i(\theta) \rangle$$

showing that

$$\ker \nu_i = \langle p \rangle + \langle f_i(\theta) \rangle$$

Let

$$\mathfrak{p}_i = \langle p \rangle + \langle f_i(\theta) \rangle$$

Then for each $\bar{f}_i$ the ideal $\mathfrak{p}_i$ is prime and satisfies $\langle p \rangle \subseteq \mathfrak{p}_i$, that is, $\mathfrak{p}_i | \langle p \rangle$.

For any ideals $\mathfrak{a}$, $\mathfrak{b}_1$, $\mathfrak{b}_2$,

$$(\mathfrak{a} + \mathfrak{b}_1)(\mathfrak{a} + \mathfrak{b}_2) \subseteq \mathfrak{a} + \mathfrak{b}_1\mathfrak{b}_2$$

so by induction

$$\begin{aligned}\mathfrak{p}_1^{e_1}\dots\mathfrak{p}_r^{e_r} &\subseteq \langle p\rangle + \langle f_1(\theta)^{e_1}\dots f_r(\theta)^{e_r}\rangle\\ &\subseteq \langle p\rangle + \langle f(\theta)\rangle\\ &= \langle p\rangle\end{aligned}$$

Thus $\langle p\rangle \mid \mathfrak{p}_1^{e_1}\dots\mathfrak{p}_r^{e_r}$, and the only prime factors of $\langle p\rangle$ are $\mathfrak{p}_1,\dots,\mathfrak{p}_r$, showing that

$$\langle p\rangle = \mathfrak{p}_1^{k_1}\dots\mathfrak{p}_r^{k_r} \tag{12.2}$$

where

$$0 < k_i \leq e_i \quad (1 \leq i \leq r) \tag{12.3}$$

The norm of $\mathfrak{p}_i$ is, by definition, $|\mathfrak{O}/\mathfrak{p}_i|$, and the isomorphisms

$$\mathfrak{O}/\mathfrak{p}_i = \mathbb{Z}[\theta]/\mathfrak{p}_i \cong \mathbb{Z}_p[\theta_i]$$

imply that

$$\mathrm{N}(\mathfrak{p}_i) = |\mathbb{Z}_p[\theta_i]| = p^{d_i}$$

where $d_i = \partial\bar{f}_i$, or equivalently $d_i = \partial f_i$. Also

$$\mathrm{N}\left(\langle p\rangle\right) = |\mathbb{Z}[\theta]/\left\langle p\right\rangle| = p^n$$

so, taking norms in Equation (12.2),

$$p^n = \mathrm{N}(\langle p\rangle) = \mathrm{N}(\mathfrak{p}_1)^{k_1}\dots\mathrm{N}(\mathfrak{p}_r)^{k_r} = p^{d_1k_1+\cdots+d_rk_r}$$

which implies that

$$d_1k_1 + \cdots + d_rk_r = n = d_1e_1 + \cdots + d_re_r \tag{12.4}$$

Equation (12.3) now leads to $k_i = e_i$ $(1 \leq i \leq r)$. □

This result is not always applicable, since in general $\mathfrak{O}$ need not be of the form $\mathbb{Z}[\theta]$; see Example 2.43. But for quadratic or cyclotomic fields we have already shown that $\mathfrak{O} = \mathbb{Z}[\theta]$, so the theorem applies in these cases—and in many others. It also has the advantage of computability. Since there is only a finite number of polynomials over $\mathbb{Z}_p$ of given degree, the factorisation of $\bar{f}$ can be performed in a finite number of steps. A little native wit helps, but, if the worst comes to the worst, there is only a finite number of polynomials of lower degree than $\bar{f}$ to try as factors.

Example 12.2. The number field $\mathbb{Q}(\mathrm{i})$ has ring of integers $\mathfrak{O} = \mathbb{Z}[\theta]$, where θ has minimal polynomial $t^2 + 1$.

To find the factorisation of $\langle 2\rangle$ we look at this polynomial (mod 2), where

$$t^2+1=(t+1)^2$$

Hence $\langle 2\rangle=\mathfrak{p}^2$ where

$$\mathfrak{p}=\langle 2\rangle+\langle \mathrm{i}+1\rangle=\langle 1+\mathrm{i}\rangle$$

because $2=(1+\mathrm{i})(1-\mathrm{i})$. Thus we recover Equation (12.1).

Example 12.3. More generally, consider the factorisation in $\mathbb{Z}[\mathrm{i}]$ of a prime $p\in\mathbb{Z}$, which we addressed in Theorem 5.34. We check that this result is consistent with Theorem 12.1.

There are three cases to consider:

(1) t^2+1 is irreducible (mod p).

(2) $t^2+1\equiv(t-\lambda)(t+\lambda)$ (mod p), (where $\lambda^2\equiv-1$ (mod p)) and $\lambda\not\equiv-\lambda$ (that is, $p\neq 2$).

(3) $t^2+1\equiv(t+1)^2$ (mod 2) when $p=2$.

In case (1) $\langle p\rangle$ is prime; in case (2) $\langle p\rangle=\mathfrak{p}_1\mathfrak{p}_2$ for distinct prime ideals $\mathfrak{p}_1,\mathfrak{p}_2$; in case (3) $\langle p\rangle=\mathfrak{p}_1^2$ for a prime ideal $\mathfrak{p}_1$.

The distinction between cases (1) and (2) is whether -1 is congruent to a square (mod p). The proof of the Two-Squares Theorem, Theorem 8.4, shows that (1) applies if p is of the form $4k+1$ ($k\in\mathbb{Z}$). A similar argument shows that (2) applies if p is of the form $4k+3$ ($k\in\mathbb{Z}$). (The Quadratic Reciprocity Law, Section 22.3, leads to the same conclusion.)

The results in this section are the tip of the iceberg of a large and significant portion of algebraic number theory. Given a prime ideal $\mathfrak{p}$ in the ring $\mathfrak{O}_K$ of integers in a number field K, we may consider the extension ideal $\mathfrak{O}_L\mathfrak{p}$ in the ring of integers $\mathfrak{O}_L$ of an extension algebraic number field. We find

$$\mathfrak{O}_L\mathfrak{p}=\mathfrak{P}_1^{e_1}\dots\mathfrak{P}_g^{e_r}$$

where $\mathfrak{P}_1,\dots,\mathfrak{P}_g$ are distinct prime ideals in $\mathfrak{O}_L$. We take up this idea in Chapter 21.

12.2 Minkowski Constants

The proof of Theorem 11.6 leaves room for improvement, because it is based on Lemma 10.7, which is far stronger than we really need. What we want is a point α such that

$$|\sigma_1(\alpha)|\dots|\sigma_s(\alpha)|\;|\sigma_{s+1}(\alpha)|^2\dots|\sigma_{s+t}(\alpha)|^2<c_1\dots c_{s+t} \tag{12.5}$$

but what we actually find is a point α satisfying the considerably stronger restriction

$$\begin{array}{c} |\sigma_1(\alpha)| < c_1, \ldots, |\sigma_s(\alpha)| < c_s \\ |\sigma_{s+1}(\alpha)|^2 < c_{s+1}, \ldots, |\sigma_{s+t}(\alpha)|^2 < c_{s+t} \end{array} \tag{12.6}$$

Certainly the inequalities (12.6) imply (12.5), but not the reverse.

The reason for using (12.6) is that we wish to employ Minkowski's Theorem. For (12.6) the relevant set of points in $\mathbb{L}^{st}$ is convex and symmetric, so the theorem applies; but for (12.5) the relevant set, though symmetric, is not convex; see Figure 12.1 below. This means we cannot use (12.6) directly. The gap between (12.5) and (12.6) is so great, however, that we might hope to find another set of inequalities, corresponding to a convex subset of $\mathbb{L}^{st}$, and implying (12.5); this would lead to improved estimates in Theorem 11.6 and Corollary 11.7.

To do this, we use the well-known inequality between arithmetic and geometric means:

$$(a_1 \ldots a_n)^{1/n} \leq \frac{1}{n}(a_1 + \cdots + a_n) \tag{12.7}$$

The result is:

Theorem 12.4. *If $\mathfrak{a} \neq 0$ is an ideal of $\mathfrak{O}$ then $\mathfrak{a}$ contains an element α with*

$$|\mathrm{N}(\alpha)| \leq \left(\frac{4}{\pi}\right)^t \cdot \frac{n!}{n^n}\sqrt{|d_K|}\,\mathrm{N}(\mathfrak{a})$$

where n is the degree of K and d_K is the discriminant.

Before giving the proof, we isolate one calculation:

Lemma 12.5. *Let X_c be the set of all $x \in \mathbb{L}^{st}$ such that*

$$|x_1| + \cdots + |x_s| + 2\sqrt{(y_1^2 + z_1^2)} + \cdots + 2\sqrt{(y_t^2 + z_t^2)} < c$$

where c is a positive real number. Then the volume of X_c is

$$v(X_c) = 2^s \left(\frac{\pi}{2}\right)^t \cdot \frac{1}{n!}c^n \tag{12.8}$$

Proof: Denote the volume by $V(r, s, c)$ and use double induction on s and t. Clearly

$$V(1, 0, c) = 2c \qquad V(0, 1, c) = \pi c^2/4$$

which starts the induction.

Assume the result for $V(s,t,c)$.

First, consider $V(s+1,t,c)$. The relevant set $X_c \subseteq \mathbb{R}^s \times \mathbb{R} \times \mathbb{C}^t$ with coordinates $(x_1,\ldots,,x_s,x,y_1+\mathrm{i}z_1,\ldots,y_t+\mathrm{i}z_t)$ is defined by

$$|x| + \sum_{i=1}^{s} |x_i| + 2\sum_{j=1}^{t} |y_j + \mathrm{i}z_j| \le c$$

The set $X_c = \emptyset$ if $|x| > c$, so, integrating over the x-coordinate,

$$\begin{aligned} V(s+1,t,c) &= \int_{-c}^{c} V(s,t,c-|x|)\mathrm{d}x \\ &= 2\int_0^c 2^s \left(\frac{\pi}{2}\right)^t \frac{(c-x)^n}{n!}\mathrm{d}x \quad \text{by induction} \\ &= 2^{s+1}\left(\frac{\pi}{2}\right)^t \frac{c^{n+1}}{(n+1)!} \end{aligned}$$

in accordance with (12.8).

Next, consider $V(s,t+1,c)$. The relevant set $X_c \subseteq \mathbb{R}^s \times \mathbb{C}^t \times \mathbb{C}$ with coordinates $(x_1,\ldots,,x_s,y_1+\mathrm{i}z_1,\ldots,y_t+\mathrm{i}z_t,z)$ is defined by

$$\sum_{i=1}^{s} |x_i| + 2\sum_{j=1}^{t} |y_j + \mathrm{i}z_j| + 2|z| \le c$$

Now

$$V(s,t+1,c) = \int_{|z|\le c/2} V(s,t,c-2|z|)\mathrm{d}z$$

Change to polar coordinates, so $z = r\mathrm{e}^{\mathrm{i}\theta}$ and $\mathrm{d}z = r\,\mathrm{d}r\,\mathrm{d}\theta$. Then

$$\begin{aligned} V(s,t+1,c) &= \int_0^{c/2}\int_0^{2\pi} 2^s \left(\frac{\pi}{2}\right)^t \frac{(c-2r)^n}{n!} r\,\mathrm{d}r\,\mathrm{d}\theta \quad \text{by induction} \\ &= 2^s\left(\frac{\pi}{2}\right)^t \frac{2\pi}{n!}\int_0^{c/2} (c-2r)^n r\,\mathrm{d}r \\ &= 2^s\left(\frac{\pi}{2}\right)^t \frac{2\pi}{n!}\left(\frac{c^{n+2}}{4(n+1)(n+2)}\right) \quad \text{by integration by parts} \\ &= 2^s\left(\frac{\pi}{2}\right)^{t+1} \frac{c^{n+2}}{(n+2)!} \end{aligned}$$

in accordance with (12.8). □

Proof of Theorem 12.4 Let X_c be the set of all $x \in \mathbb{L}^{st}$ such that

$$|x_1| + \cdots + |x_s| + 2\sqrt{(y_1^2+z_1^2)} + \cdots + 2\sqrt{(y_t^2+z_t^2)} < c$$

where c is a positive real number. Then X_c is convex and centrally symmetric. By Lemma 12.5,

$$v(X_c) = 2^s \left(\frac{\pi}{2}\right)^t \cdot \frac{1}{n!} c^n$$

By Minkowski's Theorem, X_c contains a point $\alpha \neq 0$ of $\sigma(\mathfrak{a})$ provided that

$$v(X_c) > 2^{s+2t} v(T)$$

where T is a fundamental domain for $\sigma(\mathfrak{a})$. By Theorem 9.4

$$v(T) = 2^{-t} \mathrm{N}(\mathfrak{a}) \sqrt{|d_K|}$$

so the condition on X_c becomes

$$2^s \left(\frac{\pi}{2}\right)^t \cdot \frac{1}{n!} c^n > 2^{s+2t} 2^{-t} \mathrm{N}(\mathfrak{a}) \sqrt{|d_K|}$$

which is

$$c^n > \left(\frac{4}{\pi}\right)^t n! \mathrm{N}(\mathfrak{a}) \sqrt{|d_K|}$$

For such an α

$$|\mathrm{N}(\alpha)| = |\sigma_1(\alpha) \ldots \sigma_s(\alpha) \sigma_{s+1}(\alpha)^2 \ldots \sigma_{s+t}(\alpha)^2| \leq \left(\frac{c}{n}\right)^n$$

by the inequality (12.7) between arithmetic and geometric means.

Using ε's as in Theorem 11.6, we may assume that α can be found for

$$c^n = \left(\frac{4}{\pi}\right)^t n! \mathrm{N}(\mathfrak{a}) \sqrt{|d_K|}$$

and then

$$|\mathrm{N}(\alpha)| \leq \left(\frac{4}{\pi}\right)^t \cdot \frac{n!}{n^n} \mathrm{N}(\mathfrak{a}) \sqrt{|d_K|}$$

□

The geometric considerations involved in the choice of X_c in this proof are illustrated in Figure 12.1 for the case $n = 2$, $s = 2$, $t = 0$. The three regions

A: $|xy| \leq 1$

B: $\dfrac{|x| + |y|}{2} \leq 1$

C: $|x| \leq 1$, $|y| \leq 1$

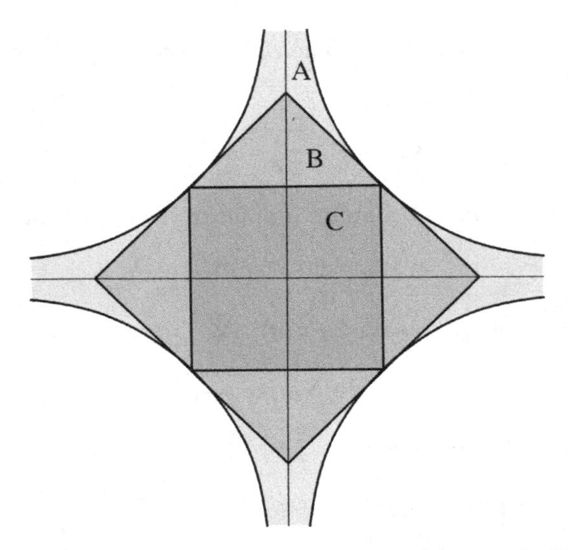

Figure 12.1. Geometry suggests the choice of X_c in the proof of Theorem 12.4, here illustrated for $n = s = 2, t = 0$. Region B is convex, lies within region A, and is larger than the more obvious region C. Therefore the use of B, in conjunction with Minkowski's Theorem, yields a better bound.

correspond respectively to the inequality (12.5), the region chosen in the proof of Theorem 12.4, and the inequality (12.6). The set A is not convex, although B, C are; $C \subseteq B \subseteq A$; and B is much larger than C (which is why it leads to a better estimate).

Corollary 12.6. *Every class of fractional ideals of K contains an ideal $\mathfrak{a}$ with*

$$N(\mathfrak{a}) \leq \left(\frac{4}{\pi}\right)^t \cdot \frac{n!}{n^n}\sqrt{|d_K|}$$

Proof: As for Corollary 11.7. □

This result suggests the introduction of special constants, which we now define:

Definition 12.7. The *Minkowski constants* are the numbers

$$M_{st} = \left(\frac{4}{\pi}\right)^t \frac{(s+2t)!}{(s+2t)^{s+2t}}$$

For future use, we give a short table of their values. The numbers in

Table 12.1. Minkowski constants.

n	s	t	M_{st}
2	0	1	0.63663
2	2	0	0.50000
3	1	1	0.28295
3	3	0	0.22223
4	0	2	0.15199
4	2	1	0.11937
4	4	0	0.09375
5	1	2	0.06226
5	3	1	0.04890
5	5	0	0.03400
6	0	3	0.03186
6	2	2	0.02502
6	4	1	0.01965
6	6	0	0.01544

the last column have all been rounded *upwards* in the fifth decimal place, to avoid underestimates.

We can now give a criterion for a number field to have class-number 1, for which the calculations required are often practicable.

Theorem 12.8. *Let $\mathfrak{O}$ be the ring of integers of a number field K of degree $n = s + 2t$ as in (9.1), and let d_K be the discriminant of K. Suppose that for every prime $p \in \mathbb{Z}$ with*

$$p \leq M_{st}\sqrt{|d_K|}$$

every prime ideal dividing $\langle p \rangle$ is principal. Then $\mathfrak{O}$ has class-number $h = 1$.

Proof: Every class of fractional ideals contains an ideal $\mathfrak{a}$ with $\mathrm{N}(\mathfrak{a}) \leq M_{st}\sqrt{|d_K|}$. Now

$$\mathrm{N}(\mathfrak{a}) = p_1 \dots p_k$$

where $p_1, \dots, p_k \in \mathbb{Z}$ and $p_i \leq M_{st}\sqrt{|d_K|}$. Further, $\mathfrak{a}|\mathrm{N}(\mathfrak{a})$, so $\mathfrak{a}$ is a product of prime ideals, each dividing some p_i. By hypothesis these prime ideals are principal, so $\mathfrak{a}$ is principal. Therefore every class of fractional ideals is equal to $[\mathfrak{O}]$, and $h = 1$. □

Specific numerical applications of this theorem, and related methods, are given in the next section.

12.3 Some Class-Number Calculations

Theorem 12.4 combines with Theorem 12.1 to provide a useful computational technique for fields of small degree and with small discriminant. The following examples show what is meant by 'small' in these circumstances.

1. $\mathbb{Q}(\sqrt{-19})$: The ring of integers is $\mathbb{Z}[\theta]$ where θ is a zero of

$$f(t) = t^2 - t + 5$$

and the discriminant is -19. Then $M_{st}\sqrt{|d_K|} \leq 0.63663\sqrt{19}$, so Theorem 12.4 applies if we know the factors of primes ≤ 2. Now we use Theorem 12.1: modulo 2, $f(t)$ is irreducible, so $\langle 2 \rangle$ is prime in $\mathfrak{O}$ (and hence every prime ideal dividing $\langle 2 \rangle$ is equal to $\langle 2 \rangle$ so is principal); modulo 3, $f(t)$ is also irreducible, so $\langle 3 \rangle$ is prime and the same argument applies.

2. $\mathbb{Q}(\sqrt{-43})$: This is similar, but now

$$f(t) = t^2 - t + 11$$

and $M_{st}\sqrt{|d_K|} \leq 0.63663\sqrt{43}$ which involves looking at primes ≤ 4. But $f(t)$ is irreducible modulo 2 and 3.

3. $\mathbb{Q}(\sqrt{-67})$: For this,

$$f(t) = t^2 - t + 17$$

and $M_{st}\sqrt{|d_K|} \leq 0.63663\sqrt{67}$ which involves looking at primes ≤ 5. But $f(t)$ is irreducible modulo 2, 3, and 5.

4. $\mathbb{Q}(\sqrt{-163})$: Now

$$f(t) = t^2 - t + 41$$

and $M_{st}\sqrt{|d_K|} \leq 0.63663\sqrt{163}$ which involves looking at primes ≤ 8. But $f(t)$ is irreducible modulo 2, 3, 5, and 7.

Combining these results with Theorem 5.28 (or using the above methods for the other values of d_K) we have:

Theorem 12.9. *The class-number of $\mathbb{Q}(\sqrt{d})$ is equal to 1 for $d = -1, -2, -3, -7, -11, -19, -43, -67, -163$.* □

As remarked in Section 5.3, these are the only values of $d < 0$ for which $\mathbb{Q}(\sqrt{d})$ has unique factorisation, or equivalently class-number 1.

Comparing with Theorem 5.29 we obtain the interesting:

Corollary 12.10. *There exist rings with unique factorisation that are not Euclidean; for example, the rings of integers of* $\mathbb{Q}(\sqrt{d})$ *for* $d = -19, -43, -67, -163$. □

The same method works easily for a few real quadratic fields (and for others with more effort). For later use, we prove:

Theorem 12.11. *The class-number of* $\mathbb{Q}(\sqrt{d})$ *is equal to* 1 *for* $d = 2, 3, 5$.

Proof: (a) $\mathbb{Q}(\sqrt{2})$: The discriminant is 8. We have $n = 2, t = 0$. By corollary 12.6 every ideal class contains an ideal with norm at most $\frac{2!}{2^2}\sqrt{8} = \frac{1}{2}\sqrt{8} \sim 1.414$, so every ideal class contains an ideal with norm at most 1. The only such ideal is $\mathfrak{O}$, so $h = 1$.

(b) $\mathbb{Q}(\sqrt{3})$: The discriminant is 12. Every ideal class contains an ideal with norm at most $\frac{1}{2}\sqrt{12} \sim 1.732$. So again every ideal class contains an ideal with norm at most 1 and $h = 1$.

(c) $\mathbb{Q}(\sqrt{5})$: The discriminant is 5. Every ideal class contains an ideal with norm at most $\frac{1}{2}\sqrt{5} \sim 1.118$. So again every ideal class contains an ideal with norm at most 1 and $h = 1$. □

We can also deal with a few cyclotomic fields. If $K = \mathbb{Q}(\zeta)$ where $\zeta^p = 1$, p prime, then the degree of K is $p - 1$, and the ring of integers is $\mathbb{Z}[\zeta]$. For $p = 3$, $K = \mathbb{Q}(\sqrt{-3})$ and we already know $h = 1$ in this case.

5. $\mathbb{Q}(\zeta)$ where $\zeta^5 = 1$: Here $n = 4$, $s = 0$, $t = 2$; and $d_K = 125$ by Theorem 3.6. Hence $M_{st}\sqrt{|d_K|} \leq 0.15199\sqrt{125}$ so we must look at primes ≤ 1. Since there are no such primes, Theorem 12.4 applies at once to give $h = 1$.

6. $\mathbb{Q}(\zeta)$ where $\zeta^7 = 1$: Here $n = 6$, $s = 0$, $t = 3$, and $d_K = -7^5$. We have to look at primes ≤ 3. The ring of integers is $\mathbb{Z}[\zeta]$ where ζ is a zero of

$$f(t) = t^6 + t^5 + t^4 + t^3 + t^2 + t + 1$$

Modulo 2, this factorises as

$$(t^3 + t^2 + 1)(t^3 + t + 1)$$

so $\langle 2 \rangle = \mathfrak{p}_1\mathfrak{p}_2$ where $\mathfrak{p}_1, \mathfrak{p}_2$ are distinct prime ideals, by Theorem 12.1. In fact

$$(\zeta^3 + \zeta^2 + 1)(\zeta^3 + \zeta + 1)\zeta^4 = 2$$

so

$$\langle 2 \rangle = \langle \zeta^3 + \zeta^2 + 1 \rangle \langle \zeta^3 + \zeta + 1 \rangle$$

and $\mathfrak{p}_1, \mathfrak{p}_2$ are principal.

Modulo 3, $f(t)$ is irreducible (by trying all possible divisors, or more enlightened methods), so $\langle 3\rangle$ is prime.

Hence, by Corollary 12.6, $h = 1$.

We now show that similar methods often allow us to compute h, even when it is not 1.

7. $\mathbb{Q}(\sqrt{10})$: The discriminant $d = 40$, $n = 2$, $s = 2$, $t = 0$. Every class of ideals contains one with norm

$$\leq M_{2,0}\sqrt{|d_K|} \leq 0.5\sqrt{40} = \sqrt{10} \sim 3.162$$

so we must factorise the primes ≤ 3. Now $\mathfrak{O} = \mathbb{Z}[\theta]$ where θ is a zero of

$$f(t) = t^2 - 10$$

$f(t) \equiv (t+1)(t-1) \pmod 3$, so $\langle 3\rangle = \mathfrak{g}_1\mathfrak{g}_2$ where $\mathfrak{g}_1 = \langle 3, 1+\sqrt{10}\rangle$, $\mathfrak{g}_2 = \langle 3, 1-\sqrt{10}\rangle$. Modulo 2, $f(t) = t \cdot t$, so $\langle 2\rangle = \mathfrak{p}^2$ for a prime ideal $\mathfrak{p}$. If $\mathfrak{p}$ is principal, say $\mathfrak{p} = \langle a + b\sqrt{10}\rangle$, then the equation

$$\mathrm{N}(\mathfrak{p})^2 = \mathrm{N}(\langle 2\rangle) = 4$$

implies that $\mathrm{N}(\mathfrak{p}) = 2$. Hence

$$a^2 - 10b^2 = \pm\, 2$$

The latter, considered modulo 10, is impossible; hence $\mathfrak{p}$ is not principal.

We have $\mathfrak{p}\mathfrak{g}_1 = \langle -2+\sqrt{10}\rangle$ and $[\mathfrak{g}_1] = [\mathfrak{p}]^{-1}$. Therefore every class of fractional ideals either contains a principal ideal or $\mathfrak{p}$, hence equals $[\mathfrak{O}]$ or $[\mathfrak{p}]$. Since $\mathfrak{p}$ is not principal, these two classes are distinct, so $h = 2$. The class-group is cyclic of order 2, and as verification

$$[\mathfrak{p}]^2 = [\mathfrak{p}^2] = [\langle 2\rangle] = [\mathfrak{O}]$$

As we said in Section 5.3, all the imaginary quadratic fields $\mathbb{Q}(\sqrt{d})$ with unique factorisation are now known, verifying a conjecture of Gauss. But Gauss also stated a more general conjecture, the *Class-Number Problem.* This states that for any given class-number h, the set of $d < 0$ for which $\mathbb{Q}(\sqrt{d}) = h$ is finite. It was proved in 1934 by Hans Heilbronn. A stronger result was proved in 1983 by Dorian Goldfeld, Benedict Gross, and Don Zagier, and it is described in a masterly survey by Goldfeld [51].

12.4 Table of Class-Numbers

To give an idea of how irregularly the class-number of $\mathbb{Q}(\sqrt{d})$ depends upon d, Table 12.2 shows, for squarefree d with $0 < d < 100$, the class-numbers h of $\mathbb{Q}(\sqrt{d})$ and h' of $\mathbb{Q}(\sqrt{-d})$.

Table 12.2. Class-numbers h of $\mathbb{Q}(\sqrt{d})$ and h' of $\mathbb{Q}(\sqrt{-d})$.

d	h	h'	d	h	h'	d	h	h'
1	—	1	34	2	4	69	1	8
2	1	1	35	2	2	70	2	4
3	1	1	37	1	2	71	1	7
5	1	2	38	1	6	73	1	4
6	1	2	39	2	4	74	2	10
7	1	1	41	1	8	77	1	8
10	2	2	42	2	4	78	2	4
11	1	1	43	1	1	79	3	5
13	1	2	46	1	4	82	4	4
14	1	4	47	1	5	83	1	3
15	2	2	51	2	2	85	2	4
17	1	4	53	1	6	86	1	10
19	1	1	55	2	4	87	2	6
21	1	4	57	1	4	89	1	12
22	1	2	58	2	2	91	2	2
23	1	3	59	1	3	93	1	4
26	2	6	61	1	6	94	1	8
29	1	6	62	1	8	95	2	8
30	2	4	65	2	8	97	1	4
31	1	3	66	2	8			
33	1	4	67	1	1			

12.5 Analytic Methods: Class-Number Formula

Efficient computational methods to compute class-numbers are known. For example, Theorem 6.32 lists all cyclotomic fields $\mathbb{Q}(\zeta_n)$ with class-number 1. Kummer's case $n = 23$ has class-number 3. The class-numbers of cyclotomic fields are highly irregular: for example $\mathbb{Q}(\zeta_{70})$ has class-number 1, $\mathbb{Q}(\zeta_{71})$ has class-number 3882809, and $\mathbb{Q}(\zeta_{72})$ has class-number 3. For larger n, compare $\mathbb{Q}(\zeta_{150})$, with class-number 11, to $\mathbb{Q}(\zeta_{151})$, with class-number 2333546653547742584439257.

It is now known that the cyclotomic field $\mathbb{Q}(\zeta_m)$ has class-number 1 if and only if m is one of the following: 1—22, 24, 25, 26, 27, 28, 30, 32, 33, 34, 35, 36, 38, 40, 42, 44, 45, 48, 50, 54, 60, 66, 70, 84, 90. See Washington [141].

One of the triumphs of classical algebraic number theory was a remarkable link with complex analysis. This is beyond our present scope, but we briefly give the general flavour. See Borevič and Šafarevič [10] p. 342 onwards for details.

Suppose that K is a number field, of degree $[K : \mathbb{Q}] = s + 2t$ where s is the number of real embeddings and $2t$ is the number of complex embeddings. Let Reg_K be an invariant called the *regulator* of K, w_K = the number of roots of unity in K, d_K = the discriminant of $K/\mathbb{Q}$, h_K = the class-number of K. Define the *Dedekind zeta function* of K to be

$$\zeta_K(z) = \sum_{\mathfrak{a}} \frac{1}{\mathrm{N}(\mathfrak{a})^z}$$

where $\mathfrak{a}$ ranges over all ideals of $\mathfrak{O}$. This series converges absolutely if $\mathrm{re}(z) > 1$. The *Class-Number Formula* states that

$$\lim_{z\to 1}(z-1)\zeta_K(z) = \frac{2^r(2\pi)^t \mathrm{Reg}_K h_K}{w_K\sqrt{d_K}} \tag{12.9}$$

The left-hand side is the *residue* of $\zeta_K(z)$ at $z = 1$, in the sense of complex analysis.

If all terms of this formula except the class-number h_K are known, we can solve for h_K. (The definition of the regulator is stated in Exercise 10.3. It is hard to compute.)

12.6 Exercises

12.1 Let $K = \mathbb{Q}(\sqrt{3})$. Use Theorem 12.1 to factorise the following principal ideals in the ring $\mathfrak{O}$ of integers of K:

$$\langle 2\rangle \qquad \langle 3\rangle \qquad \langle 5\rangle \qquad \langle 10\rangle \qquad \langle 30\rangle$$

12.2 Factorise the following principal ideals in the ring of integers of $\mathbb{Q}(\sqrt{5})$:

$$\langle 2\rangle \qquad \langle 3\rangle \qquad \langle 5\rangle \qquad \langle 12\rangle \qquad \langle 25\rangle$$

12.3 Factorise the following ideals in $\mathbb{Z}[\zeta]$ where $\zeta = e^{2\pi i/5}$:

$$\langle 2\rangle \qquad \langle 5\rangle \qquad \langle 20\rangle \qquad \langle 50\rangle$$

12.4 If K is a number field of degree n, prove that

$$|d_K| \geq \left(\frac{\pi}{4}\right)^n \left(\frac{n^n}{n!}\right)^2$$

where d_K is the discriminant.

12.5 Using the methods of this chapter, compute the class-numbers of fields $\mathbb{Q}(\sqrt{d})$ for $-10 \leq d \leq 10$.

12.6 Show that $t^5 - t + 1$ is irreducible over $\mathbb{Q}$, and let θ be a zero. Calculate the integers s, t for the field $\mathbb{Q}(\theta)$.

Calculate $\Delta[1, \theta, \ldots, \theta^4]$. Deduce that the ring of integers $\mathfrak{O}$ of $\mathbb{Q}(\theta)$ is $\mathbb{Z}(\theta)$.

Use Theorems 12.1 and 12.4 to show that $\mathbb{Z}(\theta)$ is a principal ideal domain.

12.7 Let d be squarefree with $-10 \leq d \leq 10$. Show that Dedekind's Criterion implies that if p is a rational prime then, in the ring of integers of $\mathbb{Q}(\sqrt{d})$, the ideal $\langle p\rangle$ either:

(a) Remains a prime ideal.

(b) Factorises as $\mathfrak{p}\mathfrak{q}$ for distinct prime ideals $\mathfrak{p}\mathfrak{q}$.

(c) Is the square $\mathfrak{p}^2$ of a prime ideal $\mathfrak{p}$.

Decide which case applies for $p = 2, 3$, and 5.

13

Kummer's Special Case of Fermat's Last Theorem

We now have sufficient machinery at our disposal to tackle Fermat's Last Theorem when the exponent n in the equation $x^n + y^n = z^n$ is a so-called 'regular prime', and in the 'first case' when n does not divide any of x, y, or z. The 'second case' occurs when n does divide one of x, y, or z. The second case turns out to be much harder than the first.

We begin with a short historical survey to set this version of the problem in perspective. Following this, we show how elementary methods dispose of the exponents $n = 3$ and $n = 4$, in both the first and second case. Next we discuss Sophie Germain's theorems about the first case for many exponents. We observe that in general the problem reduces to odd prime values of n.

We then give Kummer's proof of the first case of Fermat's Last Theorem for regular prime exponents. This proof, which is the main point of this chapter, uses ideals in a cyclotomic field. We do not deal with the second case, where one of x, y, or z is divisible by n; neither do we deal with irregular prime exponents. These cases are discussed, without proofs, in Chapters 16 and 17. In a final discursive section we consider the regularity property and some related matters.

13.1 Some History

The origins of Fermat's Last Theorem have been explained in the Introduction. Useful references for background reading are Cox [25], Edwards [38], Ribenboim [105, 106], Singh [123], and Stewart [127]. Fermat himself is considered to have disposed of the cases $n = 3, 4$, because he issued these specific cases as mathematical challenges to others. In fact, he produced only one written proof in the whole of his mathematical career. This states that the area of a right-angled triangle with rational sides cannot be a perfect square. Algebraically, this statement translates to the assertion that there are no (nonzero) integer solutions x, y, z of the equation $x^2 + y^2 = z^2$ where $xy/2$ is a square. Fermat deduced this from a proof by 'infinite descent' for the case $n = 4$ of what we now call Fermat's Last Theorem; see Exercise 13.3. The method of infinite descent proceeds by proving that given any solution, there is a smaller one. Since any descending sequence of natural numbers must stop, this gives a contradiction. Today this is rephrased as a proof by induction, or the assumption (for an eventual contradiction) of a minimal counterexample.

Euler produced his own proof for the case $n = 3$ in his *Algebra* of 1770. However, his proof had a subtle gap. He needed to find cubes of the form $p^2 + 3q^2$, and ingeniously showed that, for any integers a and b, if we define

$$p = a^3 - 9ab^2 \qquad q = 3(a^2b - b^3)$$

then

$$p^2 + 3q^2 = (a^2 + 3b^2)^3$$

However, he then tried to show the reverse process also works, namely, if $p^2 + 3q^2$ is a perfect cube, then there exist integers a, b satisfying the above relationships. Here he worked with algebraic numbers of the form $x + y\sqrt{-3}$, with x, y integers, believing that these numbers possess the same properties as ordinary integers—including uniqueness of factorisation. (As it happens, factorisation *is* unique in this case, but Euler did not realise that this needed proving. In Section 13.3 we use unique factorisation to fill the gap in Euler's proof.) This omission went unnoticed at the time. However, other results that Euler published gave an alternative proof for $n = 3$, without logical gaps, which justifies giving him full credit for this case.

Germain was one of the very few women doing research in mathematics around 1800. As a woman, she was not permitted to attend the École Polytechnique when it opened in Paris in 1794. Instead she assumed the identity of a student, 'Monsieur Antoine-Auguste Le Blanc', who had left the course without giving formal notice. The student's written solutions of weekly problems that Lagrange had set were elegant and insightful. Impressed by Monsieur Le Blanc's abilities, Lagrange insisted on a meeting, which revealed her subterfuge. He gave her positive encouragement—which most prominent male academics would not have done at that time—and she developed a serious interest in Fermat's Last Theorem.

Early in her work, she found it convenient to divide the problem (for prime n) into two cases:

First Case: None of x, y, z is divisible by n.

Second Case: Only one of x, y, z is divisible by n.

If two of x, y, z are divisible by n then all three are, and if all three are, then the common factor n^n can be divided out, so only these two cases are needed for a proof. Legendre credits her with a proof of the 'first case' whenever there exists an 'auxiliary prime' satisfying two technical conditions. He states that she verified these conditions for all $n \leq 97$, so she proved the first case of Fermat's Last Theorem for all n in that range. See Section 13.4.

Attention then turned to the second case. Dirichlet presented a partial proof for this case when $n = 5$ to the Paris Academy in July 1825. Legendre filled in the remaining details in September 1825, completing the full proof for $n = 5$. Dirichlet continued to work on the case $n = 7$, only to realise that the closely related case $n = 14$ was more amenable to his methods. He published the proof for $n = 14$ in 1832. The case $n = 7$ was finally proved in 1839 by Lamé. It required far subtler computations than those of earlier cases and gave the impression that further progress would be unlikely unless a completely different line of attack was found. But the next major step forward was followed by an immediate retreat.

On 1 March 1847, Lamé addressed the Paris Academy and outlined a complete proof of Fermat's Last Theorem, in which he introduced the complex nth roots of unity. Using these, he factorised the equation $x^n + y^n = z^n$ into linear terms:

$$x^n + y^n = (x + y)(x + \zeta y) \ldots (x + \zeta^{n-1} y)$$

where $\zeta = e^{2\pi i/n}$ and n is odd. Lamé acknowledged that he was indebted to Liouville for this idea.

Then Liouville took the stage. He acknowledged his contribution, but pointed out that Lamé's argument depends on unique factorisation, adding that he suspected this property might fail. Immediately the focus turned to unique factorisation. A fortnight later Pierre Wantzel announced a proof of unique factorisation for some cases, providing arguments for $n = 3, 4$, which are quadratic fields. He also stated that his method of proof fails for $n = 23$ (Cauchy [16] p. 308). (We now know that unique factorisation fails in this case.) On 24 May Liouville informed the Academy that Kummer had already shown the failure of unique factorisation three years before, but had developed a technical alternative that worked by introducing what he called 'ideal numbers'.

In 1850 Kummer produced his sensational proof of Fermat's Last Theorem for what he termed 'regular' primes, which include all primes less than 100 except for 37, 59, 67. Kummer asserted that there is an infinite number of regular primes, but this has never been proved (although Johan Jensen proved that there is an infinite number of *irregular* primes in 1915). The same year, Kummer attended to the three cases 37, 59, 67, but he made errors that went unnoticed until Harry Vandiver found them in 1920. Dimitri Mirimanoff gave a proof for $n = 37$ in 1893, and he extended this to $n \leq 257$ in 1905.

Vandiver laid down methods that made a computational approach possible, which led to proofs for $n \leq 25,000$ by Selfridge and Pollock [117], then $n \leq 125,000$ by Wagstaff [140]. By 1993 the record was $n \leq 4,000,000$, by Buhler *et al.* [14]. The first case of Fermat's last Theorem is easier than the second case. The first case was proved true for prime exponents up to 714,591,416,091,839 in 1988 by Granville and Monagan [55]. This was improved in 1990 to 7.57×10^{17} by Coppersmith [24].

Finally, Andrew Wiles (with a vital contribution from Richard Taylor) proved Fermat's Last Theorem without any special restrictions.

Kummer's proof was a watershed moment, even though it does not deal with all exponents n. Before 1850 the focus was on individual values of n and a variety of special methods. After 1850, in the late 19th century and throughout the 20th century, it shifted to seeking more general proofs for a wide variety of values of n. Wiles introduced radically new techniques in the 21st century. Before moving on to his work, we take a detailed look at Kummer's proof for the first case, and the methods behind it. We consider only the first case, but Kummer also dealt with the second case by similar but more complicated methods. The proof is too long to give here and requires new concepts; see Washington [141] chapter 9.

13.2 Fermat's Last Theorem for Fourth Powers

In this section we give what essentially is Fermat's original proof for fourth powers. We first consider what can be said about the *Fermat equation*

$$x^n + y^n = z^n \tag{13.1}$$

from an elementary point of view. If a solution to (13.1) exists, there must be a solution in which x, y, z are coprime in pairs. For if a prime q divides

x and y, then $x = qx'$, $y = qy'$,

$$q^n(x'^n + y'^n) = z^n$$

so that q also divides z, say $z = qz'$, and then $x'^n + y'^n = z'^n$. Similarly if q divides x and z, or y and z. In this way we can remove all common factors from any pair of x, y, z.

Next, if (13.1) is impossible for an exponent n then it is impossible for all multiples of n. For if $x^{mn} + y^{mn} = z^{mn}$ then $(x^m)^n + (y^m)^n = (z^m)^n$. Now any integer ≥ 3 is divisible either by 4 or by an odd prime. Hence to prove (or disprove) the conjecture *it is sufficient to consider the cases* $n = 4$ *and* n *an odd prime.*

We start with Fermat's proof for $n = 4$. It is based on the general solution of the Pythagorean equation $x^2 + y^2 = z^2$, known at least from the time of Euclid, given by:

Lemma 13.1. *The solutions of* $x^2 + y^2 = z^2$ *with pairwise coprime integers* x, y, z *are given parametrically by*

$$\pm x = 2rs \qquad \pm y = r^2 - s^2 \qquad \pm z = r^2 + s^2$$

(or with x, y *interchanged) where* r, s *are coprime and exactly one of them is odd.*

Proof: It is sufficient to consider x, y, z positive. They cannot all be odd, for this gives the contradiction 'odd + odd = odd'. Since x, y, z are pairwise coprime, precisely one of them is even. It cannot be z, for then $z = 2k$, $x = 2a + 1$, $y = 2b + 1$ where k, a, b are rational integers, and

$$(2a + 1)^2 + (2b + 1)^2 = 4k^2$$

This cannot occur since the left-hand side is clearly not divisible by 4 but the right-hand side is. So one of x, y is even. We can suppose that this is x. Then

$$x^2 = z^2 - y^2 = (z + y)(z - y)$$

Because x, $z + y$, $z - y$ are all even and positive, we can write $x = 2u$, $z + y = 2v$, $z - y = 2w$, whence

$$(2u)^2 = 2v \cdot 2w$$

or

$$u^2 = vw \qquad (13.2)$$

Now v, w are coprime, for a common factor of v, w would divide their sum $v + w = z$ and their difference $v - w = y$, which have no proper common factors. Factorising u, v, w into prime factors, (13.2) implies that v and w are both squares, say $v = r^2$, $w = s^2$. Moreover r and s are coprime, because v and w are.

Thus

$$\begin{aligned} z &= v + w = r^2 + s^2 \\ y &= v - w = r^2 - s^2 \end{aligned}$$

Because y, z are both odd, precisely one of r, s is odd. Finally

$$x^2 = z^2 - y^2 = (r^2 + s^2)^2 - (r^2 - s^2)^2 = 4r^2 s^2$$

so

$$x = 2rs$$ □

Now we can prove a theorem even stronger than the impossibility of Equation (13.1) for $n = 4$, namely:

Theorem 13.2. *The equation $x^4 + y^4 = z^2$ has no integer solutions with $x, y, z \neq 0$.*

This statement *is* stronger, since if $x^4 + y^4 = z^4$ then x, y, z^2 satisfy the above equation.

Proof: Suppose a solution of

$$x^4 + y^4 = z^2 \qquad (13.3)$$

exists. We may assume x, y, z are positive. Among such solutions there exists one for which z is smallest: assume we have this one in (13.3). Then x, y, z are coprime (or else we can cancel a common factor and make z smaller), so by Lemma 13.1

$$x^2 = r^2 - s^2 \qquad y^2 = 2rs \qquad z = r^2 + s^2 \qquad (13.4)$$

where x, z are odd and y is even. The first of these implies

$$x^2 + s^2 = r^2$$

with x, s coprime. Hence by Equation (13.1) again, since x is odd

$$x = a^2 - b^2 \qquad s = 2ab \qquad r = a^2 + b^2$$

Now we substitute from (13.4) to get

$$y^2 = 2rs = 2 \cdot 2ab(a^2 + b^2)$$

so y is even, say $y = 2k$, and

$$k^2 = ab(a^2 + b^2)$$

Since a, b and $a^2 + b^2$ are pairwise coprime

$$a = c^2 \qquad b = d^2 \qquad a^2 + b^2 = e^2$$

so that

$$c^4 + d^4 = e^2$$

This is an equation of type (13.3), but $e \leq a^2 + b^2 = r < z$, contradicting minimality of z. □

The same proof leads to Fermat's notion of 'infinite descent' if we do not assume the initial solution is minimal, and this is basically what Fermat did. The method then proves that given any (hypothetical) solution there is a smaller one. This leads to an infinitely descending sequence of ever smaller solutions, which is absurd. Therefore the hypothetical solution does not exist.

13.3 Fermat's Last Theorem for Cubes

In a similar spirit, we can prove Fermat's Last Theorem for cubes. The proof is essentially that of Euler, but we use algebraic number theory to fill the gap mentioned in Section 13.1.

The main idea is to use prime factorisation in the ring of integers $\mathbb{Z}[\omega]$ of $\mathbb{Q}(\sqrt{-3})$, where $\omega = \frac{1}{2}(-1 + \sqrt{-3})$. We mainly work with ideals rather than elements to avoid complications with units, but we revert to elements for some purposes. Theorem 5.28 proves that $\mathbb{Z}[\omega]$ is a Euclidean domain, so every ideal is principal and prime factorisation (of elements or ideals) is unique. The subring $\mathbb{Z}[\sqrt{-3}]$ also plays an important role, but it does not contain ω. However, $2\omega = -1 + \sqrt{-3} \in \mathbb{Z}[\sqrt{-3}]$, and we exploit this later in the proof of the crucial Lemma 13.6.

If $\alpha = a + b\omega$ we denote the conjugate by $\bar{\alpha} = a - b\omega$. Then

$$\overline{a + b\sqrt{-3}} = a - b\sqrt{-3}$$

The conjugate of an ideal $\mathfrak{a}$ is defined to be

$$\bar{\mathfrak{a}} = \{\bar{\alpha} : \alpha \in \mathfrak{a}\}$$

We begin by analysing how positive primes in $\mathbb{Z}$, that is, primes in $\mathbb{N}$, behave in $\mathbb{Z}[\omega]$. The usual approach exploits quadratic reciprocity, but we use more elementary ideas. There is a technical issue: our main interest is in the norm $c^2 + 3d^2$ of $c + d\sqrt{-3}$, but the norm of $a + b\omega$ is $a^2 - ab + b^2$. We cannot work directly in $\mathbb{Z}[\sqrt{-3}]$ because this is not a principal ideal domain:

$$4 = 2.2 = (1 + \sqrt{-3})(1 - \sqrt{-3}) \tag{13.5}$$

and the factors are irreducible in $\mathbb{Z}[\sqrt{-3}]$ since 2 is not of the form $c^2 + 3d^2$ for $c, d \in \mathbb{Z}$. So we have to work in $\mathbb{Z}[\omega]$. Fortunately, the units of $\mathbb{Z}[\omega]$ come to the rescue.

Lemma 13.3. *Every ideal of $\mathbb{Z}[\omega]$ has the form $\langle\alpha\rangle$ where $\alpha \in \mathbb{Z}[\sqrt{-3}]$.*

Proof: Let $\alpha = a_0 + b_0\omega$. There are rational integers a_1, a_2, b_1, b_2 such that $\omega\alpha = a_1 + b_1\omega$ and $\omega^2\alpha = a_2 + b_2\omega$. Now

$$0 = \alpha + \omega\alpha + \omega^2\alpha = (a_0 + a_1 + a_2) + (b_0 + b_1 + b_2)\omega$$

Therefore $b_0 + b_1 + b_2 = 0$, which implies that at least one of b_0, b_1, b_2 is even. If this is b_i then $b_i = 2c$ for $c \in \mathbb{Z}$, and $a_i + b_i\omega = a_i + c(2\omega) \in \mathbb{Z}[\sqrt{-3}]$. But ω is a unit, so $\langle\alpha\rangle = \langle\alpha\omega^i\rangle = \langle a_i + b_i\omega\rangle$.

□

We mention (but do not use) an intriguing corollary:

Corollary 13.4. *If a rational prime $p = a^2 - ab + b^2$ for $a, b \in \mathbb{Z}$, then there exist $c, d \in \mathbb{Z}$ such that $p = c^2 + 3d^2$.* □

Next, we classify rational primes into three classes, depending on how they factorise in the ring $\mathbb{Z}[\omega]$:

Theorem 13.5. *Prime numbers $p \in \mathbb{N}$ fall into precisely one of three classes, determined by their prime ideal factorisations in $\mathbb{Z}[\omega]$:*

(a) *$p = 3$: here $\langle 3\rangle = \mathfrak{p}^2$, where $\mathfrak{p} = \langle\sqrt{-3}\rangle$.*

(b) *$p \neq a^2 + 3b^2$ for $a, b, \in \mathbb{Z}$, where $\langle p\rangle$ remains prime in $\mathbb{Z}[\omega]$.*

(c) *$p \neq 3$ and $p = a^2 + 3b^2$ for $a, b, \in \mathbb{Z}$, where $\langle p\rangle = \mathfrak{p}\bar{\mathfrak{p}}$ for $\mathfrak{p} = \langle a + b\sqrt{-3}\rangle$, and $\bar{\mathfrak{p}} \neq \mathfrak{p}$.*

(In case (b) we know that $p \neq 3$: choose $a = 0, b = 1$.)

Proof: (a) Observe that $\sqrt{-3}^{\,2} = -3$, so $\langle 3\rangle = \mathfrak{p}^2$.

(b) $\mathrm{N}(p) = p^2$. Therefore either $\langle p\rangle$ remains prime, or $\langle p\rangle = \mathfrak{p}\mathfrak{q}$ where $\mathrm{N}(\mathfrak{p}) = \mathrm{N}(\mathfrak{q}) = p$, so $\mathfrak{p}, \mathfrak{q}$ are prime. Then

$$\bar{\mathfrak{p}}\bar{\mathfrak{q}} = \langle\bar{p}\rangle = \langle p\rangle = \mathfrak{p}\mathfrak{q}$$

By unique factorisation, either $\bar{\mathfrak{p}} = \mathfrak{p}$ and $\bar{\mathfrak{q}} = \mathfrak{q}$, or $\mathfrak{q} = \bar{\mathfrak{p}}$. In the first case $\mathfrak{p} = \langle a \rangle$ for $a \in \mathbb{Z}$, so $\mathrm{N}(\mathfrak{p}) = a^2$, a contradiction. Therefore $\mathfrak{q} = \bar{\mathfrak{p}}$ and $\langle p \rangle = \mathfrak{p}\bar{\mathfrak{p}}$. By Lemma 13.3, $\mathfrak{p} = \langle a + b\sqrt{-3} \rangle$ for $a, b \in \mathbb{Z}$. Then $p = a^2 + 3b^2$, a contradiction. Therefore $\mathfrak{p}$ is prime.

(c) If $p = a^2 + 3b^2$, let $\mathfrak{p} = \langle a + b\sqrt{-3} \rangle$. Then $\mathrm{N}(\mathfrak{p}) = p$ so $\mathfrak{p}$ is a prime ideal, and $\langle p \rangle = \mathfrak{p}\bar{\mathfrak{p}}$.

It remains to show that $\bar{\mathfrak{p}} \neq \mathfrak{p}$ in case (c). Suppose that $\mathfrak{p} = \langle a + b\omega \rangle$ for $a, b, \in \mathbb{Z}$. Then $\bar{\mathfrak{p}} = \mathfrak{p}$ if and only if $a + b\omega^2 \in \mathfrak{p}$. Since $\mathfrak{p}$ is an ideal it contains $(a + b\omega)\omega = a\omega + b\omega^2$, hence also $a - a\omega = a(1 - \omega)$. This has norm $3a^2$, so $a = \pm 1$ and $\mathfrak{p}^2 = \langle 3 \rangle$. But $p \neq 3$ in case (c). □

Here, the rational prime $p = 3$ in case (a) plays a dual role: it factorises, but the two factors are equal *and* conjugate. In case (b) $\langle p \rangle$ does not factorise, and in (c) it is the product of two *distinct* conjugate ideals. An analogous threefold classification applies to any number field, and we say more about it in Chapter 21.

We can now fill the gap in Euler's proof:

Lemma 13.6. *If $s \in \mathbb{Z}$ is odd and $s^3 = a^2 + 3b^2$ for $a, b \in \mathbb{Z}$ then $s = c^2 + 3d^2$ for $c, d \in \mathbb{Z}$.*

Proof: This proof can be streamlined to some extent, but we give details for clarity.

We may assume $s > 0$, since $(-s)^3 = -s^3$, and we may also assume that a, b are coprime. Work in the number field $\mathbb{Q}(\sqrt{-3})$, whose integers are $Z[\omega]$. Since $\mathbb{Q}(\sqrt{-3})$ is a unique factorisation domain, every ideal of $Z[\omega]$ is principal.

Let $\alpha = a + b\omega \in \mathbb{Z}[\omega]$. Recall that the conjugate of α is $\bar{\alpha} = a + b\omega^2$. Then $\mathrm{N}(\alpha) = \alpha\bar{\alpha}$.

We factorise s in $\mathbb{Z}$ according to this classification of Lemma 13.5:

$$s = (3^i)(p_1^{j_1} \cdots p_r^{j_r})(q_1^{k_1} \cdots q_s^{k_s}) \tag{13.6}$$

where 3 is in class (a), the p_i are in class (b), and the q_j are in class (c).

Passing to ideals of $\mathbb{Z}[\omega]$, define prime ideals $\mathfrak{r}, \mathfrak{p}_i, \mathfrak{q}_j$ by

$$\langle 3 \rangle = \mathfrak{r}^2 \qquad \langle p_i \rangle = \mathfrak{p}_i \qquad \langle q_i \rangle = \mathfrak{q}_i\bar{\mathfrak{q}}_i$$

(so in particular $\mathfrak{r} = \langle \sqrt{-3} \rangle$). The ideals $\mathfrak{r}, \mathfrak{p}_i, \mathfrak{q}_j, \bar{\mathfrak{q}}_k$ are all distinct: either they have distinct norms or case (c) of Theorem 13.5 applies. Then (13.6) becomes:

$$\langle s \rangle = (\mathfrak{r}^{2i})(\mathfrak{p}_1^{j_1} \cdots \mathfrak{p}_r^{j_r})((\mathfrak{q}_1\bar{\mathfrak{q}}_1)^{k_1} \cdots (\mathfrak{q}_s\bar{\mathfrak{q}}_s)^{k_s}) \tag{13.7}$$

We claim that the exponents $j_1, \dots, j_r$ are even. The required result is then an easy consequence, as we show next. After that, we prove the claim.

Suppose the claim is true, so $j_i = 2l_i$ where $l_i \in \mathbb{Z}$. Then $\pm s = A^2B$ where

$$A = 3^i p_1^{l_1} \cdots p_r^{l_r} \qquad B = \mathrm{N}(\mathfrak{q}_1^{k_1} \cdots \mathfrak{q}_s^{k_s})$$

are rational integers. Lemma (c) implies that $\mathrm{N}(\mathfrak{q}_i) = a_i^2 + 3b_i^2$ for $a, b, \in \mathbb{Z}$, so $B = (a_1^2 + 3b_1^2) \cdots (a_s^2 + 3b_s^2)$. Brahmagupta's formula (multipicativity of the norm) implies that $B = C^2 + 3D^2$ for some $C, D \in \mathbb{Z}$. Then $\pm s = A^2(C^2 + 3D^2) = (AC)^2 + 3(AD)^2$. Clearly the sign is $+$, so s has the form $c^2 + 3d^2$ for $c, d \in \mathbb{Z}$, and we are done.

It remains to prove the claim.

We are given that $s^3 = a^2 + 3b^2$. Let $\alpha = a + b\sqrt{-3}$. Then $s^3 = \alpha\bar{\alpha}$. Factorise $\langle\alpha\rangle$ into ideals, according to the classification:

$$\langle\alpha\rangle = (\mathfrak{r}^{2l})(\mathfrak{p}_1^{m_1} \cdots \mathfrak{p}_r^{m_g})(\mathfrak{q}_1^{n_1} \cdots \mathfrak{q}_r^{n_h})$$

Then

$$\langle\bar{\alpha}\rangle = (\mathfrak{r}^{2l})(\mathfrak{p}_1^{m_1} \cdots \mathfrak{p}_r^{m_g})(\bar{\mathfrak{q}}_1^{n_1} \cdots \bar{\mathfrak{q}}_r^{n_h})$$

(The only prime ideals that can occur are factors of $\langle s\rangle$.) Now

$$\langle s^3\rangle = \langle\alpha\rangle\langle\bar{\alpha}\rangle = (\mathfrak{r}^{4l})(\mathfrak{p}_1^{2m_1} \cdots \mathfrak{p}_r^{2m_g})((\mathfrak{q}_1\bar{\mathfrak{q}}_1)^{n_1} \cdots (\mathfrak{q}_1\bar{\mathfrak{q}}_r)^{n_h}))$$

But (13.7) gives an alternative expression:

$$\langle s\rangle^3 = (\mathfrak{r}^{6i})(\mathfrak{p}_1^{3j_1} \cdots \mathfrak{p}_r^{3j_r})((\mathfrak{q}_1\bar{\mathfrak{q}}_1)^{3k_1} \cdots (\mathfrak{q}_s\bar{\mathfrak{q}}_s)^{3k_s}) \qquad (13.8)$$

Comparing these factorisations, $3j_i = 2m_i$, so j_i is even for all i. This completes the proof of the claim, and with it, the lemma. □

Having disposed of this lemma, we state and prove Fermat's Last Theorem for cubes in a more symmetric form, obtained by changing z to $-z$:

Theorem 13.7. *The equation $x^3 + y^3 + z^3 = 0$ has no solution in nonzero integers.*

Proof: Euler used infinite descent, which we rephrase as a proof by contradiction for a supposed minimal counterexample. We therefore start by assuming that (x, y, z) is a counterexample with $|x|, |y|, |z|$ minimal. Except for this tactical change, the proof is basically his.

We may assume that x, y, z are pairwise coprime. Not all of them are positive. Working modulo 2, one of them is even and the other two odd.

By symmetry we may assume z is even. Moreover, $x \neq y$, since if $x = y$ then $2x^3 = -z^3$ so x is even. There exist integers u, v such that

$$x + y = 2u \qquad x - y = 2v$$

where u, v are coprime and have opposite parity. Then

$$-z^3 = x^3 + y^3 = 2u(u^2 + 3v^2) \tag{13.9}$$

Now $u^2 + 3v^2$ is odd, so u is even and v is odd. We distinguish two cases: either $\gcd(2u, u^2 + 3v^2) = 1$ or $\gcd(2u, u^2 + 3v^2) = 3$. (These are the only possible common factors.)

Case 1: $\gcd(2u, u^2 + 3v^2) = 1$.

The two factors of $-z^3$ in (13.9) are coprime, so each is the cube of a smaller number (in absolute value) a, b. Moreover, $3 \not| u$. So there exist $a, b \in \mathbb{Z}$ such that

$$2u = a^3 \qquad u^3 + 3v^2 = b^3$$

and b is odd. By Lemma 13.6, there exist $c, d \in \mathbb{Z}$ such that $b = c^2 + 3d^2$. Now

$$u = c(c^2 - 9d^2) \qquad v = 3d(c^2 - d^2)$$

It is easy to show that c, d are coprime, c is even, and d is odd. Then

$$a^3 = 2u = 2c(d - d)(c + 3d)$$

with all three factors coprime. Thus each factor is the cube of an integer:

$$-2c = k^3 \qquad c - 3d = l^3 \qquad c + 3d = m^3$$

giving a smaller solution $k^3 + l^3 + m^3 = 0$, which contradicts minimality.

Case 2: $\gcd(2u, u^2 + 3v^2) = 3$.

Now $3|u$, so $u = 3w$ for an integer w. Also $4|u$, so w is even. The numbers v, w are coprime. Neither 3 nor 4 divides v. The equation for $-z^3$ becomes

$$-z^3 = 18w(3w^2 + v^2)$$

and the two factors are coprime. Their product is a cube, so there exist smaller integers a, b such that

$$18w = a^3 \qquad 3w^2 + v^2 = b^3$$

By Lemma 13.6 there exist smaller integers c, d such that $b = c^2 + 3d^2$. Now

$$v = c(c^2 - df^2) \qquad w = 3d(c^2 - d^2)$$

Here c is odd and d is even. Now

$$a^3 = 18w = 54d(c+d)(c-d)$$

Thus $3|a$ so $(\frac{a}{3})^3 = 2d(c+d)(c-d)$. The three factors are coprime, so each is a cube:

$$-2d = k^3 \qquad c+d = l^3 \qquad c-d = m^3$$

giving a smaller solution $k^3 + l^3 + m^3 = 0$, contradicting minimality. □

13.4 Germain's Theorem

In his *Théorie des Nombres* of 1827, Legendre mentioned the work of Germain on the first case of Fermat's Last Theorem, see Section 13.1. We first prove one special case of her result, and then state two more general theorems that she proved using similar ideas.

Theorem 13.8. *If x, y, z are integers and $x^5 + y^5 = z^5$, then one of x, y, z is divisible by* 5.

Proof: Fermat's Last Theorem is normally stated for positive integers, but it is easy to see that for odd exponents, any solution involving negative integers can be converted into one with positive integers. So we can change z to $-z$, rewrite the equation in the more symmetric form

$$x^5 + y^5 + z^5 = 0$$

and prove there are no integer solutions, positive or negative, unless 5 divides one of x, y, z. (We do not need to rule out solutions involving 0, since $5|0$.) As usual we can assume that x, y, z are pairwise coprime.

Convert the equation to

$$(-x)^5 = (y+z)(y^4 - y^3z + y^2z^2 - yz^3 + z^4)$$

We claim that the two factors on the right are coprime. To see why, suppose that a prime r divides both factors. Then $y \equiv -z \pmod{r}$, so $y^4 - y^3z + y^2z^2 - yz^3 + z^4 \equiv 5y^4 \pmod{r}$. If r divides both factors then either $r = 5$, so 5 divides x and we are done, or r divides both y and $y+z$, so y, z are not coprime.

Since the product is a fifth power and the factors are coprime, each is a fifth power. (Bear in mind that $-1 = (-1)^5$.) By symmetry, the same argument applies to the equations $-y^5 = x^5 + z^5$ and $-x^5 = x^5 + y^5$. Thus there exist integers a, b, c and α, β, γ such that

$$\begin{array}{lll} y+z=a^5 & y^4-y^3z+y^2z^2-yz^3+z^4=\alpha^5 & x=-a\alpha \\ z+x=b^5 & z^4-z^3x+z^2x^2-zx^3+x^4=\beta^5 & y=-b\beta \\ x+y=c^5 & x^4-x^3y+x^2y^2-xy^3+y^4=\gamma^5 & z=-c\gamma \end{array}$$

We claim this is impossible. To see why, we deduce a series of contradictions that together eliminate all possibilities.

Observe that (mod 11) the fifth powers are $-1, 0, 1$ only. This can be proved by direct computation, or more slickly using Fermat's 'little' Theorem $a^{s-1} \equiv 1 \pmod{s}$ for prime s and a prime to s. Thus $(a^5)^2 \equiv 1 \pmod{11}$, so $a^5 \equiv \pm 1 \pmod{11}$, or $a^5 \equiv 0 \pmod{11}$.

Thus $x^5 + y^5 + z^5 \equiv 0 \pmod{11}$ implies that one of x, y, z is divisible by 11. Without loss of generality, $11|x$. Then $2x = b^5 + c^5 + (-a)^5$ is divisible by 11 and one of a, b, c is divisible by 11. However, if $11|b$ then since $11|x$ we deduce that $11|z$, contrary to x, z being coprime. Similarly 11 cannot divide c. Therefore $11|a$. But this is also impossible since then $y \equiv -z \pmod{11}$ and $\alpha^5 \equiv 5y^4 \pmod{11}$, but on the other hand $x \equiv 0, \gamma^5 \equiv y^4$, so $\alpha^5 = 5\gamma^5$. Since the fifth powers (mod 11) are $0, \pm 1$, this implies that $\alpha \equiv \gamma \equiv 0 \pmod{11}$, contrary to x, z being coprime.

This contradiction proves the claim, and with it, the theorem. □

The same argument, replacing 5 by p, proves:

Theorem 13.9. *If p is an odd prime and $2p+1$ is prime then $x^p + y^p = z^p$ implies that one of x, y, z is divisible by p.* □

More generally, this idea leads to:

Theorem 13.10. (Germain's Theorem) *Let p be an odd prime and suppose that there is another 'auxiliary' prime q such that*

(a) *$x^p + y^p + z^p \equiv 1 \pmod{q}$ implies that one of x, y, z is divisible by q, and*

(b) *$x^p \equiv p \pmod{q}$ is impossible.*

Then the Fermat equation $x^p + y^p = z^p$ implies that one of x, y, z is divisible by p. □

Germain observed that there exists such an auxiliary prime for all primes p less than 100. Legendre extended this to all primes less than 197, and many others. Historically, Germain's result focused attention on the second case (2) of Fermat's Last Theorem, which turned out to be distinctly harder than the first case.

13.5 Kummer's Lemma

This section begins the build-up to Kummer's proof of a special case of Fermat's Last Theorem, with a detailed study of the cyclotomic field $K = \mathbb{Q}(\zeta)$ where $\zeta = e^{2\pi i/p}$ for an odd prime p. As in Chapter 3 we write

$$\lambda = 1 - \zeta$$

Further we define

$$\mathfrak{l} = \langle \lambda \rangle$$

which is the ideal generated by λ in the ring of integers $\mathbb{Z}[\zeta]$ of K. We start with some properties of $\mathfrak{l}$.

Lemma 13.11. (a) $\mathfrak{l}^{p-1} = \langle p \rangle$.
(b) $\mathrm{N}(\mathfrak{l}) = p$.

Proof: For $j = 1, \ldots, p-1$ the numbers $1-\zeta$ and $1-\zeta^j$ are associates in $\mathbb{Z}[\zeta]$. Clearly $(1-\zeta)|(1-\zeta^j)$. But if we choose t such that $jt \equiv 1 \pmod{p}$ then $1-\zeta = 1-\zeta^{jt}$ so that $(1-\zeta^j)|(1-\zeta)$. Hence they are associates.

Now Equation (3.10) leads to

$$\langle p \rangle = \prod_{j=1}^{p-1} \langle 1 - \zeta^j \rangle$$

but the above remarks show that $\langle 1-\zeta^j \rangle = \langle 1-\zeta \rangle = \mathfrak{l}$, so $\langle p \rangle = \mathfrak{l}^{p-1}$ and (a) is proved. Part (b) is immediate on taking norms. □

Part (b) of Lemma 13.11 has a useful consequence:

Theorem 13.12. *Every element of* $\mathbb{Z}[\zeta]$ *is congruent modulo* $\mathfrak{l}$ *to precisely one of* $0, 1, 2, \ldots, p-1$.

Proof: Let $\phi : \mathbb{Z} \to \mathbb{Z}[\zeta]/\mathfrak{l}$ be the ring homomorphism $\phi(a) = a + \mathfrak{l}$. The kernel is $\{a \in \mathbb{Z} : \phi(a) = \mathfrak{l}\}$, which is $\{a \in \mathbb{Z} : a \in \mathfrak{l}\}$. Now $a \in \mathfrak{l}$ if and only if $a = x\lambda$ for $x \in \mathbb{Z}[\zeta]$. This implies that

$$a^{p-1} = \mathrm{N}(a) = \mathrm{N}(x)\mathrm{N}(\lambda) = p\mathrm{N}(x)$$

by (3.11). This implies that $p|a$. Conversely, if $p|a$ then $\mathfrak{l}|a$ since $\mathrm{N}(\mathfrak{l}) = p$ and $\mathfrak{l}|\mathrm{N}\mathfrak{l}$. Thus $\ker\phi = p\mathbb{Z}$ and $|\mathbb{Z}/(\ker\phi)| = |\mathbb{Z}/p\mathbb{Z}| = p$. But part (b) of Lemma 13.11 implies that $|\mathbb{Z}[\zeta]/\mathfrak{l}| = p$. Therefore ϕ is surjective, and it induces an isomorphism $\mathbb{Z}/p\mathbb{Z} \to |\mathbb{Z}[\zeta]/\mathfrak{l}$. In more concrete terms, this is what the theorem states. □

Units of $\mathbb{Z}[\zeta]$

The main aim of the rest of this section is to give a useful, though incomplete, description of the units of $K = \mathbb{Z}[\zeta]$. We start by finding which roots of unity occur, showing that there are no 'accidental' occurrences:

Lemma 13.13. *The only roots of unity in K are $\pm\zeta^s$ for integers s.*

Proof: First we show that $\mathrm{i} \notin K$ by arguing for a contradiction. If, on the contrary, $\mathrm{i} \in K$, then $2 = \mathrm{i}(1-\mathrm{i})^2$, so $\langle 2\rangle = \langle 1-\mathrm{i}\rangle^2$. Hence when $\langle 2\rangle$ is resolved into prime factors in $\mathbb{Z}[\zeta]$, it has repeated factors. Theorem 12.1 implies that the polynomial

$$f(t) = \frac{t^p - 1}{t - 1}$$

has a repeated irreducible factor modulo 2, hence that $t^p - 1$ has a repeated irreducible factor modulo 2. Then Theorem 1.6 tells us that $t^p - 1$ and $\mathrm{D}(t^p - 1) = pt^{p-1}$ are not coprime. However, p is odd, so these polynomials modulo 2 take the form $t^p + 1$, t^{p-1} which are obviously coprime. This is a contradiction.

In exactly the same way we can show that $e^{2\pi\mathrm{i}/q} \notin K$ for any odd prime $q \neq p$. We just use

$$\langle q\rangle = \left\langle 1 - e^{2\pi\mathrm{i}/q}\right\rangle^{q-1}$$

Next we remark that

$$e^{2\pi\mathrm{i}/p^2} \notin K$$

because $e^{2\pi\mathrm{i}/p^2}$ satisfies $t^{p^2} - 1 = 0$, but not $t^p - 1 = 0$, so it is a zero of

$$f(t) = (t^{p^2} - 1)/(t^p - 1) = \sum_{r=0}^{p-1} t^{rp}$$

Applying Eisenstein's criterion to $f(t+1)$, a little arithmetic shows that $f(t+1)$, hence also $f(t)$, is irreducible. Thus f is the minimal polynomial of $e^{2\pi\mathrm{i}/p^2}$. Since $[K : \mathbb{Q}] = p - 1$, Theorems 1.15 and 1.18 imply that $e^{2\pi\mathrm{i}/p^2} \notin K$.

Suppose now that $e^{2\pi i/m} \in K$ for an integer m. Then the above results show that

$$4 \nmid m \qquad q \nmid m \qquad p^2 \nmid m.$$

Hence $m|2p$ which leads at once to the required result. □

Lemma 13.14. *For each $\alpha \in \mathbb{Z}[\zeta]$ there exists $a \in \mathbb{Z}$ such that*

$$\alpha^p \equiv a \pmod{\mathfrak{l}^p}$$

Proof: By Theorem 13.12 there exists $b \in \mathbb{Z}$ such that $\alpha \equiv b \pmod{l}$. Now

$$\alpha^p - b^p = \prod_{j=0}^{p-1} (\alpha - \zeta^j b)$$

and since $\zeta \equiv 1 \pmod{\mathfrak{l}}$ each factor on the right is congruent to $\alpha - b \equiv 0 \pmod{\mathfrak{l}}$. Multiplying these factors, $\alpha^p - b^p \equiv 0 \pmod{\mathfrak{l}^p}$. □

Now we prove the main result of this section. It is known as *Kummer's Lemma*, although historically it had nothing what so ever to do with Kummer (see Long [82] p. 89).

Lemma 13.15. (Kummer's Lemma) *Every unit of $\mathbb{Z}[\zeta]$ is of the form $r\zeta^g$ where r is real and g is an integer.*

Proof: Let ε be a unit in $\mathbb{Z}[\zeta]$. There exists a polynomial $e(t) \in \mathbb{Z}[t]$ such that $\varepsilon = e(\zeta)$. For $s = 1, \ldots, p-1$ the elements

$$\varepsilon_s = e(\zeta^s)$$

are conjugate to ε, and $\varepsilon_1 = \varepsilon$. Now $1 = \pm\, \mathrm{N}(\varepsilon) = \pm\, \varepsilon_1 \ldots \varepsilon_{p-1}$, so each ε_s is also a unit. Further, if bars denote complex conjugation,

$$\varepsilon_{p-s} = e(\zeta^{p-s}) = e(\zeta^{-s}) = e(\overline{\zeta^s}) = \overline{e(\zeta^s)} = \overline{\varepsilon_s}$$

Therefore

$$\varepsilon_s \varepsilon_{p-s} = |\varepsilon_s|^2 > 0$$

Then

$$\pm 1 = \mathrm{N}(\varepsilon) = (\varepsilon_1 \varepsilon_{p-1})(\varepsilon_2 \varepsilon_{p-2}) \ldots > 0$$

so $\mathrm{N}(\varepsilon) = 1$.

Now each $\varepsilon_s/\varepsilon_{p-s}$ is a unit, of absolute value 1, and by the usual symmetric polynomial argument

$$\prod_{s=1}^{p-1}\left(t-\frac{\varepsilon_s}{\varepsilon_{p-s}}\right)$$

has coefficients in $\mathbb{Z}$. By Lemma 10.2 its zeros are roots of unity. Lemma 13.13 yields

$$\varepsilon/\varepsilon_{p-1} = \pm\zeta^u$$

for integer u. Since p is odd either u or $u+p$ is even, so

$$\varepsilon/\varepsilon_{p-1} = \pm\zeta^{2g} \tag{13.10}$$

for $0 < g \in \mathbb{Z}$.

The crucial step now is to find out whether the sign in (13.10) is positive or negative. To do this we work out the left-hand side modulo $\mathfrak{l}$, as follows. We know that for some $v \in \mathbb{Z}$

$$\zeta^{-g}\varepsilon \equiv v \pmod{\mathfrak{l}}$$

Taking complex conjugates,

$$\zeta^{g}\varepsilon_{p-1} \equiv v \pmod{\langle\bar{\lambda}\rangle}$$

But $\bar{\lambda} = 1-\zeta^{p-1}$ is an associate of λ, so in fact $\langle\bar{\lambda}\rangle = \mathfrak{l}$. Eliminate v to get

$$\varepsilon/\varepsilon_{p-1} \equiv \zeta^{2g} \pmod{\mathfrak{l}}$$

A negative sign in Equation (13.10) leads to $\mathfrak{l}|2\zeta^{2g}$. Taking norms, $\mathrm{N}(\mathfrak{l})|2^{p-1}$ which contradicts Lemma 13.11(b). So the sign in (13.10) is positive. Hence

$$\zeta^{-g}\varepsilon = \zeta^{g}\varepsilon_{p-1}$$

The two sides of this equation are complex conjugates, so they are in fact real. Therefore $\zeta^{-g}\varepsilon = r \in \mathbb{R}$. □

13.6 Kummer's Theorem

In order to state Kummer's special case of Fermat's Last Theorem, we need a technical definition.

Definition 13.16. A prime p is *regular* if it does not divide the class-number of $\mathbb{Q}(\zeta)$, where $\zeta = e^{2\pi i/p}$.

By Section 12.3, $p = 3, 5, 7$ are regular. Further discussion of the regularity property is postponed until Section 13.7, for we are now in a position to state and prove:

Theorem 13.17. *If p is an odd regular prime then the equation*

$$x^p + y^p = z^p$$

has no solutions in integers x, y, z satisfying

$$p \nmid x \qquad p \nmid y \qquad p \nmid z$$

Proof: Consider instead the equation

$$x^p + y^p + z^p = 0 \tag{13.11}$$

which exhibits greater symmetry. Since we can pass from this to the Fermat equation by changing z to $-z$, it suffices to work on (13.11). Assume, for a contradiction, that there exists a solution (x, y, z) of (13.11) in integers prime to p. We may as usual assume further that x, y, z are pairwise coprime. Factorise (13.11) in $\mathbb{Q}(\zeta)$ to obtain

$$\prod_{j=0}^{p-1} (x + \zeta^j y) = -z^p$$

and pass to ideals:

$$\prod_{j=0}^{p-1} \langle x + \zeta^j y \rangle = \langle z \rangle^p \tag{13.12}$$

First we establish that all factors on the left of this equation are pairwise coprime. For suppose $\mathfrak{p}$ is a prime ideal dividing $\langle x + \zeta^k y \rangle$ and $\langle x + \zeta^l y \rangle$ with $0 \leq k < l \leq p - 1$. Then $\mathfrak{p}$ contains

$$(x + \zeta^k y) - (x + \zeta^l y) = y\zeta^k(1 - \zeta^{l-k})$$

Now $1 - \zeta^{l-k}$ is an associate of $1 - \zeta = \lambda$, and ζ^k is a unit, so $\mathfrak{p}$ contains $y\lambda$. Since $\mathfrak{p}$ is prime either $\mathfrak{p}|y$ or $\mathfrak{p}|\lambda$. In the first case $\mathfrak{p}$ also divides z by (13.12). Now y and z are coprime integers, so there exist $a, b \in \mathbb{Z}$ such that $az + by = 1$. But $y, z \in \mathfrak{p}$ so $1 \in \mathfrak{p}$, a contradiction. On the other hand, since $\mathrm{N}(\mathfrak{l}) = p$, Theorem 6.23(a) implies that $\mathfrak{l}$ is prime, so if $\mathfrak{p}|\lambda$ then $\mathfrak{p} = \mathfrak{l}$. Then $\mathfrak{l}|z$ so

$$p = \mathrm{N}(\mathfrak{l})|\mathrm{N}(z) = z^{p-1}$$

and $p|z$ contrary to hypothesis.

Uniqueness of prime factorisation of ideals now implies that each factor on the left of Equation (13.12) is a pth power of an ideal, since the right-hand side is a pth power and the factors are pairwise coprime. In particular there is an ideal $\mathfrak{a}$ such that

$$\langle x + \zeta y \rangle = \mathfrak{a}^p$$

Thus $\mathfrak{a}^p$ is principal. Regularity of p means that $p \nmid h$, the class-number of $\mathbb{Q}(\zeta)$, and then Proposition 11.10(b) tells us that $\mathfrak{a}$ is principal, say $\mathfrak{a} = \langle \delta \rangle$. Therefore

$$x + \zeta y = \varepsilon \delta^p$$

where ε is a unit.

Now we use Lemma 13.15 to conclude that

$$x + \zeta y = r\zeta^g \delta^p$$

where r is real. By Lemma 13.14 there exists $a \in \mathbb{Z}$ such that

$$\delta^p \equiv a \pmod{\mathfrak{l}^p}$$

Hence

$$x + \zeta y \equiv ra\zeta^g \pmod{\mathfrak{l}^p}$$

Lemma 13.11(a) shows that $\langle p \rangle \mid \mathfrak{l}^p$, so

$$x + \zeta y \equiv ra\zeta^g \pmod{\langle p \rangle}$$

Now ζ^{-g} is a unit, so

$$\zeta^{-g}(x + \zeta y) \equiv ra \pmod{\langle p \rangle}$$

Take complex conjugates:

$$\zeta^g(x + \zeta^{-1} y) \equiv ra \pmod{\langle p \rangle}$$

Eliminate ra to obtain the important congruence

$$x\zeta^{-g} + y\zeta^{1-g} - x\zeta^g - y\zeta^{g-1} \equiv 0 \pmod{\langle p \rangle} \tag{13.13}$$

Observe that $1 + \zeta$ is a unit (put $t = -1$ in (3.3)). We investigate possible values for g in (13.13).

Suppose that $g \equiv 0 \pmod{p}$. Then $\zeta^g = 1$, the terms with x cancel, and (13.13) becomes

$$y(\zeta - \zeta^{-1}) \equiv 0 \pmod{\langle p \rangle}$$

so

$$y(1+\zeta)(1-\zeta) \equiv 0 \pmod{\langle p \rangle}$$

Since $1+\zeta$ is a unit,

$$y\lambda \equiv 0 \pmod{\langle p \rangle}$$

Now $\langle p \rangle = \langle \lambda \rangle^{p-1}$ and $p-1 \geq 2$, so we have $\lambda | y$. Taking norms, $p|y$, contrary to hypothesis. Hence $g \not\equiv 0 \pmod{p}$. A similar argument shows that $g \not\equiv 1 \pmod{p}$.

Rewrite (13.13) in the form

$$\alpha p = x\zeta^{-g} + y\zeta^{1-g} - x\zeta^{g} - y\zeta^{g-1}$$

for some $\alpha \in \mathbb{Z}[\zeta]$. By the previous paragraph no exponent $-g, 1-g, g, g-1$ is divisible by p. Now

$$\alpha = \frac{x}{p}\zeta^{-g} + \frac{y}{p}\zeta^{1-g} - \frac{x}{p}\zeta^{p} - \frac{y}{p}\zeta^{g-1} \tag{13.14}$$

Moreover, $\alpha \in \mathbb{Z}[\zeta]$ and $\{1, \zeta, \ldots, \zeta^{p-2}\}$ is a $\mathbb{Z}$-basis. Hence if all four exponents are incongruent modulo p we have $x/p \in \mathbb{Z}$, contrary to hypothesis. So some pair of exponents must be congruent modulo p. Since $g \not\equiv 0, 1 \pmod{p}$ the only possibility left is that $2g \equiv 1 \pmod{p}$.

But now (13.14) can be rewritten as

$$\alpha p\zeta^{g} = x + y\zeta - x\zeta^{2g} - y\zeta^{2g-1} = (x-y)\lambda$$

Taking norms we get $p|(x-y)$, so

$$x \equiv y \pmod{p}$$

By the symmetry of (13.11),

$$y \equiv z \pmod{p}$$

and hence

$$0 \equiv x^p + y^p + z^p \equiv 3x^p \pmod{p}$$

Since $p \nmid x$ we must have $p = 3$.

It remains to deal with the possibility $p = 3$, which we do by reduction modulo 9. Modulo 9, cubes of numbers prime to p (namely 1, 2, 4, 5, 7, 8) are congruent either to 1 or to -1. Hence modulo 9 a solution of (13.11) in integers prime to 3 takes the form

$$\pm 1 \pm 1 \pm 1 \equiv 0 \pmod{9}$$

which is impossible. Hence finally $p \neq 3$, a contradiction. □

A complete solution of Fermat's Last Theorem for regular primes is thus reduced to the case where one of x, y, or z is a multiple of p. Kummer dealt with this case too; his proof again depends heavily on ideal theory. Although long, it would be accessible at this stage, except for one fact. We need to know that (still for p regular) if a unit in $\mathbb{Q}(\zeta)$ is congruent modulo p to a rational integer, then it is a pth power of another unit in $\mathbb{Q}(\zeta)$. The proof of this requires new methods. We refer to Borevič and Šafarevič [10] pp. 378–381, or Washington [141] chapter 9, for the missing details.

13.7 Regular Primes

Theorem 13.17 is, of course, useless without a test for regularity. There is, in fact, quite a simple test, but once more the proofs are far beyond our present methods. We nonetheless sketch what is involved, and again refer the reader to Borevič and Šafarevič [10] or Washington [141] chapter 5 for details.

Everything rests on the Class-Number Formula (12.9), which involves complex analysis. The point is that nearly everything on the right of this equation, except h, is quite easy to compute, though the regulator Reg_K is much harder than the rest. If we could evaluate the limit on the left we could then work out h. To evaluate this limit we first extend the definition of $\zeta_K(x)$ to allow complex values of x, and then use powerful techniques from complex function theory. These involve another gadget known as a *Dirichlet L-series.*

In the case $K = \mathbb{Q}(\zeta)$ for $\zeta = e^{2\pi i/p}$, p prime, the analysis leads to an expression for h in the form of a product

$$h = h_1 h_2 \tag{13.15}$$

In this, h_2 is the class-number of the related number field $\mathbb{Q}(\zeta+\zeta^{-1})$, which is a subfield of $\mathbb{Q}(\zeta)$ and is contained in $\mathbb{R}$, and h_1 is a computable integer. This would not be very helpful, except that it can be proved that *if h_1 is prime to p, then so is h_2.* Therefore h is prime to p, or equivalently p is regular, if and only if h_1 is prime to p.

Analysis of h_1 leads to an explicit criterion: h_1 is divisible by p if and only if one of the numbers

$$S_k = \sum_{n=1}^{p-1} n^k \quad (k = 2, 4, \dots, p-3) \tag{13.16}$$

is divisible by p^2. These numbers have long been associated with the *Bernoulli numbers* B_k defined by the series expansion

$$\frac{t}{1-e^t} = \sum_{m=0}^{\infty} \frac{B_m}{m!} t^m$$

Their values behave very irregularly: for m odd $\neq 1$ they are zero, for $m = 1$ we have $B_1 = \frac{1}{2}$, and for even m the first few are:

$$B_2 = \tfrac{1}{6} \quad B_4 = -\tfrac{1}{30} \quad B_6 = \tfrac{1}{42} \quad B_8 = -\tfrac{1}{30}$$

$$B_{10} = \tfrac{5}{66} \quad B_{12} = -\tfrac{691}{2730} \quad B_{14} = \tfrac{7}{6} \quad B_{16} = -\tfrac{3617}{510}$$

The connection between the S_k and the B_k (set up in Exercise 13.12) may be shown to give:

Criterion 13.18. *A prime p is regular if and only if it does not divide the numerators of the Bernoulli numbers $B_2, B_4, \ldots, B_{p-3}$.* □

The first 10 irregular primes, found from this criterion, are 37, 59, 67, 101, 103, 131, 149, 157, 233, 257. As a check, it is possible to compute the number h_1, with the results in Table 13.1. Observe that h_1 is divisible by p exactly in the cases $p = 37, 59, 67$ (marked in bold type) as expected.

p	h_1	p	h_1
3	1	43	211
5	1	47	$5 \cdot 139$
7	1	53	4889
11	1	59	$3 \cdot \mathbf{59} \cdot 233$
13	1	61	$41 \cdot 1861$
17	1	67	$\mathbf{67} \cdot 12739$
19	1	71	$7^2 \cdot 79241$
23	3	73	$89 \cdot 134353$
29	2^3	79	$5 \cdot 53 \cdot 377911$
31	3^2	83	$3 \cdot 279405653$
37	**37**	89	$113 \cdot 118401449$
41	11^2	97	$577 \cdot 3457 \cdot 206209$

Table 13.1. Values of the class-number h_1.

13.8 Exercises

13.1 If x, y, z are integers such that $x^2 + y^2 = z^2$, prove that at least one of x, y, z is a multiple of 3, at least one is a multiple of 4, and at least one is a multiple of 5.

13.2 Show that the smallest value of z for which there exist four distinct solutions to $x^2 + y^2 = z^2$ with x, y, z pairwise coprime (not counting sign changes or interchanges of x, y as distinct) is 1105, and find the four solutions.

13.3 Fermat observed that if there are no integer solutions of the equation $X^2 + Y^4 = Z^4$ for nonzero X, Y, Z, then there are no (nonzero) integer solutions x, y, z of the equation $x^2 + y^2 = z^2$ such that $xy/2$ is a square. Prove this implication.

13.4 Show that the general solution in rational numbers of the equation

$$x^3 + y^3 = u^3 + v^3$$

is

$$\begin{aligned} x &= k(1 - (a - 3b)(a^2 + 3b^2)) \\ y &= k((a + 3b)(a^2 + 3b^2) - 1) \\ u &= k((a + 3b) - (a^2 + 3b^2)^2) \\ v &= k((a^2 + 3b^2)^2 - (a - 3b)) \end{aligned}$$

where a, b, k are rational and $k \neq 0$; or $x = y = 0$, $u = -v$; or $x = u$, $y = v$, or $x = v$, $y = u$. (*Hint*: Write $x = X - Y$, $y = X + Y$, $u = U - V$, $v = U + V$, and factorise the resulting equation in $\mathbb{Q}(\sqrt{-3})$.)

13.5 A *triangular number* has the form $\frac{1}{2}n(n + 1)$ for $n \in \mathbb{Z}$ and $n \geq 0$. Show that infinitely many integers are both triangular and square.

13.6 Modify the arguments used in the proof of Lemma 13.1 to find an expression for all solutions of the Pythagorean equation $x^2 + y^2 = z^2$ where x, y, z are Gaussian integers.

(*Hint*: The Gaussian integer $1 + \mathrm{i}$ plays the role of the prime 2. A complete discussion is in Cross [28].)

13.7 Prove Theorem 13.9.

13.8 Prove Theorem 13.10.

13.9 Find suitable auxiliary primes q in Theorem 13.10 when $p = 7$, 13, 17, 19, 31. (These are the first few cases when $2p+1$ is not prime.)

13.10 Let p be an odd prime, $\zeta = e^{2\pi i/p}$. Kummer's Lemma says that the units of $\mathbb{Z}[\zeta]$, thought of in the complex plane $\mathbb{C}$, lie on equally spaced radial lines through the origin, passing through the vertices of a regular p-gon (namely the powers ζ^s). By (3.12), $1+\zeta$ is a unit. So why does Figure 13.1 not contradict Kummer's Lemma?

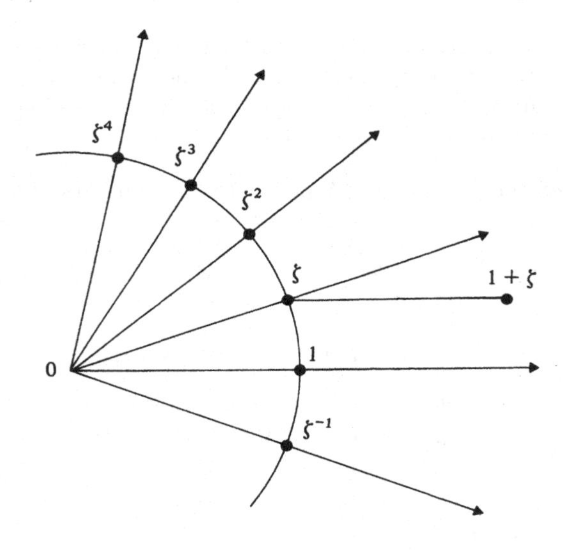

Figure 13.1. Why does this not contradict Kummer's Lemma?

13.11 For p an odd prime, show that if $\zeta = e^{2\pi i/p}$, then

$$\sqrt{\left(\frac{1-\zeta^s}{1-\zeta}\cdot\frac{1-\zeta^{-s}}{1-\zeta^{-1}}\right)}$$

is a real unit in $\mathbb{Q}(\zeta)$ for $s = 1, 2, \ldots, p-1$.

(*Hint*: It is easy to prove that the square of this expression is a real unit and is positive. Relate the expression inside the square root to the square of a polynomial in ζ.)

13.12 Relate S_k in (13.16) to the Bernoulli numbers by proving *Faulhaber's formula*:

$$S_p(n) = \sum_{k=1}^{n} k^p = \frac{1}{p+1} \sum \binom{p+1}{k} B_k n^{p+k-1}$$

(*Hint*: Define the exponential generating functions

$$\begin{aligned} G(z,n) &= \sum_{p=0}^{\infty} S_p(n) \frac{z^p}{p!} = \sum_{k=1}^{n} \mathrm{e}^{kz} = \mathrm{e}^z \frac{1-\mathrm{e}^{nz}}{1-\mathrm{e}^z} \\ \frac{z\mathrm{e}^z x}{1-\mathrm{e}^z} &= \sum_{j=0}^{\infty} B_j^+(x) \frac{z^j}{j!} \end{aligned}$$

Show that $B_j^+(0) = B_j$ and that

$$S_p(n) = \frac{B_{p+1}^+(n) - B_{p+1}^+(0)}{p+1}$$

and deduce Faulhaber's formula.)

The history of this formula is complicated. Johann Faulhaber proved the formula up to $n = 17$. Jacob Bernoulli published a version of the formula in 1713 in his combinatorics text *Ars Conjectandi* (The Art of Conjecturing) under the title 'Summae Potestatem' (Sums of Powers). The first complete proof was given by Jacobi in 1834.

IV

Elliptic Curves and Elliptic Functions

14

Elliptic Curves

We now head towards a discussion of Wiles's proof of Fermat's Last Theorem. The details are beyond the scope of this book, but many of the topics involved are accessible and important in their own right in algebraic number theory and elsewhere. For this reason we provide technical details where appropriate.

In this chapter we introduce the notion of an 'elliptic curve'. (The name is only indirectly linked to the familiar ellipse, see Exercise 15.10.) Elliptic curves are a natural class of plane curves that generalise the straight lines and conic sections studied in nearly all university mathematics courses, and in many high school courses. However, the study of elliptic curves involves two new ingredients. First, it is useful to consider *complex* curves, not just real ones. Second, for some purposes it is more satisfactory to work in complex projective space rather than the complex plane $\mathbb{C}^2$. (Algebraic geometers call $\mathbb{C}$ the complex *line* because it is 1-dimensional over $\mathbb{C}$. So $\mathbb{C}^2$ becomes the complex *plane*.) We introduce these refinements in simple stages.

The main topics discussed in this chapter are:

- Lines and conic sections in the plane.
- The 'secant process' on a conic section and its relation to Diophantine equations.
- The definition and elementary properties of elliptic curves.
- The 'tangent/secant' process on an elliptic curve and the associated group structure.

Our point of view emphasises analogies between conic sections, where the key ideas take on an especially familiar form, and elliptic curves. This should help to explain the origin of the ideas involved in the theory of elliptic curves, and make them appear more natural.

14.1 Review of Conics

The simplest real plane curves are straight lines, which can be defined as the set of solutions $(x, y) \in \mathbb{R}^2$ to a *linear* (or degree 1) polynomial equation

$$Ax + By + C = 0 \tag{14.1}$$

where $A, B, C \in \mathbb{R}$ are constants and $(A, B) \neq (0, 0)$.

Next in order of complexity come the *conic sections* or *conics*, defined by a general quadratic (or degree 2) polynomial equation

$$Ax^2 + Bxy + Cy^2 + Dx + Ey + F = 0 \tag{14.2}$$

where $A, B, C, D, E, F \in \mathbb{R}$ are constants and $(A, B, C) \neq (0, 0, 0)$.

It is well known that conic sections can be classified into seven different types: ellipse, hyperbola, parabola, two distinct lines, one 'double' line, a point, or empty. A good way to see this is to transform (14.2) into a simpler form, usually known as a *normal form*, by a change of coordinates. In fact, a general invertible linear change of coordinates

$$\begin{aligned} X &= ax + by \\ Y &= cx + dy \end{aligned}$$

(with $ad - bc \neq 0$ for invertibility) transforms (14.2) into one or other of the forms

$$\begin{aligned} \varepsilon_1 X^2 + \varepsilon_2 Y^2 + P &= 0 \\ X^2 + Y + P &= 0 \end{aligned}$$

where $P \in \mathbb{R}$ and $\varepsilon_1, \varepsilon_2 = 0, 1$, or -1.

The usual proof of this (see for example Loney [81] p. 323, Anton [1] p. 359, or Roe [111] p. 251) begins by rotating coordinates orthogonally to diagonalise the quadratic form $Ax^2 + Bxy + Cy^2$, which changes (14.2) to the slightly simpler form

$$\lambda_1 x'^2 + \lambda_2 y'^2 + \alpha x' + \beta y' + \gamma = 0$$

If $\lambda_1 \neq 0$ then the term $\alpha x'$ can be eliminated by 'completing the square', and similarly if $\lambda_2 \neq 0$ then the term $\beta y'$ can be eliminated. The coefficients of x'^2 and y'^2 can be scaled to $0, 1$, or -1 by multiplying them by a nonzero constant; furthermore, x' and y' can be interchanged if necessary. Finally, the entire equation can be multiplied throughout by -1. The result is the following catalogue of normal forms:

Theorem 14.1. *By an invertible linear coordinate change, every conic can be put in one of the following normal forms:*

(a) $X^2 + Y^2 + P = 0$

(b) $X^2 - Y^2 + P = 0$

(c) $X^2 + Y + P = 0$

(d) $X^2 + P = 0$ □

In case (a) we get an ellipse (indeed a circle) if $P < 0$, a point if $P = 0$, and the empty set if $P > 0$. In case (b) we get a (rectangular) hyperbola if $P \neq 0$ and two distinct intersecting lines if $P = 0$. Case (c) is a parabola. Case (d) is a pair of parallel lines if $P < 0$, a 'double' line if $P = 0$, and empty if $P > 0$.

Transforming back into the original (x, y) coordinates, circles transform into ellipses, rectangular hyperbolas transform into general hyperbolas, parabolas transform into parabolas, lines transform into lines, and points transform into points.

Even the conics, then, exhibit a rich set of possibilities when viewed as curves in the real plane $\mathbb{R}^2$. The situation simplifies somewhat if we consider the same equations, but in *complex* variables; it simplifies even more if we work in projective space. In complex coordinates, the map $Y \mapsto \mathrm{i}Y$ sends Y^2 to $-Y^2$, a scaling that cannot be performed over the reals. This coordinate transformation sends normal form (a) to normal form (b) and thereby abolishes the distinction between hyperbolas and ellipses.

14.2 Projective Space

We now show that in projective space, all the different types of conic section other than the double line and the point can be transformed into each other. This is the case even in *real* projective space. First, we recall the basic notions of projective geometry. For further details, see Coxeter [27], Loney [81], or Roe [111].

Definition 14.2. The *real projective plane* $\mathbb{RP}^2$ is the set of lines L through the origin in $\mathbb{R}^3$. Each such line is referred to as a *projective point.* Each plane through the origin in $\mathbb{R}^3$ is called a *projective line.* A projective point is *contained in* a projective line if and only if the corresponding line through the origin is contained in the corresponding plane through the origin.

A *projective transformation* or *projection* is a map from $\mathbb{RP}^2$ to itself of the form $L \mapsto \phi(L)$, where ϕ is an invertible linear transformation of $\mathbb{R}^3$.

Two configurations of projective lines and projective points are *projectively equivalent* if one can be mapped to the other by a projection.

This definition may seem strange when first encountered, but it represents the distillation of a considerable effort on the part of geometers to 'complete' the ordinary (or *affine*) plane $\mathbb{R}^2$ by adding 'points at infinity' at which parallel lines can be deemed to meet. We explain this idea in a moment, but first we record:

Proposition 14.3. *In the projective plane, any two projective lines meet in a unique projective point, and any two projective points can be joined by a unique projective line.*

Proof: These properties follow from the analogous properties of lines and planes through the origin in $\mathbb{R}^3$. □

Points at Infinity

Now we describe the interpretation of the projective plane in terms of points at infinity. One way to see how this comes about is to consider the plane $\mathcal{P} = \{(x, y, z) : z = 1\} \subseteq \mathbb{R}^3$. Each point $(x, y, 1) \in \mathcal{P}$ can be identified with a point (x, y) in the affine plane $\mathbb{R}^2$. We write $(x, y) \equiv (x, y, 1)$. Alternatively, the point $(x, y, 1)$ can be identified with the line through the origin in $\mathbb{R}^3$ that passes through it. Nearly every line through the origin in $\mathbb{R}^3$ is of this form: the exceptions are precisely the lines that lie in the plane $\mathcal{Q} = \{(x, y, z) : z = 0\}$, that is, the lines parallel to $\mathcal{P}$.

In the same way, any straight line M in $\mathcal{P}$ can be identified with either a straight line in $\mathbb{R}^2$, or with a plane through the origin in $\mathbb{R}^3$—namely, the unique plane that contains both the origin of $\mathbb{R}^3$ and M. Precisely *one* plane through the origin of $\mathbb{R}^3$ is not of this form, namely, the plane $\mathcal{Q}$ that is parallel to $\mathcal{P}$. See Figure 14.1.

These identifications therefore *embed* the affine plane $\mathbb{R}^2$ in the projective plane $\mathbb{RP}^2$, in such a way that points embed as projective points and lines embed as projective lines. However, $\mathbb{RP}^2$ contains exactly one extra projective line, called the *line at infinity*: namely, the projective line

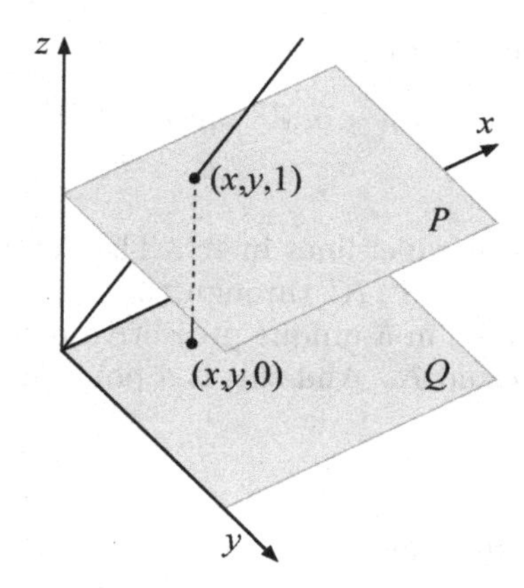

Figure 14.1. Construction of the projective plane.

that corresponds to the plane $\mathcal{Q}$ through the origin of $\mathbb{R}^3$. Moreover, $\mathbb{RP}^2$ contains extra projective points that do not correspond to points in $\mathbb{R}^2$; indeed, these are precisely the projective points that lie on the line at infinity, since they correspond to lines through the origin in $\mathbb{R}^3$ that lie in the plane $\mathcal{Q}$. These are called 'points at infinity'.

Each point at infinity corresponds to a unique direction in the plane $\mathbb{R}^2$, that is, a set of parallel lines, because any such set is parallel to precisely one line in the plane $\mathcal{Q}$. In these terms, a direction and its exact opposite, a 180° rotation, are identical. See Figure 14.2.

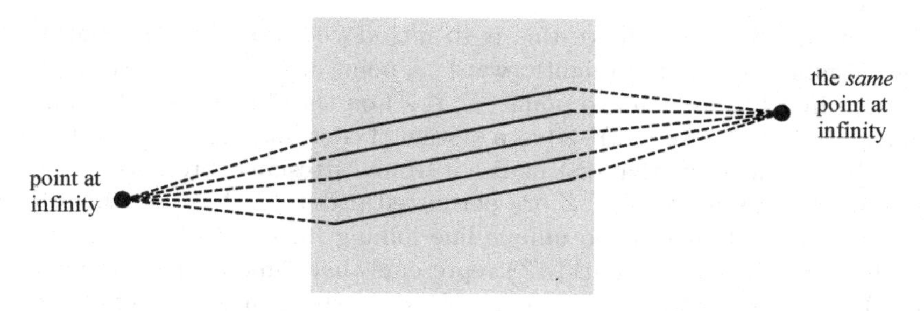

Figure 14.2. Points at infinity correspond to directions in the affine plane; that is, sets of parallel lines. Rotating by 180° does not alter the direction in this sense.

The key feature of this setup is:

Lemma 14.4. *Any two parallel lines in $\mathbb{R}^2$ meet in $\mathbb{RP}^2$ at exactly one point at infinity.*

Proof: Suppose J, K are parallel lines in $\mathbb{R}^2$. They correspond to projective lines, namely, the planes J', K' through the origin that contain J, K respectively. But these meet in a unique projective point, namely the line in $\mathcal{Q}$ that is parallel to J and K. And this is a point at infinity in $\mathbb{RP}^2$. □

Complex Projective Plane

There is also a complex analogue:

Definition 14.5. The *complex projective plane* $\mathbb{CP}^2$ is the set of lines (that is, 1-dimensional vector subspaces over $\mathbb{C}$) through the origin in $\mathbb{C}^3$.

Each such line is referred to as a *complex projective point.*

Each plane through the origin in $\mathbb{R}^3$ is called a *complex projective line.*

For notational convenience and simplicity we break with tradition and use x, y to denote complex variables, when convenient. The convention that $z = x + \mathrm{i}y$ is abandoned in this chapter and the next.

In complex projective space it is also the case that any two projective lines meet in a unique projective point, and any two projective points can be joined by a unique projective line.

The geometry of projective space, real or complex, is richer than mere lines. Any curve in $\mathbb{R}^2$ or $\mathbb{C}^2$ can be embedded in the corresponding $\mathbb{RP}^2$ or $\mathbb{CP}^2$. If the equation of the curve is polynomial, then this can be done in a systematic manner, so that 'points at infinity' on the curve can also be defined.

The easiest way to achieve this is to introduce 'homogeneous coordinates'. Again, the idea is straightforward. A point in $\mathbb{RP}^2$ is a line through the origin in $\mathbb{R}^3$. Any nonzero point (X, Y, Z) on that line defines the line uniquely. So we can use (X, Y, Z) as a system of coordinates. However, this system has two features that distinguish it from Cartesian coordinates. The first is that all values of X, Y, Z are permitted *except* $(X, Y, Z) = (0, 0, 0)$. The reason is that there is no unique line joining $(0, 0, 0)$ to the origin in $\mathbb{R}^3$. The second is that (aX, aY, aZ) represents that same projective point as (X, Y, Z) for any nonzero constant a, since clearly both points define the same line through the origin of $\mathbb{R}^3$. We therefore define an equivalence relation $\sim$ by $(X, Y, Z) \sim (aX, aY, aZ)$ for any nonzero constant a. In other words, it is not the *values* of (X, Y, Z) that determine the corresponding projective point, but their ratios.

The embedding of $\mathbb{R}^2$ into $\mathbb{RP}^2$ defined above, which identifies $(x, y) \in \mathbb{R}^2$ with $(x, y, 1) \in \mathcal{P} \subseteq \mathbb{R}^3$, also identifies the usual coordinates (x, y) on $\mathbb{R}^2$ with the corresponding coordinates $(x, y, 1)$ on $\mathbb{RP}^2$. This represents the same projective point as (ax, ay, a) for any $a \neq 0$. In other words, when $Z \neq 0$ the projective point (X, Y, Z) is the same as $(X/Z, Y/Z, Z/Z) = (X/Z, Y/Z, 1) \in \mathcal{P}$. On the other hand, when $Z = 0$ the projective point (X, Y, Z) lies in the plane $\mathcal{Q}$ and hence represents a point at infinity.

This system of coordinates (X, Y, Z) on $\mathbb{RP}^2$ is known as *homogeneous coordinates*. Homogeneous coordinates are really $\sim$-equivalence classes of triples (X, Y, Z), but it is more convenient to work with representatives and remember to take the equivalence relation $\sim$ into account. The choice of the line at infinity is conventional: in principle any line in $\mathbb{RP}^2$ can be deemed to be the line at infinity, and there is then a corresponding embedding of $\mathbb{R}^2$ in $\mathbb{RP}^2$. Indeed, any projective line in $\mathbb{RP}^2$ can be mapped to any other projective line by a projection, since any plane through the origin in $\mathbb{R}^3$ can be transformed into any other plane by an invertible linear map. For the purposes of this book, however, we employ the convention that $Z = 0$ defines the line at infinity.

The way to transform a polynomial equation in affine coordinates (x, y) into homogeneous coordinates is to replace x by X/Z and y by Y/Z, and then to multiply through by the smallest power of Z that makes the result a polynomial. For example the Cartesian equation $y - x^2 = 0$ becomes:

$$\begin{aligned} (Y/Z) - (X/Z)^2 &= 0 \\ YZ^{-1} - X^2Z^{-2} &= 0 \\ YZ - X^2 &= 0 \end{aligned}$$

Points 'at infinity' now come into play. As well as the usual points $(x, x^2) \equiv (x, x^2, 1)$, this projective curve also contains the point at infinity given by $Z = 0$, which forces $X = 0$ but any nonzero Y. Since $(0, Y, 0) \sim (0, 1, 0)$ the parabola contains exactly one new point at infinity, in addition to the usual points in $\mathbb{R}^2$. It is easy to check that this point lies in the direction towards which the arms of the affine parabola 'diverge', namely the y-axis. See Figure 14.3.

Moreover, adding this point at infinity to the parabola causes it to close up (since the point at infinity lies on *both* arms). It is now plausible that the parabola is just an ellipse in disguise—that is, that they are projectively equivalent. We can verify this by means of the projection $\phi(X, Y, Z) = (X, Y + Z, Y - Z)$, which transforms $YZ - X^2 = 0$ into $Y^2 - Z^2 - X^2 = 0$ or $X^2 + Z^2 = Y^2$. Compose with $\psi(X, Y, Z) = (X, Z, Y)$ to turn this into $X^2 + Y^2 = Z^2$. Finally restrict back to the plane $\mathbb{R}^2$ by setting $(X, Y, Z) = (x, y, 1)$ and we get $x^2 + y^2 = 1$, a circle. Which, of course, is just a special type of ellipse.

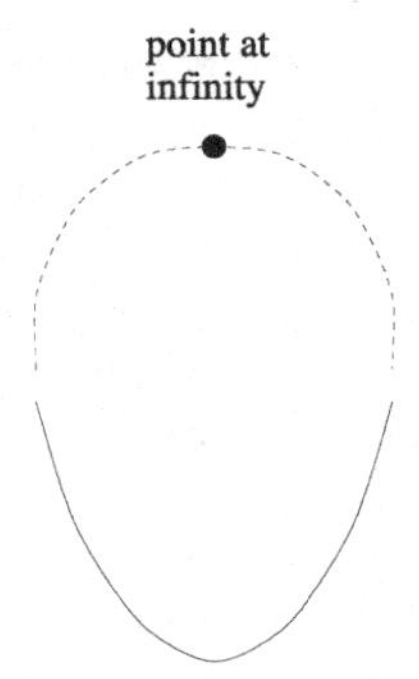

Figure 14.3. Adding a point at infinity to a parabola.

If we had not interchanged Y and Z, the result would have been $x^2 - y^2 = 1$, a hyperbola. So in fact the ellipse, hyperbola, and parabola are all projectively equivalent over $\mathbb{R}$. So in real projective space the list of conics collapses to a smaller one. Namely: ellipse (= parabola = hyperbola), intersecting lines (= pair of parallel lines), double-line, point.

What about complex projective space? Think about it. *Hint*: The real surprise is 'point'.

14.3 Rational Conics and the Pythagorean Equation

There is an interesting link between the geometry of conics and solutions of quadratic Diophantine equations.

Definition 14.6. A *rational point* in $\mathbb{R}^2$ or $\mathbb{C}^2$ is a point whose coordinates are rational numbers.

A *rational line* in $\mathbb{R}^2$ or $\mathbb{C}^2$ is a line

$$ax + by + c = 0 \tag{14.3}$$

whose coefficients a, b, c are rational numbers with $(a, b) \neq (0, 0)$.

A *rational conic* in $\mathbb{R}^2$ or $\mathbb{C}^2$ is a conic

$$f(x, y) = ax^2 + bxy + cy^2 + dx + ey + f = 0 \tag{14.4}$$

whose coefficients a, b, c, d, e, f are rational numbers, with $(a, b, c) \neq (0, 0, 0)$.

There are similar definitions for real and complex projective planes, by converting to homogeneous coordinates.

An intersection point of two rational lines is obviously a rational point. However, an intersection point of a rational line and a rational conic need not be rational—for example, consider the intersection of $x - y = 0$ with $x^2 + y^2 - 2 = 0$, which consists of the two points $(\pm\sqrt{2}, \pm\sqrt{2})$.

Not all rational conics possess rational points. For an example see Exercise 14.2. A necessary and sufficient condition for a rational conic to possess at least one rational point was proved by Legendre and can be found in Goldman [52] p. 318. However, many rational conics do possess rational points, and from now on we work with such a conic.

Proposition 14.7. *Let p be a rational point on a rational conic C. Then any rational line through p intersects C in rational points.*

Proof: We discuss real conics: the complex case is similar. Let $f(x, y)$ be a rational conic as in (14.4) and let $Ax + By + C = 0$ be a rational line. Suppose that $B \neq 0$; if not, then $A \neq 0$ and a similar argument applies. Their intersection is the set of all (x, y) for which $y = (Ax - C)/B$ and x satisfies the quadratic equation $f(x, (Ax - C)/B) = 0$. Suppose that $f(x, (Ax - C)/B) = Kx^2 + Lx + M$: then K, L, M are rational. This equation has at least one real root, given by p, so it has two real roots (which are identical if and only if the line is tangent to the conic at p). The sum of those roots is $-L/K \in \mathbb{Q}$, and the root given by p is rational; therefore the second root is also rational. □

This result immediately leads to a method for parametrising all rational points on a rational conic (provided it possesses at least one rational point). Let C be a rational conic with a rational point p and let L be any rational line. For any point $q \in L$ the line joining q to p meets C at p, and at some other point (which is distinct from p unless the line concerned is tangent to C). Define a map $\pi : L \to C$ by letting $\pi(q)$ be this second point of intersection of the line joining q to p. (See Figure 14.4.)

Theorem 14.8. *With the above notation, $\pi(q)$ is rational if and only if q is rational.*

Proof: Clearly $\pi(q)$ is rational if and only if the line joining p to q is rational. But this is the case if and only if q is a rational point. □

Example 14.9. Suppose that C is the unit circle $x^2 + y^2 - 1 = 0$, which is a rational conic. It contains the rational point $p = (-1, 0)$. Let L be the rational line $x = 0$. The rational points on L are the points $(0, t)$ where $t \in Q$.

The line joining p to $(0, t)$ has equation

$$y = t(x + 1)$$

and this meets the circle at

$$(x, y) = \left(\frac{1 - t^2}{1 + t^2}, \frac{2t}{1 + t^2}\right)$$

Thus we have the identity

$$\left(\frac{1 - t^2}{1 + t^2}\right)^2 + \left(\frac{2t}{1 + t^2}\right)^2 = 1$$

or equivalently

$$(1 - t)^2 + (2t)^2 = (1 + t)^2$$

providing solutions of the Diophantine equation $x^2 + y^2 = z^2$ in rational numbers. Indeed by Theorem 14.8 every rational solution is of this form.

This is very close to the parametrisation of Pythagorean triples obtained in Lemma 13.1. Indeed, if we set $t = r/s$ we can easily obtain the result of that lemma.

14.4 Elliptic Curves

Elliptic curves arise from the study of plane cubic curves

$$\sum_{i+j \leq 3} A_{ij} x^i y^j = 0 \tag{14.5}$$

where the A_{ij} are constants and $(A_{30}, A_{21}, A_{12}, A_{03}) \neq (0, 0, 0, 0)$.

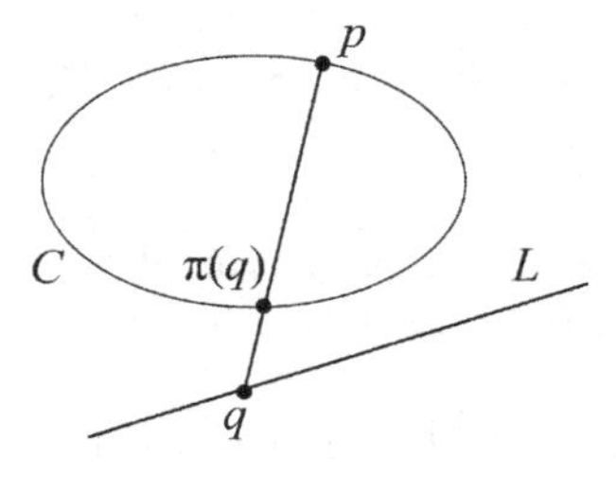

Figure 14.4. Parametrisation of the rational points of a rational conic in terms of the rational points on a rational line.

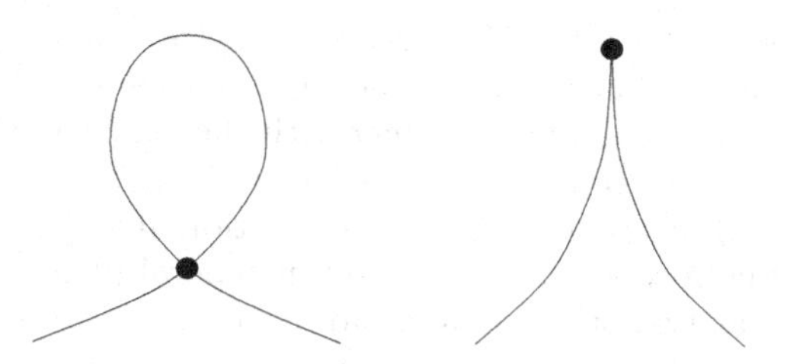

Figure 14.5. Typical singular points: (left) self-intersection, (right) cusp.

Over the reals, such curves were classified by Newton in (probably) 1668: he distinguished 58 different kinds. See Westfall [144] p. 200. As for conics, the key to such a classification is to transform coordinates so that (14.5) takes some simpler form. There are several ways to do this. In order to state the first, we need two definitions:

Definition 14.10. Let C be a curve in the plane (real or complex, affine or projective). A point $x \in C$ is *regular* if there is a unique tangent to C at x. Otherwise, x is *singular*.

The curve C is *nonsingular* if it has no singular points; that is, every point $x \in C$ is regular.

Typical singular points are *self-intersections* and *cusps* as in Figure 14.5. Although it looks as though the tangent at a cusp point is unique, actually there are three 'coincident' tangents—that is, a tangent of multiplicity 3.

Definition 14.11. Two curves $C, D \in \mathbb{CP}^2$ (or $\mathbb{RP}^2$) are *projectively equivalent* if there is a projection ϕ such that $\phi(C) = D$.

Theorem 14.12. (Weierstrass Normal Form) *Every nonsingular cubic curve in $\mathbb{CP}^2$ is projectively equivalent to a curve which in affine coordinates takes the form*

$$y^2 = 4x^3 - g_2x - g_3 \tag{14.6}$$

where $g_2, g_3 \in \mathbb{C}$ are constants.

We call (14.6) the *Weierstrass normal form* of the curve. The coefficient 4 on the x^3 term in Weierstrass normal form is traditional; it could be made equal to 1, but we will shortly see that the 4 is more convenient in some circumstances. The notation g_2, g_3 for the linear and constant coefficients is also traditional.

Proof: We sketch the proof. The first step is to establish that every nonsingular cubic curve C has at least one *inflexion point.* This is a point at which the tangent line has *triple contact* with the curve, in the following sense. Suppose that the equation of the curve is $f(x, y) = 0$, and let $(\xi, \eta) \in C$. A general line through (ξ, η) has equation $a(x-\xi)+b(y-\eta) = 0$ for $a, b, \in \mathbb{C}$. This line meets C at (ξ, η), and in general (since the equation is cubic) it meets it at two other points. However, the cubic equation that determines these intersection points may have multiple zeros. The line is a tangent at (ξ, η) if that point corresponds to a double zero. The point (ξ, η) is an inflexion if it corresponds to a triple zero.

By writing down the equations for an inflexion point, it can be shown that any nonsingular cubic curve in $\mathbb{CP}^2$ has exactly nine inflexion points, if multiplicities are taken into account. In particular, it has at least one. See Brieskorn and Knörrer [12] p. 291 for details.

By a projection, we may assume that $(0, 0, 1)$ is an inflexion point and that the tangent there has equation $X = 0$. This implies that the cubic curve has a homogeneous equation of the form

$$Y^2Z + AXYZ + BYZ^2 + CX^3 + DX^2Z + EXZ^2 + FZ^3 = 0$$

which in affine coordinates becomes

$$y^2 + (Ax + B)y + g(x) = 0 \tag{14.7}$$

where $g(x)$ is a cubic polynomial. Define new affine coordinates (x', y') by

$$y' = y + \frac{A}{2}x + \frac{B}{2} \qquad x' = x$$

Then (14.7) transforms into $y'^2 + h(x') = 0$ where h is a cubic polynomial. There exists a linear change of coordinates $x'' = px' + q$ that puts $h(x')$ into the form $4x''^3 - g_2x'' - g_3$ while leaving y' unchanged. □

We may now define an elliptic curve.

Definition 14.13. An *elliptic curve* is the set of points $(x, y) \in \mathbb{C}^2$ that satisfy the equation

$$y^2 = Ax^3 + Bx^2 + Cx + D \tag{14.8}$$

where $A, B, C, D \in \mathbb{C}$ are constants.

Strictly speaking, this defines a complex affine elliptic curve. There is projective analogue $Y^2Z = AX^3 + BX^2Z + CXZ^2 + DZ^3$, where (X, Y, Z)

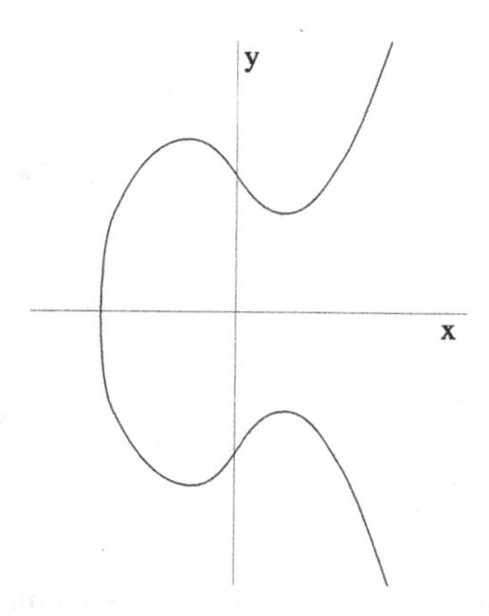

Figure 14.6. The real elliptic curve $y^2 = 4x^3 - 3x + 2$.

are homogeneous coordinates on $\mathbb{CP}^2$. When $A, B, C, D \in \mathbb{R}$ we can restrict attention to real variables, getting a *real elliptic curve.* Moreover, in the real case we can draw the graph of (14.8) in the plane to illustrate certain features of the geometry: we do this frequently below. Figure 14.6 shows the real elliptic curve $y^2 = 4x^3 - 3x + 2$ for which $g_2 = 3, g_3 = -2$.

The most important elliptic curves are those for which A, B, C, D are rational. We call these *rational elliptic curves*, and omit 'rational' whenever the context permits.

Remark. As already remarked, the name 'elliptic curve' is related only indirectly to the ellipse. As we discuss in Chapter 15, elliptic curves are closely associated with a special class of complex functions, called elliptic functions. The name of these functions comes from their connection with the arc-length of an ellipse; see Exercise 15.10.

14.5 Tangent/Secant Process

In Section 14.1 we showed that the rational points on a rational conic can be parametrised by the rational points on a rational line, once we know *one* rational point on the conic. A similar approach to rational points on a rational elliptic curve does not lead to such a definitive result, but in some respects the partial result that is thereby obtained is more interesting.

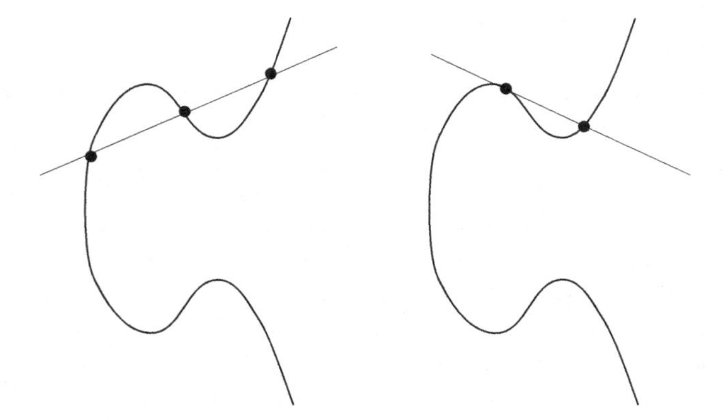

Figure 14.7. Constructing new rational points on an elliptic curve.

Proposition 14.14. *Over* $\mathbb{CP}^2$ *a rational line cuts a rational elliptic curve in three points (counting multiplicities). If two of these points are rational, then so is the third.*

Proof: Suppose that the affine equation of the elliptic curve is $f(x, y) = 0$ and let the line have affine equation $ax + by + c = 0$. Without loss of generality assume that $b \neq 0$, and solve for y, to get $y = -(ax + c)/b$. Substitute in f to get $f(x, -(ax + c)/b) = 0$. This is a cubic polynomial $px^3 + qx^2 + rx + s$ with rational coefficients, and its zeros determine the x-coordinates of the intersection points. The corresponding y-coordinates are equal to $-(ax+c)/b$, hence they are rational if and only if x is rational. Since the sum of the roots of the cubic is equal to the rational number $-p/q$, if two roots are rational then so is the third. $\square$

Incidentally, we stated Proposition 14.14 in projective form because in the affine case there are occasions when the third root of the cubic is at infinity. That is, the cubic actually reduces to a quadratic. We slid over this point in the proof. The proposition implies that once we have found two rational points on a rational elliptic curve, we can find another by drawing the line through those points and seeing where else it cuts the curve, as in Figure 14.7 (left). In fact, we can do slightly better: find *one* rational point and see where else the tangent at that point cuts the curve, Figure 14.7 (right). This construction is the *tangent/secant process.*

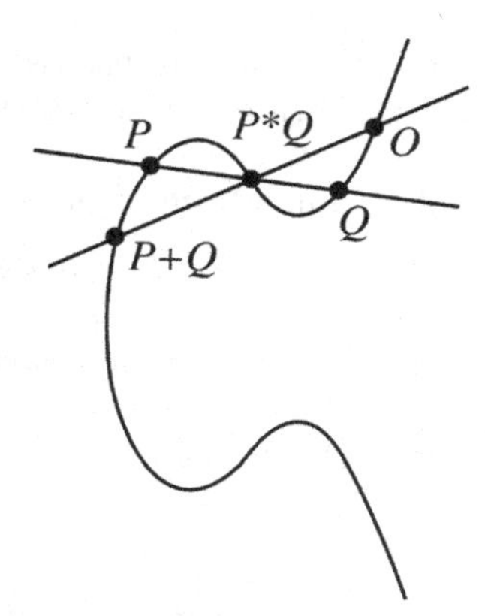

Figure 14.8. The group operation on the rational points of an elliptic curve.

14.6 Group Structure on an Elliptic Curve

We now show that the rational points on a rational elliptic curve form an abelian group, under an operation of 'addition' closely related to Proposition 14.14. This remarkable fact forms the basis of the arithmetical theory of elliptic curves.

Assume that an elliptic curve C in $\mathbb{CP}^2$ contains at least one rational point, which we denote by $\mathbb{O}$ for reasons soon to become apparent.

Definition 14.15. Let P and Q be rational points on C. Define $P * Q$ to be the third point in which the line through P and Q meets C.

Let $\mathcal{G}$ be the set of all rational points in C. For some fixed but arbitrary choice $\mathbb{O}$ of a rational point on C, define the operation $+$ on $\mathcal{G}$ by

$$P + Q = (P * Q) * \mathbb{O} \tag{14.9}$$

See Figure 14.8.

We now prove that the operation $+$ gives $\mathcal{G}$ the structure of an abelian group. In order to achieve this, we require the following fundamental theorem in algebraic geometry.

Theorem 14.16. (Bézout's Theorem) *Let $P(X,Y,Z)$ be a homogeneous polynomial of degree p over $\mathbb{C}$, let $Q(X,Y,Z)$ be a homogeneous polynomial of degree q over $\mathbb{C}$, and suppose that P, Q have no common factor of degree > 1. Then the number of intersection points of the curves in $\mathbb{CP}^2$ defined by $P = 0, Q = 0$ is precisely pq (provided multiplicities are taken into account).*

Proof: A detailed proof, along with a careful discussion of how to count multiplicities, can be found in Brieskorn and Knörrer [12] p. 227. □

Next we state without proof a lemma from algebraic geometry:

Lemma 14.17. *Let two curves of degree n meet in exactly n^2 distinct points, and let $0 \leq m \leq n$. If exactly mn of these points lie on an irreducible curve of degree mn, then the remaining $n(n-m)$ lie on a curve of degree $n-m$.*

Proof: See Brieskorn and Knörrer [12] p. 245. □

We may now prove:

Theorem 14.18. *The set $\mathcal{G}$ of rational points on a rational elliptic curve forms an abelian group under the operation $+$. The identity element is $\mathbb{O}$.*

Proof: First, observe that $P * Q = Q * P$, since the process of constructing the third point on the line through P and Q does not depend on the order in which we consider P and Q. So

$$P + Q = (P * Q) * \mathbb{O} = (Q * P) * \mathbb{O} = Q + P$$

and the operation $+$ is commutative.

We claim that $P + \mathbb{O} = P$, so $\mathbb{O}$ is the identity element. This follows since

$$P + \mathbb{O} = (P * \mathbb{O}) * \mathbb{O}$$

If $Q = P * \mathbb{O}$, then $P, \mathbb{O}, Q$ are collinear. Assume for a moment that these are distinct. Then $(P * \mathbb{O}) * \mathbb{O} = Q * \mathbb{O}$, and this must be P. If they are not distinct, then either $P = \mathbb{O}$ or $P = Q$, and in either case it is easy to complete the calculation.

The inverse of P is easily seen to be $P * (\mathbb{O} * \mathbb{O})$.

The most complicated part of the proof is the associative law

$$(P + Q) + R = P + (Q + R)$$

Figure 14.9 indicates the associated geometry.

First, we define

$$S = P + Q \qquad T = S + R \qquad U = Q + R$$

We have to prove that $P + U = T$. Denote the auxiliary points used to construct R, S, T by R', S', T', as in the figure. It suffices to show that P, U, T' are collinear. To do so, let L_1 be the line through P, Q, S', let L_2 be the line through S, R, T', let L_3 be the line through $\mathbb{O}, U', U$, let G_1 be the line through $\mathbb{O}, S', S$, and let G_2 be the line through Q, R, U'.

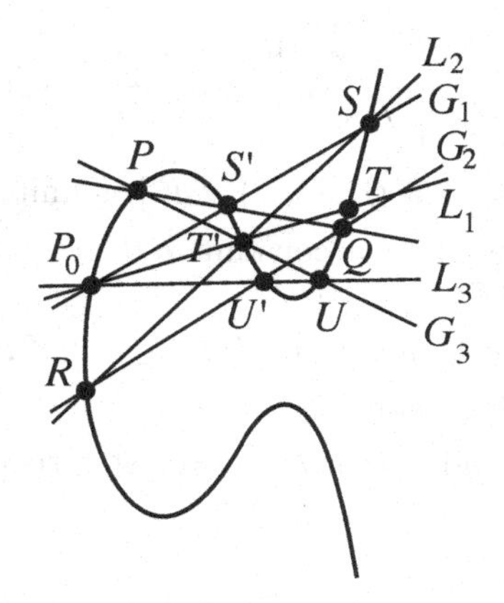

Figure 14.9. Geometry for the proof of the associative law on $\mathcal{G}$.

Recall from Lemma 14.17 that if two curves C, D of order n meet in exactly n^2 points, nm of which lie on an irreducible curve E of order m, then the remainder lie on a curve F of order $n - m$.

Apply this to the cubic curves C, D, where $D = L_1 \cup L_2 \cup L_3$, and take $E = G_1$. First, suppose $C \cap D$ contains exactly nine distinct points. Then it follows that Q, R, U', P, T', U lie on a conic. But Q, R, U' lie on the line G_2, so G_2 is a component of this conic. Let the other component, which must also be a line, be G_3. Then if P, T', U do not lie on G_2, they must lie on G_3, and the proof is complete.

If $C \cap D$ contains less than nine distinct points, then some intersection points are multiple. A suitable small perturbation of the curve C splits these apart, and a limiting argument completes the proof. For a more algebro-geometric approach, replacing the limiting procedure by Zariski continuity, see Brieskorn and Knörrer [12] p. 310. □

The operation $+$ can be defined in exactly the same way for any elliptic curve, rational or not. The same proof shows that $(C, +)$ is an abelian group. When C is rational, $\mathcal{G}$ is a subgroup.

One of the most important theorems in this area is:

Theorem 14.19. (Mordell's Theorem) *Suppose that C is a nonsingular rational cubic curve in $\mathbb{CP}^2$ having a rational point. Then the group $\mathcal{G}$ of rational points is finitely generated.*

Proof: The original proof is due to Mordell [96]. A sketch, based on a version due to Weil [142], is described in Goldman [52]. The main idea is to define a function $H(x)$, for $x \in \mathcal{G}$, that measures the 'complexity' of x, and use H as the basis of an inductive argument. This function, called the *height* has the following properties:

(a) For any $K > 0$ the set $\{P \in \mathcal{G} : H(P) < K\}$ is finite.

(b) For each $Q \in \mathcal{G}$ there exists a constant c depending only on Q such that $H(P + Q) \leq c(H(P))^2$.

(c) There exists a constant d such that $H(P) \leq d(H(2P))^{1/4}$.

(d) The quotient group $\mathcal{G}/2\mathcal{G}$ is finite.

In fact, if $P = (x, y)$ and $x = m/n$ in lowest terms, we take $H(P) = \max(|m|, |n|)$. □

Recall from Proposition 1.31 that a finitely generated abelian group is of the form $F \oplus \mathbb{Z}^k$ where F is a finite abelian group, hence a direct sum of finite cyclic groups. The group F, which is unique, consists of the elements of finite order and is called the *torsion subgroup*. The groups $\mathcal{G}$ determined by elliptic curves are very special, as is shown by the following theorem of Mazur:

Theorem 14.20. *Let $\mathcal{G}$ be the group of rational points on an elliptic curve. Then the torsion subgroup of $\mathcal{G}$ is isomorphic either to Z_l where $1 \leq l \leq 10$, or $Z_2 \oplus \mathbb{Z}_{2l}$ where $1 \leq l \leq 4$.*

Proof: The proof is very technical; see Mazur [90, 91]. □

14.7 Applications to Diophantine Equations

We now describe an application of the above ideas to an equation very similar to Fermat's. This application is due to Elkies [39].

We know that it is impossible for two cubes to sum to a cube, but might it be possible for three cubes to sum to a cube? It is; in fact $3^3+4^3+5^3 = 6^3$. Euler conjectured that in general n nth powers can sum to an nth power, but not $n-1$. It has been proved that Euler's conjecture is false. In 1966 Lander and Parkin [73] found the first counterexample: four fifth powers whose sum is a fifth power. In fact

$$27^5 + 84^5 + 110^5 + 133^5 = 144^5$$

As a check:

$$
\begin{array}{rcr}
27^5 & = & 14348907 \\
84^5 & = & 4182119424 \\
110^5 & = & 16105100000 \\
133^5 & = & 41615795893 \\
\hline
144^5 & = & 61917364224
\end{array}
\tag{14.10}
$$

They found this example by exhaustive computer search.

In 1988 Noam Elkies found another counterexample by applying the theory of elliptic curves: three fourth powers whose sum is a fourth power.

$$
\begin{array}{rcr}
2682440^4 & = & 51774995082902409832960000 \\
15365639^4 & = & 55744561387133523724209779041 \\
18796760^4 & = & 124833740909952854954805760000 \\
\hline
20615673^4 & = & 180630077292169281088848499041
\end{array}
\tag{14.11}
$$

Instead of looking for integer solutions to the equation $x^4 + y^4 + z^4 = w^4$, Elkies divided out by w^4 and looked at the surface $r^4 + s^4 + t^4 = 1$ in coordinates (r, s, t). An integer solution to $x^4 + y^4 + z^4 = w^4$ leads to a rational solution $r = x/w, s = y/w, z = t/w$ of $r^4+s^4+t^4 = 1$. Conversely, given a rational solution of $r^4+s^4+t^4 = 1$, we can assume that r, s, t all have the same denominator w by putting them over a common denominator, and that leads directly to a solution to $x^4+y^4+z^4 = w^4$. Demjanenko [32] had found a rather complicated condition for a rational point (r, s, t) to lie on the closely related surface $r^4 + s^4 + t^2 = 1$. Namely, such a rational point exists if and only if there exist x, y, u such that

$$
\begin{aligned}
r &= x + y \\
s &= x - y \\
(u^2 + 2)y^2 &= -(3u^2 - 8u + 6)x^2 - 2(u^2 - 2)x - 2u \\
(u^2 + 2)t &= 4(u^2 - 2)x^2 + 8ux + (2 - u^2)
\end{aligned}
$$

To solve Elkies's problem it is enough to show that t can be made a square. A series of simplifications shows that this can be done provided the equation

$$Y^2 = -31790X^4 + 36941X^3 - 56158X^2 + 28849X + 22030$$

has a rational solution. This equation defines an elliptic curve. (Despite the presence of a fourth power on the right-hand side, it can be transformed into a cubic. A similar transformation can be found in Section 15.2. See also McKean and Moll [94] p. 254.) Conditions are known under which no solution can exist, but these conditions did not hold in this case, which showed that such a solution might possibly exist. At this stage Elkies tried a computer search, and found the solution

$$X = -\frac{31}{467} \qquad Y = \frac{30731278}{467^2}$$

From this he deduced the rational solution

$$r = -\frac{18796760}{20615673} \qquad s = \frac{2682440}{20615673} \qquad t = \frac{2682440}{20615673}$$

This led directly to a counterexample to Euler's conjecture for fourth powers, namely

$$2682440^4 + 15365639^4 + 187960^4 = 20615673^4$$

In fact, there are infinitely many solutions. The theory of elliptic curves provides a general procedure for constructing new rational points from old ones—the tangent/secant process of Section 14.5. Using a version of this, Elkies proved that infinitely many rational solutions exist. In fact he proved that rational points are dense on the surface $r^4 + s^4 + t^4 = 1$, that is, any patch of the surface, however small, must contain a rational point. The second solution generated by the tangent/secant process is

$$\begin{aligned}
x &= 1439965710648954492268506771833175267850201426615300442218292336336633 \\
y &= 4417264698994538496943597489754952845854672497179047898864124209346920 \\
z &= 9033964577482532388059482429398457291004947925005743028147465732645880 \\
w &= 9161781830035436847832452398267266038227002962257243662070370888722169
\end{aligned}$$

After Elkies had discovered there was a solution, Roger Frye of the Thinking Machines Corporation did an exhaustive computer search. He found a smaller solution, indeed the smallest possible solution:

$$\begin{array}{rcr}
95800^4 & = & 84229075969600000000 \\
217519^4 & = & 2238663363846304960321 \\
414560^4 & = & 29535857400192040960000 \\
\hline
422481^4 & = & 31858749840007945920321
\end{array} \tag{14.12}$$

14.8 Exercises

14.1 Suppose that a conic defined over $\mathbb{Q}$ has at least one rational point and does not consist of that point alone. Prove that it has infinitely many rational points.

14.2 Prove that the rational conic $x^2 + y^2 - 3 = 0$ contains no rational points.

(*Hint:* Rational solutions correspond to integer solutions of the equation $X^2 + Y^2 = 3Z^2$. Without loss of generality X, Y, and Z have no common factor > 1. Now consider the equation (mod 3).)

14.3 Let $a, b, c \in \mathbb{Q}$ be distinct. Consider the elliptic curve E with equation

$$y^2 = (x-a)(x-b)(x-c)$$

If $P \neq O$ denote the x-coordinate of P by x. Here O is the point (∞, ∞) at infinity (in projective geometry). Write $\bar{x} = x \,(\text{mod } (\mathbb{Q}^*)^2)$ and let $E(\mathbb{Q})$ be the group of rational points of E. Define a map

$$\theta : E(\mathbb{Q}) \to (\mathbb{Q}^*/((\mathbb{Q}^*)^2)) \times (\mathbb{Q}^*/((\mathbb{Q}^*)^2)) \times (\mathbb{Q}^*/((\mathbb{Q}^*)^2))$$

by

$$\theta(x) = \begin{cases} (\overline{x-a}, \overline{x-b}, \overline{x-c}) & \text{if } P \neq O, (a,0), (b,0), (c,0) \\ (\overline{(a-b)(a-b)}, \overline{a-b}, \overline{a-c}) & \text{if } P = (a,0) \\ (\overline{b-a}, \overline{(b-a)(b-c)}, \overline{b-c}) & \text{if } P = (b,0) \\ (\overline{c-a}, \overline{c-b}, \overline{(c-a)(c-b)}) & \text{if } P = (c,0) \\ (1,1,1) & \text{if } P = O \end{cases}$$

Show that the map θ is a group homomorphism.

14.4 Let E be the elliptic curve with equation $y^2 = x^3 + 1$. Find the set $\{P \in E(\mathbb{C}) : 3P = O\}$.

14.5 Let E be the elliptic curve with equation $y^2 = x^3 - 4$. Show that $(5, 11)$ lies on E.

If P is a rational point with x-coordinate m/n, show that $2P$ has x-coordinate

$$\frac{(m^3 + 32n^3)m}{4(m^3 - 4n^3)n}$$

Define the *height* of a rational number m/n (in lowest terms) to be $H(m/n) = \max(|m|, |n|)$. Write $H(P)$ for the height of the x-coordinate of P. Prove that

$$144.H(2P) \geq H(P)^4$$

Deduce that E has infinitely many rational points.

14.6 Let $0 \neq k \in \mathbb{Q}$, and define

$$\begin{aligned} X &= \{(x,y) \in \mathbb{Q}^2 : x^3 + y^3 = k\} \\ Y &= \{(x,y) \in \mathbb{Q}^2 : y^2 = \tfrac{1}{3}(4kx^3 - 1), x \neq 0\} \end{aligned}$$

Show that there is a map $\rho : X \to Y$ defined by

$$\rho(x, y) = \left(\frac{1}{x+y}, \frac{x-y}{x+y}\right)$$

and it is a bijection.

14.7 Let $0 \neq k \in \mathbb{Q}$, and define

$$\begin{aligned} X &= \{(x, y) \in \mathbb{Q}^2 : y^2 = x^4 + k\} \\ Y &= \{(x, y) \in \mathbb{Q}^2 : y^2 = x^3 - 4kx, (x, y) \neq (0, 0)\} \end{aligned}$$

Show that there is a map $\rho : X \to Y$ defined by

$$\rho(x, y) = \big(2(x^2 + y), 4x(x2 + y)\big)$$

and it is a bijection.

14.8 Let $0 \neq k \in \mathbb{Q}$, and define two elliptic curves:

$$E : y^2 = x^3 + kx \qquad E' : y^2 = x^3 - 4kx$$

Let $K \subseteq \mathbb{C}$ be a field. Show that there are two maps

$$f : E(K) \to E'(K) \qquad g : E'(K) \to E'K)$$

defined by

$$f(x, y) = \begin{cases} \left(x + \frac{k}{x}, y\left(1 - \frac{k}{x^2}\right)\right) & \text{if } (x, y) \neq (0, 0) \\ O & \text{if } (x, y) = (0, 0) \text{ or } O \end{cases}$$

$$g(x, y) = \begin{cases} \left(\frac{x}{4} - \frac{k}{x}, \frac{y}{8}\left(1 + \frac{4k}{x^2}\right)\right) & \text{if } (x, y) \neq (0, 0) \\ O & \text{if } (x, y) = (0, 0) \text{ or } O \end{cases}$$

Show that there is a map $\rho : X \to Y$ defined by

$$\rho(x, y) = \left(\frac{1}{x+y}, \frac{x-y}{x+y}\right)$$

and each of fg, gf is the map $P \to 2P$.

Show that the map $X \to Y \subseteq E'(K) \xrightarrow{g} E(K)$ (where $X \to Y$ is the map ρ in Exercise 14.6) sends (x, y) to (x^2, xy).

15

Elliptic Functions

So far, the discussion has been algebraic. We now introduce methods from complex analysis and the concept of a modular function. In the next chapter, we show how these classical ideas lead to the Taniyama–Shimura–Weil Conjecture, which forms the centrepiece of Wiles's approach to—and proof of—Fermat's Last Theorem.

The main topics discussed in this chapter are:

- Trigonometric functions as a link between conic sections, Diophantine equations, and complex analysis.
- Weierstrassian elliptic functions and their connection with elliptic curves.
- Elliptic modular functions and their connection with elliptic curves.

An excellent reference for the material covered in this chapter, and many related topics, is McKean and Moll [94]. See also King [72] for the basic material, plus some fascinating connections with the solution of polynomial equations.

15.1 Trigonometry Meets Diophantus

In this section we explore a rich area of interconnections between trigonometric functions, complex analysis, algebraic geometry, and the Pythagorean equation

$$X^2 + Y^2 = Z^2 \tag{15.1}$$

We take the point of view that we do not yet have the machinery of trigonometric functions available and show how to derive these functions from the aforementioned interconnections. This is a useful 'dry run' for subsequent generalisations to elliptic functions. Of course it helps to bear the standard trigonometric functions in mind throughout, since then the manipulations we perform make more sense.

Approach to Trigonometric Functions

We can consider (15.1) as the projective form of the affine equation

$$x^2 + y^2 = 1 \tag{15.2}$$

by setting $x = X/Z, y = Y/Z$. Clearly integer or rational solutions of (15.1) correspond to rational solutions of (15.2). Solving (15.2) for y we get $y = \pm\sqrt{1-x^2}$, which suggests looking at the integral

$$S(x) = \int \frac{\mathrm{d}x}{y} = \int \frac{\mathrm{d}x}{\sqrt{1-x^2}} \tag{15.3}$$

In order to evaluate this integral we assume that there exists a function $s(u)$, with derivative $c(u) = ds/du$, such that

$$c^2(u) + s^2(u) = 1 \tag{15.4}$$

To define these functions uniquely we impose 'initial conditions' $s(0) = 0, c(0) = 1$.

Given such a function, we can evaluate (15.3) by substituting $x = s(u)$. Then $dx = c(u)\mathrm{d}u$, so

$$\int \frac{\mathrm{d}x}{\sqrt{1-x^2}} = \int \frac{c(u)\mathrm{d}u}{\sqrt{1-s^2(u)}} = \int \mathrm{d}u = u$$

Therefore

$$\int \frac{\mathrm{d}x}{\sqrt{1-x^2}} = s^{-1}(x) \tag{15.5}$$

Turning all this round, we can use (15.5) as the definition of the function s, by means of the equation

$$\int_0^{s(u)} \frac{\mathrm{d}x}{\sqrt{1-x^2}} = u \tag{15.6}$$

and we deduce that s and its derivative c satisfy (15.4), together with the initial conditions. Similar arguments (Exercise 15.1) show that

$$s(-u) = -s(u) \qquad c(-u) = c(u)$$

for all u for which $s(u), c(u)$ are defined.

The abstract theory of the Riemann integral guarantees that $s(u)$ is defined for u in some neighbourhood of 0. We can extend the definitions of s, c to the whole of $\mathbb{R}$ by making use of some of their special properties. So we now deduce the standard properties of trigonometric functions from our definition. Differentiate (15.4) to get

$$2c(u)c'(u) + 2s(u)c(u) = 0$$

proving that

$$\frac{\mathrm{d}}{\mathrm{d}u}c(u) = -s(u) \tag{15.7}$$

We are now in a position to derive the standard power series for sine and cosine by invoking Taylor's Theorem. By induction, successive derivatives of s, c at the origin are given by

$$s^{(n)}(0) = \begin{cases} 0 \text{ if } n \equiv 0 \pmod 4 \\ 1 \text{ if } n \equiv 1 \pmod 4 \\ 0 \text{ if } n \equiv 2 \pmod 4 \\ -1 \text{ if } n \equiv 3 \pmod 4 \end{cases}$$

and

$$c^{(n)}(0) = \begin{cases} 1 \text{ if } n \equiv 0 \pmod 4 \\ 0 \text{ if } n \equiv 1 \pmod 4 \\ -1 \text{ if } n \equiv 2 \pmod 4 \\ 0 \text{ if } n \equiv 3 \pmod 4 \end{cases}$$

so that

$$\begin{aligned} s(x) &= \sum_{n=0}^{\infty}(-1)^n\frac{x^{2n+1}}{(2n+1)!} \\ c(x) &= \sum_{n=0}^{\infty}(-1)^n\frac{x^{2n}}{(2n)!} \end{aligned} \tag{15.8}$$

These series are absolutely convergent for all $x \in \mathbb{R}$ and therefore define s, c on the whole of $\mathbb{R}$. Indeed, we can replace $x \in \mathbb{R}$ by $z \in \mathbb{C}$ and extend the definitions to the complex plane: now s, c are complex analytic. The identity (15.4) now holds when x is replaced by any complex z, because two power series that are equal on an open set of real values x are equal throughout $\mathbb{C}$. At this stage we are entitled to replace s by sin and c by cos, but to emphasise the logical line of development we continue to use the notation s, c.

Addition Formulas and Parametrisation of the Circle

We next seek formulas for $s(u+v)$ and $c(u+v)$. Working backwards from (15.7), we see that Equation (15.4) is equivalent to $X(U) = s(u)$ and $X(u) = c(u)$ being independent solutions of the second order linear differential equation

$$\frac{\mathrm{d}^2 X}{\mathrm{d}u^2} + X = 0 \tag{15.9}$$

The general solution of (15.9), with arbitrary initial conditions, is

$$X(u) = As(u) + Bc(u)$$

for constants A, B. Let v be any constant. Clearly $X(u) = s(u+v)$ is a solution of (15.9), so

$$s(u+v) = As(u) + Bc(u) \tag{15.10}$$

and by differentiation with respect to u we also have

$$c(u+v) = Ac(u) - Bs(u) \tag{15.11}$$

where the constants A, B may depend on v. Letting $u = 0$ we see that $A = -c(v), B = s(v)$. Thus we have the addition formulas

$$\begin{aligned} s(u+v) &= s(u)c(v) + c(u)s(v) \\ c(u+v) &= c(u)c(v) - s(u)s(v) \end{aligned}$$

We are now ready to return to the Pythagorean Equation (15.2), which defines a curve in $\mathbb{R}^2$, the unit circle $\mathbb{S}^1$. By Equation (15.4) the point $(c(u), s(u))$ lies in $\mathbb{S}^1$ for any $u \in \mathbb{R}$. That is, there is a map

$$\begin{aligned} \Omega : \mathbb{R} &\to \mathbb{S}^1 \\ u &\mapsto (c(u), s(u)) \end{aligned}$$

This map is continuous, indeed infinitely differentiable.

The real line $\mathbb{R}$ has a natural abelian group operation $+$, and we can use the map Ω to transport this to $\mathbb{S}^1$ if we define $\oplus$ as follows:

$$(x_1, y_1) \oplus (x_2, y_2) = (x_1x_2 - y_1y_2, x_1y_2 + y_1x_2)$$

Under this operation $\mathbb{S}^1$ is an abelian group with identity element $(1, 0)$. The inverse of (c, s) is $(c, -s)$. The addition formulas for s, c now say that Ω is a group homomorphism, that is,

$$\Omega(u + v) = \Omega(u) \oplus \Omega(v) \tag{15.12}$$

The map Ω cannot be a group isomorphism since $\mathbb{S}^1$ has elements of all finite orders but $\mathbb{R}$ does not. Therefore Ω has a nontrivial kernel $\mathcal{K}$. We claim that there is a unique real number $\varpi > 0$ such that

$$\mathcal{K} = \varpi\mathbb{Z}$$

To see this, observe that the derivative of Ω is nonsingular near $u = 0$, so there exists $\varepsilon > 0$ such that $\Omega(u) \neq (1, 0)$ whenever $0 \neq u$ and $|u| < \varepsilon$. Therefore there exists a smallest real number $\varpi > 0$ for which $\Omega(\varpi) = (1, 0)$. Then $\Omega(n\varpi) = (1, 0)$ for all $n \in \mathbb{Z}$ since $\mathcal{K}$ is a subgroup. We claim that $\mathcal{K} = \varpi\mathbb{Z}$. If not, there exists $k \in \mathcal{K}$ such that

$$n\varpi < k < (n + 1)\varpi$$

for some $n \in \mathbb{Z}$. But then $k - n\varpi \in \mathcal{K}$ and $0 < k - n\varpi < \varpi$, a contradiction.

Numerical computations show that $\varpi \sim 6.28$, and of course $\varpi = 2\pi$.

Since Ω is a homomorphism and $\varpi \in \mathcal{K}$, we deduce that $\Omega(u + \varpi) = \Omega(u)$ for all $u \in \mathbb{R}$ (hence also for all $u \in \mathbb{C}$). That is, both s and c are ϖ-periodic.

All of the usual properties of the trigonometric functions now follow standard methods. In particular we can show (Exercise 15.2) that Ω is surjective. Geometrically, Ω wraps $\mathbb{R}$ round $\mathbb{S}^1$ infinitely many times, in such a manner that the usual distance in $\mathbb{R}$ becomes $1/2\pi$ times arc-length in $\mathbb{S}^1$.

The distance function in $\mathbb{R}$ and arc-length in $\mathbb{S}^1$ determine topologies, and Ω is continuous for these topologies. Now $\mathbb{S}^1$ is compact but $\mathbb{R}$ is not.

Pythagorean Equation

The group structure on $\mathbb{S}^1$ defined by $\oplus$ has an interesting implication for the Pythagorean Equation. Namely, given two solutions

$$x^2 + y^2 = 1 \qquad u^2 + v^2 = 1$$

it implies that there exists a further solution

$$(xu - yv)^2 + (xv + yu)^2 = 1 \tag{15.13}$$

This is a special case of Brahmagupta's Theorem 4.1.

For example, from the standard (3,4,5) right triangle we know that $x = 3/5, y = 4/5$ is a solution, and so is $u = 3/5, v = 4/5$. By (15.13) we compute a new solution $(7/25, 24/25)$, so that $7^2 + 24^2 = 25^2$. In this manner we can obtain an infinite number of rational solutions of (15.2), hence of integer solutions of (15.1).

We now seek to characterise those $u \in \mathbb{R}$ for which $\Omega(u) \in \mathbb{Q}^2$, that is, both $s(u)$ and $c(u)$ are rational. To this end we introduce

$$t = t(u) = \tan\frac{u}{2}$$

From the identities

$$\cos u = \cos^2\frac{u}{2} - \sin^2\frac{u}{2} \qquad 1 = \cos^2\frac{u}{2} + \sin^2\frac{u}{2}$$

we find that

$$\cos u = \frac{1-t^2}{1+t^2} \qquad \sin u = \frac{2t}{1+t^2}.$$

Proposition 15.1. *$\Omega(u) \in \mathbb{Q}^2$ if and only if $t \in \mathbb{Q}$.*

Proof: Clearly $t \in \mathbb{Q}$ implies $\Omega(u) \in \mathbb{Q}^2$.

Conversely, if $\Omega(u) \in \mathbb{Q}^2$ then

$$\frac{1+t^2}{2t} = p \in \mathbb{Q} \qquad \frac{1-t^2}{2t} = q \in \mathbb{Q}$$

so that

$$1 + t^2 = 2pt \qquad 1 - t^2 = 2qt$$

Adding,

$$2 = 2(p+q)t$$

so that $p + q \neq 0$ and $t = \frac{1}{p+q} \in \mathbb{Q}$. □

In other words, $\Omega(u) \in \mathbb{Q}^2$ if and only if $u \in 2\arctan\mathbb{Q}$.

The identity (15.4) is now equivalent to the rational identity

$$\left(\frac{1-t^2}{1+t^2}\right)^2 + \left(\frac{2t}{1+t^2}\right)^2 = 1$$

which we have already encountered in Example 14.9. Putting $t = u/v$ where $u, v \in \mathbb{Z}$ this yields

$$(u^2 - v^2)^2 + (2uv)^2 = (u^2 + v^2)^2$$

In Lemma 11.1 we showed that all primitive integer solutions of the Pythagorean equation (that is, solutions without common factors) are of this form. Thus we have found a link between the trigonometric functions (especially $t(u)$), the Pythagorean equation, and the unit circle in the plane.

A Curious Series

Later in this chapter we develop a profound generalisation of the above to elliptic curves. We could continue to explore the circle and its links for some time, but we content ourselves with one curious formula that makes the 2π-periodicity of the trigonometric functions 'obvious'. Its analogue for doubly periodic complex functions forms the basis of Weierstrass's approach to elliptic functions. The relevant identity is a series expansion of the cosecant function

$$\csc z = \frac{1}{z} + \sum_{n\in\mathbb{Z}\setminus\{0\}} (-1)^n \left(\frac{1}{z-n\pi} + \frac{1}{n\pi}\right) \tag{15.14}$$

This series is absolutely convergent and so may be differentiated term by term, yielding the simpler identity

$$\csc z \cot z = \sum_{n\in\mathbb{Z}} \frac{(-1)^n}{(z-n\pi)^2} \tag{15.15}$$

The series in (15.14) would itself be simpler if we could use

$$\sum_{n\in\mathbb{Z}} \frac{(-1)^n}{(z-n\pi)} \tag{15.16}$$

but unfortunately this series fails to converge, which is why the more complicated (15.14) replaces it.

If we replace z in (15.15) by $z+2\pi$ then the entire series shifts two places along, term by term. This makes it obvious that the function $\csc z \cot z$

is 2π-periodic in z. It is straightforward to parlay this result into 2π-periodicity of $\sin z$ and $\cos z$. So it is possible to *define* the trigonometric functions in terms of a series whose 2π-periodicity is immediately apparent from its form.

We outline the derivation of (15.14). We rely on standard ideas from complex analysis about residues. Suppose $f(z)$ is a complex analytic function whose only singularities in $\mathbb{C}$ are poles $z = a_j, j = 1, 2, 3, \dots$ with residues b_j, where $0 < |a_1| \leq |a_2| \leq |a_3| \leq \cdots$. Suppose that there exists a sequence of circles C_j, with centre the origin and of radius R_j which tends to infinity with j, not passing through any poles, with $f(z)$ uniformly bounded on the circles C_j; that is, $|f(z)| < M$ for all $z \in \cup_j C_j$.

An example is $f(z) = \csc z$, with $R_j = (j + \frac{1}{2})\pi$, to which we return shortly.

By the residue theorem (Stewart and Tall [130] chapter 12) if x is not a pole of f then

$$\frac{1}{2\pi \mathrm{i}} \int_{C_j} \frac{f(z)}{z-x} \mathrm{d}z = f(x) + \sum_r \frac{b_r}{a_r - x}$$

where the sum is over all poles interior to C_j. Now

$$\begin{aligned} \frac{1}{2\pi \mathrm{i}} \int_{C_j} \frac{f(z)}{z-x} \mathrm{d}z &= \frac{1}{2\pi \mathrm{i}} \int_{C_j} \frac{f(z)}{z} \mathrm{d}z + \frac{x}{2\pi \mathrm{i}} \int_{C_j} \frac{f(z)}{z(z-x)} \mathrm{d}z \\ &= f(0) + \sum_r \frac{b_r}{a_r} + \frac{x}{2\pi \mathrm{i}} \int_{C_j} \frac{f(z)}{z(z-x)} \mathrm{d}z \end{aligned}$$

Let $j \to \infty$. Then the integral

$$\int_{C_j} \frac{f(z)}{z(z-x)} \mathrm{d}z$$

tends to zero, and we get

$$f(x) = f(0) + \sum_{n=1}^{\infty} b_n \left(\frac{1}{x - a_n} + \frac{1}{a_n} \right) \tag{15.17}$$

To prove (15.14) we now set $f(z) = \csc z - \frac{1}{z}$. The singularities are at $z = a_j = j\pi$, $j \neq 0$, with residues $b_j = (-1)^j$. The conditions of the above calculation are easily checked, so (15.14) follows.

Weierstrass uses a very similar series to define a class of doubly periodic functions (and has a similar problem with convergence of the simplest form of the series, which he solves in the same manner). We now begin the development of Weierstrass's theory.

15.2 Elliptic Functions

We have seen that the trigonometric functions have a number of striking properties, including:

- Periodicity: $\sin(\theta + 2\pi) = \sin(\theta), \cos(\theta + 2\pi) = \cos(\theta)$
- Algebraic differential equation: $u'(\theta)^2 = 1 - u(\theta)^2$ where $u(\theta) = \sin(\theta)$ or $\cos(\theta)$
- Parametrisation of circle: $(\cos(\theta), \sin(\theta))$ lies on the unit circle $x^2 + y^2 = 1$ and every point on the circle is of this form
- Addition theorems:

$$\begin{aligned}\cos(\theta + \phi) &= \cos(\theta)\cos(\phi) - \sin(\theta)\sin(\phi)\\ \sin(\theta + \phi) &= \sin(\theta)\cos(\phi) + \cos(\theta)\sin(\phi)\end{aligned}$$

- Integration of algebraic functions: for example,

$$\int \frac{\mathrm{d}x}{\sqrt{1-x^2}} = \sin^{-1}(x)$$

These properties are all interconnected; moreover, they illuminate the theory of quadratic Diophantine equations, and in particular the Pythagorean Equation $X^2 + Y^2 - Z^2 = 0$, which in affine form is $x^2 + y^2 - 1 = 0$.

In 1811 Legendre published the first of a series of three volumes initiating a profound generalisation of trigonometric functions, known—for the rather peripheral reason that they lead to a formula for the arc length of an ellipse—as 'elliptic functions'. In fact Legendre worked only with 'elliptic integrals', of which the most important is

$$F(k, v) = \int_0^v \frac{\mathrm{d}z}{\sqrt{(1-z^2)(1-k^2z^2)}} \tag{15.18}$$

for a complex variable z and a complex constant k. This particular integral is the *elliptic integral of the first kind*: in Legendre's theory there are also elliptic integrals of the second and third kind. A detailed treatment can be found in Hancock [56] p. 187.

Gauss, Abel, and Jacobi all noticed something that had eluded Legendre. The history is complicated. Gauss never published the idea; Jacobi made it the basis of his monumental and influential work published in 1829; and a manuscript by Abel was submitted to the French Academy of Sciences in 1826, mislaid by Cauchy, and not published until 1841, by which time Abel was long dead. Abel also published work on elliptic functions in 1827, however. Their common idea was to consider the integral (15.18) not as defining a function, but as defining its *inverse function*. Denote this inverse function by $\mathrm{sn}\,u$: it satifies the equation

$$F(k, \mathrm{sn}\,u) = u$$

Strictly speaking, sn is a family of functions parametrised by k. Associated with it are two other functions

$$\mathrm{cn}\,u = \sqrt{1 - \mathrm{sn}^2\,u} \qquad \mathrm{dn}\,u = \sqrt{1 - k^2 \mathrm{sn}^2\,u}$$

These functions have remarkable properties reminiscent of the trigonometric functions: a sample is

$$\mathrm{sn}(x+y) = \frac{\mathrm{sn}\,x\,\mathrm{cn}\,y\,\mathrm{dn}\,y + \mathrm{sn}\,y\,\mathrm{cn}\,x\,\mathrm{dn}\,x}{1 - k^2\,\mathrm{sn}^2 x\,\mathrm{sn}^2 y}$$

A vast range of similar identities can be found in Cayley [17] and Hancock [56]. The most remarkable property of all, though, is that the functions $\mathrm{sn}, \mathrm{cn}, \mathrm{dn}$ are all *doubly periodic*. That is, there exist two complex constants ω_1, ω_2 (depending on k) that are linearly independent over $\mathbb{R}$, such that

$$\begin{aligned}
\mathrm{sn}(z+\omega_1) &= \mathrm{sn}(z+\omega_2) = \mathrm{sn}\,z \\
\mathrm{cn}(z+\omega_1) &= \mathrm{cn}(z+\omega_2) = \mathrm{cn}\,z \\
\mathrm{dn}(z+\omega_1) &= \mathrm{dn}(z+\omega_2) = \mathrm{dn}\,z
\end{aligned}$$

Legendre had recognised the equivalent property of his elliptic integrals, but its expression is far more cumbersome. Moreover, $\mathrm{sn}, \mathrm{cn}, \mathrm{dn}$ are all *meromorphic* functions of a complex variable, meaning that they are analytic except for isolated singularities, which are all poles. (See Stewart and Tall [130] or any other text on complex analysis for terminology.)

In 1882 Weierstrass developed a somewhat different approach to the whole topic of doubly periodic functions, based on a function denoted $\wp(z)$. (The symbol $\wp$ is pronounced 'pay' and is a stylised old German 'p'.) The Weierstrass $\wp$-function is closely connected with elliptic curves, and the remainder of this section is devoted to this connection.

The starting-point is to consider an arbitrary doubly periodic meromorphic function $f(z)$, where $z \in \mathbb{C}$. Then there are two complex constants ω_1, ω_2, which are linearly independent over $\mathbb{R}$, such that

$$f(z+\omega_1) = f(z+\omega_2) = f(z) \tag{15.19}$$

This implies that for all $(m, n) \in \mathbb{Z}^2$

$$f(z + m\omega_1 + n\omega_2) = f(z) \tag{15.20}$$

and opens up some interesting geometry of $\mathbb{C}$.

Definition 15.2. Let $\omega_1, \omega_2 \in \mathbb{C}$ be linearly independent over $\mathbb{R}$. Then the set

$$\mathcal{L} = \mathcal{L}_{\omega_1,\omega_2} = \{z \in \mathbb{C} : z = m\omega_1 + n\omega_2 \text{ where } m, n \in \mathbb{Z}^2\}$$

is the *lattice* generated by ω_1, ω_2.

We studied lattices in $\mathbb{R}^n$ in Chapter 7. The above definition is a special case and arises when we identify $\mathbb{C}$ with $\mathbb{R}^2$. When (15.20) holds, we say that f is *$\mathcal{L}$-periodic.*

Suppose that T is the fundamental domain of $\mathcal{L}$, see Definition 7.4. By Lemma 7.5, every $z \in \mathbb{C}$ lies in exactly one of the sets $T + l$ for $l \in \mathcal{L}$. Therefore $f(z)$ is uniquely determined once we know $f(t)$ for all $t \in T$. Classically, the topological closure $\overline{T}$ of T is called the *period parallelogram.* We can consider f to be a function on the quotient torus $\mathbb{T}^2 = \mathbb{C}/\mathcal{L}$.

The main problem is to define 'interesting' doubly periodic functions for a given lattice $\mathcal{L}$. Weierstrass's idea is a generalisation of (15.14, 15.15), namely that these can be obtained by *summing over translates by the lattice.* That is, take some (initially arbitrary) function $g(z)$ and consider

$$\hat{g}(z) = \sum_{l \in \mathcal{L}} g(z - l)$$

Then $\hat{g}$ is obviously doubly periodic, because for any $l' \in \mathcal{L}$

$$\hat{g}(z + l') = \sum_{l \in \mathcal{L}} g(z - l + l') = \sum_{l'' \in \mathcal{L}} g(z + l'') = \hat{g}(z)$$

where $l'' = l' - l$.

This is all very well, but it is necessary to choose $g(z)$ with care. Three things can go wrong:

- The series defining $\hat{g}$ fails to converge.
- The series converges but $\hat{g}$ is not meromorphic.
- The series converges but $\hat{g}$ turns out to be constant.

Avoiding these pitfalls requires a certain amount of foresight. Initially it led Weierstrass to the choice $g(z) = \frac{1}{z^3}$, for which $\hat{g}(z) = -\wp'(z)$, and from this to a suitable choice of (some constant multiple of) the integral of this particular $\hat{g}$. It turns out that taking the simpler series $g(z) = \frac{1}{z^2}$ instead is not a good idea: the series for $\hat{g}$ fails to converge. However, by adding a suitable 'arbitrary constant' (which, taken on its own, *also* fails to converge, but in the 'opposite' way) he could obtain a meromorphic function. His final choice was:

Definition 15.3. Let $\mathcal{L} \subseteq \mathbb{C}$ be a lattice. Then the associated *Weierstrass $\wp$-function* is

$$\wp(z) = \frac{1}{z^2} + \sum_{l \in \mathcal{L} \setminus 0} \left(\frac{1}{(z-l)^2} - \frac{1}{l^2} \right) \tag{15.21}$$

This series is absolutely convergent provided $z \notin \mathcal{L}$ (Exercise 15.9(a)). Evidently $\wp$ is an *even* function, that is, $\wp(-z) = \wp(z)$ (Exercise 15.9(c)). We may therefore differentiate term by term to get

$$\wp'(z) = -2 \sum_{l \in \mathcal{L}} \frac{1}{(z-l)^3} \tag{15.22}$$

This series is also absolutely convergent provided $z \notin \mathcal{L}$ (Exercise 15.9(b)).

Lemma 15.4. *The function $\wp$ is meromorphic, and doubly periodic on the lattice $\mathcal{L}$.*

Proof: By the above discussion, $\wp'$ is meromorphic, and doubly periodic on the lattice $\mathcal{L}$. Therefore

$$\wp'(z + \omega_j) = \wp'(z) \qquad (j = 1, 2)$$

Integrate, and remember to include an arbitrary constant:

$$\wp(z + \omega_j) = \wp(z) + c_j \qquad (j = 1, 2) \tag{15.23}$$

for constants $c_j \in \mathbb{C}$. Since $\wp$ is an even function, $\wp(-z) = \wp(z)$. Setting $z = -\omega_j/2$ in (15.23) we have

$$\wp(\omega_j/2) = \wp(-\omega_j/2) + c_j = \wp(\omega_j/2) + c_j$$

so $c_j = 0$. Therefore $\wp$ is $\mathcal{L}$-periodic.

By absolute convergence, $\wp(z)$ is analytic for all $z \notin \mathcal{L}$. When $z \in \mathcal{L}$ we may without loss of generality take $z = 0$. Then (15.21) shows that $\wp(z)$ has a simple pole at 0, of order 2. Therefore all singularities are poles, and $\wp$ is meromorphic. □

We now prove a useful lemma:

Lemma 15.5. *Suppose that f is a doubly periodic function that is analytic throughout $\mathbb{C}$ (that is, no poles or other singularities). Then f is constant.*

Proof: Let $\overline{T}$ be the closure of the fundamental domain of $\mathcal{L}$ (the period parallelogram). Since $\overline{T}$ is compact and f is analytic on $\overline{T}$, there exists a real constant M such that $|h(z)| < M$ for all $z \in \overline{T}$. By double periodicity, $|h(z)| < M$ for all $z \in \mathbb{C}$. By Liouville's Theorem (Stewart and Tall [130] p. 213) $h(z)$ is constant. □

Define

$$g_2 = 60 \sum_{l \in \mathcal{L} \setminus 0} \frac{1}{l^4} \qquad g_3 = 140 \sum_{l \in \mathcal{L} \setminus 0} \frac{1}{l^6}$$

(These and similar expressions are called *Eisenstein series.*) Then it may be shown (Hancock [56] p. 324) that the Laurent series (Stewart and Tall [130] chapter 11) of $\wp, \wp'$ take the following forms:

$$\wp(z) = \frac{1}{z^2} + \frac{g_2}{20} z^2 + \frac{g_3}{140} z^4 + \cdots \tag{15.24}$$

$$\wp'(z) = \frac{-2}{z^3} + \frac{g_2}{10} z + \frac{g_3}{7} z^3 + \cdots \tag{15.25}$$

Theorem 15.6. *The Weierstrass $\wp$-function satisfies the differential equation*

$$\wp'(z)^2 = 4\wp(z)^3 - g_2\wp(z) - g_3 \tag{15.26}$$

Proof: By direct computation,

$$\wp'(z)^2 = \frac{4}{z^6} - \frac{2g_2}{5z^2} + \frac{4g_3}{7} + O(z^2) \tag{15.27}$$

$$\wp(z)^3 = \frac{1}{z^6} + \frac{3g_2}{20z^2} + \frac{3g_3}{28} z^3 + O(z^2) \tag{15.28}$$

where $O(z^2)$ denotes a function whose Laurent series begins with terms in z^2 or higher—which, in particular, is an analytic function for all $z \in \mathbb{C}$. Therefore

$$\wp'(z)^2 - [4\wp(z)^3 - g_2\wp(z) - g_3] = O(z^2)$$

The left-hand side, which we denote by $h(z)$, is doubly periodic; the right-hand side is analytic throughout $\mathbb{C}$. Lemma 15.5 implies that $h(z)$ is constant. Since $h(z) = O(z^2)$, it follows that $h(z) = 0$. □

Corollary 15.7. *Let C be an elliptic curve in Weierstrass normal form $y^2 = 4x^3 - g_2x - g_3$ and let $\wp$ be the corresponding Weierstrass $\wp$-function. Then $(x, y) = (\wp'(z), \wp(z))$ lies on C for all $z \in \mathbb{C}$.*

In fact, every point on C is of the above form, so the function $\wp$ *parametrises* C. There is a close analogy with the parametrisation of a circle by $(\cos(\theta), \sin(\theta))$, especially since $\sin'(\theta) = \cos(\theta)$. Moreover, the parametrisation of C behaves naturally with respect to the group operation (14.9) on C, which for clarity we now rename $\oplus$ instead of $+$. (This is not the same $\oplus$ that arose earlier in this chapter in connection with trigonometric functions.) More precisely, let C be an elliptic curve with equation $y^2 = 4x^3 - g_2x - g_3$. Then C is nonsingular if and only if the cubic $4x^3 - g_2x - g_3$ has distinct zeros, which happens if and only if the discriminant

$$9g_3^3 + 32g_2^2 \neq 0$$

There is one point at infinity on C, namely $(0, 1, 0)$, and we choose this as the identity $\mathbb{O}$ of the group $\mathcal{G}$. Straightforward calculations in coordinate geometry then show that if

$$(x_3, y_3) = (x_1, y_1) \oplus (x_2, y_2)$$

then

$$x_3 = \frac{1}{4}\left(\frac{y_1 - y_2}{x_1 - x_2}\right)^2 - (x_1 + x_2) \tag{15.29}$$

$$y_3 = \frac{y_1 - y_2}{x_1 - x_2}x_3 + \frac{x_1y_2 - x_2y_1}{x_1 - x_2} \tag{15.30}$$

We now compare this with the addition theorem for the functions $\wp, \wp'$:

Theorem 15.8. *If $u \neq v \in \mathbb{C}$ then*

$$\wp(u+v) = \frac{1}{4}\left(\frac{\wp'(u) - \wp'(v)}{\wp(u) - \wp(v)}\right)^2 - (\wp(u) + \wp(v)) \tag{15.31}$$

$$\wp'(u+v) = \frac{\wp'(u) - \wp'(v)}{\wp(u) - \wp(v)}\wp(u+v) + \frac{\wp(u)\wp'(v) - \wp(v)\wp'(u)}{\wp(u) - \wp(v)} \tag{15.32}$$

Proof: We sketch a proof: see Hancock [56] p. 351 for details. Consider the function

$$h(u) = \wp(u+v) - \frac{1}{4}\left(\frac{\wp'(u) - \wp'(v)}{\wp(u) - \wp(v)}\right)^2$$

This is doubly periodic and meromorphic: its only poles are at points where $u + v \in \mathcal{L}$. However, by construction it is finite at $u = -v$, hence at $u = -v + l$ for all $l \in \mathcal{L}$. At $u = 0$ the function $h(u)$ goes to infinity like $-\frac{1}{u^2}$. It follows that $h(u)+\wp(u)$ is doubly periodic and analytic for all $u \in \mathbb{C}$. Lemma 15.5 implies that $h(u)+\wp(u)$ is constant. Setting $u = 0$ it follows that the constant is $-\wp(v)$.

The addition formula for $\wp'$ can be obtained by differentiation and further manipulations. □

These formulas may appear complicated, but they lead immediately to a very elegant theorem:

Theorem 15.9. *The group structure on C has the property*

$$(\wp(u), \wp'(u)) \oplus (\wp(v), \wp'(v)) = (\wp(u+v), \wp'(u+v)) \tag{15.33}$$

Proof: Compare formulas (15.29) and (15.31), and (15.30) and (15.32). □

Another way to say this is that the map $u \mapsto (\wp(u), \wp'(u))$ is a homomorphism between $(\mathbb{C}, +)$ and $(C, \oplus)$.

15.3 Legendre and Weierstrass

Legendre's theory of elliptic integrals concerns the square root of a quartic polynomial $(1-z^2)(1-k^2z^2)$. Weierstrass's theory revolves around the square root of a cubic polynomial $4z^3 - g_2z - g_3$. We briefly describe the connection between the two approaches, which is essentially that suitable birational maps transform the Legendre normal form into the Weierstrass normal form, and conversely. The discussion is summarised from Hancock [56] p. 190.

Consider the integral

$$\int \frac{\mathrm{d}z}{\sqrt{4z^3 - g_2z - g_3}} \tag{15.34}$$

By Theorem 15.26 the substitution $z = \wp(t)$ transforms this into

$$\int \mathrm{d}t = t \tag{15.35}$$

so that (15.34) is equal to $\wp^{-1}(z)$. The similarity with (15.18), which we restate here for convenience,

$$F(k, v) = \int_0^v \frac{\mathrm{d}z}{\sqrt{(1-z^2)(1-k^2z^2)}} \tag{15.36}$$

is striking. In fact, elementary computations show that if we make the substitution

$$w = \frac{a_3 + a_2}{2} + \frac{a_3 - a_2}{2} \frac{z - k}{1 - kz}$$

in (15.36), where

$$\left(\frac{1-k}{1+k}\right)^2 = \frac{a_1 - a_2}{a_1 - a_3}$$

then (15.36) becomes

$$\frac{1}{2}\sqrt{\frac{a_2 - a_3}{k}} \int \frac{\mathrm{d}w}{\sqrt{(w-a_1)(w-a_2)(w-a_3)}}$$

Moreover, we can change variables from w to $u = w + c$, for a suitable constant c, and eliminate the quadratic term in $(w-a_1)(w-a_2)(w-a_3)$, reducing the cubic to Weierstrass normal form:

$$\int \frac{\mathrm{d}w}{\sqrt{(w-a_1)(w-a_2)(w-a_3)}} = 2 \int \frac{\mathrm{d}u}{\sqrt{4u^3 - g_2 u - g_3}}$$

where

$$\begin{aligned} g_2 &= -4(a_1a_2 + a_2a_3 + a_3a_1) \\ g_3 &= 4a_1a_2a_3 \end{aligned}$$

The transformation is invertible, so we can also change an elliptic integral in Weierstrass normal form to one in Legendre normal form. In short, the two theories are equivalent.

This brief description conceals some beautiful mathematics that explains the relation between quartics and cubics that is exploited in the above transformation. This involves the invariants of cubic and quartic curves, the 'resolvent cubic' of a quartic equation, and the cross-ratio invariant from projective geometry. See Hancock [56] chapter VIII.

15.4 Modular Functions

The link between elliptic curves and Fermat's Last Theorem stems from a profound generalisation of doubly periodic complex functions. Liouville proved that every single-valued doubly periodic meromorphic function on a given lattice can be expressed as a rational function of Weierstrass's $\wp$-function and its derivative; see Hancock [56] p. 437. This theorem classifies all such doubly periodic functions, and at first sight leaves little room for generalisations. However, translations are not the only interesting transformations of the complex plane—and thereby hangs a tale.

Suppose that Γ is some group of invertible maps $\mathbb{C} \to \mathbb{C}$. Then we can seek complex functions f that are *invariant* under Γ, meaning that

$$f(\gamma(z)) = f(z)$$

for all $z \in C, \gamma \in \Gamma$. Doubly periodic functions arise in this way if we take Γ to be the group of all translations

$$z \mapsto z + m\omega_1 + n\omega_2$$

by elements $m\omega_1 + n\omega_2$ of the lattice $\mathcal{L}$. That is, $m, n \in \mathbb{Z}$.

The question is: which groups Γ lead to interesting results? Translations are a special case of an important class of transformations of $\mathbb{C}$:

Definition 15.10. Let $a, b, c, d \in \mathbb{C}$ with $ad - bc \neq 0$. The function

$$g(z) = \frac{az + b}{cz + d}$$

is called a *Möbius map* or *bilinear map.*

Classically, these maps are also called *Möbius transformations* or *bilinear transformations.*

As remarked earlier, there is a technical problem: Möbius maps take the value ∞ when $z = -d/c$. The usual way to get round this is to extend $\mathbb{C}$ to the *Riemann sphere* $\mathbb{C} \cup \{\infty\}$, see Stewart and Tall [130] p. 234. If this is done, the set of Möbius maps forms a group under composition (Exercise 15.3).

Straight lines and circles in $\mathbb{C}$ correspond to circles on $\mathbb{C} \cup \{\infty\}$. Moreover, every Möbius map sends circles on $\mathbb{C} \cup \{\infty\}$ to circles. Being complex-analytic, every Möbius map is *conformal*: it preserves the angles at which curves meet; see Stewart and Tall [130] chapter 13. So Möbius maps have several remarkable properties. Translations are Möbius maps: take $a = 1, c = 1, d = 0$ to get $z \mapsto z + b$. So we may hope to find generalisations of doubly periodic functions among the Möbius maps.

The trick now is to choose fruitful subgroups of the group of Möbius maps that generalise lattices as subgroups of the group of translations. Taking the whole group leads to nothing of interest, because it is easy to see that a function that is invariant under all translations $z \mapsto z + b$—let alone all Möbius maps—must be constant. As a clue, the translations by a lattice are defined in terms of a pair of integers (m, n), so they are discrete. This leads to the following choice of group:

Definition 15.11. The *modular group* is the group of all Möbius maps

$$g(z) = \frac{az + b}{cz + d}$$

where $a, b, c, d \in \mathbb{Z}$ and $ad - bc = 1$.

It is easily seen to be a group (Exercise 15.5). Abstractly, it can be described in terms of the group $\mathbb{SL}_2(\mathbb{Z})$ of all 2×2 matrices

$$\begin{bmatrix} a & b \\ c & d \end{bmatrix}$$

with $a, b, c, d \in \mathbb{Z}$ and determinant $ad - bc = 1$. In fact, if Z is the subgroup comprising $\pm I$ where I is the identity matrix, the modular group is isomorphic to

$$\mathbb{PSL}_2(\mathbb{Z}) = \mathbb{SL}_2(\mathbb{Z})/Z$$

which is known as the *projective special linear group*. See Exercise 15.5.

For any lattice $\mathcal{L}$, the group of lattice-translations has a fundamental domain, as described earlier. The defining property of a fundamental domain is that every point of $\mathbb{C}$ lies in exactly one of its translates by the lattice. The modular group also has a fundamental domain, but this has subtler geometry than a parallelogram. Its construction is assisted by defining two particular elements of the modular group, namely

$$S = \begin{bmatrix} 0 & -1 \\ 1 & 0 \end{bmatrix} \qquad T = \begin{bmatrix} 1 & 1 \\ 0 & 1 \end{bmatrix} \tag{15.37}$$

These correspond to the functions

$$S(z) = -\frac{1}{z} \qquad T(z) = z + 1$$

It is easy to see that $S^2 = -I$ which in $\mathbb{SL}_2(\mathbb{Z})$ represents the same element as I, that T has order ∞, and $(ST)^3 = -I$ which again represents the same element as I. See Exercise 15.6.

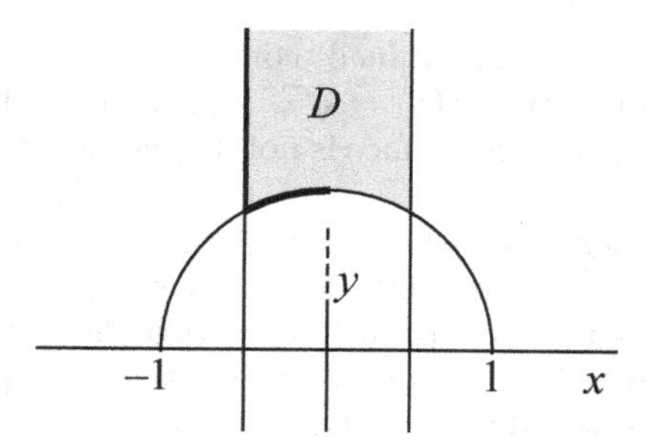

Figure 15.1. Fundamental domain for the modular group.

Let $\mathbb{H} = \{z : \mathrm{Im}(z) > 0\}$ be the upper half-plane in $\mathbb{C}$. (Despite our choice of terminology earlier, the pictures make it difficult to call this the upper half-line.) It is easy to check that the modular group maps $\mathbb{H}$ to itself. We define the *modular domain* $\mathbb{D}$ by

$$\mathbb{D} = \{z : -\tfrac{1}{2} \leq \mathrm{Re}(z) \leq 0, |z| = 1 \text{ or } -\tfrac{1}{2} \leq \mathrm{Re}(z) < \tfrac{1}{2}, |z| > 1\}$$

See Figure 15.1.

Theorem 15.12. *$\mathbb{D}$ is a fundamental domain for the modular group acting on $\mathbb{H}$.*

Proof: See Goldman [52] p. 184. □

The proof shows more: the effects of S and T on $\mathbb{D}$ are as in Figure 15.2.

One final ingredient is needed before we can define an elliptic modular function, namely: the concept of a *Riemann surface*. This concept was introduced in Riemann's Inaugural Dissertation [109] of 1851, as a general

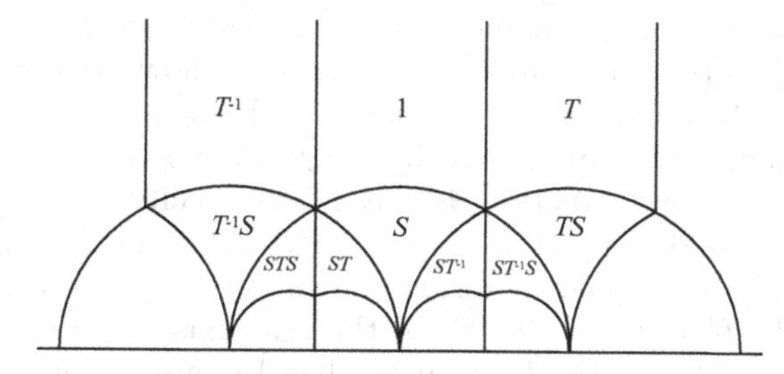

Figure 15.2. Tesselation of the upper half-plane by images of the fundamental domain.

method for making sense of 'multivalued' complex functions. We explain the idea briefly for the function $f(z) = \sqrt{z}$, and then we describe an abstract generalisation in which the surface is not associated with a previously defined function.

Every nonzero complex number, like every nonzero positive real number, has two distinct square roots. Indeed if $z = re^{i\theta}$, then $\sqrt{z} = +\sqrt{r}e^{i\theta/2}$ or $-\sqrt{r}e^{i\theta/2}$. When z is real and positive, the two choices can be distinguished by their sign, and it is reasonable to consider the positive square root as the 'natural' choice and the negative one as a secondary alternative. In the complex case, no straightforward distinction of this kind is possible, for the following reason. The complex setting reveals exactly why there are two choices of sign: the key point is that $-\sqrt{r}e^{i\theta/2} = +\sqrt{r}e^{i\theta/2+\pi} = +\sqrt{r}e^{i(\theta+2\pi)/2}$. That is, there are two alternatives *not* because the modulus r has two real square roots, but because the argument θ is defined only modulo 2π, and different choices of argument lead to two different values for the square root. The role of the argument is clearer if we consider the cube root $\sqrt[3]{z}$, which takes any of three values:

$$\sqrt[3]{r}e^{i\theta/3} \qquad \sqrt[3]{r}e^{i\theta/3+2\pi/3} \qquad \sqrt[3]{r}e^{i\theta/3+4\pi/3}$$

because the choices of argument $\theta, \theta + 2\pi, \theta + 4\pi$ lead to different results. (However, $\theta + 6\pi$ leads to the same cube root as θ.)

The impossibility of defining one choice to be the 'natural' square or cube root becomes obvious if we consider what happens as z moves along a continuous path in $\mathbb{C}$. For example, suppose that $z = e^{it}$ and t runs from 0 to 2π. When $t = 0$ the two square roots are $+1, -1$. The same is true when $t = 2\pi$. However, if we require $\sqrt{z}$ to vary continuously with z, then 1 lies on the path of square roots given by $\sqrt{e^{it}} = e^{it/2}$, and -1 lies on the path of square roots given by $\sqrt{e^{it}} = -e^{it/2}$. As t increases from 0 to 2π, the choice 1 changes continuously into -1, while the choice -1 changes continuously into 1. That is, the choices $1, -1$ swap places. Therefore neither can be considered more natural than the other.

Prior to Riemann, such phenomena were handled by declaring the function to be 'multivalued', and prescribing rules for how choices of values should be made. Riemann's idea for coping with such behaviour is radically different: define the function to be single-valued (as is now conventional whenever the word 'function' is used), but specify a domain that is different from the usual complex plane $\mathbb{C}$. For the function $\sqrt{z}$, Riemann's construction can be described in terms of two superposed copies $\mathbb{C}_1$ and $\mathbb{C}_2$ of $\mathbb{C} \setminus \{0\}$, slit from 0 to $-\infty$ along the real axis. The top left-hand quadrant of $\mathbb{C}_1$ is glued to the bottom left-hand quadrant of $\mathbb{C}_2$, and the top left-hand quadrant of $\mathbb{C}_2$ is glued to the bottom left-hand quadrant of $\mathbb{C}_1$. If we try to draw the resulting surface $\mathbb{S}$ in $\mathbb{R}^3$ then it is forced to intersect itself, but abstractly no such self-intersection is implied by the gluing recipe.

There is a canonical projection $\rho : \mathbb{S} \to \mathbb{C} \setminus \{0\}$ which identifies each copy $\mathbb{C}_j$ of $\mathbb{C}\setminus\{0\}$ with $\mathbb{C}\setminus\{0\}$. Any continuous path $\gamma(t)$ in $\mathbb{C}\setminus\{0\}$ lifts to a continuous path $\hat{\gamma}(t)$ in $\mathbb{S}$, by which we mean that $\rho(\hat{\gamma}(t)) = \gamma(t)$. Suppose, for example, that $\gamma(t)$ describes the unit circle in $\mathbb{C} \setminus \{0\}$, anticlockwise, so that $\gamma(t) = \mathrm{e}^{\mathrm{i}t}$ for $0 \leq t \leq 2\pi$. As t increases from 0 to π, this curve lifts to

$$\hat{\gamma}(t) = \mathrm{e}^{\mathrm{i}t} \in \mathbb{C}_1$$

However, because the two sheets $\mathbb{C}_1, \mathbb{C}_2$ are cross-connected along their negative real axes, the curve lifts to

$$\hat{\gamma}(t) = \mathrm{e}^{\mathrm{i}t} \in \mathbb{C}_2$$

as t increases from π to 2π. So the lifted path, unlike the original, is not a closed loop; instead, it returns to a different sheet, $\mathbb{C}_2$ rather than $\mathbb{C}_1$. The lifted loop closes up if we go round the original loop *twice*.

If the argument of z describes the path $\gamma(t)$ in $C \setminus \{0\}$, and we choose a continuously varying argument for $\sqrt{z}$, then the choices $\pm\sqrt{z}$ change in the same manner as the sheets of the surface $\mathbb{S}$. That is, we can define a single-valued square root on $\mathbb{S}$. This is Riemann's idea. The surface $\mathbb{S}$ is one of the simplest examples of a Riemann surface in the classical sense; for further examples and proofs see Stewart and Tall [130] p. 299 onwards.

The modern treatment of a Riemann surface is based on the geometry of the surface and the 'complex structure' given by its relation to $\mathbb{C}$:

Definition 15.13. A *surface* S is a topological space, covered by a countable collection of open subsets U called *patches*. Each patch U is equipped with a *local coordinate map* $\alpha_U : U \to \mathcal{D}$ where $\mathcal{D} \subseteq \mathbb{R}^2$ is the open unit disc. Finally, if $x \in S$ lies in the overlap $U \cap V$ of two patches, then the *overlap map* $\alpha_V^{-1}\alpha_U$ must be continuous where it is defined.

A *Riemann surface* is defined in the same way, but now we consider $\mathcal{D}$ to be the open unit disc in $\mathbb{C}$, and the overlap maps are required to be conformal, that is, to preserve the angles at which curves cross.

For further details see McKean and Moll [94] pp. 3–5.

We can now define an elliptic modular function:

Definition 15.14. Let $N > 0$ be an integer, and define a subgroup

$$\Gamma_0(N) \subseteq \mathbb{SL}_2(\mathbb{Z})$$

by

$$\Gamma_0(N) = \left\{ \begin{bmatrix} a & b \\ c & d \end{bmatrix} : a, b, c, d \in \mathbb{Z}, ad - bc = 1, N|c \right\}$$

This group acts on $\mathbb{H}$ and there is a compact Riemann surface $X_0(N)$ such that

$$\mathbb{H}/\Gamma_0(N) = X_0(N) \setminus \mathcal{K}$$

where $\mathcal{K}$ is some finite set of points. The points in $\mathcal{K}$ are the *cusps* of the *modular curve* $X_0(N)$ of *level* N.

Definition 15.15. An *(elliptic) modular function of level* N on $\mathbb{H}$ is a function $f(z)$ that is invariant under $\Gamma_0(N)$ and descends to a function that is meromorphic on $X_0(N)$, even at the cusps.

Here 'descends' just means that f has the same value on each orbit of $\Gamma_0(N)$ on $\mathbb{H}$ and hence defines a function on the quotient space $\mathbb{H}/\Gamma_0(N)$.

15.5 Exercises

15.1 Prove that

$$s(-u) = -s(u), \quad c(-u) = c(u)$$

for all u for which $s(u), c(u)$ are defined.

(*Hint*: $s(-u)$ and $c(-u)$ satisfy the differential Equation (15.9), hence are of the form $As(u) + Bc(u)$ for constants A, B. Which?)

15.2 Prove that the map $\Omega : \mathbb{R} \to \mathbb{S}^1$ defined by $\Omega(u) = (c(u), s(u))$ is onto.

(*Hint*: Consider the trajectory of $(s(t), c(t)$ as t increases from 0. Prove that it lies on the unit circle for all t. Calculate the tangent vector; find its length and direction.)

15.3 Let

$$f(z) = \frac{az + b}{cz + d}$$
$$g(z) = \frac{Az + B}{Cz + D}$$

be Möbius maps, so that $ad - bc \neq 0, AD - BC \neq 0$. Find a formula for $g(f(z))$, and deduce that the set of Möbius maps forms a group under composition.

15.4 Continuing Exercise 15.3, assume that $ad - bc = 1$ and $AD - BC = 1$, and prove that the modular group really is a group.

15.5 Prove that the modular group is isomorphic to

$$\mathbb{PSL}_2(\mathbb{Z}) = \mathbb{SL}_2(\mathbb{Z})/Z$$

15.6 Let S, T be as defined in (15.37). Prove that $S^2 = -I$, T has order ∞, and $(ST)^3 = -I$.

15.7 Let $f(z)$ be a doubly periodic meromorphic function, and assume that it has no poles or zeros on the boundary of the period parallelogram. Suppose that it has exactly P poles inside the period parallelogram, counted according to multiplicity. Prove that it has exactly P zeros inside the period parallelogram, counted according to multiplicity. Deduce that for any $z_0 \in \mathbb{C}$ such that $f(z) \neq z_0$ on the boundary of the period parallelogram, there are exactly P solutions of the equation $f(z) = z_0$ inside the period parallelogram, again counted according to multiplicity.

(*Hint*: Use the formula

$$N - P = \frac{1}{2\pi \mathrm{i}} \int_\gamma \frac{f'(z)}{f(z)}$$

where γ is a closed contour, f is meromorphic inside γ, and N, P are the numbers of zeros and poles of f inside γ. See Stewart and Tall [130] theorem 12.12 or Titchmarsh [137] section 3.4.)

15.8 Let $\mathcal{L}$ be a lattice and write $\wp_\mathcal{L}(z)$ for $\wp(z)$ relative to that lattice. Let $0 \leq c \in \mathbb{C}$. Prove that

$$\wp_{c\mathcal{L}}(z) = c^{-2}\wp_\mathcal{L}(\frac{z}{c})$$

15.9 Let the lattice $\mathcal{L}$ be generated by $\omega_1, \omega_2 \in \mathbb{C}$. Recall (15.21):

$$\wp(z) = \frac{1}{z^2} + \sum_{l \in \mathcal{L} \setminus 0} \left(\frac{1}{(z-l)^2} - \frac{1}{l^2} \right)$$

(a) Prove that for fixed but arbitrary $z \notin \mathcal{L}$, the series on the right is absolutely convergent. (*Hint*: Define parallelograms P_j with vertices at $\pm j\omega_1 \pm j\omega_2$, as in Figure 15.3. Estimate the size of the sum over $l \in P_j$, and consider the sum over $l \in P_1 \cup \cdots \cup P_j$. Then let $j \to \infty$.)

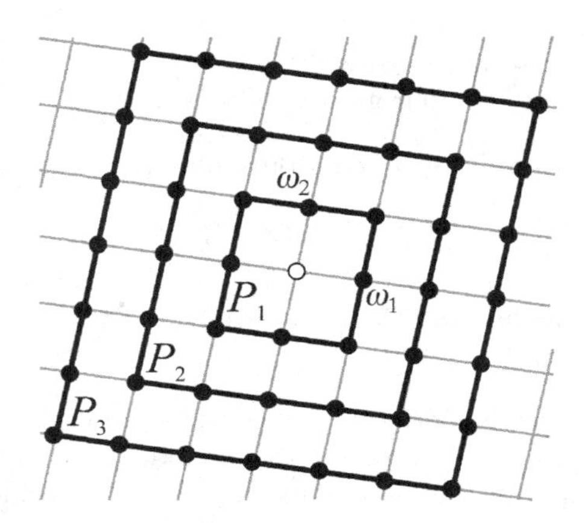

Figure 15.3. The parallelograms P_j. The origin (white circle) is omitted from the infinite series.

(b) Use the same estimates to show that the series (15.22) for $\wp'(z)$ is absolutely convergent.

(c) Prove that $\wp$ is an even function; that is, $\wp(-z) = \wp(z)$.

(d) Prove that $\wp'(z + l) = \wp'(z)$ for all $l \in \mathcal{L}$.

(e) Deduce that $\wp(z + l) = \wp(z)$ for all $l \in \mathcal{L}$.

15.10 *Why an 'elliptic function' is elliptic.*

Let $E \subseteq \mathbb{R}$ be the ellipse with equation $\frac{x^2}{a^2} + \frac{y^2}{b^2} = 1$, where $a > b$. Show that the perimeter of E has length

$$L = 4a \int_0^1 \sqrt{\frac{1 - k^2x^2}{1 - x^2}} \mathrm{d}x$$

where $k^2 = 1 - \frac{b^2}{a^2}$. The integral is known as Jacobi's complete elliptic integral of the second kind.

15.11 *Simple pendulum, exact solution.*

At school we are taught to find the period of a simple pendulum (of length l in units that make $l = 1$, under gravity g) from the equation of motion

$$\ddot{\theta} + g \sin\theta = 0$$

using the approximation $\sin\theta \sim \theta$, which leads to simple harmonic motion.

Find the exact solution, and its period, as follows:

(1) Integrate to get $I = \cos\theta - \frac{1}{2}\dot{\theta}^2$. Here I is a constant, whose interpretation is that the highest point the pendulum reaches is $\cos^{-1} I$.

(2) Substitute $x = \sqrt{\frac{2}{I-1}} \sin\frac{\theta}{2}$ to obtain

$$\dot{x} = \pm\sqrt{(1-x^2)(1-k^2x^2)} \qquad k^2 = \frac{I-1}{2}$$

(3) Deduce that

$$t = \int_0^x \frac{dy}{\sqrt{(1-y^2)(1-k^2y^2)}}$$

where the sign of the square root changes when $x = \pm 1$, the highest point.

(4) Therefore the period is $4K(k)$ where K is the complete elliptic integral.

V

Wiles's Proof of Fermat's Last Theorem

16

Path to the Final Breakthrough

We have seen that in the late 19th and early 20th centuries, the study of Fermat's Last Theorem built mainly on Kummer's methods, with the notion of ideal numbers being supplanted by Dedekind's theory of ideals in a commutative ring. The techniques required a high degree of mathematical and computational facility and were applied to more and more special cases. For instance, in 1905 Mirimanoff extended Kummer's results as far as $n \leq 257$. In 1908 Dickson generalised the theories of Germain and Legendre by investigating $x^n + y^n = z^n$ in the case where n is prime and none of x, y, z is divisible by n. Fermat's Last Theorem was proving to have a nasty sting in its tail. Despite the apparently simple statement of the problem, the proofs of special cases were becoming ever more complex, requiring the highly specialised activity of mathematical experts.

16.1 Wolfskehl Prize

In 1908 the situation changed dramatically, and the problem was opened up to other mathematicians and to a wider world of amateurs. The agent of change was Paul Friedrich Wolfskehl, the son of a wealthy Jewish banker; he was born in Darmstadt in 1856. He first studied medicine, obtaining his doctorate in 1880. However, debilitating multiple sclerosis made it impossible for him to practice surgery, and in 1880 he turned instead to mathematics. He began this activity in Bonn but moved to Berlin the following year, where he attended lectures by the 72-year-old Kummer. So fascinated did Wolfskehl become with the still unproved Last Theorem of Fermat that he left 100,000 marks in his will, to be awarded to the first person either to prove the theorem or to give a counterexample; see Barner [4]. In today's currency, this would have been worth around $2 million. The prize was announced by the Royal Society of Science in Göttingen, on 13 September 1908, exactly two years after Wolfskehl's death; it was to be claimed on or before 13 September 2007. Although hyperinflation in Germany in the 1920s greatly diminished the value of the bequest, it was still valued at 75,000 Deutsche marks at the end of the 20th century, thanks to judicious investment. (The mark was replaced by the euro, and euro banknotes were first available in 2002.)

Wolfskehl's act of altruism proved a mixed blessing to the mathematical community. In the first year alone, 621 solutions were submitted, and although the frequency slowly decreased, attempted solutions continued to flow in for the next ninety years. The total number sent to the Göttingen Academy has been estimated at over 5,000, and each attempt had to be read and considered by one of the judges. The endless succession of 'proofs' of Fermat's Last Theorem kept the staff and assistants involved continually busy. Not only did they have to deal with problems regularly, they could also become involved in protracted correspondence in addition to the initial reply. One correspondence on record extended to over sixty communications.

Other universities did not escape the burden. At the Royal Society of Science in Berlin the numerous attempted proofs were dealt with by a single individual, Albert Fleck, who courteously replied to each aspirant, highlighting the error in the manuscript and succinctly explaining the mistake.

Sometimes the solutions were put forward by eminent mathematicians. Ferdinand Lindemann (1852–1939), who is famous for his proof of the transcendence of π, published a fallacious proof of Fermat's Last Theorem in 1901. He soon withdrew it, but he continued his efforts with a 64-page paper in 1908. Fleck showed him his error on pages 23 and 24, rendering the remainder of the enterprise worthless. Fleck was a true 'amateur' who loved his work: his 'Fermat Clinic' at the Berlin Academy consisted solely of himself, at his desk in his room in the Mathematics Department. For these efforts, Fleck was awarded the Leibniz silver medal of the Berlin Mathematical Society in 1915, and he continued in this task until his death in 1943. As a Jew in Nazi Germany, his final years were blighted by persecution and humiliation. As the 20th century continued, the volume of solutions diminished, but they continued to arrive at intervals from all corners of the world. When the Berlin wall was removed, the number of solutions from Eastern Europe suddenly increased, because academics from the former Soviet Union were once more able to communicate freely with the West.

Despite the large number of attempted proofs, the actual advances in the early 20th century were prosaic and highly technical. In 1909, Arthur Wieferich focused on the 'first case', where n is a prime not dividing any of x, y, z, and proved that if there is a solution then the condition

$$2^{n-1} \equiv 1 \pmod{n^2}$$

must be satisfied. Relationships like this became much more useful with the arrival of computers in the middle of the 20th century, because it was then practical to check them for large p. The American mathematician Harry Schultz Vandiver (1882–1973) introduced methods that made a computational approach to the full theorem possible for any specific exponent n (not *too* large). He had little formal education, and left school early to work in his father's firm. In 1904 he collaborated with the 20-year-old George Birkhoff in a paper on the factorisation of integers of the form $a^n - b^n$, becoming yet another in the long line of amateurs who were fascinated with number theory. He took a university appointment in 1919 and worked extensively on Fermat's Last Theorem, leading to the award of the Cole Prize of the American Mathematical Society in 1931. His findings built on the work of Kummer and were particularly amenable to computation. In 1952, at the age of seventy, he used a computer to prove Fermat's Last Theorem for $n \leq 2,000$.

The value of the exponent continued to be raised at intervals over the years, as we saw in Section 13.1. The methods continued to involve heavy calculations, making painful step-by-step progress without any simple fundamental insight that addressed the whole problem in a truly conceptual way. The proof was remarkably elusive. It seemed that the Wolfskehl Prize would be unclaimed in the few years left before time ran out in 2007.

16.2 Other Directions

Meanwhile, mathematics was continuing to grow in other directions, which seemed at the time to have nothing whatsoever to do with Fermat's Last Theorem. However, history is littered with cases where mathematicians attempting to solve one problem ended up formulating and proving something quite different. Indeed, Kummer's original breakthrough in his proof of many cases of Fermat's Last Theorem occurred when he was working on a totally different problem in the generalised theory of quadratic reciprocity. In the same manner, the ingredients that were to lead Wiles to the final proof of Fermat's Last Theorem arose in areas which, at first, seemed to have no possible link with it.

In the last decade of the 19th century, Henri Poincaré developed the new theory of algebraic topology in his book *Analysis Situs* (1895). He invented ways to translate topological problems into algebraic form. He classified surfaces in terms of their 'fundamental group' which, among other things, gives information about the number of 'holes' in the surface and relates this number to an integer called the 'genus'. A sphere with no holes has genus 0, a torus has genus 1, and other surfaces with 'more holes' have genus $g \geq 2$.

Initially, this idea seemed to have no relationship with Fermat's Last Theorem. However, there is a connection. An integer solution of Fermat's equation, say $a^n + b^n = c^n$, corresponds to the rational solution $x = a/c, y = b/c$ of the polynomial equation

$$x^n + y^n - 1 = 0 \tag{16.1}$$

Therefore Fermat's Last Theorem is equivalent to showing that this polynomial equation has no rational solutions. The Cambridge mathematician Louis Mordell had the bright idea of looking not only at the rational solutions of a polynomial equation $Q(x, y) = 0$ with rational coefficients, but also at its complex solutions. Topologically, such solutions are related to a surface whose genus happens to be $(n-1)(n-2)/2$. For $n \geq 4$, the genus is therefore 2 or more. In 1922 Mordell formulated what is now called the Mordell Conjecture: a polynomial equation $Q(x, y) = 0$ with rational coefficients and genus $g \geq 2$ has only finitely many rational solutions. If this could be proved, then it would immediately follow that for any $n \geq 4$ the Fermat equation $a^n + b^n = c^n$ has at most a finite number of integer solutions.

At first this seemed not to carry the Fermat quest very far forward. To start with, it was an unproved conjecture. Even if it were proved, it would show only that the equation has a finite number of solutions, when what we actually wish to show is that there are none. Nevertheless, the Mordell Conjecture turned out to be an important step towards the final proof of Fermat's Last Theorem. Early work of André Weil [142] led to significant progress in special cases, and the full Mordell Conjecture was finally proved by Gerd Faltings in 1983; see Bloch [9]. The proof was immediately followed by new results. In 1985, two different papers were published to confirm that Fermat's Last Theorem is true for 'almost all' n.

Andrew Granville and 'Roger' Heath-Brown showed that the proportion of those n for which Fermat's Last Theorem is true tends to 1 as n becomes large. It is a remarkable result but still not the full proof that was so eagerly sought. Moreover, it tells us nothing about any particular exponent n.

16.3 Modular Functions and Elliptic Curves

Other ideas of Poincaré also proved to be seminal in the proof of Fermat's Last Theorem, although again the link was not obvious when they were first introduced. As a visual thinker, Poincaré loved to study systems that have symmetry. An area of particular interest was that of symmetries in complex function theory, so Poincaré studied complex functions $f(z)$ that remain invariant when their domains are operated on by an element of the modular group; see Definition 15.11. That is, Möbius maps

$$g(z) = \frac{az+b}{cz+d}$$

where $a, b, c, d \in \mathbb{Z}$ and $ad - bc = 1$.

When $z = -d/c$, the image under the transformation is infinite, so that to obtain a more satisfactory theory, it is best to adjoin the point at infinity to the complex plane to give a surface that is topologically like the surface of a sphere. Möbius maps have the delightful property of mapping circles on this sphere to circles and preserving angles. Functions that are invariant under a countably infinite group of Möbius maps are called *automorphic*.

Poincaré went further, and considered those functions transforming the upper half-plane ($z = x + iy$ where $y > 0$) to itself that remain invariant under the same kinds of map. Adding one or two technical conditions, he developed a theory of *modular functions*. We discuss these in Chapter 17. For the moment, it is sufficient to know that modular functions have properties that eventually made them a pivotal idea in the proof of Fermat's Last Theorem.

The introduction of complex numbers into the study of Fermat's Last Theorem—particularly the study of polynomial equations with rational coefficients, as in the Mordell Conjecture—played another important role. This relates to elliptic curves, which we recall are defined by a cubic equation of the form

$$y^2 = Ax^3 + Bx^2 + Cx + D$$

where A, B, C, D are all rational. The trick that opens up a route to a proof of Fermat's Last Theorem can be formulated as follows: imagine this equation for complex x and y, and to attempt to parametrise it with functions $x = f(z), y = g(z)$ satisfying the equation

$$g(z)^2 = Af(z)^3 + Bf(z)^2 + Cf(z) + D$$

However, this point of view on the ideas involved is a fairly recent one: the early formulations were stated in more technical ways. See Rubin and Silverberg [113].

16.4 Taniyama–Shimura–Weil Conjecture

In 1955, a highly significant step was taken by two Japanese mathematicians who were planning a conference in Tokyo on algebraic number theory. Yutaka Taniyama was interested in elliptic curves. He had a powerful intuitive grasp of mathematics but was prone to making errors. But his friend Goro Shimura, a much more formal mathematician, realised that Taniyama had an instinctive ability to imagine new relationships that were not available to more careful thinkers. At their conference they presented a number of problems for consideration by the participants. Four of these, proposed by Taniyama, dealt with possible relationships between elliptic curves and modular functions. From these developed what became known as the Taniyama–Shimura–Weil Conjecture. This conjecture hypothesised that every elliptic curve can be parametrised by modular functions. (Its technical statement was different, and only much later was it reinterpreted in this way as a result of other discoveries in the area.)

At the time this was a surprising idea to most workers in the field, who saw elliptic curves and modular functions as inhabiting quite different parts of mathematics, so at first the conjecture was not taken seriously. Shimura left Tokyo for Princeton in 1957, resolving to return in two years to continue work with his colleague. His plans were not realised; in November 1958 Taniyama committed suicide. A letter left beside his body explained that he did not really know why he had decided on this action: simply that he was in a frame of mind where he had lost confidence in his future. He was due to be married within a month. A few weeks later his fiancée also took her own life. They had promised each other they would never be parted and she chose to follow him in death.

Shimura reacted to this double tragedy by devoting his energies to understanding the relationship between elliptic curves and modular functions. Over the years he gathered so much supporting evidence that the Taniyama–Shimura–Weil Conjecture became more widely appreciated. It occupies a pivotal position between two different areas of mathematics.

Both of these areas had been studied intensely, but had remained separate. If the conjecture were true, then unsolved problems in one area could be translated into the language and concepts of the other and perhaps solved by the novel methods available there.

In the 1960s and 1970s, hundreds of mathematical papers appeared which showed that if the Taniyama–Shimura–Weil Conjecture were true, then other—very important—results would follow. A whole mathematical industry was being built on a principle that still eluded proof.

16.5 Frey's Elliptic Equation

In the depths of the Black Forest in Germany, near the town of Oberwolfach, is a retreat for mathematical researchers, where they can gather in a relaxed environment to share their ideas. In the summer of 1984 a group of number theorists assembled to discuss their latest ideas on elliptic equations. In a lecture at the meeting Gerhard Frey, from Saarbrucken, building on an idea of Yves Hellegouarch, forever changed the landscape in the search for a proof of Fermat's Last Theorem.

In common with almost all of the great breakthroughs in number theory, Frey's idea depended on an ingenious calculation. He made the assumption that a genuine solution to Fermat's equation exists, so that $a^n + b^n = c^n$ where a, b, c are integers and $n > 2$. Such a solution would, of course, be a counterexample to Fermat's Last Theorem. He then wrote the following elliptic equation on the board:

$$y^2 = x(x + a^n)(x - b^n) = x^3 + (a^n - b^n)x^2 - a^n b^n x \tag{16.2}$$

which later became known as the *Frey curve*. He explained that this equation has very special properties. For instance, its discriminant is

$$(x_1 - x_2)^2(x_2 - x_3)^2(x_3 - x_1)^2$$

where x_1, x_2, x_3 are the roots of (16.2). In this case $x_1 = 0, x_2 = -a^n, x_3 = b^n$, so the discriminant is

$$(-a^n - b^n)^2(b^n - 0)^2(0 - (-a)^n)^2 = c^{2n}b^{2n}a^{2n} = (abc)^{2n}$$

(using $a^n + b^n = c^n$). Frey remarked that it is highly unusual for a discriminant to be a perfect power in this way, and he went on to suggest that the equation has other equally strange properties which mean that it contradicts the Taniyama–Shimura–Weil Conjecture. He was unable to prove this in full, but he offered convincing evidence for such a connection.

So, if the Taniyama–Shimura–Weil Conjecture is true, then there cannot be any solution of the Fermat equation ... so Fermat's Last Theorem must be true.

16.6 The Amateur who Became a Model Professional

Andrew Wiles now enters the story. His love of mathematics dated from his childhood in Cambridge. As he recalled in the BBC Television Programme *Horizon* on 27 September 1997:

> I was a 10-year-old, and one day I happened to be looking in my local public library and I found a book on math and it told a bit about the history of this problem—that someone had resolved this problem 300 years ago, but no one had ever seen the proof, no one knew if there was a proof, and people ever since have looked for the proof. And here was a problem that I, a 10-year-old, could understand, but none of the great mathematicians in the past had been able to resolve. And from that moment of course I just tried to solve it myself. It was such a challenge, such a beautiful problem.

This problem was Fermat's Last Theorem. It became an obsession. As a teenager, Wiles reasoned that Fermat would have had only limited resources, which did not include the subtler theories that came after him. Wiles therefore felt that it was worthwhile to attack Fermat's Last Theorem using only the knowledge that he already had from school. As his interest developed, though, he began to read the literature on the subject, and to delve more and more deeply into it.

In 1971 he went to Merton College, Oxford, to study mathematics. After graduating in 1974 he moved to Clare College, Cambridge, to study for a doctorate. At the time he wanted to pursue his quest for a proof of Fermat's Last Theorem, but his PhD supervisor John Coates advised against this, because it was possible to spend many years working on the problem but getting nowhere. Instead, Wiles worked in his supervisor's area of expertise which happened to be the Iwasawa theory of elliptic curves—a fortuitous choice, given how the story would later turn out.

Wiles was a Junior Research Fellow at Clare College from 1977 to 1980, including a period at Harvard. In 1980 he was awarded his doctorate, and then he spent a time at Bonn before taking a post at the Princeton Institute for Advanced Study in 1981. He became a professor at Princeton University in 1982. He was awarded a Guggenheim Fellowship to visit the Institut des Hautes Études Scientifiques and the École Normale Supérieure in Paris during 1985–86, and there events occurred that were to change his life.

In 1986, Ken Ribet completed a chain of arguments that began with the Frey curve and used ideas of Jean-Pierre Serre on modular Galois groups, to prove Frey's contention that the Taniyama–Shimura–Weil Conjecture implies Fermat's Last Theorem. Wiles took this as an opportunity to begin work in earnest, for two reasons: First, Fermat's Last Theorem, despite its evident allure, had seemed to be just an isolated curiosity; indeed, Gauss had long ago dismissed it for that very reason. But now it was linked to the Taniyama–Shimura–Weil Conjecture, which was at the forefront of modern algebraic number theory. Second, if he could prove the Taniyama–Shimura–Weil Conjecture, he would finally crack the problem that had defeated the entire mathematical community for nearly three hundred and fifty years and haunted him from childhood.

He soon learned that so many people continued to have an interest in Fermat's Last Theorem that talking about it would lead to wide-ranging discussions that would use up valuable time. So for the next seven years he worked on the problem in secret. Only his wife and later his young children and his Head of Department were aware of what he was doing. He spent his life on his mathematics and with his family. When he got stuck, he took a walk down the road to the lake near the Princeton Institute. Sometimes the combination of relaxation and deep incubation of ideas would suddenly come together in a new revelation. He found it necessary always to have a pencil and paper with him, to write down anything that occurred before it slipped his mind.

In 1988 he was stunned to see an announcement in *The Washington Post* and *The New York Times* that Fermat's Last Theorem had been proved by Yoichi Miyaoka of Tokyo University. Miyaoka also translated the number-theoretic problem into one in another area of mathematics, but he related the problem to differential geometry, not elliptic curves. He presented the first outline of his proof at a seminar in Bonn. Two weeks later he released a five-page algebraic proof, and close scrutiny by other mathematicians began. But soon his 'theorem' was seen to contradict a result in geometry that had been proved conclusively several years before. A fortnight later, Gerd Faltings pinpointed the fatal flaw in Miyaoka's proof. Within two months the consensus was that Miyaoka had failed. Wiles could breathe again and continue his work.

In the next three years he made considerable progress with various parts of the proof. As he explained later,

> Perhaps I can best describe my experience of doing mathematics in terms of a journey through a dark unexplored mansion. You enter the first room of the mansion, and it's completely dark. You stumble

> around bumping into the furniture, but gradually you learn where each piece of furniture is. Finally, after six months or so, you find the light switch, you turn it on, and suddenly it's all illuminated. You can see exactly where you were. Then you move into the next room and spend another six months in the dark. So each of these breakthroughs, while sometimes they're momentary, sometimes over a period of a day or so, they are the culmination of—and couldn't exist without—the many months of stumbling around in the dark that preceded them.

He tried to use the Iwasawa theory that he had studied for his PhD. He knew that the theory as it stood would be of little help, so he tried to generalise it and fix it up to attack his difficulties. It didn't work. In 1991, after a period of getting nowhere, he met Coates at a conference, who told him about something that appeared to bridge the gap. A brilliant young student, Mattheus Flach, had just written a beautiful paper analysing elliptic equations. Wiles took a look at the work and concluded that it was exactly what he needed. Progress thereafter was more rapid.

In 1993 Wiles gave a series of three lectures at the Isaac Newton Institute in Cambridge, England, on Monday, Tuesday, and Wednesday 21–23 June. The title of the series was 'Modular forms, elliptic curves and Galois representations'. With typical modesty, he made no advance announcements of his recent activity. Even so, many of the giants of number theory realised that something special was about to happen, and they attended the lectures—some with cameras ready to record the event for posterity. In the course of his lectures, Wiles proved a partial version of the Taniyama–Shimura–Weil Conjecture. It was sufficiently powerful to have a very special corollary. At 10:30 am, at the end of his third lecture, he wrote this corollary on the blackboard. It was the statement of Fermat's Last Theorem. At this point he turned to the audience, and as he sat down he said 'I will stop here'.

16.7 Technical Hitch

Wiles was not allowed to stop there. His proof now had to be subject to the usual reviewing process, and soon doubts began to arise. In response to a query from a colleague, Nick Katz, he realised that there was a hole in his use of the Flach technique, employed in the final stages of his proof. On 4 December 1993 he issued a statement that in the reviewing process a number of issues had arisen, most of which had been resolved. However, in view of the speculation buzzing around at the time, he acknowledged that a certain problem had occurred, and he wished to withdraw his claim that he had a proof. Despite this, he said that he remained confident that he could repair the difficulty using the methods he had announced in his Cambridge lectures.

His life was suddenly in turmoil. Instead of being able to work in secret, his difficulties were now public knowledge. In view of the many false proofs of Fermat's Last Theorem that had preceded Wiles's announcement, many mathematical colleagues began to voice doubts about the validity of his proof. In March 1994, in *Scientific American*, Faltings wrote:

> If it were easy, he would have solved it by now. Strictly speaking, it was not a proof when it was announced.

In the same magazine, Weil was even more damning:

> I believe he has had some good ideas in trying to construct the proof, but the proof is not there. To some extent, proving Fermat's Theorem is like climbing Everest. If a man wants to climb Everest and falls short of it by 100 yards, he has not climbed Everest.

From the beginning of 1994, Wiles began to collaborate with his former student Richard Taylor in an attempt to fill the gaps in the proof. They concentrated on the step based on Flach's method, which was now seen to be inadequate, but they were unable to find an alternative argument. In August, Wiles addressed the International Congress of Mathematicians and had to announce that he was no nearer to a solution. Taylor suggested that they revisit Flach's method to see if another approach were possible, but Wiles was sure it would never work. Nevertheless, he agreed to give it another try to convince Taylor that it was hopeless.

16.8 Flash of Inspiration

They worked on alternative approaches for a couple of weeks, with no result. Then Wiles suddenly had a blinding inspiration as to why the Flach technique failed:

> In a flash I saw that the thing that stopped it working was something that would make another method I had tried previously work.

His inspiration cut through the final difficulties. On 6 October he sent the new proof to three mathematicians primed for the job, and all three reviewers found the new ideas satisfactory. The new method was simpler than his earlier failed attempt, and one of the three, Faltings, even suggested a further simplification of part of the argument. By the following year there was general agreement that the proof was correct and complete. When

Taylor lectured at the British Mathematical Colloquium in Edinburgh in April 1995, there were no longer any real doubts about its validity.

The proof was finally published in May 1995 in two papers in the journal *Annals of Mathematics*. The first, from pages 443 to 551, was Wiles's paper on 'Modular elliptic curves and Fermat's Last Theorem', the second, from pages 553 to 572, was the final step by Taylor and Wiles, entitled 'Ring theoretic properties of Hecke algebras'. See Wiles [145], Taylor and Wiles [134], and the survey by Darmon *et al.* [30].

In the years that followed, Wiles was fêted around the world. In 1995 he received the Schock Prize in Mathematics from the Royal Swedish Academy of Sciences and the Prix Fermat from the Université Paul Sabatier. The American Mathematical Society awarded him the Cole Prize in Number Theory, worth $4,000. He was presented with a $50,000 share in the 1995/6 Wolf Prize by the Israeli President Ezer Weizman for his 'spectacular contributions to number theory and related fields, major advances on fundamental conjectures, and for settling Fermat's Last Theorem'. The other recipient, Robert Langlands, was honoured for his own work in number theory, automorphic forms, and group representations. In 1996 Wiles received the National Academy of Sciences Award ($5,000) followed in 1997 by a five-year MacArthur Fellowship ($275,000). On 27 June 1997, after his proof had been published for the statutory two years laid down in the rules, he received the Wolfskehl Prize. A decade later, the hundred-year period laid down in the original bequest would have run out.

There is no Nobel Prize in mathematics. Until recently, the equivalent honour was the Fields Medal, awarded to up to four mathematicians every four years at the International Congress of Mathematicians. But by tradition the medal is limited to individuals under the age of forty, and Wiles was just over this age when he proved Fermat's Last Theorem. So in August 1998 the Congress celebrated this event at the Fields Medal Ceremony by awarding Wiles a special Silver Plaque—a unique honour in the history of the organisation. In 1999 he won the King Faisal International Prize for Science ($200,000), being nominated for this honour by the London Mathematical Society. In addition he has been awarded honorary degrees at many universities around the world and, in the New Years Honours List in 2000, he became Sir Andrew Wiles, Knight Commander of the British Empire. Many other honours followed, among them the Abel Prize—explicity modelled on the Nobel prize but awarded by Norway—in 2016. The ten-year-old boy had grown to achieve his lifetime ambition, and had been lionised around the world for his success. He had conquered a problem that had foiled the world of mathematicians for three hundred fifty-eight years.

16.9 Exercises

These exercises really are intended to be taken seriously. Don't just be amused (or not, according to taste); do them.

16.1 Think about how creative mathematics needs hard work (first) and relaxation. Note how the great insights mentioned in this chapter occurred. Does this suggest any way to help yourself understand difficult mathematics?

16.2 Why can some mathematical questions be stated in very simple terms that almost anyone can understand, yet when they are finally solved the proof is highly technical and comprehensible only to experts in very specialised fields—and even for them, very difficult?

16.3 Use the Internet to look up the history of other great problems in mathematics, such as the Four Colour Theorem, the Poincaré Conjecture, and the Riemann Hypothesis. How do they relate to the two previous questions?

16.4 Find some other great unsolved problems in mathematics for which even the statement of the problem requires deep and difficult theories. (*Hint*: Try the Birch–Swinnerton-Dyer Conjecture, the Hodge Conjecture, and the Langlands Programme.)

16.5 Find a lake or other idyllic setting; relax and think great thoughts.

16.6 Prepare yourself for the rigours to come. Subtler details will be outlined in the next two chapters.

17

Wiles's Strategy and Subsequent Developments

We continue sketching Wiles's proof of Fermat's Last Theorem, which we saw in Chapter 16 depends on the theory of modular functions. With this theory as the necessary context, we state the Taniyama–Shimura–Weil Conjecture, which forms the centrepiece of Wiles's approach to—and proof of—Fermat's Last Theorem. We then develop the circle of new ideas that leads to the proof of a special case of that conjecture. Wiles, building on the work of Frey and others, realised that this special case, the semistable Taniyama–Shimura–Weil Conjecture, immediately implies the truth of Fermat's Last Theorem. A vital first step is the definition of the Frey elliptic curve, which links Fermat's Last Theorem to elliptic curves.

We conclude this chapter with a summary of several more recent results and conjectures that either develop the theory further, or provide insight into related questions. The topics discussed are the full Taniyama–Shimura–Weil Conjecture, the Langlands Programme, the Catalan Conjecture, the Pillai Conjecture, the Fermat–Catalan Conjecture, the ABC Conjecture, and the Beal Conjecture.

17.1 Frey's Elliptic Curve

A major reason why problems like Fermat's Last Theorem remain unsolved for centuries is that it is difficult to find a reasonable line of attack—a place to start from. As we have seen, the 'big idea' of the 1840s and 1850s was to reformulate the problem in a cyclotomic field. This idea led to significant progress, and although in the end it failed to prove Fermat's Last Theorem, it left a legacy that was far more important than the theorem itself: the whole machinery of ideals in algebraic number theory. This kind of development is quite common in mathematics; the significance of a notorious unsolved problem often lies not in its answer (nothing of great importance would follow easily or directly from knowing whether Fermat's Last Theorem is true or false) but in the methods that the search for an answer can open up. Such problems serve as glorious reminders of areas of massive ignorance and quell any belief that mathematics is 'pretty much worked out'. As it turned out, Fermat's Last Theorem has stimulated the creation of *several* major mathematical theories, whose far-reaching consequences are still being discovered. Wiles's ideas, leading to a complete proof of the Taniyama–Shimura–Weil Conjecture—which *is* important in its own right because it opens up new lines of attack on all kinds of questions—is the latest addition to the list.

When the approach to Fermat's Last Theorem by way of cyclotomic fields ground to a halt in the 1980s, no plausible line of attack seemed to be visible. That situation changed dramatically with the work of Hellegouarch [62] and Frey [44, 45], who indulged in some major lateral thinking that revealed a startling link between Fermat's Last Theorem and elliptic curves. Bearing in mind that elliptic curves form one of the deepest areas of number theory, one equipped with a vast array of powerful machinery, the significance of this breakthrough was immediately evident to number-theorists.

It was Frey, above all, who made the link solid and complete. He started in the obvious way: assume there is a counterexample and derive a contradiction. That is, assume there exist three pairwise coprime nonzero integers a, b, c that satisfy the Fermat Equation

$$a^p + b^p = c^p \tag{17.1}$$

for a prime p, and then...

A lot of people have tried this, but the big problem is what comes after 'then...'. Namely, to derive a contradiction. This is where everyone either made a mistake or got stuck. Frey's key idea is to consider what we now call the *Frey elliptic curve* $\mathcal{E}$, whose equation is

$$y^2 = x(x - a^p)(x + b^p) \tag{17.2}$$

Since the above solution gives rise to two further solutions $b^p + a^p = c^p$ and $a^p + (-c)^p = (-b)^p$ we can arrange for b to be even and $a \equiv -1 \pmod 4$. These conditions are among those needed to make $\mathcal{E}$ 'semistable', a notion that we discuss below. For technical reasons, it is also useful to assume

$p > 3$, which involves no loss of generality since Euler proved Fermat's Last Theorem for $p = 3$.

Frey's chief contribution was to recognise that the curve $\mathcal{E}$ has such strange properties that, intuitively, it cannot possibly exist. If this could be *proved*, then by contradiction there are no solutions to the Fermat equation, and Fermat's Last Theorem is proved.

Moreover, Frey [45] provided strong but incomplete evidence *why* $\mathcal{E}$ cannot exist. Namely, if it does, then it contradicts the Taniyama–Shimura–Weil Conjecture. The main gap in his argument was filled in by Serre, but only by invoking a conjecture of his own, the Special Level Reduction Conjecture. In 1986 Ribet [107] proved this conjecture, although the proof was not published until 1990. At this stage the hoped-for proof of Fermat's Last Theorem rested only on the Taniyama–Shimura–Weil Conjecture. It was a sufficiently powerful special case of this conjecture that Wiles attacked, over a period of seven years, and (not without hiccups, as we have seen) demolished.

17.2 Taniyama–Shimura–Weil Conjecture

The Taniyama–Shimura–Weil Conjecture (often also attributed to some subset of those three mathematicians) can be stated in numerous forms, which look very different but are equivalent given the state of knowledge in the field in the 1980s. In essence, it is this:

Conjecture 17.1. (Taniyama–Shimura–Weil Conjecture) *Every elliptic curve over $\mathbb{Q}$ is modular.*

However, in order for this to make sense we have to explain what it means for an elliptic curve to be modular, and this is where the different alternatives arise.

The best known approach involves a technique we have used repeatedly: reduction modulo p, where $p \in \mathbb{Z}$ is a prime. Suppose that E is an elliptic curve over $\mathbb{Q}$, and let $\mathbb{F}_p$ be the field with p elements. We can write E in projective form as a homogeneous cubic with integer coefficients, and we can then reinterpret those coefficients as integers modulo p. We then get a cubic equation over $\mathbb{F}_p$, also in projective form; and we define b_p to be the number of distinct solutions over $\mathbb{F}_p$, including any that lie at infinity. For instance, suppose that E has affine equation

$$y^2 = x^3 + 22$$

so that in projective form it becomes

$$y^2z = x^3 + 22z^3$$

and let $p = 5$. In $\mathbb{F}_5$ we have $22 \equiv 2$, so we are trying to count the projectively distinct solutions of

$$y^2z = x^3 + 2z^3$$

with $x, y, z \in \mathbb{F}_5$. By trial and error we find that there are exactly six of them, namely

$$(x, y, z) = (0, 1, 0) \quad (1, 0, 3) \quad (1, 1, 2) \quad (1, 1, 4) \quad (1, 4, 2) \quad (1, 4, 4)$$

The first two involve 0 so are at infinity. For example, when $(x, y, z) = (1, 4, 4)$ then $y^2z - (x^3 + 2z^3) = -65$, which is congruent to 0 modulo 5.

Remember that if any solution is multiplied throughout by a nonzero constant, then projectively this is the same solution—so that, for instance, $(1, 1, 2)$ is the same as $(3, 3, 1)$. We conclude that $b_5 = 6$ for this elliptic curve E.

The numbers b_p, for various p, encode useful information about E. When E is modular, there is a formula relating *all* the numbers b_p, for all primes p, to a single function. This function is called an *eigenform*. It can be written

$$f(z) = \sum_{n=1}^{\infty} a_n e^{2\pi i n z} \tag{17.3}$$

and has some very specific properties (technically, it is a normalised cusp form of weight 2 for the $\Gamma_0(N)$ of Definition 15.15, and it is an eigenfunction for all Hecke operators). We then have:

Definition 17.2. An elliptic curve E over $\mathbb{Q}$ is *modular* if there exists an eigenform (17.3) such that for all but finitely many primes p,

$$b_p = p + 1 - a_p$$

It is now known that this definition leads to an alternative, equivalent formulation of the Taniyama–Shimura–Weil Conjecture, which states that E can be parametrised by modular functions of a certain kind—much as the circle can be parametrised by trigonometric functions and every elliptic curve can be parametrised by the Weierstrass $\wp$-function and its derivative:

Conjecture 17.3. (Taniyama–Shimura–Weil Conjecture, alternative formulation) *Let* $y^2 = Ax^3 + Bx^2 + Cx + D$ *be an elliptic curve, where* $A, B, C, D \in$

Q. Then there exist modular functions $f(z), g(z)$, both of the same level N, such that

$$g(z)^2 = Af(z)^3 + Bf(z)^2 + Cf(z) + D$$

See Cox [25] and Mazur [92] for further information on this alternative formulation.

Wiles did not prove the full Taniyama–Shimura–Weil Conjecture, although this has now been done as a consequence of Wiles's ideas; see Section 17.4. He realised that a more accessible special case would be sufficient: the Semistable Taniyama–Shimura–Weil Conjecture. In order to explain what 'semistable' means, we must discuss some numerical invariants of elliptic curves.

Consider a rational elliptic curve $y^2 = Ax^3 + Bx^2 + Cx + D$. Over the complex numbers, the cubic equation $p(x) = Ax^3 + Bx^2 + Cx + D = 0$ has three roots x_1, x_2, x_3. Classically, the discriminant of $p(x)$ is defined to be $(x_1 - x_2)^2(x_1 - x_3)^2(x_2 - x_3)^2$. It is the forerunner of the discriminant of an algebraic number field, which we defined in Section 2.2. We may now define four invariants of the Frey curve:

The *discriminant*, which equals $(abc)^{2p}$. Already we see that the Frey curve is special, since this is a perfect $2p$th power, a highly unusual circumstance.

The *minimal discriminant*, equal to $2^{-8}(abc)^{2p}$. Since b is even and $p \geq 5$, this is an integer.

The *conductor*, which is the product of all primes that divide a, b, or c. If an elliptic curve is modular, then it can be parametrised by modular functions whose level is equal to the conductor of the curve.

The *j-invariant*, which equals $\frac{2^8(a^{2p}+b^{2p}+a^pb^p)^3}{a^{2p}b^{2p}c^{2p}}$ and is a complete invariant in the sense that any two elliptic curves with the same j-invariant are isomorphic over $\mathbb{C}$.

Now we can define semistability.

Definition 17.4. An elliptic curve is *semistable* if whenever a prime $l > 3$ divides the discriminant, only two of the three roots of $p(x)$ are congruent modulo l; *and* similar but more technical conditions on the primes 2 and 3 hold, which we omit.

Lemma 17.5. *The Frey curve $\mathcal{E}$ is semistable.*

Proof: The discriminant is $(abc)^{2p}$ and the zeros of the cubic are $0, a^p, -b^p$ where a^p, b^p are coprime. When $l = 2, 3$ the conditions b even, $a \equiv -1 \pmod 4$ are also required in the proof. □

Wiles's main result is:

Theorem 17.6. (Semistable Taniyama–Shimura–Weil Conjecture) *Every semistable elliptic curve over* $\mathbb{Q}$ *is modular.* □

Thus every semistable elliptic curve over $\mathbb{Q}$ can be parametrised by modular functions of some level N (equal to the conductor). This leads to:

Lemma 17.7. *If l is an odd prime dividing N, then the j-invariant of $\mathcal{E}$ can be written in the form $l^{-mp}q$ where $m > 0$ and q is a rational number whose numerator and denominator in lowest terms are not divisible by l.*

Proof: The j-invariant of $\mathcal{E}$ is

$$\frac{2^8(a^{2p}+b^{2p}+a^pb^p)^3}{a^{2p}b^{2p}c^{2p}} = \frac{2^8(c^{2p}-b^pc^p)^3}{(abc)^{2p}}$$

The power of l dividing the denominator is a multiple of p. Since a, b, c are pairwise coprime, the above fraction is in lowest terms. The result follows since N is the product of the primes dividing abc. □

This lemma fails for $l = 2$ because of the factor 2^8.

17.3 Sketch Proof of Fermat's Last Theorem

We can now sketch Wiles's proof of Fermat's Last Theorem. We need one further ingredient from complex analysis: that of a modular form of weight 2. We begin with an elliptic integral of the first kind

$$\int \frac{\mathrm{d}x}{\sqrt{Ax^3+Bx^2+Cx+D}}$$

Setting $y^2 = Ax^3+Bx^2+Cx+D$, defining an elliptic curve, this becomes $\int \frac{\mathrm{d}x}{y}$. If the elliptic curve is modular, then there exist modular functions $f(z), g(z)$ such that $x = f(z), y = g(z)$ parametrises the curve. In this case

$$\frac{\mathrm{d}x}{y} = \frac{\mathrm{d}f}{g} = \frac{f'(z)\mathrm{d}z}{g(z)} = F(z)\mathrm{d}z$$

where

$$F(z) = \frac{f'(z)}{g(z)}$$

It is not hard to see that although F is not a modular function, it comes close:

$$F\left(\frac{az+b}{cz+d}\right) = (cz+d)^2 F(z)$$

See Exercise 17.2.

In this case we say that F is a *modular form* of *weight* 2 and *level* N. Modular forms of this type have some distinctive features; in particular, if the parametrisation $(f(z), g(z))$ is chosen carefully, the function $F(z)$ is analytic and its Fourier expansion has zero constant term. It is then called a *cusp form*. Moreover, it is possible to work out $F(z)$ from arithmetic information about the elliptic curve, namely, the number of solutions to the congruences $y^2 \equiv Ax^3 + Bx^2 + Cx + D \pmod{p}$ for all primes p. It is this connection, rather than anything to do with Fermat's Last Theorem, that makes the Taniyama–Shimura–Weil Conjecture so important.

And now for the climax:

Theorem 17.8. (Fermat's Last Theorem) *If p is an odd prime then the Fermat equation*

$$x^p + y^p = z^p$$

has no solutions in nonzero integers x, y, z.

Proof: The full proof can be found in Wiles [145] and Wiles and Taylor [134] and is several hundred pages long. We can, however, sketch how the proof follows from the concepts introduced above. For a longer, more technical sketch, see Ribet [108].

We aim for a proof by contradiction and suppose that there exists a solution $a^p + b^p = c^p$ for nonzero integers a, b, c. Let $\mathcal{E}$ be the corresponding Frey elliptic curve (17.2). By Theorem 17.6 $\mathcal{E}$ is modular, hence has a cusp form F of weight 2 and level N, where N is the conductor of $\mathcal{E}$.

Lemma 17.7 now allows us to invoke Serre's Level Reduction Conjecture, proved by Ribet, and this implies that for any odd prime l dividing N there exists a cusp form F' of weight 2 and level N/l, which inherits various useful properties of F. (It would be too technical to say which.) Inductively, we can consider an odd prime l' dividing N/l and repeat the argument to get a cusp form of level N/ll', and so on. The conductor is divisible by 2 since b is even, and by definition the conductor is a product of distinct primes. We may therefore remove all odd prime factors of N and deduce that there exists a cusp form of level 2.

The dimension of the space of such cusp forms is equal to the genus of a compact Riemann surface denoted by $X_0(N)$. But by direct calculation, the genus of $X_0(2)$ can be shown to equal 0. (Indeed, the genus of $X_0(N)$ is zero for $N \leq 10$.) That is, there are no cusp forms of weight 2 and level

2. This is a contradiction, and Fermat's Last Theorem is therefore true. □

17.4 Recent Developments

Wiles's proof of Fermat's Last Theorem is important, but not because it solved that problem and thereby closed down that line of research. On the contrary, it is important because it introduced new methods and made new connections, opening up many new areas for future work. In this section we indicate some of these developments, including appropriate background material.

Full Taniyama–Shimura–Weil Conjecture

Wiles required (and proved) only the semistable case of this conjecture. The full conjecture would be even more important. Wiles's methods opened up the entire area, and in 2001 Christophe Breuil, Brian Conrad, Fred Diamond, and Richard Taylor published a proof of the full Taniyama–Shimura–Weil Conjecture. It is now called the *Modularity Theorem.* See Conrad *et al.* [23], Diamond [34], Breuil *et al.* [11], and the summary by Darmon [29].

Their methods are firmly in the spirit of Wiles's pioneering work, and we content ourselves with two observations. The first is that they prove a more general theorem than the Taniyama–Shimura–Weil Conjecture by rephrasing it in algebraic form. This more general conjecture has technical advantages, which make the proof possible. The second is that their methods make heavy use of Galois Theory. We describe 'classical' Galois Theory, which considers finite field extensions, in Chapter 18. These breakthroughs require the more general theory for infinite-dimensional algebraic field extensions, and they involve a number of technical concepts that are beyond our scope here.

Langlands Programme

The Modularity Theorem is one of the simplest (!) special cases of a set of far-reaching (and highly technical) conjectures, made by Robert Langlands in the winter of 1966-67 and outlined in a letter [76] to Weil in January 1967. In retrospect, Hilbert's class field theory (classifying abelian Galois extensions) can be viewed as a special case: it is equivalent to the Langlands conjectures for the multiplicative group K^* of nonzero elements of a number field K. The Langlands Programme aims to associate number-theoretic objects that generalise elliptic curves over any number field with 'automorphic forms', which generalise modular forms. It involves Galois theory, generalised reciprocity laws, infinite-dimensional group representations, and much else. Broadly speaking, it postulates the existence of deep connections between algebraic geometry, algebraic number theory, and analytic number theory.

Analytic number theory is the application of complex analysis to number theory. It originated in attempts to prove the Prime Number Theorem, conjectured by Lagrange in 1797 and in unpublished work by Gauss in 1792. This states that the number $\pi(x)$ of primes less than x is asymptotic to $x/\log x$. Dirichlet pioneered the use of complex analysis in number theory, and in 1837 he used it to prove that any arithmetic progression $an+b$ with a, b coprime contains infinitely many primes. In this work he introduced new complex functions called L-functions, which are of fundamental importance in number theory. In 1848 and 1850 Pafnuty Chebyshev attempted to use Dirichlet's ideas to prove the Prime Number Theorem, and almost succeeded. He proved that if

$$\lim_{x\to\infty} \frac{\pi(x)}{x/\log x}$$

exists, then it must equal 1. His proof uses Euler's zeta-function

$$\zeta(x) = \sum_{n=1}^{\infty} n^{-x}$$

In 1959 Riemann extended the definition of the zeta-function to $\zeta(z)$ where $z \in \mathbb{C}$, opening up powerful methods of complex analysis. He used it to investigate the distribution of prime numbers and stated the famous Riemann Hypothesis that all nontrivial zeros of $\zeta(z)$ are of the form $\frac{1}{2}+\mathrm{i}y$, currently one of the major unsolved problems in mathematics.

Completion of the Langlands Programme would lead to huge advances on rational points of elliptic curves, the topological construction of algebraic varieties, and even the Riemann Hypothesis. At the moment almost all of the Langlands Programme remains firmly in the realm of conjecture. However, in 1996 Freitas, Le Hung, and Siksek [42] proved that elliptic curves defined over real quadratic fields are modular. As with this one, many advances verify the programme in special cases. Such results are important in their own right, but they carry extra weight because potentially any special case might disprove the programme's conjectures. At the very least, this would require modifications of the programme.

In 1998, Laurent Lafforgue proved the Langlands conjectures for the general linear group $\mathbb{GL}(n, K)$ over a function field K, generalising Vladimir

Drinfeld's proof for $\mathbb{GL}(2, K)$ published in 1980. In 2008, Ngô Bao Châu proved the 'fundamental lemma', conjectured by Langlands and Diana Shelstad in 1983, which is basic to the Langlands Programme, and in 2018 Vincent Lafforgue set up a special case of the conjectured link from automorphic forms to Galois representations. The state of play in 2004 was surveyed by Frenkel [43]; Friedberg [47] published a brief explanation of the basic principles in 2018.

Catalan Conjecture

Fermat's Last Theorem is just one of many famous questions in number theory about integer powers. In 1844 the Belgian mathematician Eugène Catalan published a short letter in the *Journal für die Reine und Angewandte Mathematik* (Journal for Pure and Applied Mathematics), universally known as Crelle's journal after its founder and first editor August Crelle. It read:

> I beg you, sir, to please announce in your journal the following theorem that I believe true although I have not yet succeeded in completely proving it; perhaps others will be more successful. Two consecutive whole numbers, other than 8 and 9, cannot be consecutive powers; otherwise said, the equation $x^m - y^n = 1$ in which the unknowns are positive integers only admits a single solution.

This statement became known as the Catalan Conjecture.

The earliest significant explicit application of algebraic numbers to the Catalan Conjecture occurred in 1850, when Victor Lebesgue proved the result when the smaller of the two powers is a square; that is, the equation $x^a - y^2 = 1$ has no solutions. It took another one hundred eleven years before anyone could prove that there are no integer solutions to the deceptively similar equation $x^2 - y^a = 1$. Very little was known about this equation; for example in 1961 it was proved that if there is a solution then x must have at least 3 billion digits. In the same year Chao Ko proved that no solutions exist except the one that started the whole game: $3^2 - 2^3 = 1$. However, it took another three years before his proof became known to the mathematical community.

The upshot of these discoveries is a useful simplification of the Catalan Conjecture. Its proof or disproof reduces to a special case, in which both x and y occur to odd prime powers. In this case the distinction between $+1$ and -1 ceases to be relevant because -1 to an odd power is -1. So the equation under consideration becomes

$$x^p - y^q = 1 \tag{17.4}$$

where p and q are odd primes. From now on we call this the *Catalan equation*. The Catalan Conjecture is now equivalent to the assertion that for all odd primes p, q the Catalan equation has no nonzero whole number solutions.

Meanwhile, a different line of attack suggested that perhaps the Catalan Conjecture could be reduced to a computer calculation. Suppose a solution exists. Suppose moreover that the sizes of x, y, p and q are bounded; that is, if a solution exists, then there must be one for which these four numbers have a specific, limited size, called a bound. Then a computer could, in principle, try every possibility up to that bound. If no solutions were found in this range, the proof would be complete. Alternatively, a solution would appear, and that would prove the conjecture false.

Proving that such a bound exists, and working out what it is, would be difficult, but there were encouraging precedents. Bit by bit, pieces of the puzzle seemed to be slotting into place. In 1929 Karl Siegel proved a general theorem on Diophantine equations, and a direct consequence is that for fixed odd primes p and q, the Catalan equation has a finite number of solutions. The hope was that this finite number could be proved to be zero, but at least number theorists now knew that there cannot be infinitely many solutions. This, in turn, implies that for fixed p and q the sizes of x and y are bounded. However, Siegel's theorem does not say how big that bound is. It also fails to specify any bound on the sizes of p and q. So, while it is a step in the right direction, it does not open the way for a computer attack.

In 1955 Davenport and Klaus Roth proved another general theorem, and this one did provide an explicit bound on the sizes of x and y. However, the bound was so gigantic that a computer search would take much longer than the lifetime of the universe, so again the practical implications of their theorem were nil. Nonetheless, number theorists felt they were edging just a little bit closer to a proof, and further evidence for this belief soon showed up. In 1966–67 Alan Baker proved that in general, an integer combination of logarithms of rational numbers cannot be small; indeed, it has to be larger than some function of the integers and rationals concerned. In 1976 Robert Tidjeman [136] applied Baker's theorem to the Catalan equation and proved that the exponents p and q must both be less than some explicitly computable bound.

Later work showed that the largest solution to (17.4) satisfies

$$|x| \leq \exp\exp\exp\exp 730$$

At this point, a computer could in principle solve the problem, by checking all numbers in this range. But again, the range was much too great for any practical computation to be possible—even allowing for probable improvements in computer power. In 1999, Maurice Mignotte proved that if a nontrivial solution exists then $p < 7.15 \times 10^{11}$ and $q < 7.78 \times 10^{16}$. Modern computers can handle numbers that big—but those are the exponents, not the numbers x and y, let alone their powers x^p and y^q. So Mignotte's theorem, although an exciting theoretical result, did not reduce the problem to a feasible calculation.

Nevertheless, mathematicians were slowly whittling away at the Catalan Conjecture. A few more improvements of the same kind, and a computer solution would be within reach. But when a solution finally appeared, it came from a different direction altogether. And it made hardly any use of a computer. Instead, it went back to the tried and tested strategy of algebraic number theory—but with better tactics.

In 2002, Preda Mihăilescu startled the mathematical community by publishing a proof of the Catalan Conjecture [95]. His ingenious, highly technical, proof is based on cyclotomic integers. The methods come from his PhD thesis 'Cyclotomy of Rings and Primality Testing', an area with no obvious connection to the Catalan Conjecture. An expanded description of the proof can be found in Schoof [116]. See also Bilu [7].

The first step is to rewrite the Catalan equation as $x^p - 1 = y^q$ and borrow a trick from Gauss. The left-hand side factorises as $(x-1)(x^{p-1} + x^{p-2} + \ldots + x + 1)$. The right-hand side is a qth power; therefore the left hand side is also a qth power. If the two numbers $x-1$ and $x^{p-1}+x^{p-2}+\ldots+x+1$ have no common factor, then each factor must also be a qth power. In 1960 John Cassels (usually known as 'Ian') proved that this assumption leads to a contradiction. However, there might be a common factor. He made progress on this case too, showing that such a common factor must be p. In fact, p divides the second term but p^2 does not. Cassels was unable to derive a contradiction from this result, but it convinced number theorists that further progress on the Catalan Conjecture might be possible by pursuing a similar strategy. There were some standard tricks that might apply, but none of them seemed to work.

Using the cyclotomic integer $\zeta = e^{2\pi \mathrm{i}/p}$, the Catalan equation splits into linear factors:

$$y^q = x^{p-1} - 1 = (x-1)(x-\zeta)(x-\zeta^2)\ldots(x-\zeta^{p-1})$$

If the factors on the right-hand side have no common factor, and prime factorisation is unique in the cyclotomic integers for the prime p, then each term $x - \zeta^k$ must itself be a unit times a prime power. From this it is not too hard to derive a contradiction. However, neither of these statements is true. Undaunted, Mihăilescu decided to find out what happens instead. The resulting analysis is quite long, and too complicated to do more than summarise. It relied on a number of ideas about the ring of cyclotomic integers.

The current proof, a simplification of Mihăilescu's original version that includes more recent work—some by him—has several main components. The first step, which he completed in 2000, is to prove that the exponents p and q satisfy a very stringent condition: they are *double Wieferich primes.* This means that $p^{q-1} - 1$ is divisible by q^2 and $q^{p-1} - 1$ is divisible by p^2. We saw in Section 16.1 that Wieferich introduced a related condition in 1909 when working on Fermat's last theorem: a *Wieferich prime* p is one for which p^2 divides $2^{p-1} - 1$. The only known Wieferich primes are 1093 and 3511, and their rarity suggests that double Wieferich primes might also be rare. Indeed, if it could be proved that they do not exist; that would polish off the Catalan Conjecture. However, they do exist: an example is $p = 83, q = 4871$, and only five more such pairs are currently known. Nevertheless, this result is a good start.

The next step, accomplished in 2002, is to derive a much simpler condition. Namely, either $p - 1$ is divisible by q or $q - 1$ is divisible by p. A key idea is to analyse the structure of 'annihilators', a technical concept that turns statements about ideals back into useful statements about numbers. Another is to exploit known results about the units of cyclotomic number fields. A deep and difficult theorem proved by Francisco Thaine in 1988 relates annihilators of units to ideals, and this is a key ingredient in proving the second step.

The third step relates the sizes of the primes p and q. Neither can be too large compared to the other; specifically, $p < 4q^2$ and $q < 4p^2$. Mihăilescu proved this in 2003, and it simplifies parts of his original proof. The method fails to work for some smallish values of p and q, so these have to be handled by other methods. In this manner, he proved that no new solutions of the Catalan equation exist if either p or q is 41 or more. The improved proof replaces 41 by 5.

By putting all these things together, Mihăilescu was able to prove that if the Catalan equation has a solution, then the two primes p and q that occur must satisfy the condition $p < q$. However, because p and q are odd, the equation is symmetric in p and q, in the sense that if $x^p - y^q = 1$ then $(-y)^q - (-x)^p = 1$. So we can swap p and q. Therefore, by the same argument, $q < p$. This is a contradiction, so no new solution to the Catalan equation exists.

Whenever someone solves a major problem in mathematics, an immediate reflex is to try the same method on other similar problems. However, Mihăilescu's proof uses so many special features of the Catalan equation that it is difficult to get any significant generalisations working. So, for the moment at least, it is a 'one-off', carefully steering its way through a host of difficulties. That makes Mihăilescu's achievement all the more remarkable.

Pillai Conjecture

A result that is in some respects stronger was conjectured in 1945 by Subbayya Sivasankaranarayana Pillai as a brief remark in [103]. The Pillai Conjecture states that each positive integer occurs only finitely many times as a difference of perfect powers. That is, for any integer $c > 0$ the equation $x^m - y^n = c$ has finitely many integer solutions. This problem remains open, but it would be a consequence of the ABC Conjecture 17.14 below.

Fermat–Catalan Conjecture

More generally, consider the Diophantine equation

$$x^a + y^b = z^c \tag{17.5}$$

The 'surprising' solutions occur when a, b, c are 'large' in some sense. For the Pythagorean equation ($a = b = c = 2$), with its infinite family of solutions, the exponents a, b, c should clearly be considered small. We explain below why a sensible interpretation of the 'size' of a solution, in this context, is the number

$$s = \frac{1}{a} + \frac{1}{b} + \frac{1}{c}$$

The *smaller* s is, the larger a, b, c must be. The crucial distinction is that 'large' solutions have $s < 1$ but 'small' ones have $s > 1$. The *only* known integer solutions of (17.5) for large a, b, c (see Mazur [93]) are:

$$\begin{aligned} 1 + 2^3 &= 3^2, \\ 2^5 + 7^2 &= 3^4 \\ 7^3 + 13^2 &= 2^9 \\ 2^7 + 17^3 &= 71^2 \\ 3^5 + 11^4 &= 122^2 \\ 17^7 + 76271^3 &= 21063928^2 \\ 1414^3 + 2213459^2 &= 65^7 \\ 9262^3 + 15312283^2 &= 113^7 \\ 43^8 + 96222^3 &= 30042907^2 \\ 33^8 + 1590342^2 &= 15613^3 \end{aligned} \tag{17.6}$$

By convention, 1 is treated as 1^∞ and $\frac{1}{\infty} = 0$. So $s = \frac{5}{6}$ for the first solution above.

The first five of these solutions have been known for centuries; the last five are due to Frits Beukers and Don Zagier. The main conjecture here is:

Conjecture 17.9. (Fermat–Catalan Conjecture) *In total, for all large* (a, b, c) *(that is,* $s < 1$*) there exists only a finite number of coprime integer solutions of Equation* (17.5).

The name of the conjecture is modern, due to Henri Darmon and Granville; it reflects the fact that a positive solution would imply both the Catalan Conjecture and Fermat's Last Theorem.

The main positive result is a recent theorem of Darmon and Loïc Merel, who prove that there are no solutions with $(a, b, c) = (g, g, 3)$ for $g > 3$. Darmon and Granville have proved that for each individual triple (a, b, c) with $s < 1$ there exist only finitely many coprime integer solutions x, y, z of (17.5). The Fermat–Catalan Conjecture says more than this: the number of triples (a, b, c) with $s < 1$ for which coprime integer solutions exist is also finite.

ABC Conjecture

The problems above lead to a far-reaching and potentially enormously powerful conjecture. Its proof would revolutionise number theory. In a way, it concerns the Fermat equation for the exponent 1; that is, $a + b = c$. Of course there are infinitely many integer solutions, and for any a, b there is a unique c; namely $a + b$. So this equation looks distinctly trivial. However, there's a twist: what matter now are the number-theoretic properties of a, b, and c.

In order to formulate the relevant conjectures and theorems with some precision, we require a few simple concepts:

Definition 17.10. Let N be an integer. Then the *radical* $\operatorname{rad} N$ is the product of all distinct prime factors of N.

If $N \neq 0, \pm 1$ then the *power function* of N is

$$P(N) = \frac{\log |N|}{\log \operatorname{rad} N}$$

By convention,

$$P(\pm 1) = \infty$$

Obviously $P(N) = 1$ if and only if N is squarefree. If N is a perfect kth power, then $P(N) \geq k$. We define a *k-powered number* to be a number N for which $P(N) \geq k$. Roughly speaking, the larger k becomes, the rarer k-powered numbers are. For example, when $k = 2$:

Proposition 17.11. *As $x \to \infty$ the number of squarefree integers between* 0 *and x is of the form*

$$\frac{6}{\pi^2}x + O(\sqrt{x})$$

Proof: See Exercise 17.1. □

Informally, this proposition tells us that roughly 60% of integers (up to a given size) are squarefree. So at most 40% are 2-powered, and this is presumably an overestimate.

Fermat's Last Theorem asserts that a particular 'linear relation' between perfect nth powers is rare—indeed, so rare that it never happens. That is, the sum of two nth powers is never an nth power. More generally, we can look at linear relations between three k-powered numbers and ask how rare those are. To be precise, choose three real numbers $a, b, c \geq 1$ and a real number x. Let $S(a, b, c; x)$ be the set of all triples (A, B, C) of integers (assumed relatively prime and nonzero) such that

$$\begin{gathered} A + B + C = 0 \\ |A| \leq x \qquad |B| \leq x \qquad |C| \leq x \\ P(A) \geq a \qquad P(B) \geq b \qquad P(c) \geq c \end{gathered}$$

Given a, b, c, how rapidly do we expect the cardinality of $S(a, b, c; x)$ to grow as $x \to \infty$? That is, how rare (or how common) are solutions to $A + B + C = 0$ in a-, b-, and c-powered numbers A, B, C?

A heuristic argument leads to a striking guess. Ignore the condition that A, B, C be relatively prime, because this does not change the likely result much. Then there are roughly $x^{1/a}$ choices for A, $x^{1/b}$ choices for B, and $x^{1/c}$ choices for C. Since $|A + B + C| \leq 3x$, the probability (whatever that means here) that $A + B + C = 0$ is of the order $\frac{1}{a} + \frac{1}{b} + \frac{1}{c} - 1$, so the cardinality of $S(a, b, c; x)$ should be comparable to $x^{\frac{1}{a}+\frac{1}{b}+\frac{1}{c}-1}$. This argument is very rough-and-ready, but it focuses attention on the *basic exponent*

$$d = \frac{1}{a} + \frac{1}{b} + \frac{1}{c} - 1 = s - 1$$

and leads us to consider three different cases:

$$d < 0 \qquad d = 0 \qquad d > 0$$

where we expect very different results. Roughly speaking, if $d > 0$ then we expect a proportion x^d of solutions with $|A|, |B|, |C| \leq x$, so that in particular if we allow x to take on all values, then we expect there to exist infinitely many solutions to $A+B+C=0$. When $d<0$, however, we expect there to be only finitely many solutions (allowing x to range over all values). When $d=0$ we have a delicate transitional case and caution is needed even in formulating a sensible conjecture.

Suppose that $a \le b \le c \in \mathbb{N}$. Table 17.1 shows all cases for which $d \ge 0$. This table is familiar from other areas of mathematics, notably polyhedra and tilings in higher dimensions; see Coxeter [26]. This raises the hope that the approach being adopted here is significant.

a	b	c	d
1	*	*	*
2	2	*	*
2	3	3	$\frac{1}{6}$
2	3	4	$\frac{1}{12}$
2	3	5	$\frac{1}{30}$
2	3	6	0
3	3	3	0

Table 17.1. Integer a, b, c for which $d \ge 0$. Entries * can be any positive integer.

When $d>0$ and $a,b,c \in N$ we can get large numbers of (a,b,c') solutions with c' close to c from single Diophantine equations, such as $x^a + y^b = Ez^c$ for some fixed integer $E \ne 0$. Thus, for instance, to obtain lots of $(2,2,c)$ solutions we might consider $x^2+y^2=z^c$. When $c=2$ the problem is that of Pythagorean triples, and these occur in sufficient profusion to establish the conjectured asymptotics, in the sense that the number of solutions with $|A|,|B|,|C|<x$ is at least $x^{d-\varepsilon}$ for any $\varepsilon>0$, however small.

When $d<0$ the problem becomes, if anything, more interesting. Extensive numerical experiments are consistent with:

Conjecture 17.12. ((a,b,c) Conjecture) *If $s<1$ then the number of solutions A,B,C of the equation $A+B+C=0$ with $P(A) \ge a, P(B) \ge b, P(C) \ge c$ is finite.*

Indeed, there is a stronger conjecture:

Conjecture 17.13. (Uniform (a,b,c) Conjecture) *Let $d_0<0$ be real. If $s<1$ then the number of solutions A,B,C of the equation $A+B+C=0$ with $P(A) \ge a, P(B) \ge b, P(C) \ge c$ and $s-1 \le d_0$ is finite.*

The Uniform (a,b,c) Conjecture implies the (a,b,c) Conjecture. However, at the time of writing neither conjecture has been proved in any case whatsoever. However, both would be consequences of the so-called ABC-Conjecture of Masser and Oesterlé [88, 101], one of the biggest open questions in current number theory. We state this conjecture after setting up one necessary concept.

Define an *ABC solution* to be a triple (A,B,C) of nonzero coprime integers such that $A+B+C=0$. Define the *power* of (A,B,C) to be

$$P(A,B,C) = \frac{\log \max(|A|,|B|,|C|)}{\log \operatorname{rad}(ABC)}$$

Then the conjecture is:

Conjecture 17.14. (Masser–Oesterlé ABC Conjecture) *For any real $\rho > 1$ there exist only finitely many ABC solutions with $P(A,B,C) \geq \rho$.*

Beal Conjecture

Finally, we mention the Beal Conjecture; see also Mauldin [89]. Andrew Beal is a number theory enthusiast living in Dallas, Texas. When Fermat's Last Theorem was proved, he decided to follow the example of Paul Wolfskehl. He offered a prize of \$5,000 (increasing annually by \$5,000 up to a total of \$50,000) for a proof of:

Conjecture 17.15. (Beal Conjecture) *Let x,y,z,a,b,c be positive integers with $a,b,c > 2$. If $x^a + y^b = z^c$, then x,y,z have a common factor > 1.*

This conjecture is currently open, and the prize is now \$1 million.

All ten 'large' solutions (17.6) of $x^a + y^b = z^c$ have one exponent equal to 2, so the Beal Conjecture is consistent with the known data related to the Fermat–Catalan Conjecture. One related result is that in any solution, x, y, and z must be 3-powerful (every prime factor occurs to a power 3 or higher). In 1995 Abderrahmane Nitaj [100] proved that there are infinitely many sums $X+Y=Z$ where all of X,Y,Z are 3-powerful; an example is

$$271^3 + 2^3 3^5 73^3 = 919^3$$

However, no known case has all of X,Y,Z perfect powers (of higher degree than 1).

It is clear that plenty of unsolved questions about sums of powers remain to keep the next generation of number theorists busy.

17.5 Exercises

17.1 Give a heuristic argument to show that the number $S(x)$ of squarefree integers less than x is approximately $\frac{6}{\pi^2}x$.

(*Hint:* Let p_j be the primes in increasing order and consider the sequence of integers $1, 2, 3, ..., x$. Remove from this sequence all multiples of p_1^2, leaving approximately $(1 - \frac{1}{p_1^2})x$ integers. From these, remove all mutiples of p_2^2, then p_3^2, and so on. Continue until $p_j \sim \sqrt{x}$, so that the number of integers left (which are the squarefree ones) is approximately

$$x \prod_{p_j \leq \sqrt{x}} \left(1 - \frac{1}{p_j^2}\right)$$

Now use Euler's result that

$$\frac{\pi^2}{6} = \sum_{n=1}^{\infty} \frac{1}{n^2} = \prod_{p_j} \left(1 - \frac{1}{p_j^2}\right)^{-1}$$

to complete the estimate.)

17.2 Use estimates based on the Möbius function $\mu(n)$, which is 1 when n is squarefree and 0 otherwise, to show that $S(x) = \frac{6}{\pi^2}x + O(\sqrt{x})$ for $x \to \infty$.

(*Hint:* Write $S(x) = \sum_{n \leq x} \sum_{d^2 | n} \mu(d)$ and estimate errors.)

17.3 If $f(z), g(z)$ are modular functions and

$$F(z) = \frac{f'(z)}{g(z)}$$

prove that

$$F\left(\frac{az+b}{cz+d}\right) = (cz+d)^2 F(z)$$

17.4 Show that over the Gaussian integers $\mathbb{Z}[\mathrm{i}]$ the equation

$$(78 + 78\mathrm{i})^2 = (-23\mathrm{i})^3 + \mathrm{i}$$

is valid. That is, two perfect powers (other than 8 and 9) differ by a unit.

17.5 Prove that Table 17.1 lists all solutions of the inequality $d \geq 0$, where

$$d = \frac{1}{a} + \frac{1}{b} + \frac{1}{c} - 1$$

and a, b, c are positive integers.

17.6 Verify the following Fermat-Catalan solutions in Gaussian integers:

$$
\begin{aligned}
(8+5i)^2+(5+3i)^3 &= (1+2i)^7 \\
(20+9i)^2+(1+8i)^3 &= (1+i)^{15} \\
(1+2i)^7+(49+306i)^2 &= (27+37i)^3 \\
(44+83i)^2+(31+39i)^3 &= (5+2i)^7 \\
(238+72i)^3+(7+6i)^8 &= (7347-1240i)^2 \\
(3i)^{13}+(2761-2761i)^2 &= (239i)^3 \\
(2797i)^3+(13i)^9 &= (75090-75090i)^2
\end{aligned}
$$

See

math.stackexchange.com/questions/69291/complex-solutions-for-fermat-catalan-conjecture

for their originators and further discussion.

17.7 Search the Internet to find the latest information on the status of the unsolved problems discussed in this chapter.

VI

Galois Theory and Other Topics

18

Extensions and Galois Theory

One of the most famous incidents in the history of mathematics took place on 31 May 1832, when the 20-year-old French mathematician and revolutionary Évariste Galois was killed in a duel over a young woman. He worked in several areas of mathematics, but his main interest was in the solution of polynomial equations by *radicals*, that is, expressions built up from the coefficients using only the usual operations of algebra, together with nth roots for any integer n. A familiar example is the usual formula for the solutions of a quadratic equation, $at^2 + bt + c = 0$:

$$t = \frac{-b \pm \sqrt{b^2 - 4ac}}{2a}$$

which can be traced back to ancient Babylon. In the 1500s, Italian mathematicians found similar, but more complicated, expressions for solutions of cubic and quartic equations.

For a long time it was widely assumed that some radical formula must exist for the quintic equation, but no one could discover it. Gauss thought otherwise; in his *Disquisitiones Arithmeticae* he wrote 'there is little doubt that this problem does not so much defy modern methods of analysis as that it proposes the impossible', but gave no reason for this opinion. Eventually Paolo Ruffini and Niels Henrik Abel proved that no such formula exists. Ruffini's proof had a gap, which was filled later; Abel's was very terse.

It is often said that Galois was also aiming to prove the insolubility of the quintic, but this distorts the history. He was well aware of the work of Ruffini and Abel, and he had little interest in impossibility proofs. He was after bigger game: a general understanding of polynomial equations that *can* be solved by radicals. He obtained this understanding by studying what we now call the Galois group of a polynomial. The insolubility of the quintic is a direct corollary of one of his main results, since its Galois group does not have the correct order, but Galois does not mention this in his surviving papers, probably because it seemed obvious to him.

Galois led a difficult life and his attempts to gain recognition for these important results made little progress. What we know of his ideas comes from a small quantity of rather untidy manuscripts that passed to one of his friends after his death (see Neumann [98]) and were neglected for decades before Joseph Liouville figured out what Galois was saying. Although there were forerunners such as Lagrange, Ruffini, and Cauchy, we now recognise Galois as the person who took the first *significant* steps towards what we now call group theory. In modern terms, he realised that complex solutions of a polynomial equation (*zeros* of the polynomial) are closely related to its *symmetry group*: the permutations of those solutions that preserve all algebraic relations between them. These permutations form a subgroup of the group all permutations of the solutions.

The modern abstract approach to group theory—indeed, to algebra in general—evolved from these concrete beginnings. The modern treatment, known as *Galois Theory*, considers not the polynomial as such, but the subfield of the complex numbers generated by its zeros. The Fundamental Theorem of Galois Theory (Theorem 18.34 below) states that under certain technical conditions on this field there is a bijection between subgroups of the Galois group and subfields of the field concerned. This is the *Galois correspondence*. Galois Theory has been generalised to arbitrary fields; see for example Fenrick [40], Garling [48], and Stewart [128]. In this chapter we restrict attention to number fields, whose concrete nature simplifies the presentation. Most of this chapter is extracted from the general presentation in Stewart [128], with some notational changes.

18.1 A Simple Example

We illustrate the basic structures of Galois Theory with a simple example.

Let $\theta = \sqrt{2} + \sqrt{3} \in \mathbb{C}$. Theorem 2.2 implies that θ is an algebraic number. It is easy to show this directly, and we find that $\theta^4 - 10\,\theta^2 + 1 = 0$. The polynomial $f(t) = t^4 - 10\,t^2 + 1$ is irreducible over $\mathbb{Q}$, so $K = \mathbb{Q}(\theta)$ has degree 4 over $\mathbb{Q}$ by Theorem 1.18. That is, $[K : \mathbb{Q}] = 4$.

It is easy to show that K contains $\sqrt{2}$ and $\sqrt{3}$, hence also $\sqrt{6}$, and that the numbers $1, \sqrt{2}, \sqrt{3}, \sqrt{6}$ are linearly independent over $\mathbb{Q}$. The conjugates of θ are the four numbers $\pm\sqrt{2} \pm \sqrt{3}$. By Theorem 2.7 there are four automorphisms of K, all of which are the identity on $\mathbb{Q}$, defined by:

$$\begin{array}{ll} \sigma_1(\theta) = \sqrt{2} + \sqrt{3} & \sigma_2(\theta) = \sqrt{2} - \sqrt{3} \\ \sigma_3(\theta) = -\sqrt{2} + \sqrt{3} & \sigma_4(\theta) = -\sqrt{2} - \sqrt{3} \end{array}$$

These automorphisms form a group G of order 4, with identity element σ_1, which we write as $\sigma_1 = 1$. By direct calculation the σ_i commute, $\sigma_2^2 = 1, \sigma_3^2 = 1$, and $\sigma_2\sigma_3 = \sigma_4$. Therefore $G \cong \mathbb{Z}_2 \times \mathbb{Z}_2$. This is the Galois group $\mathrm{Gal}(K/\mathbb{Q})$.

We now ask: what are the subfields of K? We can guess some of them:

$$\mathbb{Q} \qquad \mathbb{Q}(\sqrt{2}) \qquad \mathbb{Q}(\sqrt{3}) \qquad \mathbb{Q}(\sqrt{6}) \qquad K$$

Are there any others? A direct approach is complicated; we have to take arbitrary elements of the form

$$\alpha = p + q\sqrt{2} + r\sqrt{3} + s\sqrt{6} \tag{18.1}$$

with $p, q, r, s \in \mathbb{Q}$, calculate their sums and products, and find out when some subset is closed under such operations. (There are shortcuts, but even then it is rather messy.)

An indirect approach using the Galois correspondence answers this question in two steps:

(a) Write down all subgroups H of the Galois group.

(b) For each such H calculate the *fixed field* of H; that is, the set of elements α such that $\sigma(\alpha) = \alpha$ for all $\sigma \in H$.

The group G has five subgroups:

$$\{1\} \qquad \{1, \sigma_2\} \qquad \{1, \sigma_3\} \qquad \{1, \sigma_4\} \qquad G$$

Using (18.1) we easily find the corresponding fixed fields:

$$K \qquad \mathbb{Q}(\sqrt{2}) \qquad \mathbb{Q}(\sqrt{3}) \qquad \mathbb{Q}(\sqrt{6}) \qquad \mathbb{Q}$$

The Fundamental Theorem of Galois Theory states that these fixed fields comprise *all* subfields of K. This shows that we made a good guess.

The subgroups of G are partially ordered by inclusion $\subseteq$. So are the subfields of K. However, making a subgroup larger imposes more conditions on the fixed field, so it gets smaller. Thus the Galois correspondence reverses the inclusion order. Writing $\mathrm{Fix}(H)$ for the fixed field of a subgroup H, we can state this more precisely:

$$\text{If } H_1 \subseteq H_2 \text{ then } \mathrm{Fix}(H_1) \supseteq \mathrm{Fix}(H_2)$$

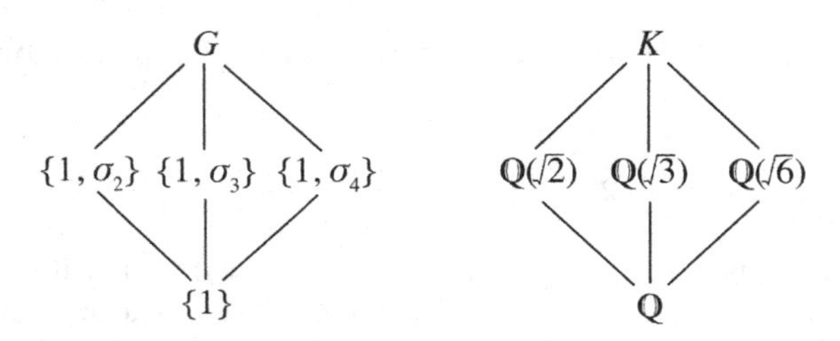

Figure 18.1. *Left*: Lattice of subgroups of $G = \mathbb{Z}_2 \times \mathbb{Z}_2$. *Right*: Lattice of fixed subfields of K.

The inclusion relations between the subgroups of G can be summed up by the *lattice diagram* of Figure 18.1 (left). In such diagrams, $X \subseteq Y$ if there is a sequence of upward-sloping lines from X to Y.

The lattice of subfields is shown in Figure 18.1 (right). This particular lattice looks the same turned upside down, but $\mathrm{Fix}(\{1\}) = K$ and $\mathrm{Fix}(G) = \mathbb{Q}$, so this is misleading.

Bearing this example in mind may help to clarify the general theory, which we now develop. A (slightly) more complicated example is discussed in Section 18.9; for others see Stewart [128] chapter 13.

18.2 Normality and Separability

Let L/K be an extension, as defined in Section 1.3, so that in particular K and L are number fields.

Definition 18.1. Let L/K be an extension. Then M is an *intermediate field* if $K \subseteq M \subseteq L$.

The Galois correspondence relates intermediate fields M to subgroups of the Galois group, which is the group of automorphisms of L that are the identity on K. For this correspondence to be a bijection, two key properties must hold: L/K should be 'normal' and 'separable'. For number fields, contained in $\mathbb{C}$, separability is automatic, but normality is not. Galois worked with subfields of $\mathbb{C}$ and used the separability property without comment, but he emphasised the importance of normality.

Definition 18.2. A polynomial $f(t) \in K[t]$ *splits* over K if

$$f(t) = \alpha(t - \theta_1) \cdots (t - \theta_n)$$

where $\alpha, \theta_1 \ldots \theta_n \in K$.

Definition 18.3. An extension L/K is *normal* if every polynomial $f \in K[t]$ that is irreducible over K, and has at least one zero in L, splits in L.

Equivalently, every zero in $\mathbb{C}$ of every such f belongs to L.

Examples 18.4. (a) Let $\zeta = e^{2\pi i/5}$, a primitive 5th root of unity. The extension $\mathbb{Q}(\zeta)/\mathbb{Q}$ is normal, because it is a splitting field for the irreducible polynomial $f(t) = t^4 + t^3 + t^2 + t + 1$.

Indeed, the zeros of f are ζ^k for $1 \leq k \leq 4$, and they all belong to $\mathbb{Q}(\zeta)$.

(b) Let $\alpha = \sqrt[3]{2} \in \mathbb{R}$. The extension $\mathbb{Q}(\alpha)/\mathbb{Q}$ is not normal. The irreducible polynomial $t^3 - 2$ has a zero in $\mathbb{Q}(\alpha)$, namely α, but it does not split in $\mathbb{Q}(\alpha)$. The other zeros are $\omega\alpha$ and $\omega^2\alpha$, where $\omega = e^{2\pi i/3}$, and these do not lie in $\mathbb{Q}(\alpha)$ because they are not real.

One way to deal with non-normal extensions is to enlarge them so they become normal. There is a canonical way to achieve this:

Definition 18.5. Let K be a subfield of $\mathbb{C}$ and suppose that f is a nonzero polynomial in $K[t]$. A subfield $\Sigma \subseteq \mathbb{C}$ is a *splitting field* if $K \subseteq \Sigma$ and:

(a) f splits over Σ.

(b) If $K \subseteq \Sigma' \subseteq \Sigma$ and f splits over Σ', then $\Sigma' = \Sigma$.

Condition (b) is clearly equivalent to:

(c) $\Sigma = K(\theta_1, \ldots, \theta_n)$ where the θ_j are the zeros of f in Σ.

Every polynomial over a subfield K of $\mathbb{C}$ has a splitting field:

Theorem 18.6. *If K is a number field and f is any nonzero polynomial over K, then there exists a unique splitting field Σ for f over K. Moreover, Σ is also a number field.*

Proof: We can take $\Sigma = K(\alpha_1, \ldots, \alpha_n)$, where the α_j are the zeros of f in $\mathbb{C}$. In fact, this is the only possibility, so Σ is unique. The degree $[\Sigma : K]$ is finite by Corollary 2.4, and $[K : \mathbb{Q}]$ is finite, so $[\Sigma : \mathbb{Q}]$ is finite. $\square$

Recall that Theorem 1.19, the classification theorem for simple algebraic extensions, shows that $K(\alpha)/K \cong (K[t]/\langle m \rangle)/K$, where m is the minimal polynomial of α. Also, m is irreducible over K. Using this, we prove:

Theorem 18.7. *An extension L/K (of number fields) is normal if and only if L is a splitting field for some polynomial over K.*

Proof: Let L/K be normal. By Corollary 2.4, $L = K(\alpha_1, \dots, \alpha_s)$ for finitely many α_j that are algebraic over K. Let m_j be the minimal polynomial of α_j over K and let $f = m_1 \dots m_s$. Each m_j is irreducible over K and has a zero $\alpha_j \in L$, so by normality each m_j splits over L. Hence f splits over L. Since L is generated by K and the zeros of f, Definition 18.5(c) shows that it is the splitting field for f over K.

To prove the converse, suppose that L is the splitting field for some polynomial g over K. The extension L/K is finite. To prove it is normal, let f be an irreducible polynomial over K with a zero in L. Then we must show that f splits in L.

Let $M \supseteq L$ be a splitting field for fg over K. Suppose that θ_1 and θ_2 are zeros of f in M. By irreducibility, f is the minimal polynomial of both θ_1 and θ_2 over K. Therefore

$$[K(\theta_1) : K] = \partial f = [K(\theta_2) : K] \tag{18.2}$$

by Theorem 1.18. We now prove that

$$[L(\theta_1) : L] = [L(\theta_2) : L] \tag{18.3}$$

To do so, consider two towers of subfields:

$$\begin{aligned} K \subseteq K(\theta_1) \subseteq L(\theta_1) \subseteq M \\ K \subseteq K(\theta_2) \subseteq L(\theta_2) \subseteq M \end{aligned}$$

For $j = 1$ or 2 the Tower Law, Theorem 1.15, implies:

$$[L(\theta_j) : L][L : K] = [L(\theta_j) : K] = [L(\theta_j) : K(\theta_j)][K(\theta_j) : K] \tag{18.4}$$

By (18.2), $[K(\theta_1){:}K] = [K(\theta_2){:}K]$. Clearly $L(\theta_j)$ is the splitting field for g over $K(\theta_j)$, and by Theorem 1.19 $K(\theta_1)$ is isomorphic to $K(\theta_2)$. Therefore by Corollary 1.20 the extensions $L(\theta_j)/K(\theta_j)$ are isomorphic for $j = 1, 2$, so they have the same degree. Substituting in (18.4) and cancelling, we obtain (18.3).

But now, if $\theta_1 \in L$ then $[L(\theta_1) : L] = 1$, so $[L(\theta_2) : L] = 1$, so $\theta_2 \in L$ also. Hence L/K is normal. □

We can dispose of separability far more easily:

Definition 18.8. An irreducible polynomial f over a subfield K of $\mathbb{C}$ is *separable* over K if it has simple zeros in $\mathbb{C}$, or equivalently, simple zeros in its splitting field.

An extension L/K is *separable* if every irreducible f over K is separable.

This means that over its splitting field, or over $\mathbb{C}$, f takes the form

$$f(t) = k(t - \alpha_1) \dots (t - \alpha_n)$$

where the α_j are all different.

Separability is not a great concern in this book, because it is automatic for number fields. From now on we use, usually without comment, the next proposition:

Proposition 18.9. *Every extension of number fields is separable.*

Proof: For subfields K of $\mathbb{C}$, Corollary 1.7 implies that every extension of K is separable. □

More generally, field extensions in characteristic zero are always separable. In characteristic p they may or may not be, depending on the fields involved. We state one important case:

Theorem 18.10. *Every extension L/K where L and K are finite fields is separable.*

Proof: See Garling [48] Corollary to Theorem 10.8.

We record a commonly used term:

Definition 18.11. A general field extension L/K is *Galois* (or a *Galois extension*) if it is finite, normal, and separable.

We do not employ this terminology, but it is standard in much of the literature. Every normal number field extension is Galois; so is every extension of finite fields.

18.3 Galois Group

We now define the Galois group of any normal extension L/K of number fields. First, we need:

Definition 18.12. Let L/K be an extension. Then a *K-automorphism* of L is an automorphism ϕ of L such that $\phi(k) = k$ for all $k \in K$.

Theorem 18.13. *The set of all K-automorphisms of L is a group under the operation of composition of functions.*

Proof: Suppose that for $j = 1, 2$ the map $\gamma_j : L \to L$ is a K-automorphism. It is routine to check that:

(a) The identity map on L is a K-automorphism.

(b) The composition $\gamma_1\gamma_2$ is a K-automorphism.

(c) The inverse γ_1^{-1} is a K-automorphism.

Thus the K-automorphisms of L form a group. □

This leads to:

Definition 18.14. Let L/K be a normal extension of number fields. Then the *Galois group* of L/K, written

$$\mathrm{Gal}(L/K)$$

is the set of all K-automorphisms of L under the operation of composition.

Section 2.2 introduced the conjugates of an algebraic number θ and related them to the monomorphisms $\sigma : \mathbb{Q}(\theta) \to \mathbb{C}$. We can characterise the elements of the Galois group using conjugates and the Primitive Element Theorem 2.5:

Theorem 18.15. *Let L/K be a normal extension of number fields, so that $L = K(\theta)$ for some algebraic number $\theta \in \mathbb{C}$. Let the conjugates of θ be $\theta_1, \ldots, \theta_n$. Then*

(a) *There is a unique K-automorphism $\gamma_j : L \to L$ such that $\gamma_j(\theta) = \theta_j$.*

(b) *Every K-automorphism of L is equal to some γ_j.*

Proof: Let $p \in K[t]$ be the minimal polynomial of θ over K. This is irreducible over K. The conjugates θ_j of θ are distinct by Corollary 1.7.

Proof of (a): Since L/K is normal, every conjugate $\theta_j \in L$. Each θ_j has the same minimal polynomial p over K, so there is a K-automorphism γ_j of L with $\gamma_j(\theta) = \theta_j$ by Corollary 1.20.

Proof of (b): Let γ be a K-automorphism of L. Then

$$p(\gamma(\theta)) = \gamma(p(\theta)) = \gamma(0) = 0$$

so $\gamma(\theta) = \theta_j$ for some j. Therefore $\gamma(f(\theta)) = f(\theta_j)$ for all $f \in K[t]$ since γ is a K-automorphism. Now $K(\theta) = \{f(\theta) : f \in K[t]\}$, so $\gamma = \gamma_j$. □

Corollary 18.16. *If L/K is a normal extension of number fields,*

$$|\text{Gal}(L/K)| = [L : K]$$

□

The next result is needed later.

Lemma 18.17. *Suppose that L/K is a normal extension, M is an intermediate field, and τ is a K-automorphism of L. Then $\tau(\text{Gal}(L/M)) = \tau\text{Gal}(L/M)\tau^{-1}$.*

Proof: Let $M' = \tau(M)$, and take $\gamma \in \text{Gal}(L/M), x_1 \in M'$. Then $x_1 = \tau(x)$ for some $x \in M$. Now

$$(\tau\gamma\tau^{-1})(x_1) = \tau\gamma(\tau^{-1}(x_1)) = \tau\gamma(x) = \tau(x) = x_1$$

so

$$\tau\text{Gal}(L/M)\tau^{-1} \subseteq \text{Gal}(M') \tag{18.5}$$

Similarly $\tau^{-1}\text{Gal}(M')\tau \subseteq \text{Gal}(L/M)$ so

$$\tau\text{Gal}(L/M)\tau^{-1} \supseteq \text{Gal}(M') \tag{18.6}$$

Now (18.5) and (18.6) imply that

$$\tau\text{Gal}(L/M)\tau^{-1} = \text{Gal}(M')$$

□

18.4 Fixed Field

The Galois group lets us pass from an extension to a group, while preserving useful structure. Now we want to go in the opposite direction, from a group to a field extension. To do so, we require:

Definition 18.18. Let L/K be a normal extension of number fields with Galois group $G = \text{Gal}(L/K)$, and let H be a subgroup of G. Then the *fixed field* of H is

$$\text{Fix}(H) = \{x \in L : \gamma(x) = x \text{ for all } \gamma \in H\}$$

We now justify the 'field' part of the name:

Proposition 18.19. *Let L/K be a normal extension of number fields. Then* $\text{Fix}(H)$ *is a field and* $K \subseteq \text{Fix}(H) \subseteq L$.

Proof: By definition $\text{Fix}(H) \subseteq L$. We must prove it is a field. Let $x, y \in \text{Fix}(H)$ and $\gamma \in H$. Since γ is a K-automorphism of L, we have:

$$\begin{aligned}
\gamma(1) &= 1, \text{ so } 1 \in \text{Fix}(H) \\
\gamma(x \pm y) &= \gamma(x) \pm \gamma(y) = x \pm y, \text{ so } x \pm y \in \text{Fix}(H) \\
\gamma(xy) &= \gamma(x)\gamma(y) = xy, \text{ so } xy \in \text{Fix}(H) \\
\gamma(1/x) &= 1/\gamma(x) = 1/x \text{ if } x \neq 0, \text{ so } 1/x \in \text{Fix}(H)
\end{aligned}$$

Therefore $\text{Fix}(H)$ is a field.

By definition, $\gamma(K)$ is the identity on K for any $\gamma \in G$, hence for any $\gamma \in H$. But this states that $\gamma(x) = x$ for all $x \in K$, so $x \in \text{Fix}(H)$ for all $x \in K$. That is, $K \subseteq \text{Fix}(H)$. □

Again with L/K a normal extension of number fields, let $\mathcal{G}$ be the set of all subgroups of $\text{Gal}(L/K)$ and let $\mathcal{F}$ be the set of all intermediate fields M (see Definition 18.1). Then we can interpret Gal and Fix as two maps:

$$\begin{aligned}
\text{Gal} : \mathcal{F} \to \mathcal{G} \quad & M \mapsto \text{Gal}(L/M) \\
\text{Fix} : \mathcal{G} \to \mathcal{F} \quad & H \mapsto \text{Fix}(H)
\end{aligned}$$

These maps constitute the *Galois correspondence* between $\mathcal{F}$ and $\mathcal{G}$. In Theorem 18.34 below we prove that they are mutual inverses. The next few subsections lead up to this result, together with other properties of the Galois correspondence.

18.5 Linear Independence of Monomorphisms

Let L/K be an extension. In the terminology of Section 1.1, a *monomorphism* $\lambda : K \to L$ is an injective ring homomorphism. We now prove a theorem of Dedekind, who was the first to make a systematic study of field monomorphisms:

Theorem 18.20. (Dedekind) *If K and L are subfields of $\mathbb{C}$, then every set of distinct monomorphisms $K \to L$ is linearly independent over L.*

Proof: Let $\lambda_1, \ldots, \lambda_n$ be distinct monomorphisms $K \to L$. Assume for a contradiction that these are linearly dependent over L. Then there exist $a_1, \ldots, a_n \in L$, not all zero, such that

$$a_1\lambda_1(x) + \cdots + a_n\lambda_n(x) = 0 \tag{18.7}$$

for all $x \in K$.

Remove all zero terms and renumber, so now

$$a_1\lambda_1(x) + \cdots + a_k\lambda_k(x) = 0 \tag{18.8}$$

and all $a_i \neq 0$ for $1 \leq i \leq k$. At least one of the a_i is nonzero, so $k \geq 1$; since the λ_i are nonzero, in fact $k \geq 2$.

Among all equations of the form (18.7) with all $a_i \neq 0$, consider one for which the number k of nonzero a_j is minimal. We have just seen that $k \geq 2$.

By minimality, no equation like (18.8) has fewer than k nonzero terms.

Since $\lambda_1 \neq \lambda_k$ there exists $y \in K$ such that $\lambda_1(y) \neq \lambda_k(y)$, so $y \neq 0$. Now (18.8) holds with yx in place of x, so

$$a_1\lambda_1(yx) + \cdots + a_k\lambda_k(yx) = 0$$

for all $x \in K$. Therefore

$$a_1\lambda_1(y)\lambda_1(x) + \cdots + a_k\lambda_k(y)\lambda_k(x) = 0 \tag{18.9}$$

for all $x \in K$. Multiply (18.8) by $\lambda_1(y)$ and subtract (18.9), to get

$$a_2[\lambda_2(x)\lambda_1(y) - \lambda_2(x)\lambda_2(y)] + \cdots + a_k[\lambda_k(x)\lambda_1(y) - \lambda_k(x)\lambda_k(y)] = 0$$

The coefficient of $\lambda_k(x)$ is $a_k[\lambda_1(y) - \lambda_k(y)] \neq 0$, so (if necessary removing any terms that vanish) we have an equation of the form (18.8) with fewer nonzero terms. This is a contradiction. □

We now come to the main theorem of this chapter, whose proof is similar to that of Theorem 18.20. Recall that $|G|$ denotes the order of a group G.

Theorem 18.21. *Let G be a finite subgroup of the group of automorphisms of a field K, and let K_0 be the fixed field of G. Then $[K : K_0] = |G|$.*

Proof:

Let $n = |G|$, and suppose that the elements of G are $g_1, \ldots, g_n$, where $g_1 = 1$. We prove separately that $[K : K_0] < n$ and $[K : K_0] > n$ are impossible.

(1) Suppose that $[K : K_0] = m < n$. Let $\{x_1, \ldots, x_m\}$ be a basis for K over K_0. By standard linear algebra there exist $y_1, \ldots, y_n \in K$, not all zero, such that

$$y_1 g_1(x_i) + \cdots + y_n g_n(x_i) = 0 \tag{18.10}$$

for $i = 1, \ldots, m$. Let x be any element of K. Then

$$x = \alpha_1 x_1 + \cdots + \alpha_m x_m$$

where $\alpha_1, \ldots, \alpha_m \in K_0$. Now

$$\begin{aligned} y_1 g_1(x) + \cdots + y_n g_n(x) &= y_1 g_1\left(\sum_l \alpha_l x_l\right) + \cdots + y_n g_n\left(\sum_l \alpha_l x_l\right) \\ &= \sum_l \alpha_l [y_1 g_1(x_l) + \cdots + y_n g_n(x_l)] \\ &= 0 \end{aligned}$$

by (18.10). Hence the distinct monomorphisms $g_1, \ldots, g_n$ are linearly dependent, contrary to Theorem 18.20. Therefore $m \geq n$.
(2) Next, suppose for a contradiction that $[K : K_0] > n$. Then there exists a set $\{x_1, \ldots, x_{n+1}\}$ of $n+1$ elements of K that are linearly independent over K_0. By standard linear algebra there exist $y_1, \ldots, y_{n+1} \in K$, not all zero, such that for $j = 1, \ldots, n$

$$y_1 g_j(x_1) + \cdots + y_{n+1} g_j(x_{n+1}) = 0 \tag{18.11}$$

Choose $y_1, \ldots, y_{n+1}$ so that as few as possible are nonzero, and renumber so that

$$y_1, \ldots, y_r \neq 0 \qquad y_{r+1}, \ldots, y_{n+1} = 0$$

Equation (18.11) now becomes

$$y_1 g_j(x_1) + \cdots + y_r g_j(x_r) = 0 \tag{18.12}$$

Let $g \in G$, and operate on (18.12) with g. This gives a system of equations

$$g(y_1) g g_j(x_1) + \cdots + g(y_r) g g_j(x_r) = 0 \ (1 \leq j \leq n)$$

As the g_j run through all elements of G, so do the gg_j. Thus this system of equations is equivalent to the system

$$g(y_1) g_j(x_1) + \cdots + g(y_r) g_j(x_r) = 0 \ (1 \leq j \leq n) \tag{18.13}$$

Multiply (18.12) by $g(y_1)$ and (18.13) by y_1 and subtract to get

$$[y_2 g(y_1) - g(y_2) y_1] g_j(x_2) + \cdots + [y_r g(y_1) - g(y_r) y_1] g_j(x_r) = 0 \ (1 \leq j \leq n)$$

This system of equations has fewer nonzero terms than (18.12). We obtain a contradiction unless all the coefficients

$$y_i g(y_1) - y_1 g(y_i)$$

are zero. If this happens then

$$y_i y_1^{-1} = g(y_i y_1^{-1})$$

for all $g \in G$, so that $y_i y_1^{-1} \in K_0$. Thus there exist $z_1, \ldots, z_r \in K_0$ and an element $k \in K$ such that $y_i = kz_i$ for all i. Then (18.12), with $j = 1$, becomes

$$x_1 k z_1 + \cdots + x_r k z_r = 0$$

and since $k \neq 0$ we may divide by k, which shows that the x_i are linearly dependent over K_0. This is a contradiction.

Therefore $[K : K_0] \not< n$ and $[K : K_0] \not> n$, so $[K : K_0] = n = |G|$. □

Corollary 18.22. *If G is the Galois group of the finite extension L/K, and H is a finite subgroup of G, then*

$$[\mathrm{Fix}(H) : K] = [L : K]/|H|$$

Proof: By the Tower Law, $[L : K] = [L : \mathrm{Fix}(H)][\mathrm{Fix}(H) : K]$, so $[\mathrm{Fix}(H) : K] = [L : K]/[L : \mathrm{Fix}(H)]$. But this equals $[L : K]/|H|$ by Theorem 18.21. □

Example 18.23. Let $K = \mathbb{Q}(\zeta)$ where $\zeta = \exp(2\pi \mathrm{i}/5) \in \mathbb{C}$. Now $\zeta^5 = 1$ and $\mathbb{Q}(\zeta)$ consists of all elements

$$p + q\zeta + r\zeta^2 + s\zeta^3 + t\zeta^4 \tag{18.14}$$

where $p, q, r, s, t \in \mathbb{Q}$. The Galois group of $\mathbb{Q}(\zeta)/\mathbb{Q}$ is easy to find, because the conjugates of ζ are $\zeta, \zeta^2, \zeta^3, \zeta^4$. This follows because they all are zeros of $t^4 + t^3 + t^2 + t + 1$, which is irreducible by Lemma 3.8, so it is the minimal polynomial of each ζ^j for $1 \leq j \leq 4$. By Theorem 18.15 there are precisely four $\mathbb{Q}$-automorphisms:

$$\begin{array}{ll} 1: p + q\zeta + r\zeta^2 + s\zeta^3 + t\zeta^4 & \mapsto p + q\zeta + r\zeta^2 + s\zeta^3 + t\zeta^4 \\ \alpha: & \mapsto p + s\zeta + q\zeta^2 + t\zeta^3 + r\zeta^4 \\ \beta: & \mapsto p + r\zeta + t\zeta^2 + q\zeta^3 + s\zeta^4 \\ \gamma: & \mapsto p + t\zeta + s\zeta^2 + r\zeta^3 + q\zeta^4 \end{array} \tag{18.15}$$

It is easy to check that all of these are $\mathbb{Q}$-automorphisms. The only point to bear in mind is that $1, \zeta, \zeta^2, \zeta^3, \zeta^4$ are *not* linearly independent over $\mathbb{Q}$. However, their linear relations are generated by just one: $\zeta+\zeta^2+\zeta^3+\zeta^4 = -1$, and this relation is preserved by all of the candidate $\mathbb{Q}$-automorphisms.

Alternatively, observe that $\zeta, \zeta^2, \zeta^3, \zeta^4$ all have the same minimal polynomial $t^4 + t^3 + t^2 + t + 1$ and use Corollary 1.20.

We deduce that the Galois group of $\mathbb{Q}(\zeta)/\mathbb{Q}$ has order 4. It is easy to find the fixed field of this group: it turns out to be $\mathbb{Q}$. Therefore, by Theorem 18.21, $[\mathbb{Q}(\zeta) : \mathbb{Q}] = 4$, as expected from Corollary 18.16.

18.6 K-Automorphisms and K-Monomorphisms

We begin by generalising the concept of a K-automorphism of a subfield L of $\mathbb{C}$, by relaxing the condition that the map should be onto. We continue to require it to be injective.

Definition 18.24. Suppose that a number field K is a subfield of each of the subfields M and L of $\mathbb{C}$. Then a *K-monomorphism* of M into L is a field monomorphism $\phi : M \to L$ such that $\phi(k) = k$ for every $k \in K$.

We say that ϕ *fixes* K.

Example 18.25. Suppose that $K = \mathbb{Q}, M = \mathbb{Q}(\alpha)$ where α is a real cube root of 2, and $L = \mathbb{C}$. Define $\phi : M \to L$ by

$$\phi(p + q\alpha + r\alpha^2) = p + q\omega\alpha + r\omega^2\alpha^2 \qquad p, q, r \in \mathbb{Q}$$

where $\omega = \mathrm{e}^{2\pi \mathrm{i}/3}$. Since α and $\omega\alpha$ have the same minimal polynomial, namely $t^3 - 2$, Corollary 1.20 implies that ϕ is a K-monomorphism.

There are two other K-monomorphisms $M \to L$ in this case. One is the identity, and the other takes α to $\omega^2\alpha$.

In general if $K \subseteq M \subseteq L$ then any K-automorphism of L restricts to a K-monomorphism $M \to L$. We are particularly interested in when this process can be reversed.

Theorem 18.26. *Suppose that L/K is a finite normal extension and $K \subseteq M \subseteq L$. Let τ be any K-monomorphism $M \to L$. Then there exists a K-automorphism σ of L such that $\sigma|_M = \tau$.*

Proof: By Theorem 18.7, L is the splitting field over K of some polynomial f over K. Hence it is simultaneously the splitting field over M for f and over $\tau(M)$ for $\tau(f)$. But $\tau|_K$ is the identity, so $\tau(f) = f$. Consider the diagram

$$\begin{array}{ccc} M & \longrightarrow & L \\ \downarrow\tau & & \sigma\downarrow \\ \tau(M) & \longrightarrow & L \end{array}$$

where σ is yet to be found. By Theorem 1.19 there is an isomorphism $\sigma : L \to L$ such that $\sigma|_M = \tau$. Therefore σ is an automorphism of L, and since $\sigma|_K = \tau|_K$ is the identity, σ is a K-automorphism of L. □

This result can be used to construct K-automorphisms:

Proposition 18.27. *Suppose that L/K is a finite normal extension, and α, β are zeros in L of the irreducible polynomial p over K. Then there exists a K-automorphism σ of L such that $\sigma(\alpha) = \beta$.*

Proof: By Corollary 1.20 there is an isomorphism $\tau : K(\alpha) \to K(\beta)$ such that $\tau|_K$ is the identity and $\tau(\alpha) = \beta$. By Theorem 18.26, τ extends to a K-automorphism σ of L. □

18.7 Normal Closures

An extension that is not normal can be enlarged to give a normal extension. Among such enlargements there is a unique minimal one:

Definition 18.28. Let L be a finite extension of K. A *normal closure* of L/K is an extension N/L such that

(a) N/K is normal;

(b) If $L \subseteq M \subseteq N$ and M/K is normal, then $M = N$.

Thus N is the smallest extension of L that is normal over K.

Theorem 18.29. *If L/K is a finite extension in $\mathbb{C}$, then there exists a unique normal closure $N \subseteq \mathbb{C}$ of L/K, and N/K is a finite extension.*

Proof: By the Primitive Element Theorem $K = K[\theta]$ for some $\theta \in L$. Let the conjugates of θ be $\theta_1, \ldots, \theta_n$. Let $N = K(\theta_1, \ldots, \theta_n)$. Then L/N is normal since N is a splitting field for the minimal polynomial of θ over K.

If $N \subseteq M \subseteq N$ and M/K is normal, then M must contain $\theta_1, \ldots, \theta_n$. Therefore $M \supseteq N$. Therefore N is a normal closure.

If M is any other normal closure then $M \subseteq N$, but by the same argument $M \supseteq N$. Therefore $M = N$. □

Example 18.30. Consider $\mathbb{Q}(\alpha)/\mathbb{Q}$ where α is the real cube root of 2. By Example 18.4(b) this extension is not normal. If we let K be the splitting field for $t^3 - 2$ over $\mathbb{Q}$, contained in $\mathbb{C}$, then $K = \mathbb{Q}(\alpha, \alpha\omega, \alpha\omega^2)$ where $\omega = e^{2\pi i/3} = (-1 + i\sqrt{3})/2$. This is the same as $\mathbb{Q}(\alpha, \omega)$. Now K is the normal closure for $\mathbb{Q}(\alpha)/\mathbb{Q}$.

Normal closures place restrictions on the image of a monomorphism.

Lemma 18.31. *Suppose that $K \subseteq L \subseteq N \subseteq M$ where L/K is finite and N is the normal closure of L/K. Let τ be any K-monomorphism $L \to M$. Then $\tau(L) \subseteq N$.*

Proof: Let $\alpha \in L$. Let m be the minimal polynomial of α over K. Then $m(\alpha) = 0$ so $\tau(m(\alpha)) = 0$. But $\tau(m(\alpha)) = m(\tau(\alpha))$ since τ is a K-monomorphism, so $m(\tau(\alpha)) = 0$ and $\tau(\alpha)$ is a zero of m. Therefore $\tau(\alpha)$ lies in N since N/K is normal. Therefore $\tau(L) \subseteq N$. □

The next theorem provides a sort of converse.

Theorem 18.32. *For a finite extension L/K the following are equivalent:*

(1) *L/K is normal.*

(2) *There exists a finite normal extension N of K containing L such that every K-monomorphism $\tau : L \to N$ is a K-automorphism of L.*

(3) *For every finite extension M of K containing L, every K-monomorphism $\tau : L \to M$ is a K-automorphism of L.*

Proof: We show that (1) $\Rightarrow$ (3) $\Rightarrow$ (2) $\Rightarrow$ (1).

(1) $\Rightarrow$ (3). If L/K is normal then L is the normal closure of L/K, so by Lemma 18.31, $\tau(L) \subseteq L$. But τ is a K-linear map defined on the finite-dimensional vector space L over K, and is a monomorphism. Therefore $\tau(L)$ has the same dimension as L, whence $\tau(L) = L$ and τ is a K-automorphism of L.

(3) $\Rightarrow$ (2). Let N be the normal closure for L/K. Then N exists by Theorem 18.29 and has the requisite properties by (3).

(2) $\Rightarrow$ (1). Suppose that f is any irreducible polynomial over K with a zero $\alpha \in L$. Then f splits over N by normality, and if β is any zero of f in N, then by Proposition 18.27 there exists an automorphism σ of N such that $\sigma(\alpha) = \beta$. By hypothesis, σ is a K-automorphism of L, so $\beta = \sigma(\alpha) \in \sigma(L) = L$. Therefore f splits over L and L/K is normal. □

Theorem 18.33. *Suppose that L/K is a finite extension of degree n. Then there are precisely n distinct K-monomorphisms of L into the normal closure N of L/K, and hence into any given normal extension M of K containing L.*

Proof: By the Primitive Element Theorem there exists $\theta \in L$ such that $L = K(\theta)$. Let $p(t) \in K[t]$ be the minimal polynomial of θ over K. Then $p(t)$ is irreducible and $\partial p = [L : K] = n$. Over $\mathbb{C}$, we have

$$p(t) = (t - \theta_1) \cdots (t - \theta_n)$$

where θ_i are the conjugates of θ, and are distinct. (One of them, say θ_1, is equal to θ.)

The minimal polynomial of each θ_i is also $p(t)$, so there are n K-monomorphisms $\sigma_i : K \to L$ determined by $\sigma_i(\theta) = \theta_i$. Moreover, any K-monomorphism $\sigma : K \to L$ must be one of the σ_i, since $\sigma(\theta)$ has the same minimal polynomial as θ. □

18.8 Fundamental Theorem of Galois Theory

Recall from Sections 18.3 and 18.4 that whenever K/L is a normal extension of number fields, there are two maps

$$\mathrm{Gal} : \mathcal{F} \to \mathcal{G} \qquad \mathrm{Fix} : \mathcal{G} \to \mathcal{F}$$

where $\mathcal{F}$ is the set of intermediate fields for L/K and $\mathcal{G}$ is the set of subgroups of $\mathrm{Gal}(L/K)$.

We are now ready to prove Galois's great discovery, which for simplicity we state and prove only for extensions of number fields.

Theorem 18.34. (Fundamental Theorem of Galois Theory) *If L/K is a finite normal number field extension with Galois group G, then:*

(a) *The Galois group* $\mathrm{Gal}(L/K)$ *has order* $[L : K]$.

(b) *The maps* Gal *and* Fix *are mutual inverses, and set up an order-reversing bijection between* $\mathcal{F}$ *and* $\mathcal{G}$.

(c) *If M is an intermediate field, then*

$$[L : M] = |\mathrm{Gal}(L/M)| \qquad [M : K] = |G|/|\mathrm{Gal}(L/M)|$$

(d) *An intermediate field M is a normal extension of K if and only if* $\mathrm{Gal}(L/M)$ *is a normal subgroup of* G.

(e) *If an intermediate field M is a normal extension of K, then the Galois group of M/K is isomorphic to the quotient group* $G/\mathrm{Gal}(L/M)$.

Proof: Let $G = \mathrm{Gal}(L/K)$.

Part (a) is a restatement of Corollary 18.16.

For part (b), suppose that M is an intermediate field, and let $[L : M] = d$. Then $|\mathrm{Gal}(L/M)| = d$ by Theorem 18.21. On the other hand, if H is a subgroup of G of order d, then $[L : \mathrm{Fix}(H)] = d$ by Corollary 18.16. Hence the composite functions Gal Fix and Fix Gal preserve $[L : M]$ and $|H|$, respectively.

From their definitions, $\mathrm{Fix}\,\mathrm{Gal}(L/M) \supseteq M$ and $\mathrm{Gal}\,\mathrm{Fix}(H) \supseteq H$. Therefore these inclusions are equalities, proving (b).

For part (c), again note that L/M is normal. Corollary 18.16 states that $[L : M] = |\mathrm{Gal}(L/M)|$, and the other equality follows immediately.

We now prove part (d). If M/K is normal, let $\tau \in G$. Then $\tau|_M$ is a K-monomorphism $M \to L$, so is a K-automorphism of M by Theorem 18.32. Hence $\tau(M) = M$. By Lemma 18.17, $\tau\mathrm{Gal}(L/M)\tau^{-1} = \mathrm{Gal}(L/M)$, so $\mathrm{Gal}(L/M)$ is a normal subgroup of G.

Conversely, suppose that $\mathrm{Gal}(L/M)$ is a normal subgroup of G. Let σ be any K-monomorphism $M \to L$. By Theorem 18.26, there is a K-automorphism τ of L such that $\tau|_M = \sigma$. Now $\tau\mathrm{Gal}(L/M)\tau^{-1} = \mathrm{Gal}(L/M)$ since $\mathrm{Gal}(L/M)$ is a normal subgroup of G, so by Lemma 18.17, $\mathrm{Gal}(\tau(M))$ $= \mathrm{Gal}(L/M)$. By part (b) above, Gal is a bijection, so $\tau(M) = M$. Hence $\sigma(M) = M$ and σ is a K-automorphism of M. By Theorem 18.32, M/K is normal.

Finally we prove part (e). Let G' be the Galois group of M/K. Define $\phi : G \to G'$ by

$$\phi(\tau) = \tau|_M \qquad \tau \in G$$

This is clearly a group homomorphism $G \to G'$, for by Theorem 18.32 $\tau|_M$ is a K-automorphism of M. By Theorem 18.26, ϕ is onto. The kernel of ϕ is obviously $\mathrm{Gal}(L/M)$, so by standard group theory

$$G' = \mathrm{im}(\phi) \cong G/\ker(\phi) = G/\mathrm{Gal}(L/M)$$

where im is the image and ker the kernel. □

Remark. The Fundamental Theorem of Galois Theory holds for any finite normal separable (that is, Galois) field extension, with essentially the same proof, although the Primitive Element Theorem cannot be used for infinite extensions in prime characteristic, so different arguments are occasionally needed. The Primitive Element Theorem holds for all extensions L/K when K, L are finite fields, so the Fundamental Theorem of Galois Theory holds for such extensions. See for example Fenrick [40], Garling [48], and Stewart [128].

18.9 Examples of Galois Groups

We discussed one example of the Galois correspondence in Section 18.1. We now consider two more examples. In general, computing the Galois group of a specific polynomial is quite complicated, even for degree 3, because there is no simple relation between the coefficients of the polynomial and its zeros (which is what the Galois group permutes). Indeed, it was this difficulty that led to Galois's ideas being rejected by the French Academy of Sciences. The task becomes rapidly more difficult for larger degrees. However, computer programs can carry out this task fairly efficiently for polynomials over $\mathbb{Q}$. The Maple command galois computes Galois groups for irreducible polynomials over $\mathbb{Q}$ of degree ≤ 8. GAP also has an algorithm to compute Galois groups.

Example 18.35.

We revisit Example 18.23. Let $f(t) = t^4 + t^3 + t^2 + t + 1$ over $\mathbb{Q}$, which is irreducible by Lemma 3.8. Its zeros are $\zeta, \zeta^2, \zeta^3, \zeta^4$ where $\zeta = e^{2\pi i/5}$ is a primitive 5th root of 1. In particular $\zeta^5 = 1$, so the splitting field of f over $\mathbb{Q}$ is $K = \mathbb{Q}(\zeta)$. The degree $[K : \mathbb{Q}] = 4$, so the Galois group $G = \mathrm{Gal}(K/\mathbb{Q})$ has order 4. A basis for K over $\mathbb{Q}$ is $\{1, \zeta, \zeta^2, \zeta^3\}$.

The Galois group comprises the four $\mathbb{Q}$-automorphisms $\{1, \alpha, \beta, \gamma\}$ listed in (18.15). Apply α repeatedly to ζ. What happens is:

$$\zeta \mapsto \zeta^2 \mapsto \zeta^4 \mapsto \zeta^8 = \zeta^3 \mapsto \zeta^6 = \zeta$$

so $\alpha^4 = 1$ and $G \cong \mathbb{Z}_4$ is cyclic. Moreover, $\alpha^2 = \gamma$ and this has order 2. Therefore G has three subgroups:

Order 4:	G	$G \cong \mathbb{Z}_4$
Order 2:	$\{1, \gamma\}$	$A \cong \mathbb{Z}_2$
Order 1:	$\{1\}$	$I \cong \mathbf{1}$

The inclusion relations between the subgroups of G are shown in Figure 18.2 (left).

The Galois correspondence gives the intermediate fields, which are the fixed fields of the subgroups of G; see Figure 18.2 (right). The fixed fields

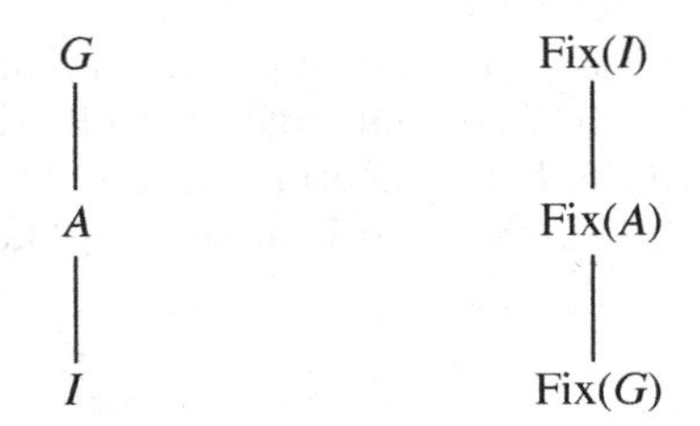

Figure 18.2. *Left*: Lattice of subgroups of $G = \mathbb{Z}_4$. *Right*: Lattice of intermediate fields of $K = \mathbb{Q}(\zeta)$.

are obvious except perhaps for A. Consider a general element of K:

$$x = a + b\zeta + c\zeta^2 + d\zeta^3$$

Then

$$\gamma(x) = a + b\zeta^4 + c\zeta^3 + d\zeta$$

so $x = \gamma(x)$ if and only if $a = d, b = c$. Therefore Fix(A) is spanned by $\zeta + \zeta^4$ and $\zeta^2 + \zeta^3$. Since $1 + \zeta + \zeta^2 + \zeta^3 + \zeta^4 = 0$, we have

$$\zeta^2 + \zeta^3 = 1 - (\zeta + \zeta^4)$$

so Fix$(A) = \mathbb{Q}(\zeta + \zeta^4)$. In summary:

$$\begin{aligned} \text{Fix}(I) &= K \\ \text{Fix}(A) &= \mathbb{Q}(\zeta + \zeta^4) \\ \text{Fix}(G) &= \mathbb{Q} \end{aligned}$$

Example 18.36.

Let $f(t) = t^4 - 2$ over $\mathbb{Q}$, and let K be the splitting field for f in $\mathbb{C}$.
(1) Factorising,

$$f(t) = (t - \xi)(t + \xi)(t - \mathrm{i}\xi)(t + \mathrm{i}\xi)$$

where $\xi = \sqrt[4]{2}$ is real and positive, so $K = \mathbb{Q}(\xi, \mathrm{i})$. Since K is the splitting field, $K/\mathbb{Q}$ is finite and normal; it is separable because we are working in $\mathbb{C}$.
(2) We find the degree of $K/\mathbb{Q}$. By the Tower Law,

$$[K : \mathbb{Q}] = [\mathbb{Q}(\xi, \mathrm{i}) : \mathbb{Q}(\xi)][\mathbb{Q}(\xi) : \mathbb{Q}]$$

The minimal polynomial of i over $\mathbb{Q}(\xi)$ is $t^2 + 1$, since $\mathrm{i}^2 + 1 = 0$ but $\mathrm{i} \notin \mathbb{R} \supseteq \mathbb{Q}(\xi)$. So $[\mathbb{Q}(\xi, \mathrm{i}) : \mathbb{Q}(\xi)] = 2$.

Now ξ is a zero of f over $\mathbb{Q}$, and f is irreducible by Eisenstein's Criterion, Theorem 1.11. Hence $[\mathbb{Q}(\xi):\mathbb{Q}] = 4$, and $[K:\mathbb{Q}] = 2.4 = 8$.
(3) The Galois group of $K/\mathbb{Q}$ therefore has order 8 by Theorem 18.34(c). By Corollary 1.20 there are $\mathbb{Q}$-automorphisms σ, τ of K such that

$$\begin{array}{ll} \sigma(\mathrm{i}) = \mathrm{i} & \sigma(\xi) = \mathrm{i}\xi \\ \tau(\mathrm{i}) = -\mathrm{i} & \tau(\xi) = \xi \end{array}$$

Products of these yield eight distinct $\mathbb{Q}$-automorphisms of K, which we list in a table:

Automorphism	Effect on ξ	Effect on i
1	ξ	i
σ	$\mathrm{i}\xi$	i
σ^2	$-\xi$	i
σ^3	$-\mathrm{i}\xi$	i
τ	ξ	$-\mathrm{i}$
$\sigma\tau$	$\mathrm{i}\xi$	$-\mathrm{i}$
$\sigma^2\tau$	$-\xi$	$-\mathrm{i}$
$\sigma^3\tau$	$-\mathrm{i}\xi$	$-\mathrm{i}$

Table 18.1. The eight $\mathbb{Q}$-automorphisms of $K/\mathbb{Q}$.

These must be all the elements of $\mathrm{Gal}(K/\mathbb{Q})$ since its order is 8.
(4) The generator-relation presentation

$$G = \langle \sigma, \tau : \sigma^4 = \tau^2 = 1,\ \tau\sigma = \sigma^3\tau \rangle$$

shows that G is the dihedral group of order 8, which we write as $\mathbb{D}_4$ since it is the symmetry group of the square (4-gon) and contains $\mathbb{Z}_4$.

(Some writers prefer the notation $\mathbb{D}_8$ since it has order 8; common alternatives are $\mathbf{D}_4$ and $\mathbf{D}_8$.)
(5) A routine but lengthy exercise yields the subgroups of G:

Order 8:	G	$G \cong \mathbb{D}_4$
Order 4:	$\{1, \sigma, \sigma^2, \sigma^3\}$	$S \cong \mathbb{Z}_4$
	$\{1, \sigma^2, \tau, \sigma^2\tau\}$	$T \cong \mathbb{Z}_2 \times \mathbb{Z}_2$
	$\{1, \sigma^2, \sigma\tau, \sigma^3\tau\}$	$U \cong \mathbb{Z}_2 \times \mathbb{Z}_2$
Order 2:	$\{1, \sigma^2\}$	$A \cong \mathbb{Z}_2$
	$\{1, \tau\}$	$B \cong \mathbb{Z}_2$
	$\{1, \sigma\tau\}$	$C \cong \mathbb{Z}_2$
	$\{1, \sigma^2\tau\}$	$D \cong \mathbb{Z}_2$
	$\{1, \sigma^3\tau\}$	$E \cong \mathbb{Z}_2$
Order 1:	$\{1\}$	$I \cong 1$

(6) The inclusion relations between the subgroups of G are shown in the lattice diagram of Figure 18.3 (left).

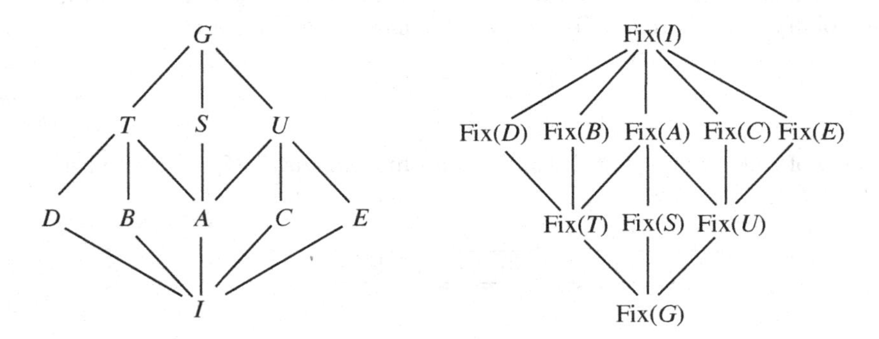

Figure 18.3. *Left*: Lattice of subgroups. *Right*: Lattice of subfields.

(7) Under the Galois correspondence we obtain the intermediate fields. Since the correspondence reverses inclusions, we obtain the lattice diagram in Figure 18.3 (right).

Remark. The lattice diagrams for $\mathcal{F}$ and $\mathcal{G}$ do *not* look the same unless one of them is turned upside down. Hence there does not exist a correspondence like the Galois correspondence but *preserving* inclusion relations.

(8) We now describe the elements of these intermediate fields. There are three obvious subfields of K of degree 2 over $\mathbb{Q}$, namely $\mathbb{Q}(\mathrm{i})$, $\mathbb{Q}(\sqrt{2})$, $\mathbb{Q}(\mathrm{i}\sqrt{2})$. These are clearly the fixed fields $\mathrm{Fix}(S)$, $\mathrm{Fix}(T)$, and $\mathrm{Fix}(U)$, respectively. The other fixed fields can be found by taking an arbitrary element of K, expressed uniquely in the form

$$x = a_0 + a_1\xi + a_2\xi^2 + a_3\xi^3 + a_4\mathrm{i} + a_5\mathrm{i}\xi + a_6\mathrm{i}\xi^2 + a_7\mathrm{i}\xi^3$$

where $a_0, \ldots, a_7 \in \mathbb{Q}$. Then we apply all elements γ in the subgroup, compute $\gamma(x)$, and equate it to x to find the required restrictions on the a_j. The upshot is that:

$$\begin{array}{lll} \mathrm{Fix}(A) = \mathbb{Q}(\mathrm{i}, \sqrt{2}) & \mathrm{Fix}(B) = \mathbb{Q}(\xi) & \mathrm{Fix}(C) = \mathbb{Q}((1+\mathrm{i})\xi) \\ \mathrm{Fix}(D) = \mathbb{Q}(\mathrm{i}\xi) & \mathrm{Fix}(E) = \mathbb{Q}((1-\mathrm{i})\xi) & \end{array}$$

It is now easy to verify the inclusion relations of Figure 18.3 (right).

18.10 Galois Group of a Finite Field

In Chapter 21 we need to know the Galois group of the finite field $\mathbb{F}_q$ over its prime subfield $\mathbb{F}_p$, where $q = p^n$. To state the required result, we first need:

Theorem 18.37. *Let $\mathbb{F}_q$ be a finite field of prime characteristic p. Then the map $\alpha : \mathbb{F}_q \to \mathbb{F}_q$ defined by $\alpha(\theta) = \theta^p$, for $\theta \in \mathbb{F}_q$ is an automorphism.*

Proof: Clearly $\alpha(\theta_1\theta_2) = \alpha(\theta_1)\alpha(\theta_2)$. To show that addition is preserved, use the binomial theorem:

$$\alpha(\theta_1 + \theta_2) = \sum_{i=0}^{p} \binom{p}{i} \theta_1^{p-i}\theta_2^i = \theta_1^p + \theta_2^p = \alpha(\theta_1) + \alpha(\theta_2)$$

where we use the well-known result that $\binom{p}{i}$ is divisible by p when $i \neq 0, p$. □

Definition 18.38. The map $\alpha : \mathbb{F}_q \to \mathbb{F}_q$ defined by $\alpha(\theta) = \theta^p$, for $\theta \in \mathbb{F}_q$ is the *Frobenius automorphism* of $\mathbb{F}_q$.

We can now state:

Theorem 18.39. *Let $q = p^n$ where p is prime. The Galois group of $\mathbb{F}_q/\mathbb{F}_p$ is cyclic of order n. It is generated by the Frobenius automorphism α defined by $\alpha(\theta) = \theta^p$, for $\theta \in \mathbb{F}_q$.*

Proof: See Garling [48] theorem 12.4. Exercise 18.21 also describes the proof. □

18.11 Norm and Trace

We end this chapter by generalising Section 2.7, defining the norm and trace relative to an extension L/K and proving a few key properties. By Theorem 2.5 there is a primitive element α such that $L = K(\alpha)$, with conjugates $\alpha_1, \ldots, \alpha_n$. The field L has basis $\{\alpha, \alpha^2, \ldots, \alpha^m\}$ over K where m is the degree of the minimal polynomial of α over K. There are n K-monomorphisms $\sigma_j : L \to \mathbb{C}$, determined by $\sigma_j(\alpha) = \alpha_j$.

Definition 18.40. The *(relative) trace* of α in L/K is

$$T_{L/K}(\alpha) = \sigma_1(\alpha) + \cdots + \sigma_n(\alpha) = \alpha_1 + \cdots + \alpha_n$$

The *(relative) norm* of α in L/K is

$$N_{L/K}(\alpha) = \sigma_1(\alpha) \cdots \sigma_n(\alpha) = \alpha_1 \cdots \alpha_n$$

The field polynomial or *characteristic polynomial* of α with respect to L/K is

$$c_\alpha(t) = \prod_{j=1}^{n} (t - \sigma_j(\alpha)) = \prod_{j=1}^{n} (t - \alpha_j)$$

The trace and norm are (up to sign) the coefficient of t^{n-1} and the constant term is $c_\alpha(t)$.

Example 18.41. Let $\alpha = \sqrt[3]{2} \in \mathbb{R}$, $K = \mathbb{Q}, L = \mathbb{Q}(\alpha)$. The conjugates of α are $\alpha_1 = \alpha, \alpha_2 = \omega\alpha, \alpha_3 = \omega^2\alpha$. The trace is $\alpha_1+\alpha_2+\alpha_3 = \alpha(1+\omega+\omega^2) = 0$. The norm is $\alpha_1\alpha_2\alpha_3 = \alpha^3(1.\omega.\omega^2) = 2$.

This example is not a normal extension, but both the norm and trace of α lie in $\mathbb{Q}$. This is a general phenomenon:

Theorem 18.42. *Every coefficient of c_α lies in K. In particular, both $T_{L/K}(\alpha)$ and $N_{L/K}(\alpha)$ lie in K.*

Proof: The coefficients of c_α are symmetric polynomials in the α_j, hence are fixed by $\mathrm{Gal}(L/K)$. Therefore they lie in K. □

Theorem 18.43. (Transitivity Formulas) *Suppose that L/K and K/J are extensions and $\alpha \in J$. Then*

$$\begin{aligned} T_{L/J}(\alpha) &= T_{K/J}(T_{L/K}(\alpha)) \\ N_{L/J}(\alpha) &= N_{K/J}(N_{L/K}(\alpha)) \end{aligned}$$

Proof: Let $\sigma_j : L \to \mathbb{C}$ be the K-monomorphisms of L, and let $\sigma_j : K \to \mathbb{C}$ be the J-monomorphisms of K. Define

$$\mu_{jk} = \sigma_j \tau_k$$

Then the μ_{jk} are distinct, each μ_{jk} is a J-monomorphism $L \to \mathbb{C}$, and all J-monomorphisms $L \to \mathbb{C}$ arise in this manner. Now

$$T_{L/J}(\alpha) = \sum_{j,k} \sigma_j \tau_k(\alpha) = \sum_j \sigma_j \left(\sum_k \tau_k(\alpha) \right) = T_{K/J}(T_{L/K}(\alpha))$$

The proof for the norm is similar, replacing the sum by the product. □

For later use we interpret the norm, trace, and field polynomial in terms of linear algebra:

Lemma 18.44. *Let L/K be an extension of number fields. Viewing L as a vector space over K, define linear maps $l_a : L \to L$ by $l_a(x) = ax$ for $a, x \in L$. Then*

$$\begin{aligned} \operatorname{tr}(l_a) &= \mathrm{T}_{L/K}(a) \\ \det(l_a) &= \mathrm{N}_{L/K}(a) \\ \det(tI - l_a) &= c_\alpha(t) \end{aligned}$$

where $c_\alpha(t)$ is the field polynomial of a.

Proof: Let $\{u_1, \ldots, u_n\}$ be a basis for L over K. Then

$$au_j = \sum_{k=1}^{n} c_{jk} u_k$$

Let $\{\sigma_i\}$ be the set of all K-monomorphisms $K \to L$. Then

$$\sigma_i(a)\sigma_i(u_j) = \sum_{k=1}^{n} c_{jk}\sigma_i(u_k)$$

so we have a matrix equation

$$(\sigma_i(u_j))(\operatorname{diag}_i(\sigma_i(a)) = (c_{jk})(\sigma_i(u_k))$$

By Theorem 18.20 $\det(\sigma_i(u_j)) \neq 0$, so the trace, determinant, and characteristic polynomials of the matrices (c_{jk}) and $(\operatorname{diag}_i(\sigma_i(a))$ coincide. Compute them using the diagonal matrix. □

Proposition 18.45. *Let K be a number field, with $a \in K^*$. Then*

$$\mathrm{N}(a\mathfrak{O}_K) = |N_{K/\mathbb{Q}}(a)| \qquad (18.16)$$

In particular, if $a \in \mathbb{Q}^*$ then

$$\mathrm{N}(a\mathbb{Z}) = |a|$$

Proof: Both sides of (18.16) are multiplicative, so it is enough to consider the case when $a \in \mathfrak{O}_K$. Now

$$[\mathfrak{O}_K : a\mathfrak{O}_K] = |\det l_a|$$

where $l_a : K \to K$ is defined by $l_a(x) = ax$. Apply Lemma 18.44. □

18.12 Exercises

18.1 Determine splitting fields over $\mathbb{Q}$ for the polynomials t^3-1, t^4+5t^2+6, t^6-8, in the form $\mathbb{Q}(\alpha_1, \ldots, \alpha_k)$ for explicit α_j.

18.2 Find the degrees of these fields as extensions of $\mathbb{Q}$.

18.3 Show that every degree-2 extension in $\mathbb{C}$ is normal. Is this true if the degree is greater than 2?

18.4 If Σ is the splitting field for f over K and $K \subseteq L \subseteq \Sigma$, show that Σ is the splitting field for f over L.

18.5 Let ζ be a primitive 6th root of unity. Find all $\mathbb{Q}$-automorphisms of $\mathbb{Q}(\zeta)$ and prove directly that they are linearly independent over $\mathbb{Q}$.

18.6 Construct the normal closure N for the following extensions:

(a) $\mathbb{Q}(\alpha)/\mathbb{Q}$ where α is the real fifth root of 3

(b) $\mathbb{Q}(\beta)/\mathbb{Q}$ where β is the real seventh root of 2

(c) $\mathbb{Q}(\sqrt{2}, \sqrt{3})/\mathbb{Q}$

(d) $\mathbb{Q}(\alpha, \sqrt{2})/\mathbb{Q}$ where α is the real cube root of 2

18.7 Find the Galois groups of the extensions in Exercise 18.6.

18.8 Find the Galois groups of the extensions $N/\mathbb{Q}$ for the normal closures N of the extensions in Exercise 18.6.

18.9 Prove 'bare hands' that the degree of the extension

$$\mathbb{Q}(\sqrt{3}, \sqrt{5}, \sqrt{7})/\mathbb{Q}$$

is 8. Find its Galois group.

18.10 Let $\gamma = \sqrt{2+\sqrt{2}}$. Show that $\mathbb{Q}(\gamma)/\mathbb{Q}$ is normal, with cyclic Galois group. Show that $\mathbb{Q}(\gamma, \mathrm{i})$=$\mathbb{Q}(\mu)$ where $\mu^4 = \mathrm{i}$; that is, μ is a primitive 10th root of unity.

18.11 Find the Galois group of $t^6 - 7$ over $\mathbb{Q}$.

18.12 Find the Galois group of $t^6 - 2t^3 - 1$ over $\mathbb{Q}$.

18.13 Let $\zeta = \mathrm{e}^{\mathrm{i}\pi/6}$ be a primitive 12th root of unity. Find the Galois group $\mathrm{Gal}(\mathbb{Q}(\zeta)/\mathbb{Q})$ as follows.

(a) Prove that ζ is a zero of the polynomial $t^4 - t^2 + 1$, and that the other zeros are $\zeta^5, \zeta^7, \zeta^{11}$.

(b) Prove that $t^4 - t^2 + 1$ is irreducible over $\mathbb{Q}$, and is the minimal polynomial of ζ over $\mathbb{Q}$.

(c) Prove that $\text{Gal}(\mathbb{Q}(\zeta)/\mathbb{Q})$ consists of four $\mathbb{Q}$-automorphisms ϕ_j, defined by

$$\phi_j(\zeta) = \zeta^j \qquad j = 1, 5, 7, 11$$

(d) Prove that $\text{Gal}(\mathbb{Q}(\zeta)/\mathbb{Q}) \cong \mathbb{Z}_2 \times \mathbb{Z}_2$.

18.14 Using the subgroup structure of $\mathbb{Z}_2 \times \mathbb{Z}_2$, find all intermediate fields between $\mathbb{Q}$ and $\mathbb{Q}(\zeta)$ in Exercise 18.13.

18.15 The group $\mathbb{D}_4$ is isomorphic to the symmetry group of a square. Use the geometry of the square to find the list of subgroups in (5) of Example 18.36 and interpret them geometrically.

18.16 Show that $K = \mathbb{Q}(\sqrt{5}, \mathrm{i})$ is a normal separable extension of $\mathbb{Q}$ of degree 4. Find its Galois group.

18.17 Let θ be the real zero of $t^3 - t + 1$, and let $\phi, \bar{\phi}$ be the other two zeros. Let $K = \mathbb{Q}(\theta)$. Prove that

$$\begin{aligned} \phi + \bar{\phi} &= -\theta \\ \phi.\bar{\phi} &= -1/\theta \\ ((\phi - \theta)(\bar{\phi} - \theta)(\phi - \bar{\phi}))^2 &= -23 \end{aligned}$$

Deduce that the normal closure L is $K(\sqrt{-23}) = \mathbb{Q}(\theta, \phi, \bar{\phi})$, a separable normal extension of degree 6 over $\mathbb{Q}$. Find its Galois group G. List all subgroups of G and find their fixed fields. Which subfields are normal over $\mathbb{Q}$?

18.18 Let $K = \mathbb{Q}(\zeta)$ where ζ is a primitive 9th root of unity. Show that it is a cyclic extension of $\mathbb{Q}$ of degree 6. Using the Galois group, construct an intermediate field M of degree 3 over $\mathbb{Q}$. Show that $M = \mathbb{Q}(\theta)$ where $\theta = 2\cos 2\pi/9$ and find the minimal polynomial of θ over $\mathbb{Q}$.

18.19 Let L/K be a number field extension and let $f(t) \in K[t]$ be irreducible of degree d. Let the normal closure of L/K be N/K. Prove that the order of the Galois group $G = \text{Gal}(N/K)$ is divisible by d, and G is isomorphic to a subgroup of the symmetric group $\mathbb{S}_d$.

18.20 Let $f(t) = t^5 - 6t + 3 \in \mathbb{Q}[t]$, and let N be a splitting field for f over K. Show that f has three real zeros and one complex conjugate pair of zeros. Use this fact and Exercise 18.19 to prove that $\text{Gal}(N/\mathbb{Q}) \cong \mathbb{S}_5$.

18.21 Prove Theorem 18.39. (*Hint*: The Galois group G has order n, and it contains the Frobenius automorphism $\phi(x) = x^p$. Let the order of ϕ be d. Prove that every element of $\mathbb{F}_q$ is a zero of the polynomial $t^{p^d} - t$. Deduce that $d = n$ so ϕ generates G and G is cyclic.)

18.22 Use a computer algebra package to find two distinct monic irreducible cubic polynomials $p(t) \in \mathbb{Z}[t]$ with the property that if α is a zero of p then so is $\alpha^2 - 8$. Find their Galois groups over $\mathbb{Q}$.

18.23 Let $p(t) = t^6 + t^5 + 16t^4 + 11t^3 + 91t^2 + 36t + 181$. Use a computer algebra package to calculate the zeros of p and show that none of them is real. Show that if α is a zero of p then so is $\alpha^2 + 5$. Find the Galois group of p over $\mathbb{Q}$.

(*Hint*: Let $f(t) = t^2 + 5$. Expand the polynomial $f(f(f(t))) - t$ and show that it is equal to $(t^2 - t + 5)p(t)$.)

18.24 Use the same idea as in Exercises 18.22 and 18.23 to construct an irreducible monic quintic polynomial in $\mathbb{Z}[t]$ with Galois group $\mathbb{Z}_5$.

19

Cyclotomic and Cubic Fields

In this chapter we revisit cyclotomic fields to establish some further important properties. These results are not needed in the proof of Kummer's Theorem, but they are basic to algebraic number theory and illustrate the use of Galois theory. We also take a brief look at cubic fields, concentrating on a theorem of Artin concerning cubic fields in which the minimal polynomial has only one real zero.

The results on cyclotomic fields are elegant and relatively complete. Those on cubic fields cover only a small part of that area and rely on complicated calculations. The theory of cubic fields is well developed, but many simple questions remain unanswered. One reason is that in general the Galois group is not abelian; the best-understood areas of algebraic number theory focus on abelian extensions—those with an abelian Galois group. The Kronecker–Weber Theorem, which is beyond the scope of this book, shows that all abelian extensions of $\mathbb{Q}$ are subfields of cyclotomic fields. See Long [82] chapter 6 theorem (1.12).

19.1 Cyclotomic Polynomials

Until now our main concern with cyclotomic fields has been with pth roots of unity for prime p. The composite case is more complicated, but of considerable interest.

As motivation we consider $\mathbb{Q}(\zeta)$ where $\zeta = e^{i\pi/6}$ is a primitive 12th root of unity. We classify the powers ζ^j according to the minimal power d such that $(\zeta^j)^d = 1$. That is, we consider when they are *primitive* dth roots of unity. We can then factorise $t^{12} - 1$ by grouping zeros according to d:

$$\begin{aligned} t^{12} - 1 = &(t-1)\times \\ &(t-\zeta^6)\times \\ &(t-\zeta^4)(t-\zeta^8)\times \\ &(t-\zeta^3)(t-\zeta^9)\times \\ &(t-\zeta^2)(t-\zeta^{10})\times \\ &(t-\zeta)(t-\zeta^5)(t-\zeta^7)(t-\zeta^{11}) \end{aligned}$$

This expression simplifies to

$$t^{12} - 1 = (t-1)(t+1)(t^2+t+1)(t^2+1)(t^2-t+1)(t^4-t^2+1)$$

(The explicit form of the final factor $t^4 - t^2 + 1$ is not immediately obvious. One way to find it is to divide $t^{12} - 1$ by all the other factors.)

Defining $\Phi_d(t)$ to be the factor corresponding to primitive dth roots of unity, we have

$$t^{12} - 1 = \Phi_1(t)\Phi_2(t)\Phi_3(t)\Phi_4(t)\Phi_6(t)\Phi_{12}(t)$$

These computations show that in this case every factor Φ_d lies in $\mathbb{Z}[t]$. In fact, it turns out that the factors are all *irreducible* over $\mathbb{Z}$. This is obvious for all factors here except $t^4 - t^2 + 1$, where it can be proved by considering the factorisation $(t-\zeta)(t-\zeta^5)(t-\zeta^7)(t-\zeta^{11})$. See Exercise 19.7.

This calculation motivates:

Definition 19.1. The polynomial

$$\Phi_n(t) = \prod_{a\in\mathbb{Z}_n,\ (a,n)=1} (t-\zeta^a) \tag{19.1}$$

is the nth *cyclotomic polynomial* over $\mathbb{C}$. (We shortly prove it is a polynomial over $\mathbb{Z}$.)

The most basic property of cyclotomic polynomials is the identity

$$t^n - 1 = \prod_{d|n} \Phi_d(t) \tag{19.2}$$

which is a direct consequence of their definition.

Recall that $\mathbb{Z}_n^*$ denotes the group of units of $\mathbb{Z}_n$. Since the units are the numbers between 0 and $n-1$ that are prime to n, the order of $\mathbb{Z}_n^*$ is given by an important number-theoretic function:

Definition 19.2. The *Euler function* $\phi(n)$ is the number of integers a, with $1 \leq a \leq n-1$, such that a is prime to n.

We immediately deduce:

Proposition 19.3. (a) *The order of* $\mathbb{Z}_n^*$ *is equal to* $\phi(n)$.
(b) *The cyclotomic polynomial* $\Phi_n(t)$ *has degree* $\phi(n)$.

Proof: The order of $\mathbb{Z}_n^*$ is the number of $a \in \mathbb{Z}_n$ such that $(a, n) = 1$. This is the definition of $\phi(n)$.

The degree is the number of linear factors in (19.1), which again is the number of $a \in \mathbb{Z}_n$ such that $(a, n) = 1$. □

The Euler function $\phi(n)$ has numerous interesting properties. In particular

$$\phi(p^k) = (p-1)p^{k-1} \tag{19.3}$$

if p is prime, and

$$\phi(r)\phi(s) = \phi(rs) \tag{19.4}$$

when r, s are coprime. Together these equations determine $\phi(n)$ for all $n \geq 1$. See Exercise 19.8.

Monomorphisms $\mathbb{Q}(\zeta) \to \mathbb{C}$

Suppose that $\sigma : \mathbb{Q}(\zeta) \to \mathbb{C}$ is a monomorphism. Since ζ is a primitive nth root of unity, so is $\sigma(\zeta)$. Therefore $\sigma(\zeta) = \zeta^i$ where i is prime to n. Moreover, this equation determines σ on the whole of $\mathbb{Q}(\zeta)$. So we can define σ_i to map ζ to ζ^i, and every monomorphism σ is one of the σ_i.

We will show that each σ_i is a monomorphism, but at this stage we do not know that, because we do not know the minimal polynomial of ζ. (It turns out to be $\Phi_n(t)$, but we do not yet know this is irreducible over $\mathbb{Q}$, so we must avoid circular logic.) However, we do know that the field polynomial $c_\zeta(t)$ satisfies

$$c_\zeta(t) = \prod_{i \in X}(t - \sigma_i(\zeta)) = \prod_{i \in X}(t - \zeta^i)$$

for some subset X of $\{\zeta^i : (i, n) = 1\}$ such that $1 \in X$. Moreover, $c_\zeta(t)$ lies in $\mathbb{Q}[t]$ by Theorem 2.9, and its zeros in $\mathbb{C}$ are the ζ^i for $i \in X$, which are distinct. This is enough to prove:

Theorem 19.4. *The field polynomial* c_ζ *is irreducible over* $\mathbb{Q}$ *and is the minimum polynomial of* ζ.

Proof: The field polynomial $c_\zeta(t)$ lies in $\mathbb{Q}[t]$ by Theorem 2.9. Let the minimum polynomial of ζ over $\mathbb{Q}$ be $m(t) \in \mathbb{Q}[t]$. By Theorem 2.11, c_ζ is a power of m. But the zeros of c_ζ are distinct, so $m = c_\zeta$. Now m is irreducible over $\mathbb{Q}$ by Proposition 1.17, so c_ζ is irreducible over $\mathbb{Q}$. □

Despite this result, we still have not proved that $c_\zeta = \Phi_n$. Certainly it divides Φ_n, but if Φ_n happened to be reducible, then c_ζ could be one of its factors. To complete the proof that Φ_n is irreducible, we turn to Galois theory.

19.2 Galois Group of Cyclotomic Field

In this section we calculate the Galois group of $\mathbb{Q}(\zeta)/\mathbb{Q}$. The key step is to find its order; once we do that everything is wide open.

Theorem 19.5. *Let ζ be a primitive nth root of unity in $\mathbb{C}$, and let $G = \mathrm{Gal}(\mathbb{Q}(\zeta)/\mathbb{Q})$ be the Galois group of $\mathbb{Q}(\zeta)$ over $\mathbb{Q}$. Then $|G| = \phi(n)$.*

Proof: We prove:

(a) $|G| = \phi(n)$ when $n = p^k$ is a prime power.

(b) Suppose that a, b are coprime. If $|G| = \phi(n)$ when $n = a$ and $n = b$, then $|G| = \phi(n)$ for $n = ab$.

The result then follows by induction on the number of distinct prime-power factors of n, using (19.3) and (19.4).

Proof of (a): Suppose that $n = p^k$. If $k = 0$ then $\Phi_n(t) = \Phi_1(t) = t - 1$, which is irreducible.

If $k > 0$ we use the Eisenstein trick. We know that

$$\Phi_{p^k}(t) = \frac{t^{p^k} - 1}{t^{p^{k-1}} - 1} = 1 + t^{p^{k-1}} + t^{2p^{k-1}} + \cdots + t^{(p-1)p^{k-1}}$$

Set $t = u + 1$. Now

$$\begin{aligned}\Phi_{p^k}(t) &= 1 + (u+1)^{p^{k-1}} + (u+1)^{2p^{k-1}} + \cdots + (u+1)^{(p-1)p^{k-1}} \\ &= ph(t) + u^{(p-1)p^{k-1}}\end{aligned}$$

where $h(t)$ has constant term 1. By Eisenstein's Criterion, $\Phi_n(t)$ is irreducible over $\mathbb{Q}$.

Proof of (b): Let $n = ab$ where $(a, b) = 1$. Now ζ^a is a primitive bth root of unity, and ζ^b is a primitive ath root of unity.

In the field extension $\mathbb{Q}(\zeta)/\mathbb{Q}$, consider the intermediate fields $M_a = \mathbb{Q}(\zeta^a)$ and $M_b = \mathbb{Q}(\zeta^b)$. The corresponding Galois groups are

$$\begin{aligned} \mathrm{Gal}(\mathbb{Q}(\zeta)/\mathbb{Q}) &= G \\ \mathrm{Gal}(\mathbb{Q}(\zeta)/\mathbb{Q}(\zeta^a)) &= G_a \\ \mathrm{Gal}(\mathbb{Q}(\zeta)/\mathbb{Q}(\zeta^b)) &= G_b \\ \mathrm{Gal}(\mathbb{Q}(\zeta)/\mathbb{Q}(\zeta)) &= 1 \end{aligned}$$

Since G is abelian, all extensions of subfields of $\mathbb{Q}(\zeta)$ are normal. Therefore

$$\begin{aligned} \mathrm{Gal}(\mathbb{Q}(\zeta^a)/\mathbb{Q}) &\cong G/G_a \\ \mathrm{Gal}(\mathbb{Q}(\zeta^b)/\mathbb{Q}) &\cong G/G_b \end{aligned}$$

We are assuming that the theorem is true for $n = a, n = b$, so we know that

$$|G/G_a| = \phi(b) \qquad |G/G_b| = \phi(a)$$

Also, $G_a \cap G_b = 1$, because any $\mathbb{Q}$-automorphism fixing ζ^a and ζ^b must fix ζ. So the subgroup $G_aG_b \subseteq G$ is isomorphic to $G_a \times G_b$, which has order $\phi(a)\phi(b) = \phi(n)$. Therefore $|G| \geq \phi(n)$.

However, we already know that $|G| \leq \phi(n)$, since any $\gamma \in G$ must equal some σ_i with $(i, n) = 1$. Therefore $|G| = \phi(n)$. That is, condition (1) holds for $n = ab$. □

By now the next result is not a surprise, but it still requires proof.

Theorem 19.6. (a) *The Galois group* $\mathrm{Gal}(\mathbb{Q}(\zeta)/\mathbb{Q})$ *consists of the* $\mathbb{Q}$*-automorphisms* σ_k *defined by*

$$\sigma_k(\zeta) = \zeta^k$$

where $0 \leq k \leq n-1$ *and* k *is prime to* n.

(b) $\mathrm{Gal}(\mathbb{Q}(\zeta)/\mathbb{Q})$ *is isomorphic to* $\mathbb{Z}_n^*$ *and in particular it is an abelian group.*

Proof: Let $f(t) = t^n - 1 \in \mathbb{Q}[t]$. Then the splitting field of f is $\mathbb{Q}(\zeta)$. Theorem 18.7 implies that the extension $\mathbb{Q}(\zeta)/\mathbb{Q}$ is normal. It is separable since the characteristic is zero. So we can apply the Galois Correspondence to $\mathbb{Q}(\zeta)/\mathbb{Q}$.

(a) Let $\gamma \in G = \mathrm{Gal}(Q(\zeta)/\mathbb{Q})$. We have already shown that $\gamma = \sigma_i$ for some i. But there are $\phi(n)$ choices for i, and $|G| = \phi(n)$. Therefore every σ_i lies in the Galois group, and these give the whole of G.

(b) Clearly $\sigma_j\sigma_k = \sigma_{jk}$, so the map $\sigma_k \mapsto k$ is an isomorphism from $\mathrm{Gal}(\mathbb{Q}(\zeta)/\mathbb{Q})$ to $\mathbb{Z}_n^*$. □

We saw in Theorem 2.11 that the field polynomial of any element is a power of its minimal polynomial. Galois theory adds some extra detail. Recall the definition of the field polynomial c_α for an extension L/K, Definition 18.40:

$$c_\alpha(t) = \prod_{\gamma \in \Gamma} (t - \gamma(\alpha))$$

The coefficients of $c_\alpha(t)$ clearly lie in $\mathrm{Fix}(\Gamma)$; by the Galois correspondence applied to N/K where N is a normal closure, this is K. We now generalise Theorem 2.11(a):

Proposition 19.7. (a) *Let L/K be a normal extension and let $\alpha \in L$. Let $p(t) = c_\alpha(t)$ be the field polynomial of α and let $m(t)$ be the minimum polynomial of α. Let $H = \mathrm{Gal}(L/K(\alpha)) \subseteq \Gamma$. Then $p(t) = m(t)^d$ where $d = |H|$.*

(b) *If $L = K(\alpha)$ then $p(t)$ is the minimal polynomial of α over K. In particular, it is irreducible over K.*

Proof: (a) Let $G = \mathrm{Gal}(L/K)$ and let $\gamma_1, \gamma_2 \in G$. Then

$$\begin{aligned} \gamma_1(\alpha) = \gamma_2(\alpha) &\iff \gamma_2^{-1}\gamma_1(\alpha) = \alpha \\ &\iff \gamma_2^{-1}\gamma_1 \in \mathrm{Gal}(L/K(\alpha)) = H \\ &\iff \gamma_1 = \gamma_2 h \text{ for some } h \in H \end{aligned}$$

Decompose Γ into left cosets γH. Let $\gamma_1, \ldots, \gamma_k$ be a set if representatives, one from each such coset. Then $k = |\Gamma|/|H| = |\Gamma|/d$. By the above calculation, the γ_i are distinct. Now

$$\Gamma = \gamma_1 H \,\dot\cup\, \gamma_2 H \,\dot\cup \cdots \dot\cup\, \gamma_k H$$

where $\dot\cup$ indicates disjoint union. Therefore

$$\begin{aligned} p(t) &= \prod_{i=1}^{k} \prod_{h \in H} (t - \gamma_i h(\alpha)) \\ &= \prod_{i=1}^{k} \prod_{h \in H} (t - \gamma_i(\alpha)) \quad \text{since } h \in H \\ &= \left(\prod_{i=1}^{k} (t - \gamma_i(\alpha)) \right)^d = q(t)^d \end{aligned}$$

where

$$q(t) = \prod_{i=1}^{k} (t - g\gamma_i(\alpha))$$

and $d = |\Gamma|/k = |H|$.

We claim that $q(t)$ is the minimal polynomial of α. The Galois correspondence implies that $p(t) \in K[t]$. We claim that $q(t) \in K[t]$. We know that $p(t) = q(t)^d$ as polynomials over L. Apply an arbitrary element $\gamma \in \Gamma$ to deduce that $q^\gamma(t)^d = p^\gamma(t) = p(t)$. Therefore $q^\gamma(t)^d = q(t)^d$ as elements of $L[t]$. Both $q^\gamma(t)$ and $q(t)$ are monic. Factorising into irreducibles of $L[t]$ we deduce that $q^\gamma(t) = q(t)$. Since $g \in \Gamma$ is arbitrary, $q(t)$ is fixed by Γ, so $q(t) \in K[t]$.

Certainly $q(\alpha) = 0$, so the minimal polynomial $m(t)$ of α over K divides $q(t)$. But $q(t)$ is divisible by all distinct conjugates $\gamma(\alpha)$ of α, each of which is a zero of $m(t)$. Moreover, all zeros of $m(t)$ arise in this manner. Since $m(t)$ is irreducible and L/K is a number field extension, the zeros of $m(t)$ are simple. Therefore $q(t) = m(t)$ as required.

(b) Follows immediately because $L = K(\alpha)$ implies that $H = \mathrm{Gal}(L/L) = 1$, so $d = 1$. Therefore $p(t) = m(t)$, which is irreducible over K since it is a minimal polynomial. □

Using this result, we can finally prove:

Corollary 19.8. *The cyclotomic polynomial $\Phi_n(t)$ is irreducible over $\mathbb{Q}$ for all n.*

Proof: Consider the (normal) extension $\mathbb{Q}(\zeta)/\mathbb{Q}$ where ζ is a primitive nth root of unity. All zeros of $\Phi_n(t)$ are conjugate under the Galois group, so $\Phi_n(t)$ is the field polynomial of ζ. By Theorem 2.11(a), $\Phi_n = m^r$ for some r, where m is the minimal polynomial of ζ over $\mathbb{Q}$. If $r > 1$ then Φ_n has a multiple zero; however, the zeros of Φ_n are distinct by definition. Therefore $\Phi_n = m$. Now m is irreducible over $\mathbb{Q}$ by Proposition 19.7. □

Table 19.1 shows the first 15 cyclotomic polynomials.

At the moment all we have proved is that $\Phi_n(t)$ has rational coefficients. The table suggests that the coefficients are actually integers.

It also suggests that these integers are either 0, 1, or -1, but this is misleading; Exercise 19.19 shows that Φ_{105} is a counterexample. This is the smallest n for which Φ_n has a coefficient that is not 0, 1, or -1. As a sample for larger n, the polynomial Φ_{255255} has a coefficient with absolute value 532. The survey by Helmut Maier [85] discusses theorems by several authors showing that the coefficients of Φ_n are unbounded over all n and estimating the growth rate of the (absolute values of the) largest coefficients with respect to n. The Sister Beiter Conjecture [5, 6] states that when $n = pqr$, for primes $3 \leq p < q < r$, the largest coefficient is less than

n	$\Phi_n(t)$
1	$t-1$
2	$t+1$
3	t^2+t+1
4	t^2+1
5	$t^4+t^3+t^2+t+1$
6	t^2-t+1
7	$t^6+t^5+t^4+t^3+t^2+t+1$
8	t^4+1
9	t^6+t^3+1
10	$t^4-t^3+t^2-t+1$
11	$t^{10}+t^9+t^8+t^7+t^6+t^5+t^4+t^3+t^2+t+1$
12	t^4-t^2+1
13	$t^{12}+t^{11}+t^{10}+t^9+t^8+t^7+t^6+t^5+t^4+t^3+t^2+t+1$
14	$t^6-t^5+t^4-t^3+t^2-t+1$
15	$t^8-t^7+t^5-t^4+t^3-t+1$

Table 19.1. Cyclotomic polynomials $\Phi_n(t)$ for $1 \leq n \leq 15$.

or equal to $\frac{2}{3}p$. A proof by Branko Juran and others [70] was announced in 2023. This article includes an up-to-date history of the conjecture.

We now prove that $\Phi_n(t) \in \mathbb{Z}[t]$ using the recursive procedure of Exercise 19.5.

Theorem 19.9. *For all n, the cyclotomic polynomial $\Phi_n(t)$ has integer coefficients.*

Proof: Use induction on n. When $n = 1$ we have $\Phi_1(t) = t - 1 \in \mathbb{Z}[t]$. Now assume that the result is valid for all $m < n$; we prove it for n. Let

$$g(t) = (t^n - 1)/\Phi_n(t)$$

so that

$$t^n - 1 = g(t)\Phi_n(t)$$

By (19.1), both $g(t)$ and $\Phi_n(t)$ are monic. By (19.2),

$$g(t) = \prod_{d|n, d\neq n} \Phi_d(t)$$

By induction each factor $\Phi_d(t) \in \mathbb{Z}[t]$, so $g(t) \in \mathbb{Z}[t]$. Moreover, $g(t)$ and $\Phi_n(T)$ are monic. By Corollary 1.9, $\Phi_n(t) \in \mathbb{Z}[t]$. This completes the induction. $\square$

19.3 Subfields of Cyclotomic Fields

Theorem 19.6 shows that the Galois group of a cyclotomic field $\mathbb{Q}(\zeta_n)/\mathbb{Q}$ is isomorphic to $\mathbb{Z}_n^*$. The structure of this group is known. It takes an especially simple form when $n = p$ is prime; in this case $\mathbb{Z}_p^*$ is a cyclic group $\mathbb{Z}_{p-1}$ of order $p-1$. The subgroup structure of cyclic groups $\mathbb{Z}_m$ is well known; there is one cyclic subgroup of order d for each divisor d of m. If γ is a generator of $\mathbb{Z}_m$, this subgroup is generated by $\gamma^{m/d}$.

Using the Galois correspondence, we can then determine all subfields of $\mathbb{Q}(\zeta_p)$.

Example 19.10. Let $p = 5$, with $\zeta = e^{2\pi i/5}$. We showed in Example 18.35 that the subfields of $\mathbb{Q}(\zeta)$ are $\mathbb{Q}, \mathbb{Q}(\sqrt{5})$, and $\mathbb{Q}(\zeta)$. The Galois group G is generated by γ where $\gamma(\zeta) = \zeta^2$. Then

$$\gamma^0(\zeta) = 1 \qquad \gamma(\zeta) = \zeta^2 \qquad \gamma^2(\zeta) = \zeta^4 \qquad \gamma^3(\zeta) = \zeta^3$$

The subgroups are $\mathbf{1}, \{1, \gamma^2\}, G$. The fixed fields of $\mathbf{1}, G$ are $\mathbb{Q}(\zeta)$ and $\mathbb{Q}$, respectively. The interesting case is the fixed field of γ^2. In this case we notice that γ^2 is the same as complex conjugation κ. Symmetrising over γ^2 we see that 1, $\zeta + \zeta^4$, and $\zeta^2 + \zeta^3$ are fixed by κ. The sum of these is zero, so the fixed field is $\mathbb{Q}(\zeta + \zeta^4)$, which is the same as $\mathbb{Q}(\zeta^2 + \zeta^3)$.

By the Galois correspondence, $[\mathbb{Q}(\zeta+\zeta^4) : \mathbb{Q}] = 2$. Therefore $\theta = \zeta+\zeta^4$ satisfies a quadratic equation over $\mathbb{Q}$. Indeed,

$$\theta^2 = \zeta^2 + \zeta^3 + 2 = (-1-\theta) + 2 = 1 - \theta$$

and the minimal polynomial of θ over $\mathbb{Q}$ is $m(t) = t^2 + t - 1$.

The zeros of $m(t)$ are $\frac{-1\pm\sqrt{5}}{2}$. Therefore $\text{Fix}(\{1, \gamma^2\}) = \mathbb{Q}(\sqrt{5})$.

Example 19.11. Let $p = 11$, with $\zeta = e^{2\pi i/11}$. We show that the nontrivial proper subfields of $\mathbb{Q}(\zeta)$ are $\mathbb{Q}(\sqrt{-11})$ and a field of degree 5 over $\mathbb{Q}$ with minimal polynomial $t^5 + t^4 - 4t^3 - 3t^2 + 3t + 1$.

By Theorem 19.6 the Galois group is $\text{Gal}(\mathbb{Q}(\zeta)/\mathbb{Q}) \cong \mathbb{Z}_{11}^* \cong \mathbb{Z}_{10}$. By experiment a generator is $\alpha : \zeta \mapsto \zeta^2$.

The nontrivial proper subgroups of $\mathbb{Z}_{10}$ are $\langle\alpha^5\rangle \cong \mathbb{Z}_2$ and $\langle\alpha^2\rangle \cong \mathbb{Z}_5$.

The fixed field $\text{Fix}(\mathbb{Z}_5)$ has $\mathbb{Q}$-basis $\{u_2, u_2\}$ where

$$\begin{aligned} u_1 &= \zeta + \zeta^4 + \zeta^5 + \zeta^9 + \zeta^3 \\ u_2 &= \zeta^2 + \zeta^8 + \zeta^{10} + \zeta^7 + \zeta^6 \end{aligned}$$

and has degree 2 over $\mathbb{Q}$. Routine calculations show that

$$u_1 + u_2 = -1 \qquad u_1 u_2 = 3$$

so the minimal polynomial of either u_i is t^2+t+3, with zeros $\frac{1}{2}(-1\pm\sqrt{-11})$. Thus $\mathrm{Fix}(\mathbb{Z}_5) = \mathbb{Q}(\sqrt{-11})$.

$\mathrm{Fix}(\mathbb{Z}_2)$ has $\mathbb{Q}$-basis $\{v_1, v_2, v_3, v_4, v_5\}$ where

$$v_1 = \zeta + \zeta^{10} \quad v_2 = \zeta^2 + \zeta^9 \quad v_3 = \zeta^3 + \zeta^8 \quad v_4 = \zeta^4 + \zeta^7 \quad v_5 = \zeta^5 + \zeta^6$$

and has degree 5 over $\mathbb{Q}$. By computer algebra, the v_j have minimal polynomial $t^5 + t^4 - 4t^3 - 3t^2 + 3t + 1$.

In these examples, the quadratic subfields are relatively easy to compute. Moreover, they appear to be of the form $\mathbb{Q}(\sqrt{\pm p})$. This suggests a more general investigation of the quadratic subfields of cyclotomic fields, a topic that we return to in Section 22.2.

19.4 Class-Numbers of Cyclotomic Fields

Table 19.2 shows the class-numbers of the first 50 cyclotomic fields. In general, these are hard to compute and vary wildly in size, with no obvious patterns.

n	class-number	n	class-number
1–22	1	40	1
23	3	41	121
24–28	1	42	1
29	8	43	211
30	1	44–45	1
31	9	46	3
32–36	1	47	695
37	37	48	1
38	1	49	43
39	2	50	1

Table 19.2. Class-number of cyclotomic field $\mathbb{Q}(\zeta_n)$.

19.5 Some Cubic Fields

To get away from abelian extensions, we now dip our toes into the murky waters of cubic number fields. A full treatment would take another book, so we content ourselves with a remarkable theorem of Artin and a few examples.

Definition 19.12. An extension $K/\mathbb{Q}$ is *cubic* if $K = \mathbb{Q}(\alpha)$ where α has cubic minimal polynomial $m(t)$, that is, $\partial m = 3$.

Here we consider only the case when α is real, and its other two conjugates are nonreal. These cases are distinguished by the signs of the discriminant of the cubic. Any cubic can be transformed into the simpler form $f(t) = t^3 + at + b$ by a shift of origin and a scaling of t. The discriminant for this cubic is $-4a^3 - 27b^2$. If it is positive, there are three real zeros; if negative, one zero is real and the other two are complex conjugate.

We apply Dirichlet's Units Theorem 10.9. Let K be a splitting field for f. Since K embeds in $\mathbb{R}$, its subgroup of roots of unity is $\{\pm 1\}$. The group of units U_K has rank 1, and U_N has rank 2. So K has a fundamental unit, which generates U_K modulo $\{\pm 1\}$. There is a unique generator $u \in U_K$ such that $u > 1$ in $\mathbb{R}$.

Theorem 19.13. (Artin's Theorem) *With the above conditions on K, the discriminant d_K satisfies $|d_K| < 4u^3 + 24$.*

Before giving the proof, which is based on the proof in Fröhlich and Taylor [46], we observe a useful implication:

Corollary 19.14. *Let $1 < v \in U_K$. If $4v^{3/2} + 24 < |d_K|$, then v is a fundamental unit.*

Proof: Let $N \subseteq \mathbb{C}$ be the splitting field of m, and let U be the group generated by u. Now $U_K = \{\pm 1\} \times U$, so $v = u^n$ for some $n > 0$. But if $n \geq 2$ we would have $|d_K| \leq 4u^3 + 24 = 4v^{3/n} + 24 \leq 4v^{3/2} + 24$, a contradiction. Thus $n = 1$ and $v = u$. □

Now we prove Artin's Theorem. The proof involves some complicated estimates and is a bit of a scramble.

Proof: Since u is a unit, $N_{K/Q}(u) = \pm 1$. Let the conjugates of u, other than u itself, be $z, \bar{z}$. Now $u > 0$ and $z\bar{z} > 0$, so $uz\bar{z} = +1$. Swapping z and $\bar{z}$ if necessary, there exist $r > 1$ and $\theta \in (0, \pi)$ such that

$$u = r^2 \qquad z = r^{-1}e^{i\theta} \qquad \bar{z} = r^{-1}e^{-i\theta}$$

The ring of integers $\mathfrak{O}$ of K contains $\mathbb{Z}[u]$, so the discriminant of $\mathbb{Z}[u]$ is divisible by d_K. In particular, $|d_K| \leq |D|$. Since $4u^3 + 24 = 4(r^6 + 6)$, it is enough to prove that $|D| < 4(r^6 + 6)$.

To establish this inequality, define a function $\psi(r, \theta)$ by:

$$\psi(r,\theta) = \sqrt{D} = \begin{vmatrix} 1 & r^2 & r^4 \\ 1 & r^{-1}e^{i\theta} & r^{-2}e^{2i\theta} \\ 1 & r^{-1}e^{-i\theta} & r^{-2}e^{-2i\theta} \end{vmatrix} = -2i((r^3 + r^{-3})\sin\theta - \sin 2\theta)$$

To simplify notation, set $\rho = \frac{1}{2}(r^3 + r^{-3})$. By calculus applied to $\psi(r, \theta)$, the expression $|D|$ has a maximum only when $|\sqrt{D}|$ has a maximum, which requires $\partial\psi(r,\theta)/\partial r = 0$. This leads to the condition

$$2\rho\cos\theta - 2\cos 2\theta = 0$$

If we let $x = \cos\theta$ this becomes $\rho x - 2x^2 + 1 = 0$. Let $g(x) = 2x^2 - \rho x - 1$. Now

$$|D| \leq 16(\rho - \cos\theta)^2 \sin^2\theta = 16(\rho - x)^2(1 - x^2)$$

so

$$|D| \leq 16(\rho^2 - 2\rho x + x^2)(1 - x)^2$$

For maximality, $\rho x = 2x^2 - 1$, so $\rho^2 x^2 = 4x^4 - 4x^2 + 1$. Thus

$$\begin{aligned} |D| \leq 16(\rho^2 + 1 - x^4 - x^2) &\leq 4(r^6 + 6 + (r^{-6} - 4x^2 - 4x^4)) \\ &= 4u^3 + 24 + 4(r^{-6} - 4x^2 - 4x^4) \end{aligned}$$

It therefore suffices to show that $r^{-6} - 4x^2 - 4x^4 < 0$, or equivalently

$$4x^4 + 4x^2 - r^{-6} > 0 \tag{19.5}$$

To establish this we consider the zeros of $g(x)$. Since $2\rho = r + r^{-1} > 2$ we have

$$g(1) = 2 - \rho - 1 = 1 - \rho < 0$$

Also

$$g\left(\frac{-1}{2r^3}\right) = \frac{2}{4r^6} + \frac{r^3 + r^{-3}}{4r^3} - 1 = \frac{3}{4}(r^{-6} - 1) < 0$$

and $g(0) = 1 > 0$. The graph of g is a parabola, so g has two zeros: one with $x > 1$, which is impossible since $x = \cos\theta$, and the other with $x < -1/2r^3$. Thus $x^2 > 1/(4r^6)$, so $4x^2 - r^{-6} > 0$, and (19.5) is an obvious consequence. □

We give one application of Artin's Theorem:

Example 19.15. Let $K = \mathbb{Q}(\sqrt[3]{2})$. We prove that a fundamental unit is $u = \sqrt[3]{2} - 1$.

Clearly u is a unit since its inverse is $v = 1 + \sqrt[3]{2} + \sqrt[3]{4}$. Moreover, $v \sim 3.85 \cdots$. By Example 2.41, $\mathfrak{o}_K = \mathbb{Z}[\sqrt[3]{2}]$ so $d_k = 108$. But

$$4(3.85)^{3/2} + 24 < 108$$

and Corollary 19.14 implies that v is a fundamental unit.

The most important lesson to be drawn from Example 19.15 is that when it comes to cubic fields, the calculations become extremely intricate, even when finding objects that are known to exist. For this reason, computer algebra has become essential when studying specific examples.

Nevertheless, many general results can be proved, among them the following:

Theorem 19.16. (Ishida) *Let $a \geq 2$ be an integer such that $4a^3 + 27$ is squarefree, and let v be the unique real zero of $f(t) = t^3 + at - 1$. Then v^{-1} is a fundamental unit of $\mathbb{Q}(v)$, and $v > 1$.*

Proof: Since $\mathrm{D}f(t) = 3x^2 + a > 0$, the cubic $f(t)$ has a unique real zero, which we denote by v. The discriminant of f is $-(4a^3 + 27)$, which is squarefree. By Theorem 2.30, $\mathfrak{O}_K = \mathbb{Z}[v]$ and $d_K = -(4a^3 + 27)$.

Dirichlet's Units Theorem implies that there is a fundamental unit u. Now $v^{-1} = a + v^2$, so $v^{-1} > 1$. Therefore $v^{-1} = u^n$ for some $n \geq 1$. We claim $n = 1$. If not, $n \geq 2$, and we seek a contradiction. By Artin's Theorem 19.13,

$$4a^3 + 27 = |d_K| < 4u^3 + 24$$

whence $a < u$. Now

$$a^2 < u^2 \leq u^n = v^{-1} = v^2 + a < a + 1$$

which is impossible since $a \geq 2$. □

A detailed survey of known results on cubic number fields can be found in Marques and Ward [86]. Shanks [120] studies cyclic cubic fields—that is, having cyclic Galois group $\mathbb{Z}_3$—which are generated by a zero of the polynomial $t^3 - at^2 - (a+3)t - 1$. Godwin and Samet [50] list all the totally real cubic number fields of discriminant less than 2×10^4 and show that 764 of them have class-number 1. This is 92%, so a surprisingly high proportion of this class of cubic number fields comprises unique factorisation domains.

19.6 Exercises

19.1 The Möbius function $\mu(n)$ is defined to be 0 if n has a squared prime factor, and otherwise $(-1)^r$ where r is the number of distinct prime factors. Show that μ is multiplicative; that is, if m, n are coprime then $\mu(mn) = \mu(m)\mu(n)$.

19.2 Prove that $\sum_{d|n} \mu(d) = 1$ if $n = 1$ and 0 if $n > 1$.

19.3 Prove the *Möbius inversion formula*:

If f, g are arithmetic functions such that

$$g(n) = \sum_{d|n} f(d) \qquad (n \geq 1)$$

then

$$f(n) = \sum_{d|n} \mu(d) g(\frac{n}{d}) \qquad (n \geq 1)$$

(*Hint*: Use the result of Exercise 19.2.)

19.4 Show that the sum of all primitive nth roots of unity is $\mu(n)$.

19.5 Prove that

$$\Phi_n(t) = \prod_{d|n} (t^d - 1)^{\mu(n/d)}$$

19.6 We can use the identity (19.2) recursively to compute $\Phi_n(t)$. Thus

$$\Phi_1(t) = t - 1$$

so

$$t^2 - 1 = \Phi_2(t)\Phi_1(t)$$

which implies that

$$\Phi_2(t) = \frac{t^2 - 1}{\Phi_1(t)} = \frac{t^2 - 1}{t - 1} = t + 1$$

Similarly

$$\Phi_3(t) = \frac{t^3 - 1}{t - 1} = t^2 + t + 1$$

and

$$\Phi_4(t) = \frac{t^4 - 1}{(t - 1)(t + 1)} = t^2 + 1$$

and so on.

Use this method to verify Table 19.1.

19.7 Use the factorisation $(t-\zeta)(t-\zeta^5)(t-\zeta^7)(t-\zeta^{11})$ to show that t^4-t^2+1 is irreducible over $\mathbb{Z}$.

19.8 Prove (19.3) and (19.4).

(*Hint*: For (19.3) exclude the multiples of p between 0 and p^k.

For (19.4) use the Chinese Remainder Theorem to show that $\mathbb{U}_{rs} \cong \mathbb{U}_r \times \mathbb{U}_s$, where $\mathbb{U}_t$ is the multiplicative group of invertible elements of the ring $\mathbb{Z}_t$.)

19.9 Let $\zeta = \zeta_7$. Prove that $\mathbb{Q}(\zeta)$ has precisely four subfields, whose degrees over $\mathbb{Q}$ are $1, 2, 3, 6$. Two of these are $\mathbb{Q}$ and $\mathbb{Q}(\zeta)$. Prove that the extension of degree 3 is $\mathbb{Q}(\alpha)$ and the extension of degree 2 is $\mathbb{Q}(\beta)$ where $\alpha = \zeta + \zeta^6, \beta = \zeta + \zeta^2 + \zeta^4$.

19.10 (continued) Find the minimum polynomial $m(t)$ over $\mathbb{Q}$ of α. Let its conjugates be $\alpha_1 = \alpha = \zeta + \zeta^6, \alpha_2 = \zeta^2 + \zeta^5, \alpha_3 = \zeta^3 + \zeta^4$. Prove that $\alpha_{i+1} = \alpha_i^2 - 2$ for $i = 0, 1, 2$. Deduce that if θ is a zero of $m(t)$ then so is $\theta^2 - 2$. Prove this directly by relating $m(t^2-2)$ to $m(t)$.

19.11 (continued) Find the minimum polynomial $m(t)$ over $\mathbb{Q}$ of β.

Show that $\mathbb{Q}(\beta) = \mathbb{Q}(\sqrt{-7})$ and express β as $a + b\sqrt{-7}$ for $a, b \in \mathbb{Q}$.

What is the ring of integers of $\mathbb{Q}(\beta)$?

19.12 Let $\zeta = \zeta_p$, for p prime. By Exercise 13.1, the element $1+\zeta$ is a unit in $\mathbb{Q}(\zeta)$. Deduce that $1+\zeta^i$ is a unit in $\mathbb{Q}(\zeta)$ for $1 \le i \le p-1$.

Show that when $p > 3$ the elements $1 + \zeta^i + \zeta^{2i}$ are units in $\mathbb{Q}(\zeta)$ for $1 \le i \le p-1$.

19.13 Let $\zeta = \zeta_p$ for p prime. Let $q \neq p$ be prime and suppose that there exists $a \in \mathbb{Z}$ with $a \not\equiv 1 \pmod{q}$. Suppose that $q|(a^p-1)/(a-1)$. Show that a is a zero of the minimal polynomial of ζ when reduced $\pmod{q}$. Using computer algebra, find all such primes q for $3 \le p \le 11$ and $2 \le a \le p-1$.

19.14 What happens for larger a? *Hint*: Try $p = 3, p = 5$.

19.15 Show that $t^3 + 10t + 1$ is irreducible over $\mathbb{Q}$, and let θ be a real zero.

Use the discriminant to prove that the ring of integers is $\mathbb{Z}[\theta]$.

Use Artin's Theorem to show that $u = -1/\theta$ generates the group of positive units.

19.16 Let $K = \mathbb{Q}(\zeta)$ where ζ is a primitive 5th root of unity, and let $G = \mathrm{Gal}(K/\mathbb{Q})$. Show that:

(a) The only quadratic field contained in K is $\mathbb{Q}(\sqrt{5})$.

(b) The only roots of unity in K are the tenth roots.

(c) Use logarithmic space to show that if u is a unit of $\mathbb{Z}[\zeta]$ and $|u| = 1$ (complex absolute value) then u is a root of unity.

19.17 Use Galois Theory to find all subfields of $\mathbb{Q}(\zeta)$ when ζ is a primitive 13th root of unity. More precisely, find a generator for the Galois group, which is cyclic of order 12; then list all subgroups of the Galois group.

For each subfield K specify a $\mathbb{Q}$-basis of K, whose elements are sums of distinct powers of ζ as in Example 19.11.

Compute (using computer algebra) the common minimal polynomial of the elements of this $\mathbb{Q}$-basis.

19.18 Use cyclotomic polynomials to prove that for every integer $n > 1$ there are infinitely many primes p such that $p \equiv 1 \pmod{n}$.

(*Hint*: (a) Prove that if a prime p does not divide n then p divides $\Phi_n(a)$ for some $a \in \mathbb{Z}$ if and only if $p \equiv 1 \pmod{n}$.

(b) Assume for a contradiction that there are only finitely many primes $p \equiv 1 \pmod{n}$, and let them be $p_1, \ldots, p_r$. Let $M = p_1 \ldots p_r$ and $N \in \mathbb{Z}$. Show that $\Phi_n(NM) \equiv \pm 1 \pmod{M}$. Deduce that $\Phi_n(NM) \not\equiv 0 \pmod{p_i}$ and that no prime factor of $\Phi_n(NM)$ divides n.

(c) Prove that for large enough N we have $\Phi(NM) \neq \pm 1$, and deduce that there is a prime p dividing $\Phi_n(NM)$, that p is none of the p_i, and that $p \equiv 1 \pmod{n}$: contradiction.)

19.19 Calculate $\Phi_{105}(t)$ and show that it has a coeffficient not equal to $-1, 0, 1$.

20

Prime Ideals Revisited

20.1 Introduction

Let L/K be an extension. In preparation for Chapter 21, it is convenient to make a slight change to our standard notation, writing the ring of integers of K as $\mathfrak{o}$ and the ring of integers of L as $\mathfrak{O}$. We then use lowercase letters $\mathfrak{a}$ for ideals of $\mathfrak{o}$ and uppercase $\mathfrak{A}$ for ideals of $\mathfrak{O}$. This convention is common in the literature.

In this chapter we establish additional properties of prime ideals. We begin with the valuation map, which specifies the power to which a given prime ideal $\mathfrak{p}$ occurs in the factorisation of a fractional ideal $\mathfrak{a}$. We discuss the structure of the residue ring $\mathfrak{o}/\mathfrak{a}$ for ideals $\mathfrak{a}$. We generalise Gauss's Lemma to polynomials over a ring of integers, and we introduce the technique of localisation at a prime ideal, local rings, and discrete valuation rings.

The main aim of this chapter is to set up some useful methods and results, which we use in Chapter 21 on ramification—how prime ideals behave when the number field is extended.

20.2 Valuation Relative to a Prime Ideal

Let K be a number field with ring of integers $\mathfrak{o}$. In Section 6.2 we defined the group $\mathbb{F}$ of fractional ideals of $\mathfrak{o}$. For simplicity, we use notation such as

$$\prod_{\mathfrak{p}} f(\mathfrak{p})$$

to mean the product of the expressions $f(\mathfrak{p})$ over *all* prime ideals $\mathfrak{p}$. This makes sense provided $f(\mathfrak{p}) = 1$ for all but a finite number of $\mathfrak{p}$.

We make heavy use of Theorem 6.12: every fractional ideal $\mathfrak{a} \in \mathbb{F}$ can be expressed *uniquely* as a product

$$\mathfrak{a} = \prod_{\mathfrak{p}} \mathfrak{p}^{a(\mathfrak{p})} \tag{20.1}$$

where $\mathfrak{p}$ runs through all prime ideals of $\mathfrak{o}$, the $a(\mathfrak{p}) \in \mathbb{Z}$ (so they may be *negative*), and all but finitely many $a(\mathfrak{p})$ are zero.

We introduce some systematic notation for the decomposition of a fractional ideal into a product of powers of prime ideals.

Definition 20.1. For any prime ideal $\mathfrak{p}$, and for each fractional ideal $\mathfrak{a}$, the integer $v_{\mathfrak{p}}(\mathfrak{a})$ is the power $a(\mathfrak{p})$ occurring in (20.1). Letting $\mathfrak{a}$ vary over $\mathbb{F}$, for each prime ideal $\mathfrak{p}$ the integer $v_{\mathfrak{p}}(\mathfrak{a})$ defines a map $v_{\mathfrak{p}} : \mathbb{F} \to (\mathbb{Z}, +)$, called the *valuation map* relative to $\mathfrak{p}$.

By (20.1), we clearly have:

$$\mathfrak{a} = \prod_{\mathfrak{p}} \mathfrak{p}^{v_{\mathfrak{p}}(\mathfrak{a})} \tag{20.2}$$

We interpret $v_{\mathfrak{p}}(\mathfrak{a})$ as the power to which $\mathfrak{p}$ appears in the prime (ideal) factorisation of $\mathfrak{a}$. For fractional ideals this power may be negative.

Proposition 20.2. *If $\mathfrak{a}, \mathfrak{b}$ are fractional ideals of $\mathfrak{o}$ and $\mathfrak{p}$ is any prime ideal of $\mathfrak{o}$ then*

$$v_{\mathfrak{p}}(\mathfrak{a}\mathfrak{b}) = v_{\mathfrak{p}}(\mathfrak{a}) + v_{\mathfrak{p}}(\mathfrak{b}) \tag{20.3}$$

$$v_{\mathfrak{p}}(\mathfrak{o}) = 0 \tag{20.4}$$

$$v_{\mathfrak{p}}(\mathfrak{a}^{-1}) = -v_{\mathfrak{p}}(\mathfrak{a}) \tag{20.5}$$

Proof: All three properties follow directly from (20.2). □

In more abstract terms, these properties state that $v_{\mathfrak{p}} : \mathbb{F} \to (\mathbb{Z}, +)$ is a group homomorphism.

Proposition 20.3. *A fractional ideal* $\mathfrak{a}$ *is an ideal if and only if* $v_{\mathfrak{p}}(\mathfrak{a}) \geq 0$ *for all* $\mathfrak{p}$. □

Proposition 20.4. *Let* $\mathfrak{a}, \mathfrak{b}$ *be fractional ideals. Then*

$$\mathfrak{a}|\mathfrak{b} \Leftrightarrow \mathfrak{a} \supseteq \mathfrak{b} \Leftrightarrow v_{\mathfrak{p}}(\mathfrak{a}) \leq v_{\mathfrak{p}}(\mathfrak{b}) \text{ for all } \mathfrak{p} \tag{20.6}$$

$$v_{\mathfrak{p}}(\mathfrak{a} \cap \mathfrak{b}) = \sup(v_{\mathfrak{p}}(\mathfrak{a}), v_{\mathfrak{p}}(\mathfrak{b})) \tag{20.7}$$

$$v_{\mathfrak{p}}(\mathfrak{a} + \mathfrak{b}) = \inf(v_{\mathfrak{p}}(\mathfrak{a}), v_{\mathfrak{p}}(\mathfrak{b})) \tag{20.8}$$

$$v_{\mathfrak{p}}(\mathfrak{a}\mathfrak{b}) = v_{\mathfrak{p}}(\mathfrak{a} \cap \mathfrak{b})) + v_{\mathfrak{p}}(\mathfrak{a} + \mathfrak{b})) \tag{20.9}$$

□

Proposition 20.5. *Let* $\mathfrak{a}, \mathfrak{b}$ *be nonzero ideals. Then the following are equivalent:*

$$\mathfrak{a} + \mathfrak{b} = \mathfrak{o} \tag{20.10}$$

$$\mathfrak{a} \cap \mathfrak{b} = \mathfrak{a}\mathfrak{b} \tag{20.11}$$

$$v_{\mathfrak{p}}(\mathfrak{a})v_{\mathfrak{p}}(\mathfrak{b}) = 0 \text{ for all } \mathfrak{p} \tag{20.12}$$

□

If any condition (20.10), (20.11), or (20.12) holds, we say that $\mathfrak{a}$ and $\mathfrak{b}$ are *coprime*. Another term is '$\mathfrak{a}$ is prime to $\mathfrak{b}$'. (This is the correct historical term; see Hardy and Wright [58]. It often appears as '$\mathfrak{a}$ is coprime to $\mathfrak{b}$', which is clear but arguably ungrammatical: 'coprime *with*' is justifiable but no one uses it. Another term for 'coprime' is 'relatively prime'.)

Corollary 20.6. *If ideals* $\mathfrak{a}_1, \ldots, \mathfrak{a}_n$ *of* $\mathfrak{o}$ *are pairwise coprime, then*

$$\mathfrak{a}_1 \cap \cdots \cap \mathfrak{a}_n = \mathfrak{a}_1 \cdots \mathfrak{a}_n \tag{20.13}$$

Proof: By uniqueness of prime factorisation for ideals, $\mathfrak{a}_n$ is prime to $\mathfrak{a}_1 \cdots \mathfrak{a}_{n-1}$. Now use (20.11) and induction on n. □

20.3 Residue Rings and Fields

Let K be a number field with ring of integers $\mathfrak{o}$. Recall Definition 6.6, which we restate for convenience:

Definition 20.7. Let $\mathfrak{a}$ be a nonzero ideal of $\mathfrak{o}$. Then the *residue ring* of $\mathfrak{a}$ is the quotient ring $\mathfrak{o}/\mathfrak{a}$.

We write $x \equiv y \pmod{\mathfrak{a}}$ to mean that $x - y \in \mathfrak{a}$. In coset notation, $\mathfrak{a} + x = \mathfrak{a} + y$. Since the quotient map is a homomorphism, addition, subtraction, and multiplication are preserved modulo $\mathfrak{a}$.

Theorem 20.8. *Let $\mathfrak{a}$ be an ideal of $\mathfrak{o}$. Then the following statements are equivalent:*
(a) *The ideal $\mathfrak{a}$ is prime.*
(b) *The ideal $\mathfrak{a}$ is maximal.*
(c) *The residue ring $\mathfrak{o}/\mathfrak{a}$ is a finite field.*

Proof: Combine the results of Lemma 6.2 and Theorem 6.4(d) to show that (a) $\Leftrightarrow$ (b). Lemma 6.2(a) states that (b) holds if and only if $\mathfrak{o}/\mathfrak{a}$ is a field. By Corollary 6.5, this field is finite. □

Finite fields are discussed in Section 1.7. They are the fields $\mathbb{F}_q$ where $q = p^n$ and p is a rational prime.

Example 20.9. Let $K = \mathbb{Q}(\mathrm{i})$, with ring of integers $\mathbb{Z}[\mathrm{i}]$. By Theorem 5.34, the prime ideals are of three kinds:
(a) $\langle p \rangle$ where p is a rational prime of the form $4k + 3$.
(b) $\langle a + \mathrm{i}b \rangle$ where $p = a^2 + b^2$ is a rational prime of the form $4k + 1$, so:

$$p = (a + \mathrm{i}b)(a - \mathrm{i}b)$$

(c) The ideal $\langle 1 + \mathrm{i} \rangle$.

For example, the ideals $\langle 3 \rangle$, $\langle 7 \rangle$, $\langle 11 \rangle$, $\langle 19 \rangle$ are prime in $\mathbb{Z}[\mathrm{i}]$. However, $\langle 2 \rangle$, $\langle 5 \rangle$, $\langle 13 \rangle$, $\langle 17 \rangle$ are not:

$$\begin{aligned} \langle 2 \rangle &= \langle 1 + \mathrm{i} \rangle^2 \\ \langle 5 \rangle &= \langle 2 + \mathrm{i} \rangle \langle 2 - \mathrm{i} \rangle \\ \langle 13 \rangle &= \langle 3 + 2\mathrm{i} \rangle \langle 3 - 2\mathrm{i} \rangle \\ \langle 17 \rangle &= \langle 4 + \mathrm{i} \rangle \langle 4 - \mathrm{i} \rangle \end{aligned}$$

and the factors are prime ideals.

Now consider the residue fields $\mathfrak{o}/\mathfrak{p}$.

When $p = 4k+3$ we take $\mathfrak{p} = \langle p \rangle$. The residue field is $\mathbb{Z}[\mathrm{i}]/p\mathbb{Z}[\mathrm{i}] \cong \mathbb{F}_p[\mathrm{i}]$, which is one representation of the field $\mathbb{F}_{p^2}$ with p^2 elements. (The integer -1 is not a square in $\mathbb{F}_p$, so $t^2 + 1 \in \mathbb{F}_p[t]$ is irreducible; the quotient $\mathbb{F}_p[t]/\langle t^2 + 1 \rangle$ can be thought of as $\mathbb{F}_p[\mathrm{i}]$.)

When $p = 4k+1 = a^2+b^2$ and $\mathfrak{p} = \langle a \pm ib \rangle$ the residue field is isomorphic to $\mathbb{F}_p$.

When $\mathfrak{p} = \langle 1 + \mathrm{i} \rangle$ the residue field is isomorphic to $\mathbb{F}_2$.

Next, we prove a generalisation of the Chinese Remainder Theorem to ideals.

Theorem 20.10. (Chinese Remainder Theorem for Ideals)
Let $\mathfrak{p}_j$ $(1 \leq j \leq n)$ be finitely many prime ideals of $\mathfrak{o}$ and let r_j be positive integers. Then for any $x_j \in \mathfrak{o}$ there exists $x \in \mathfrak{o}$ such that

$$x \equiv x_j \pmod{\mathfrak{p}_j^{r_j}} \qquad (1 \leq j \leq n)$$

Moreover, x is unique modulo $\mathfrak{p}_1^{r_1} \cdots \mathfrak{p}_n^{r_n}$.

Proof: Translated into ring language, the theorem states that if $\phi_j : \mathfrak{o} \to \mathfrak{o}/p_j^{r_j}$ is the quotient map, then the map

$$f : \frac{\mathfrak{o}}{\mathfrak{p}_1^{r_1} \cdots \mathfrak{p}_n^{r_n}} \to \prod_{j=1}^{n} \frac{\mathfrak{o}}{\mathfrak{p}_j^{r_j}}$$

defined by $f(x) = (\phi_1(x), \ldots, \phi_n(x))$ induces an isomorphism $f : \frac{\mathfrak{o}}{\mathfrak{p}_1^{r_1} \cdots \mathfrak{p}_n^{r_n}} \cong \prod_{j=1}^{n} \mathfrak{o}/\mathfrak{p}_j^{r_j}$. We prove the theorem in this form. Obviously f is a ring homomorphism, with kernel $\ker f = \mathfrak{p}_1^{r_1} \cap \cdots \cap \mathfrak{p}_n^{r_n}$. By Corollary 20.6, $\ker f = \mathfrak{p}_1^{r_1} \cdots \mathfrak{p}_n^{r_n}$. Thus $\frac{\mathfrak{o}}{\mathfrak{p}_1^{r_1} \cdots \mathfrak{p}_n^{r_n}}$ is isomorphic to the image $\mathrm{im} f$. To complete the proof we must show that f is surjective. This follows inductively from Corollary 1.2, because

$$(\mathfrak{p}_1^{r_1} \cdots \mathfrak{p}_m^{r_m}) \cap \mathfrak{p}_{m+1}^{r_{m+1}} = (\mathfrak{p}_1^{r_1} \cdots \mathfrak{p}_m^{r_m})(\mathfrak{p}_{m+1}^{r_{m+1}})$$

since the prime ideals $\mathfrak{p}_i$ are distinct. □

20.4 Generalised Gauss's Lemma

The classical Gauss's Lemma (Theorem 1.8) states that if a monic polynomial over $\mathbb{Z}$ is reducible over $\mathbb{Q}$, then it is reducible over $\mathbb{Z}$. We generalise this useful result to number fields. The proof is modelled on the classical case.

Definition 20.11. The *content ideal* $\mathfrak{a}_f$ of a polynomial $f \in K[t]$ is the fractional ideal of $\mathfrak{o}$ generated by the coefficients of f. That is,

$$\mathfrak{a}_f = a_1\mathfrak{o} + \cdots + a_k\mathfrak{o} \qquad (20.14)$$

Let $f \in K[t]$. For each prime ideal $\mathfrak{p}$ of $\mathfrak{o}$ define

$$v_{\mathfrak{p}}(f) = v_{\mathfrak{p}}(\mathfrak{a}_f)$$

By (20.14) and (20.8),

$$v_{\mathfrak{p}}(f) = \inf_j v_{\mathfrak{p}}(a_j) \tag{20.15}$$

We now prove a decisive property of content ideals:

Lemma 20.12. *If $f, g \in K[t]$ then*

$$v_{\mathfrak{p}}(fg) = v_{\mathfrak{p}}(f) + v_{\mathfrak{p}}(g) \tag{20.16}$$

Proof: Let $h = fg$ and let $\mathfrak{a}_h$ be the content ideal of h. Let

$$\begin{aligned} h(t) &= a_k t^k + \cdots + a_0 \\ g(t) &= b_n t^n + \cdots + b_0 \\ h(t) &= c_m t^m + \cdots + c_0 \end{aligned}$$

where the $a_j, b_j, c_j \in K$. Let r be the smallest non-negative integer such that $v_{\mathfrak{p}}(b_r) = v_{\mathfrak{p}}(f)$, which exists by (20.15) applied to f. Then

$$\begin{aligned} v_{\mathfrak{p}}(b_r) &\leq v_{\mathfrak{p}}(b_j) \quad \text{for all } j \text{ such that } b_j \neq 0 \\ v_{\mathfrak{p}}(b_r) &< v_{\mathfrak{p}}(b_j) \quad \text{for all } j < r \text{ such that } b_j \neq 0 \end{aligned}$$

Define s for g analogously using the coefficients c_s. Then the coefficient d_{r+s} of t^{r+s} in fg is equal to $b_r c_s + d$, where d is a sum of monomials $r_k t^k$, each of which has $v_{\mathfrak{p}}(r_k) > v_{\mathfrak{p}}(b_r c_s)$. Therefore

$$v_{\mathfrak{p}}(b_r) + v_{\mathfrak{p}}(c_s) = v_{\mathfrak{p}}(f) + v_{\mathfrak{p}}(g)$$

so

$$v_{\mathfrak{p}}(fg) \leq v_{\mathfrak{p}}(f) + v_{\mathfrak{p}}(g)$$

However, $\mathfrak{a}_{fg} \subseteq \mathfrak{a}_f \mathfrak{a}_g$ since every coefficient of fg is a sum of products $b_j c_k$, which lie in $\mathfrak{a}_f \mathfrak{a}_g$. Thus

$$v_{\mathfrak{p}}(fg) \geq v_{\mathfrak{p}}(f) + v_{\mathfrak{p}}(g)$$

and (20.16) is proved. □

Theorem 20.13. (Generalised Gauss's Lemma) *If $f, g, h \in K[t]$ are monic, with $h = fg$, and $h \in \mathfrak{o}[t]$, then $f, g \in \mathfrak{o}[t]$.*

Proof: Since f, g are monic and in $\mathfrak{o}[t]$, both $v_\mathfrak{p}(f)$ and $v_\mathfrak{p}(g)$ are nonpositive. Since h is monic and in $\mathfrak{o}[t]$, we have $v_\mathfrak{p}(h) = 0$ for all $\mathfrak{p}$. By (20.16), $v_\mathfrak{p}(f) + v_\mathfrak{p}(g) = v_\mathfrak{p}(h) = 0$, so both $v_\mathfrak{p}(f)$ and $v_\mathfrak{p}(g)$ are zero, and $f, g \in \mathfrak{o}[t]$.
□

Corollary 20.14. *The minimal polynomial over K of an element $a \in L$ that is integral over $\mathfrak{o}$ lies in $\mathfrak{o}[t]$.* □

20.5 Localisation

Localisation is a construction that focuses attention on a single prime ideal $\mathfrak{p}$ in the ring of integers of a number field. It is like a partial field of fractions in which some denominators (related to $\mathfrak{p}$) are forbidden. We define it and establish some basic properties. It is used in Section 21.11 to prove an important theorem of Dedekind.

Consider a number field K with ring of integers $\mathfrak{o}$, so in particular K is the field of fractions of $\mathfrak{o}$. Let $\mathfrak{p}$ be a prime ideal of $\mathfrak{o}$. The complement $S = \mathfrak{o} \setminus \mathfrak{p}$ is a *multiplicative subset* of $\mathfrak{o}$; that is, if $x, y \in S$ then $xy \in S$. (Proof: if $xy \in \mathfrak{p}$ then x or y lies in $\mathfrak{p}$.)

Definition 20.15. The *localisation* $\mathfrak{o}_\mathfrak{p}$ is the subset $S^{-1}\mathfrak{o} \subseteq K$, where $S = \mathfrak{o} \setminus \mathfrak{p}$. This consists of all elements of K of the form x/s where $x \in \mathfrak{o}$ and $s \in S$ (that is, $s \notin \mathfrak{p}$).

Example 20.16. Let $K = \mathbb{Q}$, so $\mathfrak{o} = \mathbb{Z}$. Consider the prime ideal $\mathfrak{p} = \langle 2 \rangle$. Then the localisation $\mathfrak{o}_\mathfrak{p}$ consists of all rationals p/q in lowest terms such that q is odd.

Definition 20.17. A *local ring* is a ring with a unique maximal ideal.

In Chapter 1 we remarked that although a maximal ideal need not contain every proper ideal, every proper ideal is contained in *some* maximal ideal. We now give the proof, which requires Zorn's Lemma. It depends on the condition that the ring has a 1, which we are assuming throughout, but on this occasion we include that condition in the statement to avoid confusion.

Theorem 20.18. *Every proper ideal of a ring with* 1 *is contained in a maximal ideal.*

Proof: A proper ideal of R cannot contain 1, since $\langle 1 \rangle = R$.

Let I be a proper ideal. Consider the set $\mathcal{P}$ of all proper ideals that contain I. This set is partially ordered by inclusion. Zorn's Lemma states that if the union of every totally ordered subset $\mathcal{T}$ of elements of $\mathcal{P}$ belongs to $\mathcal{P}$, then $\mathcal{P}$ has a maximal element.

Obviously the union of all ideals in $\mathcal{T}$ is an ideal. It is also proper, since this union does not contain 1. Thus $\mathcal{P}$ has a maximal element J. Now $J \supseteq I$, and J is maximal since any proper ideal J' that contains J also contains I, hence J' lies in $\mathcal{P}$; maximality of J implies that $J' = J$.

Proposition 20.19. *The following conditions are equivalent:*

(a) *R is a local ring.*

(b) *If $a, b \in R$ and $a + b$ is a unit, then either a or b is a unit.*

(c) *The set of non-units is a maximal ideal of R.*

Proof: We show that (a)$\Rightarrow$(b)$\Rightarrow$(c)$\Rightarrow$(a).

(a)$\Rightarrow$(b): Let M be the unique maximal ideal of R. Suppose for a contradiction that both a and b are non-units but $a + b$ is a unit. Then $\langle a \rangle$ and $\langle b \rangle$ are proper ideals. By Theorem 20.18 each of them is contained in a maximal ideal, which must be M by uniqueness. Therefore both are contained in M. Now $a + b \in M$, but a proper ideal cannot contain a unit. This is the required contradiction.

(b)$\Rightarrow$(c): If a is a non-unit and $r \in R$, then ra is a non-unit. (Proof: if $(ra)s = 1$ then $a(rs) = 1$ so a is a unit.) Combined with (b) this shows that the set of non-units M is an ideal. It must be maximal since any larger ideal contains a unit, so it is equal to R.

(c)$\Rightarrow$(a) Let M be the set of non-units. By (c) it is a maximal ideal. Let I be any proper ideal. I cannot contain a unit, so $I \subseteq M$, and R has a unique maximal ideal M, so it is local. □

Next, we establish some simple basic properties of localisation.

Proposition 20.20. *With the above notation:*

(a) *$\mathfrak{o}_\mathfrak{p}$ is a subring of K.*

(b) *$\mathfrak{o} \subseteq \mathfrak{o}_\mathfrak{p}$.*

(c) *If $\mathfrak{a}$ is an ideal of $\mathfrak{o}$, then $S^{-1}\mathfrak{a}$ is an ideal of $\mathfrak{o}_\mathfrak{p}$, and every ideal of $\mathfrak{o}_\mathfrak{p}$ arises in this manner.*

(d) *If $\mathfrak{a}, \mathfrak{b}$ are ideals of $\mathfrak{o}$ then $S^{-1}(\mathfrak{a}\mathfrak{b}) = (S^{-1}\mathfrak{a})(S^{-1}\mathfrak{b})$.*

(e) *If $\mathfrak{a}$ is an ideal of $\mathfrak{o}$ and $\mathfrak{a}$ is prime to $\mathfrak{p}$, then $S^{-1}\mathfrak{a} = \mathfrak{o}_\mathfrak{p}$.*

(f) *$\mathfrak{o}_\mathfrak{p}$ is a local ring with unique maximal ideal $\mathfrak{m}_\mathfrak{p} = S^{-1}\mathfrak{p}$.*

(g) *The only nonzero proper ideals of $\mathfrak{o}_\mathfrak{p}$ are the powers $\mathfrak{m}_\mathfrak{p}^k$.*

(h) *The complement of $\mathfrak{m}_\mathfrak{p}$ in $\mathfrak{o}_\mathfrak{p}$ is precisely the set of units of $\mathfrak{o}_\mathfrak{p}$.*

(i) *If L/K is an extension and $M \subseteq L$ is an $\mathfrak{o}$-module, then $S^{-1}M$ is an $\mathfrak{o}_\mathfrak{p}$-module with the obvious action.*

Proof: (a) and (b) are obvious.

For (c), suppose that $\mathfrak{a}$ is an ideal of $\mathfrak{o}_\mathfrak{p}$. We claim that $\mathfrak{a} = S^{-1}(\mathfrak{a} \cap \mathfrak{o})$. Clearly $\mathfrak{a} \supseteq S^{-1}(\mathfrak{a} \cap \mathfrak{o})$. For the opposite inclusion, let $x \in \mathfrak{a}$. Then $x = a/s$ with $a \in \mathfrak{a}, s \in S$. Then $sx \in \mathfrak{a} \cap \mathfrak{o}$, so $x \in S^{-1}(\mathfrak{a} \cap \mathfrak{o})$.

(d) is obvious.

For (e): Since $\mathfrak{a}$ is prime to $\mathfrak{p}$, $\mathfrak{a} \cap \mathfrak{p} = 0$. Thus every nonzero element of $\mathfrak{a}$ lies in S. Let $0 \neq a \in \mathfrak{a}$. Then $a \in S$ so $S^{-1}\mathfrak{a} \ni a^{-1}a = 1$. Therefore $S^{-1}\mathfrak{a} = \mathfrak{o}_\mathfrak{p}$.

Part (f) is immediate since $\mathfrak{m}_\mathfrak{p} \supseteq \mathfrak{m}_\mathfrak{p}^2 \supseteq \mathfrak{m}_\mathfrak{p}^3 \supseteq \cdots$ is a decreasing sequence of ideals, and every nonzero proper ideal occurs in it.

(g): By (c) any ideal $\mathfrak{A}$ of $\mathfrak{o}_\mathfrak{p}$ is $S^{-1}\mathfrak{a}$ for an ideal $\mathfrak{a}$ of $\mathfrak{o}$. Write

$$\mathfrak{a} = \prod \mathfrak{p}_i^{e_i}$$

for prime ideals $\mathfrak{p}_i$, and number so that $\mathfrak{p}_1 = \mathfrak{p}$. By part (d),

$$\mathfrak{A} = S^{-1}\mathfrak{a} = \prod (S^{-1}\mathfrak{p}_i)^{e_i}$$

But now by part (e) the only term that does not reduce to $\mathfrak{o}_\mathfrak{p}$ is

$$\mathfrak{A} = S^{-1}\mathfrak{a} = (S^{-1}\mathfrak{p}_1)^{e_1} = (S^{-1}\mathfrak{p})^{e_1} = \mathfrak{m}_\mathfrak{p}^{e_1}$$

Part (h) follows from Theorem 20.19 and (g).

Part (i) is obvious. □

We see from this proof that localisation at $\mathfrak{p}$ preserves the multiplicative structure of ideals but kills off all prime ideals that are not equal to $\mathfrak{p}$. In fact, localisation at $\mathfrak{p}$ preserves many other aspects of the structure, but further discussion is beyond the scope of this book. See Borevič and Šafarevič [10], Fröhlich and Taylor [46], Lang [74, 75], and Long [82].

Definition 20.21. A *discrete valuation ring* is a Dedekind domain that is also a local ring; that is, it is a Dedekind domain that has a unique maximal (that is, nonzero prime) ideal.

In particular, the localisation $\mathfrak{o}_\mathfrak{p}$ of $\mathfrak{o}$ at a prime ideal $\mathfrak{p}$ is a discrete valuation ring.

Next, we need:

Theorem 20.22. *A Dedekind domain with only finitely many prime ideals is a principal ideal domain.*

Proof: Let the prime ideals be $\mathfrak{p}_1, \ldots, \mathfrak{p}_s$. Consider any ideal

$$0 \neq \mathfrak{a} = \mathfrak{p}_1^{r_1} \cdots \mathfrak{p}_s^{r_s}$$

For each i choose an element $\pi_i \in \mathfrak{p}_i \setminus \mathfrak{p}_i^2$. By the Chinese Remainder Theorem 20.10 there is an element $\alpha \in \mathfrak{o}$ such that

$$\alpha \equiv \pi_i^{e_i} \pmod{\mathfrak{b}} \tag{20.17}$$

where $\mathfrak{b} = (\mathfrak{p}_1 \cdots, \mathfrak{p}_s)^k$ and k is very large.

If the prime factorisation of $\langle \alpha \rangle$ is $\mathfrak{p}_1^{e_1} \cdots \mathfrak{p}_s^{e_s}$ then (20.17) implies that $e_i = r_i$ for all i. Therefore $\mathfrak{a} = \langle \alpha \rangle$ is principal. □

Corollary 20.23. *In a discrete valuation ring, the unique maximal ideal is principal. In particular, this holds for the localisation* $\mathfrak{o}_\mathfrak{p}$ *of the ring of integers* $\mathfrak{o}$ *of a number field* K *at a prime ideal* $\mathfrak{p}$.

Proof: The ring $\mathfrak{o}_\mathfrak{p}$ is a Dedekind domain, and it has only one prime ideal since prime ideals are maximal. Therefore Theorem 20.22 applies. □

20.6 Exercises

20.1 Let $\mathfrak{a}, \mathfrak{b}, \mathfrak{c}$ be ideals of the ring of integers $\mathfrak{o}$ of a number field. Prove that

$$\begin{aligned} v_\mathfrak{p}(\mathfrak{a} \cap \mathfrak{b} \cap \mathfrak{c}) &= \sup(v_\mathfrak{p}(\mathfrak{a}), v_\mathfrak{p}(\mathfrak{b}), v_\mathfrak{p}(\mathfrak{c})) \\ v_\mathfrak{p}(\mathfrak{a} + \mathfrak{b} + \mathfrak{c}) &= \inf(v_\mathfrak{p}(\mathfrak{a}), v_\mathfrak{p}(\mathfrak{b}), v_\mathfrak{p}(\mathfrak{c})) \end{aligned}$$

20.2 Generalise Exercise 20.1 to any finite number of ideals.

20.3 What about Equation (20.9) for three ideals? That is, find and prove a formula for $v_\mathfrak{p}(\mathfrak{abc})$ in terms of sums and intersections of the three ideals.

20.4 Let $K = \mathbb{Q}(\mathrm{i})$ and $L = K(\sqrt{2}) = \mathbb{Q}(\sqrt{2}, \mathrm{i})$. Then $\mathfrak{o} = \mathbb{Z}[\mathrm{i}]$, and we saw in Example 2.44 that $\mathfrak{O} = \mathbb{Z}[1, \theta, \mathrm{i}, \frac{1}{2}\theta(1+\mathrm{i})]$ where $\theta = \sqrt{2}$. Let $x = a + b\theta + c\mathrm{i} + d(\frac{1}{2}\theta(1+\mathrm{i})) \in L$. Show that

$$\begin{aligned} \mathrm{T}_{L/K}(x) &= 4a \\ \mathrm{N}_{L/K}(x) &= a^4 - 4a^2b^2 + 4b^4 + 2a^2c^2 + 4b^2c^2 + c^4 - 4a^2bd + 8b^3d \\ &\quad -8abcd + 4bc^2d + 8b^2d^2 - 4acd^2 + 4bd^3 + d^4 \end{aligned}$$

Deduce that both $\langle 3+\sqrt{2}\rangle_{\mathfrak{O}}$ and $\langle 3-\sqrt{2}\rangle_{\mathfrak{O}}$ have trace 12 and norm 49.

20.5 Find the field polynomials of $3+\sqrt{2}$ and $3-\sqrt{2}$ in Exercise 20.4.

20.6 Let R be a local ring. Prove that if $r \in R$ then either r or $1-r$ is a unit.

20.7 In the proof of Theorem 20.22 we used an ideal $\mathfrak{b} = (\mathfrak{p}_1 \cdots, \mathfrak{p}_s)^k$ and k is very large. Clarify the meaning of 'very large' and explain why this leads to the equation $e_i = r_i$ for all i.

20.8 Show that $t^3 - 3t + 1$ is irreducible over $\mathbb{Q}$, and all three zeros are real. Let θ be one of them. Let $K = \mathbb{Q}(\theta)$ with ring of integers $\mathfrak{O}$.

Compute $\Delta[1, \theta, \theta^2]$. Deduce that if $a + b\theta + c\theta^2 \in \mathfrak{O}$ for $a, b, c \in \mathbb{Q}$ then $a, b, c \in S^{-1}\mathbb{Z}$ where $S = \{3^n : n \geq 0\}$.

Let $\mathfrak{a} = \mathbb{Z}[\theta]$. Show that $\theta, \theta + 2$ are units of $\mathfrak{a}$ and of $\mathfrak{O}$. Show that

$$(\theta+1)^3 = 3\theta(\theta+2)$$

$$(\theta+1)\mathfrak{O} \text{ is a prime ideal of } \mathfrak{O}$$

$$(\theta+1)\mathfrak{O} \cap \mathfrak{a} = (\theta+1)\mathfrak{a}$$

$$\mathfrak{O} = \mathfrak{a} + (\theta+1)\mathfrak{O} = \mathfrak{a} + 3\mathfrak{O}$$

Deduce that $\mathfrak{O} = \mathfrak{a} = \mathbb{Z}[\theta]$.

20.9 With the same notation as in the previous exercise, show that $2\mathfrak{O}$ is a prime ideal of $\mathfrak{O}$, and deduce that $\mathfrak{O}$ is a principal ideal ring.

20.10 Let $K = \mathbb{Q}(\sqrt{3}, \sqrt{7})$. Find the content ideal of the polynomial

$$t^5 - \frac{419\sqrt{3} + 25\sqrt{7}}{12}t^3 + \sqrt{21}t^2 - 21t + 84 \in K[t]$$

20.11 Let $K = \mathbb{Q}(\sqrt{3}, \sqrt{7})$. Find the content ideal of the polynomial

$$(\sqrt{3} + \sqrt{21})t^5 - 17\sqrt{3}t + \sqrt{21} \in K[t]$$

20.12 Does there exist a Dedekind domain with exactly two prime ideals? If so, construct one; if not, prove it.

20.13 Generalise your result for the previous exercise to any specific finite number of prime ideals.

20.14 Is Theorem 20.18 true even when the ring does not have a 1?

21

Ramification Theory

Extensions of number fields are of central importance in algebraic number theory, and so are prime ideals. It is therefore of interest to find out what happens to a prime ideal when the field is extended. This chapter is devoted to that topic, often called *Hilbert Theory* because Hilbert made a feature of it in his *Zahlbericht* and developed many of the main results.

Let L/K be an extension. We continue to use the notation $\mathfrak{o}$ for the ring of integers of K and $\mathfrak{O}$ and the ring of integers of L, and correspondingly we use lowercase letters $\mathfrak{a}$ for ideals of $\mathfrak{o}$ and uppercase $\mathfrak{A}$ for ideals of $\mathfrak{O}$.

Suppose that $\mathfrak{p}$ is a prime ideal of $\mathfrak{o}$. Since $\mathfrak{o} \subseteq \mathfrak{O}$, we have $\mathfrak{p} \subseteq \mathfrak{O}$. It is usually not an ideal of $\mathfrak{O}$, but it generates the ideal $\langle \mathfrak{p} \rangle_{\mathfrak{O}} = \mathfrak{p}\mathfrak{O}$. Factorising into prime ideals of $\mathfrak{O}$, the ideal $\mathfrak{p}\mathfrak{O}$ decomposes as a finite product $\mathfrak{p}\mathfrak{O} = \prod \mathfrak{P}_i^{e_i}$ where the $\mathfrak{P}_i$ are prime ideals of $\mathfrak{O}$ and the $e_i > 0$. These are the prime ideals 'above' $\mathfrak{p}$, and our main aim is to elucidate the relationship between $\mathfrak{p}$ and the $\mathfrak{P}_i$. The powers e_i are the 'ramification indices' for $\mathfrak{p}$.

We begin with some simple motivating examples. Then we define prime ideals above $\mathfrak{p}$ and prove some basic properties. It is natural to ask how the Galois group $\mathrm{Gal}(L/K)$ acts in this situation. It is easy to see that it permutes the prime ideals above $\mathfrak{p}$ while preserving the powers e_i. A prime ideal $\mathfrak{p}$ is said to 'ramify' in $\mathfrak{O}$ if some $e_i > 1$. We relate the decomposition of $\mathfrak{p}\mathfrak{O}$ to norms.

The story gets properly off the ground in Section 21.4, where we set up the terminology and concepts of ramification theory, and we relate inertia degrees (degrees of certain field extensions) and ramification indices to the degree of L/K. Next we go back to the action of the Galois group

Gal(L/K), showing that when L/K is normal this group acts transitively on the prime ideals above $\mathfrak{p}$. This leads to strong constraints on ramification indices and inertia degrees. We define the decomposition group and the inertia group and state (without proof) Kummer's Criterion, which relates the prime ideals above $\mathfrak{p}$ to the minimal polynomial of a primitive element, modulo $\mathfrak{p}$. We use the trace form to prove (in the special case $K = \mathbb{Q}$) a theorem of Dedekind which shows that only finitely many $\mathfrak{p}$ ramify; indeed, a necessary and sufficient condition is that $\mathfrak{p}$ divides the discriminant.

Many of these concepts are illustrated in an extended example in Section 22.6 of the next chapter, after we have used ramification theory in cyclotomic fields to prove the Law of Quadratic Reciprocity.

21.1 Motivation

We begin with some simple examples that show there are interesting phenomena to investigate. The first proposition is obvious, but crucial:

Proposition 21.1. *The ring of integers $\mathfrak{o}$ of K is a subset of the ring of integers $\mathfrak{O}$ of L.*

Proof: If $\alpha \in \mathfrak{o}$ then $\alpha \in K \subseteq L$, so $\alpha \in L$. Since $\alpha \in \mathfrak{o}$, there is a polynomial $p \in \mathbb{Z}[t]$ such that $p(\alpha) = 0$. But since $\alpha \in L$, this implies that $\alpha \in \mathfrak{O}$. □

The fields K and L, and the rings $\mathfrak{o}$ and $\mathfrak{O}$, are therefore related by the following diagram, where arrows denote inclusion:

$$\begin{array}{ccc} K & \rightarrow & L \\ \uparrow & & \uparrow \\ \mathfrak{o} & \rightarrow & \mathfrak{O} \end{array}$$

Our aim is to relate prime ideals of $\mathfrak{O}$ to prime ideals of $\mathfrak{o}$. In general, any ideal $\mathfrak{a}$ of $\mathfrak{o}$ *generates* an ideal $\mathfrak{a}\mathfrak{O}$ of $\mathfrak{O}$, which need not equal $\mathfrak{a}$. When $\mathfrak{a}$ is principal, say $\mathfrak{a} = \langle a \rangle_{\mathfrak{o}}$, then $\mathfrak{a}\mathfrak{O} = \langle a \rangle_{\mathfrak{O}} = a\mathfrak{O}$. We set the scene with two examples.

Example 21.2. Consider the extension $L/K = \mathbb{Q}(\mathrm{i})/\mathbb{Q}$. The corresponding rings of integers are $\mathfrak{o} = \mathbb{Z}, \mathfrak{O} = \mathbb{Z}[\mathrm{i}]$. We already know that some prime ideals of $\mathbb{Z}$ cease to be prime in $\mathbb{Z}[\mathrm{i}]$.

(a) For example, consider $\mathfrak{p} = \langle 5 \rangle \subseteq \mathfrak{o}$. We can reinterpret $\mathfrak{p} = \langle 5 \rangle$ as the ideal of $\mathbb{Z}[\mathrm{i}]$ generated by 5, not as the ideal of $\mathbb{Z}$ generated by 5. Using subscripts to indicate which ring we are thinking of,

$$\langle 5\rangle_{\mathbb{Z}} = \{5n : n \in \mathbb{Z}\}$$
$$\langle 5\rangle_{\mathbb{Z}[\mathrm{i}]} = \{5a + 5b\mathrm{i} : a, b \in \mathbb{Z}\}$$

Now $5 = (2+\mathrm{i})(2-\mathrm{i})$ in $\mathbb{Z}[\mathrm{i}]$. The ideals $\mathfrak{P}_1 = \langle 2+\mathrm{i}\rangle$ and $\mathfrak{P}_2 = \langle 2-\mathrm{i}\rangle$ are both prime in $\mathfrak{O}$ since they have norm 5, and

$$\mathfrak{p}\mathfrak{O} = \mathfrak{P}_1\mathfrak{P}_2$$

(b) Consider $\mathfrak{p} = \langle 2\rangle_{\mathbb{Z}} \subseteq \mathfrak{o}$, and the related ideal $\mathfrak{P} = \langle 2\rangle_{\mathbb{Z}[\mathrm{i}]} \subseteq \mathbb{Z}[\mathrm{i}]$. Now

$$2 = -\mathrm{i}(1+\mathrm{i})^2$$

so

$$\mathfrak{P} = \langle 1+\mathrm{i}\rangle^2_{\mathbb{Z}[\mathrm{i}]}$$

Moreover, $\langle 1+\mathrm{i}\rangle_{\mathbb{Z}[\mathrm{i}]}$ has norm 2, so it is a prime ideal of $\mathfrak{O}$. Thus $\langle 2\rangle_{\mathbb{Z}[\mathrm{i}]}$ is the square of a prime ideal in $\mathbb{Z}[\mathrm{i}]$.

Example 21.3. For a more challenging case, let $K = \mathbb{Q}(\mathrm{i})$ and $L = K(\sqrt{2}) = \mathbb{Q}(\sqrt{2}, \mathrm{i})$. Then $\mathfrak{o} = \mathbb{Z}[\mathrm{i}]$, and by Example 2.44 $\mathfrak{O} = \mathbb{Z}[1, \theta, \mathrm{i}, \frac{1}{2}\theta(1+\mathrm{i})]$ where $\theta = \sqrt{2}$.

The principal ideal $\mathfrak{p} = \langle 7\rangle_{\mathfrak{o}}$ is a prime ideal of $\mathfrak{o}$ because 7 is a rational prime of the form $4k+3$. It generates an ideal $\mathfrak{p} = \langle 7\rangle_{\mathfrak{O}}$ in $\mathfrak{O}$, and this is not prime because

$$7 = 3^2 - 2.1^2 = (3+\sqrt{2})(3-\sqrt{2})$$

so in $\mathfrak{O}$ we have

$$\langle 7\rangle_{\mathfrak{O}} = \langle 3+\sqrt{2}\rangle_{\mathfrak{O}}\langle 3-\sqrt{2}\rangle_{\mathfrak{O}}$$

We have not investigated whether these two factors are prime ideals in $\mathfrak{O}$, but we can already see that $\mathfrak{p}\mathfrak{O}$ is not a prime ideal in $\mathfrak{O}$. So its prime ideal factors in $\mathfrak{O}$ have obvious mathematical interest.

21.2 Prime Ideals Above $\mathfrak{p}$

Examples 21.2 and 21.3 motivate a key issue: how ideals of $\mathfrak{o}$ relate to ideals of $\mathfrak{O}$. Unique factorisation into prime ideals lets us focus on the prime ideals $\mathfrak{p}$ of $\mathfrak{o}$. By Proposition 21.1, $\mathfrak{o} \subseteq \mathfrak{O}$, so each $\mathfrak{p}$ generates an ideal of $\mathfrak{O}$, namely

$$\mathfrak{p}\mathfrak{O} = \left\{\sum_i x_i y_i : x_i \in \mathfrak{p}, y_i \in \mathfrak{O}\right\}$$

By unique prime ideal factorisation,

$$\mathfrak{p}\mathfrak{O} = \prod_{j=1}^{g} \mathfrak{P}_j^{e_j} \tag{21.1}$$

for certain distinct prime ideals $\mathfrak{P}_j$ of $\mathfrak{O}$. Here we define g so that exactly g different $\mathfrak{P}_j$ occur with nonzero exponent. We then number the $\mathfrak{P}_j$ so that $e_j > 0$ for $1 \leq j \leq g$.

Definition 21.4. In Equation (21.1), the power e_j is the *multiplicity* of $\mathfrak{P}_j$. It is also called the *ramification index* of $\mathfrak{P}_j$.

The $\mathfrak{P}_j$ are the prime ideals of $\mathfrak{O}$ *above* $\mathfrak{p}$.

Figure 21.1 explains the mental image behind the terminology.

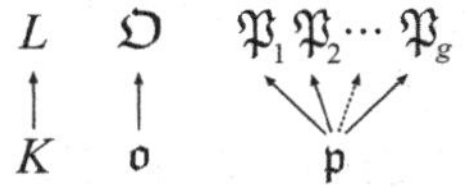

$L \quad \mathfrak{O} \quad \mathfrak{P}_1 \mathfrak{P}_2 \cdots \mathfrak{P}_g$

$K \quad \mathfrak{o} \quad \mathfrak{p}$

Figure 21.1. Image for the term 'above'. Arrows denote inclusion.

Lemma 21.5. *If $\mathfrak{P}$ is a prime ideal of $\mathfrak{O}$ then $\mathfrak{P} \cap \mathfrak{o}$ is a prime ideal of $\mathfrak{o}$.*

Proof: By the isomorphism theorems for quotient rings, there is a monomorphism $\sigma : \mathfrak{o}/(\mathfrak{o} \cap \mathfrak{P}) \to \mathfrak{O}/\mathfrak{P}$ induced by the containment

$$\frac{\mathfrak{O}}{\mathfrak{P}} \supseteq \frac{\mathfrak{o} + \mathfrak{P}}{\mathfrak{P}} \cong \frac{\mathfrak{o}}{\mathfrak{o} \cap \mathfrak{P}} \tag{21.2}$$

By Theorem 6.2 the image of $\mathfrak{o}/(\mathfrak{P} \cap \mathfrak{o})$ in $\mathfrak{O}/\mathfrak{P}$ is a subring of a domain, hence a domain. By Theorem 6.2 again, $\mathfrak{P} \cap \mathfrak{o}$ is a prime ideal of $\mathfrak{o}$. □

Proposition 21.6. *The following conditions on a prime ideal $\mathfrak{p}$ of $\mathfrak{o}$ and a prime ideal $\mathfrak{P}$ of $\mathfrak{O}$ are equivalent:*

(a) *$\mathfrak{P}$ lies above $\mathfrak{p}$.*

(b) $\mathfrak{P} \supseteq \mathfrak{p}\mathfrak{O}$.

(c) $\mathfrak{P} \cap \mathfrak{o} = \mathfrak{p}$.

Proof: Condition (b) holds if and only if $\mathfrak{P}$ divides $\mathfrak{p}\mathfrak{O} = \prod_{j=1}^{g} \mathfrak{P}_j^{e_j}$, so by unique factorisation $\mathfrak{P} = \mathfrak{P}_j$ for some j, which is (a). Thus (a) and (b) are equivalent.

Suppose (c) holds. By Lemma 21.5, $\mathfrak{P} \cap \mathfrak{o}$ is a prime ideal of $\mathfrak{o}$. But $\mathfrak{P} \cap \mathfrak{o} \supseteq \mathfrak{p}$ so maximality of $\mathfrak{p}$ implies (b), unless $\mathfrak{P} \cap \mathfrak{o} = \mathfrak{o}$. But in this case $\mathfrak{P} \supseteq \mathfrak{o}\mathfrak{O} = \mathfrak{O}$, contradicting $\mathfrak{P}$ being prime. Thus (c) $\Rightarrow$ (b).

Conversely, if (b) holds then $\mathfrak{P} \supseteq \mathfrak{p}\mathfrak{O}$, so $\mathfrak{P} \cap \mathfrak{o} \supseteq \mathfrak{p}\mathfrak{O} \cap \mathfrak{o} \supseteq \mathfrak{p}$. By Lemma 21.5, $\mathfrak{p}\mathfrak{O} \cap \mathfrak{o}$ is a prime ideal of $\mathfrak{o}$, so maximality implies (c). Therefore (b) and (c) are equivalent. □

Theorem 21.7. *Let L/K be an extension of number fields. Then*

(a) *Every prime ideal $\mathfrak{P}$ of $\mathfrak{O}$ lies above some prime ideal $\mathfrak{p}$ of $\mathfrak{o}$.*

(b) *For every prime ideal $\mathfrak{p}$ of $\mathfrak{o}$, at least one prime ideal $\mathfrak{P}$ of $\mathfrak{O}$ lies above $\mathfrak{p}$.*

(c) *For every prime ideal $\mathfrak{p}$ of $\mathfrak{o}$, only finitely many prime ideals $\mathfrak{P}$ of $\mathfrak{O}$ lie above $\mathfrak{p}$.*

Proof: Part (a) follows from Lemma 21.5. Parts (b) and (c) are immediate from (21.1). □

Proposition 21.8. *Let L/K be an extension with $\mathfrak{P}$ a prime ideal above $\mathfrak{p}$. Then there is a monomorphism of residue fields $\mathfrak{o}/\mathfrak{p} \to \mathfrak{O}/\mathfrak{P}$. Both fields have the same (prime) characteristic.*

Proof: By Proposition 21.6(c) we have $\mathfrak{p} = \mathfrak{o} \cap \mathfrak{P}$, so $\mathfrak{o}/\mathfrak{p} = \mathfrak{o}/(\mathfrak{o} \cap \mathfrak{P})$. By (21.2), there is a monomorphism $\mathfrak{o}/(\mathfrak{o} \cap \mathfrak{P}) \to \mathfrak{O}/\mathfrak{P}$. Therefore there is a monomorphism $\mathfrak{o}/\mathfrak{p} \to \mathfrak{O}/\mathfrak{P}$.

Both $\mathfrak{o}/\mathfrak{p}$ and $\mathfrak{O}/\mathfrak{P}$ are finite fields and therefore have the same characteristic, which is prime. □

Definition 21.9. Denote the degree of the field extension $(\mathfrak{O}/\mathfrak{P}) \,/\, (\mathfrak{o}/\mathfrak{p})$ by

$$f_{L/K}(\mathfrak{P}) = [\mathfrak{O}/\mathfrak{P} : \mathfrak{o}/\mathfrak{p}]$$

Theorem 21.10. *Let $\mathfrak{P}$ be a prime ideal of $\mathfrak{O}$ lying above the prime ideal $\mathfrak{p}$ of $\mathfrak{o}$. Then*

$$\mathrm{N}(\mathfrak{P}) = \mathrm{N}(\mathfrak{p})^{f_{L/K}(\mathfrak{P})}$$

Proof: By Proposition 21.8, the field $\mathfrak{O}/\mathfrak{P}$ is a finite extension of the field $\mathfrak{o}/\mathfrak{p}$, and we have defined $[\mathfrak{O}/\mathfrak{P} : \mathfrak{o}/\mathfrak{p}] = f_{L/K}(\mathfrak{P})$. Now

$$\mathrm{N}(\mathfrak{p}) = |\mathfrak{o}/\mathfrak{p}| \qquad \mathrm{N}(\mathfrak{P}) = |\mathfrak{O}/\mathfrak{P}|$$

and $\mathfrak{O}/\mathfrak{P}$ is a vector space over $\mathfrak{o}/\mathfrak{p}$ of dimension $d = f_{L/K}(\mathfrak{P})$. Therefore

$$\mathrm{N}(\mathfrak{P}) = |\mathfrak{O}/\mathfrak{P}| = |\mathfrak{o}/\mathfrak{p}|^d = \mathrm{N}(\mathfrak{p})^d = \mathrm{N}(\mathfrak{p})^{f_{L/K}(\mathfrak{P})}$$

21.3 Action of the Galois Group

Since we are working with a field extension L/K it is natural to investigate what the Galois group $\mathrm{Gal}(L/K)$ can tell us. We assume that L/K is normal, so the Galois correspondence, Theorem 18.34, applies.

Examples 21.2 and 21.3 make it clear that in this context there are interesting problems. They also hint at useful structure. For instance, in both examples the extra factors in $\mathfrak{O}$ are permuted by the Galois group of L/K. This should not be a great surprise, because:

Theorem 21.11. *Let L/K be a normal extension of number fields, with respective rings of integers $\mathfrak{O}$ for L and $\mathfrak{o}$ for K. Then:*

(a) *The Galois group $\mathrm{Gal}(L/K)$ fixes every element of $\mathfrak{o}$ and leaves $\mathfrak{O}$ invariant as a set.*

(b) *If $\mathfrak{A}$ is an ideal of $\mathfrak{O}$ and $\gamma \in \mathrm{Gal}(L/K)$, then $\gamma(\mathfrak{A})$ is an ideal of $\mathfrak{O}$.*

(c) *If $\mathfrak{P}$ is a prime ideal of $\mathfrak{O}$ and $\gamma \in \mathrm{Gal}(L/K)$, then $\gamma(\mathfrak{P})$ is a prime ideal of $\mathfrak{O}$.*

(d) *Let $\mathfrak{p}$ be a prime ideal of $\mathfrak{o}$, and let $\mathfrak{p}\mathfrak{O} = \mathfrak{P}_1^{e_1} \dots \mathfrak{P}_k^{e_k}$ be its prime factorisation in $\mathfrak{O}$, where the $\mathfrak{P}_j$ are distinct prime ideals of $\mathfrak{O}$. For any $\gamma \in \mathrm{Gal}(L/K)$, we have $\gamma(\mathfrak{P}_i) = \mathfrak{P}_{\gamma(i)}$ for some $\gamma(i) \in \{1, \dots, k\}$, and $e_{\gamma(i)} = e_i$.*

Proof: Let $\Gamma = \mathrm{Gal}(L/K)$. Everything is straightforward since Γ consists of K-automorphisms of L, but we spell out the details.

(a) The group Γ fixes every element of K, and $\mathfrak{o} \subseteq K$, so Γ fixes every element of $\mathfrak{o}$. The elements $a \in \mathfrak{O}$ are the elements of L that satisfy a polynomial equation $f(a) = 0$ with coefficients in $\mathbb{Z}$. If $\gamma \in \Gamma$ then $\gamma(a) \in L$, and $f(\gamma(a)) = \gamma(f(a)) = f(a) = 0$ since γ acts trivially on $\mathbb{Z}$.

(b) We have $\gamma(\mathfrak{A})\mathfrak{O} = \gamma(\mathfrak{A})\gamma(\mathfrak{O}) = \gamma(\mathfrak{A}\mathfrak{O}) = \gamma(\mathfrak{A})$ since $\mathfrak{A}$ is an ideal of $\mathfrak{O}$.

(c) Prime ideals are maximal. If $\gamma(\mathfrak{P}) \subsetneq \mathfrak{B}$ for an ideal $\mathfrak{B}$ of $\mathfrak{O}$, then $\mathfrak{P} \subsetneq \gamma^{-1}(\mathfrak{B})$. Therefore $\gamma^{-1}(\mathfrak{B}) = \mathfrak{O}$, so $\mathfrak{B} = \mathfrak{O}$ and $\gamma(\mathfrak{P})$ is maximal, hence prime.

(d) We have $\gamma(\mathfrak{p}\mathfrak{O}) = \gamma(\mathfrak{p})\gamma(\mathfrak{O}) = \gamma(\mathfrak{p})\mathfrak{O}$. Therefore

$$\mathfrak{P}_1^{e_1} \dots \mathfrak{P}_k^{e_k} = \gamma(\mathfrak{P}_1)^{e_1} \dots \gamma(\mathfrak{P}_k)^{e_k}$$

Now apply unique factorisation into prime ideals. □

Part (d) tells us that when L/K is normal, the Galois group permutes the prime ideals in the factorisation of $\mathfrak{p}\mathfrak{O}$ and preserves their multiplicities. In fact, we prove in Theorem 21.20 that $\mathrm{Gal}(L/K)$ acts transitively on these prime ideals. That is, if $\mathfrak{P}_i$ and $\mathfrak{P}_j$ are any two of them, there exists $\gamma \in \mathrm{Gal}(L/K)$ such that $\gamma(\mathfrak{P}_i) = \mathfrak{P}_j$. Thus all $\mathfrak{P}_j$ are 'on the same footing' up to K-automorphisms of L; that is, they have very similar properties.

21.4 Ramification

Examples 21.2 and 21.3 show that when L/K is an extension, some prime ideals of K extend to prime ideals of L, but others do not. In this section we examine in more detail the relation between a prime ideal $\mathfrak{p}$ of $\mathfrak{o}$ and the ideal $\mathfrak{p}\mathfrak{O}$ that it generates in $\mathfrak{O}$, and we seek to generalise the apparent patterns observed in those two examples. The basis for the analysis is Equation (21.1), which defines the prime ideals $\mathfrak{P}_i$ above $\mathfrak{p}$ and the positive integer g.

Recall that $\mathfrak{o}/\mathfrak{p}$ and $\mathfrak{O}/\mathfrak{P}_j$ are finite fields, by Theorem 20.8. By Proposition 21.8, there is a monomorphism from $\mathfrak{o}/\mathfrak{p}$ to $\mathfrak{O}/\mathfrak{P}_j$ for any j. This leads to:

Definition 21.12. The degree

$$f_j = [\mathfrak{O}/\mathfrak{P}_j : \mathfrak{o}/\mathfrak{p}]$$

of the field extension $(\mathfrak{O}/\mathfrak{P}_j) / (\mathfrak{o}/\mathfrak{p})$ is the *inertia degree* of $\mathfrak{P}_j$ over $\mathfrak{p}$. In Definition 21.9 we wrote this number as $f_{L/K}(\mathfrak{P}_j)$; now we abbreviate this to f_j.

The multiplicity e_j in (21.1) is the *ramification index* of $\mathfrak{P}_j$ over $\mathfrak{p}$.

If $e_j > 1$ for some j, we say that L/K is *ramified* at $\mathfrak{p}$ (or $\mathfrak{p}$ is ramified in L, or $\mathfrak{p}$ ramifies in L). If $e_j = 1$ then L/K is *unramified* at $\mathfrak{p}$.

We need a definition and a standard theorem from the theory of modules over principal ideal domains:

Definition 21.13. Let R be a domain and let M be an R-module. Then M is *torsion-free* if $0 \neq x \in M$ and $0 \neq r \in R$ imply that $rx \neq 0$.

Theorem 21.14. *If R is a principal ideal domain and M is a finitely generated torsion-free R-module, then M is a free R-module.*

Proof: See Dummit and Foote [36] chapter 12 or Hungerford [66] section VI.6. □

This theorem is the module analogue of the usual structure theorem for finitely generated abelian groups, which is the case when $R = \mathbb{Z}$.

We also restate Lemma 6.20 in the relevant notation:

Proposition 21.15. *Let $\mathfrak{P}$ be a nonzero prime ideal of $\mathfrak{O}$ and let $r \in \mathbb{Z}$. Then there is an additive group isomorphism*

$$\mathfrak{O}/\mathfrak{P} \cong \mathfrak{P}^r/\mathfrak{P}^{r+1} \tag{21.3}$$

□

We are now ready to prove an important formula:

Theorem 21.16. *With the above notation,*

$$[L : K] = \sum_{j=1}^{g} e_j f_j$$

Proof: Localise at $\mathfrak{p}$: this multiplies $\mathfrak{o}$ and $\mathfrak{O}$ by S^{-1}, where $S = \mathfrak{o} \setminus \mathfrak{p}$. To avoid complicated notation in this proof (and only here) we continue to write $\mathfrak{o}$ for the localisation $\mathfrak{o}_\mathfrak{p}$, and similarly for other rings and ideals. Now $\mathfrak{o}$ becomes a discrete valuation ring, hence a principal ideal domain, and $\mathfrak{O}$ is a finitely generated $\mathfrak{o}$-module. It is obviously torsion-free since L is a field, so it is a free $\mathfrak{o}$-module by Theorem 21.14. Its rank is $n = [L : K]$, and $\mathfrak{O}/(\mathfrak{p}\mathfrak{O})$ is a vector space of dimension n over $\mathfrak{o}/\mathfrak{p}$:

$$\dim_{\mathfrak{o}/\mathfrak{p}} \mathfrak{O}/(\mathfrak{p}\mathfrak{O}) = [L : K] \tag{21.4}$$

As in (21.1), let $\mathfrak{p}\mathfrak{O} = \prod_{j=1}^{g} \mathfrak{P}_j^{e_j}$. Since $\mathfrak{P}_j^{e_j} \supseteq \mathfrak{p}\mathfrak{O}$ for each j, there is a homomorphism

$$\mathfrak{O}/(\mathfrak{p}\mathfrak{O}) \to \mathfrak{O}/(\mathfrak{P}_j^{e_j})$$

hence a homomorphism

$$\phi : \mathfrak{O}/(\mathfrak{p}\mathfrak{O}) \to \prod_{j=1}^{g} \mathfrak{O}/(\mathfrak{P}_j^{e_j})$$

Each $\mathfrak{O}/(\mathfrak{P}_j^{e_j})$ is a vector space over $\mathfrak{o}/\mathfrak{p}$, so the direct product is also a vector space over $\mathfrak{o}/\mathfrak{p}$.

The kernel of ϕ is the intersection of all $\mathfrak{P}_j^{e_j}$, which is $\mathfrak{p}\mathfrak{O}$ modulo $\mathfrak{p}\mathfrak{O}$, so this map is injective. By the Chinese Remainder Theorem 20.10, ϕ is surjective, so it is an isomorphism.

The main step now is to find the dimension over K of $\mathfrak{O}/(\mathfrak{P}^e)$, where $\mathfrak{P}$ is any of the $\mathfrak{P}_j$ and $e = e_j$.

By Corollary 20.23 the ideal $\mathfrak{P}$ is principal; let a generator be α. Let $k \in \mathbb{Z}$, with $k \geq 1$. We can view $\mathfrak{P}^k/\mathfrak{P}^{k+1}$ as a vector space over $\mathfrak{o}/\mathfrak{p}$. Consider the map

$$\psi : \mathfrak{O}/\mathfrak{P} \to \mathfrak{P}^k/\mathfrak{P}^{k+1}$$

induced by multiplying elements of $\mathfrak{O}$ by α^k. By Equation (21.3) of Proposition 21.15, ψ is an isomorphism.

The $\mathfrak{o}/\mathfrak{p}$-vector space $\mathfrak{O}/\mathfrak{P}^e$ has a descending chain of subspaces induced by

$$\mathfrak{O} \supseteq \mathfrak{P} \supseteq \mathfrak{P}^2 \supseteq \cdots \supseteq \mathfrak{P}^e$$

By definition, $\dim_{\mathfrak{o}/\mathfrak{p}}(\mathfrak{O}/\mathfrak{P}) = f_{\mathfrak{P}}$, so $\dim_{\mathfrak{o}/\mathfrak{p}}(\mathfrak{O}/\mathfrak{P}^e) = ef_{\mathfrak{P}}$. Setting $\mathfrak{P} = \mathfrak{P}_j$, we have proved that $\dim_{\mathfrak{o}/\mathfrak{p}}(\mathfrak{O}/\mathfrak{P}_j^{e_j}) = e_j f_j$. Therefore, using (21.4),

$$[L : K] = \dim_{\mathfrak{o}/\mathfrak{p}}(\mathfrak{O}/\mathfrak{p}\mathfrak{O}) = \sum_{j=1}^{g} \dim_{\mathfrak{o}/\mathfrak{p}}(\mathfrak{O}/\mathfrak{P}_j^e) = \sum_{j=1}^{g} e_j f_j$$

□

Definition 21.17. If $f_j = e_j = 1$ for all j we say that $\mathfrak{p}$ *splits completely* in L. That is, $\mathfrak{p}\mathfrak{O} = \mathfrak{P}_1 \ldots \mathfrak{P}_g$ where the $\mathfrak{P}_j$ are distinct prime ideals of $\mathfrak{O}$.

If $g = 1$ and $f_1 = 1$, whence $e_1 = [L : K]$, we say that $\mathfrak{p}$ *ramifies completely* in L. That is, $\mathfrak{p}\mathfrak{O} = \mathfrak{P}_1^{e_1}$ is a power of a single prime ideal of $\mathfrak{O}$.

If $g = 1$ and $e_1 = 1$, so $f_1 = [L : K]$, we say that $\mathfrak{p}$ is *inert* in L. That is, $\mathfrak{p}\mathfrak{O} = \mathfrak{P}_1$ is a prime ideal of $\mathfrak{O}$.

Example 21.18. In Example 21.2, prime ideals $\langle p \rangle$ for $\mathbb{Q}$ behave like this:

If $p = 4k + 1$ then $\langle p \rangle$ splits completely in $\mathbb{Q}[\mathrm{i}]$.

If $p = 2$ then $\langle p \rangle$ ramifies completely in $\mathbb{Q}[\mathrm{i}]$.

If $p = 4k - 1$ then $\langle p \rangle$ is inert in $\mathbb{Q}[\mathrm{i}]$.

21.5 Splitting of a Prime Number in a Quadratic Field

Let $d \in \mathbb{Z}$ be squarefree, $L = \mathbb{Q}(\sqrt{d})$, with ring of integers $\mathfrak{O}$. Let $p \in \mathbb{Z}$ be a rational prime. We ask how $\langle p \rangle = p\mathfrak{O}$ factorises in $\mathfrak{O}$.

We have $\sum_1^g e_i f_i = 2$. There are three possibilities:

(a) $g = 2, e_1 = e_2 = 1, f_1 = f_2 = 1$. Then p splits in $\mathfrak{O}$ and $\mathfrak{O}\mathfrak{p} = \mathfrak{P}_1\mathfrak{P}_2$ where the factors are distinct.

(b) $g = 1, e_1 = 1, f_1 = 2$. Then p is inert in $\mathfrak{O}$; that is, $\mathfrak{O}\mathfrak{p}$ is prime in $\mathfrak{O}$.

(c) $g = 1, e_1 = 2, f_1 = 1$. Then p ramifies in $\mathfrak{O}$; that is, $\mathfrak{O}\mathfrak{p} = \mathfrak{P}_1^2$.

We seek further details.

Theorem 21.19. *Let* $L = \mathbb{Q}(\sqrt{d})$.

(a) *If* p *is an odd prime and* d *is a quadratic residue* (mod p) *then* p *splits. So does* 2 *if further* $d \equiv 1 \pmod 8$.

(b) *If* p *is an odd prime and* d *is a quadratic nonresidue* (mod p) *then* p *is inert. So is* 2 *if further* $d \equiv 5 \pmod 8$.

(c) *If* p *is an odd prime that divides* d *then it ramifies in* L. *So does* 2 *if* $d \equiv 2$ *or* $3 \pmod 4$.

Proof: First, suppose that p is odd.

Now $\mathfrak{O} = \mathbb{Z}[\sqrt{d}]$ or $\mathfrak{O} = \mathbb{Z}[(1+\sqrt{d})/2]$ depending on $d \pmod 4$. Passing to residue classes modulo $p\mathfrak{O}$, we see that in the second case any element $a + b(1+\sqrt{d})/2$ is congruent (mod p) to $a + (b+p)(1+\sqrt{d})/2 \in \mathbb{Z} + \mathbb{Z}\sqrt{d}$. Thus, for any d, we have

$$\mathfrak{O}/(p\mathfrak{O}) \cong (\mathbb{Z} + \mathbb{Z}\sqrt{d})/p\mathbb{Z}$$

Moreover,

$$\mathbb{Z} + \mathbb{Z}\sqrt{d} \cong \mathbb{Z}[t]/\langle t^2 - d\rangle$$

which implies that

$$\mathfrak{O}/(p\mathfrak{O}) \cong \mathbb{Z}_p[t]/\langle t^2 - \bar{d}\rangle$$

where $\bar{d}$ is the residue class of $d \pmod p$.

Now cases (a), (b), (c) correspond to:

(a′) $t^2 - d \in \mathbb{Z}_p[t]$ is a product of two distinct linear factors. That is, d is a nonzero square in $\mathbb{Z}_p$ (quadratic residue).

(b′) $t^2 - d \in \mathbb{Z}_p[t]$ is irreducible. That is, d is not a square in $\mathbb{Z}_p$ (quadratic nonresidue).

(c′) $t^2 - d \in \mathbb{Z}_p[t]$ is the square of a linear factor. That is, d is zero in $\mathbb{Z}_p$.

Next, suppose that $p = 2$.

If $d \equiv 2$ or $3 \pmod 4$, then $\mathfrak{O} = \mathbb{Z} + \mathbb{Z}\sqrt{d}$, so, as before, $\mathfrak{O}/2\mathfrak{O} \cong \mathbb{F}_2[t]/\langle t^2 - d\rangle$. Now $t^2 - d$ is either t^2 or $t^2 + 1 = (t+1)^2$, so 2 ramifies.

If $d \equiv 1 \pmod 4$, then the minimal polynomial of $(1+\sqrt{d})/2$ is $t^2 - t - (d-1)/4$, so as before $\mathfrak{O}/2\mathfrak{O} \cong \mathbb{F}_2[t]/(t^2 - t - \delta)$, where $\delta \equiv (d-1)/4 \pmod 2$.

For $d \equiv 1 \pmod 8$, $\delta = 0$ and $t^2 - t - \delta = t(t-1)$, so 2 splits.

For $d \equiv 5 \pmod 8$, $\delta = 1$ and $t^2 - t - \delta = t^2 + t + 1$, which is irreducible over $\mathbb{F}_2$, so 2 is inert. □

The calculations in (a′)–(c′) are special cases of Dedekind's Criterion, Theorem 12.1.

21.6 Ramification in Normal Extensions

When L/K is normal we can use Galois theory to obtain more detailed information on the relationship between prime ideals of $\mathfrak{o}$ and prime ideals of $\mathfrak{O}$. As a sample of this approach, we now prove that if L/K is a normal extension then the Galois group $\mathrm{Gal}(L/K)$ acts transitively on the set of all $\mathfrak{P}_j$ above $\mathfrak{p}$. That is, once we know one $\mathfrak{P}_i$ we can in principle find them all.

Theorem 21.20. *Let L/K be a normal extension of number fields, with $\mathfrak{p}$ a prime ideal in $\mathfrak{o}$. Then:*
(a) *The Galois group* $\mathrm{Gal}(L/K)$ *acts transitively on the set of prime ideals $\mathfrak{P}_i$ of $\mathfrak{O}$ that lie above $\mathfrak{p}$.*
(b)

$$e_1 = \cdots = e_g \qquad f_1 = \cdots = f_g$$

(c)

$$[L : K] = efg$$

where $e = e_i$ and $f = f_i$ are the common values for $1 \leq i \leq g$.

Proof: (a) Let $G = \mathrm{Gal}(L/K)$. We have (21.1):

$$\mathfrak{p}\mathfrak{O} = \prod_{i=1}^{g} \mathfrak{P}_i^{e_i} \tag{21.5}$$

Let $\mathfrak{P} = \mathfrak{P}_i$ for some fixed but arbitrary i. By Lemma 6.28 there exists $a \in \mathfrak{P}$ such that $\langle a \rangle \mathfrak{P}^{-1}$ is an ideal of $\mathfrak{O}$ that is prime to $\mathfrak{p}\mathfrak{O}$. Thus the product

$$\mathfrak{C} = \prod_{\sigma \in G} \sigma(\langle a \rangle \mathfrak{P}^{-1}) = \prod_{\sigma \in G} \frac{\langle \sigma(a) \rangle \mathfrak{O}}{\sigma(\mathfrak{P})} = \frac{(\mathrm{N}_{L/K}(a))\mathfrak{O}}{\prod_{\sigma \in G} \sigma(\mathfrak{P})}$$

is a nonzero ideal of $\mathfrak{O}$.

Consider the final expression

$$\frac{(\mathrm{N}_{L/K}(a))\mathfrak{O}}{\prod_{\sigma \in G} \sigma(\mathfrak{P})}$$

Since $a \in \mathfrak{P}$, we have $\mathrm{N}_{L/K}(a) \in \mathfrak{P} \cap \mathfrak{o} = \mathfrak{p}$. Thus the numerator of this expression is divisible by $\mathfrak{p}\mathfrak{O}$. Since $\langle a \rangle \mathfrak{P}^{-1}$ is prime to $\mathfrak{p}\mathfrak{O}$, every $\sigma(\langle a \rangle \mathfrak{P}^{-1})$ is also prime to $\mathfrak{p}\mathfrak{O}$. Thus the denominator of this expression is also divisible by $\mathfrak{p}\mathfrak{O}$, in order to cancel this factor of the numerator. Thus for any i we have

$$\prod_{j=1}^{g} \mathfrak{P}_j^{e_j} \quad \text{divides} \quad \prod_{\sigma \in G} \sigma(\mathfrak{P}_i)$$

In particular, every prime ideal factor $\mathfrak{P}_j$ of the left-hand side must appear as a factor in the right-hand side, so every $\mathfrak{P}_j$ is an image $\sigma(\mathfrak{P}_i)$ of the specified prime ideal $\mathfrak{P}_i$. Thus G is transitive on the $\mathfrak{P}_i$.

(b) Now choose indices $j \neq k$. By transitivity there exists $\sigma \in G$ such that $\sigma(\mathfrak{P}_k) = \mathfrak{P}_j$. Apply σ to the factorisation (21.5), to obtain

$$\prod_{i=1}^{g} \mathfrak{P}_i^{e_i} = \prod_{i=1}^{g} \sigma(\mathfrak{P}_i)^{e_i}$$

Take $v_{\mathfrak{P}_j}$ of both sides to conclude that $e_j = e_k$. Thus there exists e such that

$$e = e_1 = \cdots = e_g \tag{21.6}$$

Now, for any $\sigma \in G$ we have $\mathfrak{O}/\mathfrak{P}_i \cong \mathfrak{O}/\sigma(\mathfrak{P}_i)$, so transitivity implies that there exists f such that

$$f = f_1 = \cdots = f_g \tag{21.7}$$

(c) Denote the residue field $\mathfrak{o}/\mathfrak{p}$ (which is finite) by $\mathbb{F}_\mathfrak{p}$. Since $\mathfrak{O}$ is a lattice in L,

$$\begin{aligned}
[L:K] &= \dim_\mathfrak{o} \mathfrak{O} = \dim_{\mathbb{F}_\mathfrak{p}} \mathfrak{O}/(\mathfrak{p}\mathfrak{O}) \\
&= \dim_{\mathbb{F}_\mathfrak{p}} \left(\prod_{i=1}^{g} \mathfrak{O}/\mathfrak{P}_i^{e_i} \right) \\
&= \sum_{i=1}^{g} e_i f_i = efg
\end{aligned}$$

by (21.6) and (21.7). □

21.7 Integrality Properties

We recast some of the results of Section 2.4 in a more general setting that applies to extensions. The proofs use essentially the same ideas.

Theorem 21.21. *Let R be a ring, $A \subseteq R$ a subring, and $x \in R$. Then the following are equivalent:*

(a) *There is a monic polynomial $f \in A[t]$ such that $f(x) = 0$.*

(b) *The ring $A[x]$ is a finitely generated A-module.*

(c) *There exists a subring $B \subseteq R$ such that $A \subseteq B$ and B is a finitely generated A-module.*

Proof: (a) $\Rightarrow$ (b). Let M be the A-submodule of R generated by $x, \ldots, x^{n-1}$, where $n = \partial f$. By (a) we have

$$x^n = a_{n-1}x^{n-1} + \cdots + a_0$$

with the $a_i \in A$. Therefore $x^n \in M$. Multiplying this equation by x^i and using induction, $x^{n+i} \in M$ for all $i \geq 1$. Now $A[x] = M$ so $A[x]$ is finitely generated as an A-module.

(b) $\Rightarrow$ (c). This is clear.

(c) $\Rightarrow$ (b). Let $y_1, \ldots, y_n$ generate B as an A-module. Now $x \in B \subseteq R$, so $xy_i \in B$ for $1 \leq i \leq n$. Therefore

$$xy_i = \sum_{j=1}^{n} a_{ij}y_j \qquad (1 \leq i \leq n)$$

so

$$\sum_{j=1}^{n} (\delta_{ij}x - a_{ij})y_j = 0 \qquad (1 \leq i \leq n)$$

where δ_{ij} is the Kronecker delta. This is a set of homogeneous linear equations in the y_i. Let $d = \det(\delta_{ij}x - a_{ij})$. Then $d = 0$ since the system has a nonzero solution, and d expands to a monic polynomial in x. □

Definition 21.22. Let R be a ring, $A \subseteq R$ a subring, and $x \in R$. Then x is *integral over* A if it satisfies any of the equivalent conditions (a), (b), (c) of Theorem 21.21.

Proposition 21.23. *Let R be a ring, $A \subseteq R$ a subring, and $x_1, \ldots, x_n \in R$. If x_i is integral over $A[x_1, \ldots, x_{i-1}]$ for all i with $2 \leq i \leq n$, then $A[x_1, \ldots, x_n]$ is a finitely generated A-module.*

Proof: A straightforward induction on n using Theorem 21.21. □

Definition 21.24. Let R be a ring, contained in a field K. The *integral closure* of R in K is the set of all $\alpha \in K$ that are integral over R; this is, satisfy $f(\alpha) = 0$ for some $f \in R[t]$.

Proposition 21.25. *The integral closure of R in K is a ring.*

Proof: This is a simple consequence of Proposition 21.23. □

21.8 Decomposition and Inertia Groups

Again assume that L/K is normal, and the respective rings of integers are $\mathfrak{O}$ and $\mathfrak{o}$. Let $G = \mathrm{Gal}(L/K)$. We briefly discuss further detail on how the Galois group $\mathrm{Gal}(L/K)$ acts on the set of primes $\mathfrak{P}$ above a given $\mathfrak{p}$. The results here are too extensive to survey completely, but we can at least set up some of the concepts.

Definition 21.26. Let $G = \mathrm{Gal}(L/K)$ where L/K is normal. The *decomposition group* $D_{\mathfrak{P}_j}$ is the subgroup of those $\gamma \in G$ that map $\mathfrak{P}_j$ to itself.

By the orbit-stabiliser theorem, $g = |G|/|D_{\mathfrak{P}_j}|$ for any j. By Galois theory, $[L : K] = |G|$, so for all j,

$$|D_{\mathfrak{P}_j}| = ef \tag{21.8}$$

Definition 21.27. The group $D_{\mathfrak{P}_j}$ has a subgroup $I_{\mathfrak{P}_j}$ consisting of all elements $\gamma \in G$ that induce the identity on $\mathfrak{O}/\mathfrak{P}_j$. This group is the *inertia group* of $\mathfrak{P}_j$.

Let $\mathfrak{P}$ be above $\mathfrak{p}$. Let $\sigma \in G$. Let σ belong to the decomposition group $D = D_{\mathfrak{P}}$ and denote the inertia group by I. Then $\sigma(\mathfrak{O}) = \mathfrak{O}$ and $\sigma(\mathfrak{P}) = \mathfrak{P}$, so it induces an automorphism $\bar{\sigma}$ of $(\mathfrak{O}/\mathfrak{P})/(\mathfrak{o}/\mathfrak{p})$.

Theorem 21.28. *Let $\mathfrak{P}$ be above $\mathfrak{p}$. Then $\mathfrak{O}/\mathfrak{P}$ is a normal separable extension of $\mathfrak{o}/\mathfrak{p}$ of degree f (residue index). The reduction map $\rho(\sigma) = \bar{\sigma}$ is a surjective homomorphism*

$$\rho : D \to \mathrm{Gal}((\mathfrak{O}/\mathfrak{P})/(\mathfrak{o}/\mathfrak{p}))$$

The inertia group I is the kernel of the reduction map. Moreover, $|I| = e$.

Proof: That the inertia group I is the kernel of the reduction map is a restatement of Definition 21.27.

Let K_D be the fixed field of D, let $\mathfrak{o}_D$ be the integral closure of $\mathfrak{o}$ in K_D, and let $\mathfrak{p}_D = \mathfrak{P} \cap \mathfrak{o}_D$, which is a prime ideal. By Proposition 21.23

and the definition of D, the ideal $\mathfrak{P}$ is the only prime factor of $\mathfrak{O}\mathfrak{p}_D$. Let

$$\mathfrak{O}\mathfrak{p}_D = \mathfrak{P}^{e'}$$

and let f' be the *residual degree*

$$f' = [(\mathfrak{O}/\mathfrak{P}) : (\mathfrak{o}_D/\mathfrak{p}_D)]$$

By 21.8 we have

$$e'f' = [K : K_D] = |D| = ef$$

where the step $[K : K_D] = |D|$ follows from the Fundamental Theorem of Galois Theory, Theorem 18.34(c). Now

$$\mathfrak{o}/\mathfrak{p} \subseteq \mathfrak{o}_D/\mathfrak{p}_D \subseteq \mathfrak{O}/\mathfrak{P}$$

so $f' \leq f$. Since $\mathfrak{p}\mathfrak{o}_D \subseteq \mathfrak{p}_D$, we have $e' \leq e$. Thus $e = e'$ and $f = f'$. But this implies that

$$\mathfrak{o}/\mathfrak{p} \cong \mathfrak{o}_D/\mathfrak{p}_D$$

Let $\bar{x}$ be a primitive element for $(\mathfrak{O}/\mathfrak{P})/(\mathfrak{o}/\mathfrak{p})$ and let x be a representative for x. Let the minimal polynomial of x over K_D be

$$m(t) = t^r + a_r t^{r-1} + \cdots + a_0$$

Then $a_i \in D$ by 18.34. The zeros of m are all of the form $\sigma(x)$ with $\sigma \in D$. The reduced polynomial

$$\bar{m}(t) = t^r + \bar{a}_r t^{r-1} + \cdots + \bar{a}_0$$

has coefficients in $\mathfrak{o}/\mathfrak{p}$ and its roots are all of the form $\bar{\sigma}(\bar{x})$ with $\sigma \in D$.

Therefore $\mathfrak{O}/\mathfrak{P}$ contains all the conjugates of $\bar{x}$ over $\mathfrak{o}/\mathfrak{p}$, so the extension $(\mathfrak{O}/\mathfrak{P})/(\mathfrak{o}/\mathfrak{p})$ is normal (and separable since the fields are finite).

Moreover, every conjugate of $\bar{x}$ over $\mathfrak{o}/\mathfrak{p}$ has the form $\bar{\sigma}(\bar{x})$, so every $(\mathfrak{o}/\mathfrak{p})$-automorphism of $\mathfrak{O}/\mathfrak{P}$ is equal to some $\bar{\sigma}$. Thus the Galois group of $(\mathfrak{O}/\mathfrak{P})/(\mathfrak{o}/\mathfrak{p})$ can be identified with D/I. The order of this group is $[(\mathfrak{O}/\mathfrak{P}) : (\mathfrak{o}/\mathfrak{p}) = f$, so $|D|/|I| = f$, and $|I| = e$. □

Corollary 21.29. *Let $\mathfrak{p}$ be a prime ideal of $\mathfrak{o}$. Then the following conditions are equivalent:*

(a) $\mathfrak{p}$ *does not ramify in* $\mathfrak{O}$.
(b) *The inertia group* $I_{\mathfrak{P}}$ *of any* $\mathfrak{P}$ *over* $\mathfrak{p}$ *is trivial.* □

Example 21.30. Again the Gaussian integers provide a simple but informative example. Here we take $K = \mathbb{Q}, L = \mathbb{Q}[\mathrm{i}]$, and $\mathfrak{o} = \mathbb{Z}, \mathfrak{O} = \mathbb{Z}[\mathrm{i}]$.

Case 1: $p = 2$. By Example 21.2(b), $\langle 2 \rangle = \langle 1 + \mathrm{i} \rangle^2$. The ramification index is $e = 2$. The residue field is $\mathbb{F}_2$. The Galois group $G \cong \mathbb{Z}_2$ generated by complex conjugation. The decomposition group is G since only one prime lies above 2. The inertia group is also G, because $a + b\mathrm{i} \equiv a - b\mathrm{i}$ (mod 2), hence also (mod $1 + \mathrm{i}$).

The prime $2 \in \mathbb{Z}$ is the only prime that ramifies, since the discriminant is -4 and Theorem 21.33 implies that any ramified prime divides it.

Case 2: $p = 4k + 1 = a^2 + b^2$. Then $\langle p \rangle = \langle a + \mathrm{i}b \rangle \langle a - \mathrm{i}b \rangle$. The Galois group swaps these two ideals. Therefore the decomposition groups are trivial. The inertia groups are also trivial. Both residue fields are $\mathbb{F}_p$.

Case 3: $p = 4k + 3$. Any such prime is inert. The decomposition group is $G \cong \mathbb{Z}_2$. However, the Galois group does not act trivially on the residue field, so the inertia group is trivial.

21.9 Frobenius Map

Let $\mathfrak{p}$ be a prime ideal of $\mathfrak{o}$ that does not ramify in $\mathfrak{O}$, and let $\mathfrak{P}$ be a prime ideal above $\mathfrak{p}$.

The inertia group of $\mathfrak{P}$ is the identity by Corollary 21.29. Its decomposition group D is isomorphic to the Galois group G of $(\mathfrak{O}/\mathfrak{P})/(\mathfrak{o}/\mathfrak{p})$ by Theorem 21.28. By Theorem 18.39, G is cyclic with generator the map $\bar{\sigma} : \bar{x} \mapsto \bar{x}^q$ where $q = |\mathfrak{o}/\mathfrak{p}|$. Therefore D is cyclic with generator σ such that

$$\sigma(x) = x^q \pmod{\mathfrak{P}}$$

for $x \in \mathfrak{O}$. This is the Frobenius automorphism of $\mathfrak{O}/\mathfrak{P}$, see Definition 18.38. It is also called the *Frobenius map* of $\mathfrak{P}$ or the *Artin symbol.* In this context it is traditionally denoted by

$$(\mathfrak{P}, L/K)$$

It is easy to show that if $\tau \in G$ then

$$(\tau(\mathfrak{P}), L/K) = \tau(\mathfrak{P}, L/K)\tau^{-1}$$

In particular, if L/K is abelian then $(\mathfrak{P}, L/K)$ depends only on the ideal $\mathfrak{p}$ of $\mathfrak{o}$. If so, the traditional notation is

$$\left(\frac{L/K}{\mathfrak{p}}\right) = (\mathfrak{P}, L/K) \qquad (21.9)$$

Proposition 21.31. *Let F be an intermediate field, $K \subseteq F \subseteq L$, and let f be the residual degree of $\mathfrak{P} \cap F$ over K. Then*

(a) $(\mathfrak{P}, L/F) = (\mathfrak{P}, L/K)^f$.

(b) *If F/K is normal, the restriction of $(\mathfrak{P}, L/K)$ to F is equal to $(\mathfrak{P} \cap F, F/K)$.*

Proof: (a) Let $\sigma = (\mathfrak{P}, L/K)$. Then $\sigma(\mathfrak{P}) = \mathfrak{P}$ and $\sigma(x) \equiv x^q \pmod{\mathfrak{P}}$ for all $x \in \mathfrak{O}$, where $q = |\mathfrak{o}/\mathfrak{p}|$. Thus $\sigma^f(\mathfrak{P}) = \mathfrak{P}$ and $\sigma^f(x) \equiv x^{q^f} \pmod{\mathfrak{P}}$ for all $x \in \mathfrak{O}$. By definition of f,

$$q^f = |(\mathfrak{O} \cap F)/(\mathfrak{P} \cap F)|$$

Further, the decomposition group of $\mathfrak{P}$ over F is a subgroup of the decomposition group D of $\mathfrak{P}$ over K. Its order is

$$[(\mathfrak{O}/fP) : (\mathfrak{O} \cap F)/(\mathfrak{P} \cap F)] = \frac{1}{f}[(\mathfrak{O}/\mathfrak{P}) : (\mathfrak{o} : \mathfrak{p})] = \frac{|D|}{f}$$

by (21.8). Since D is cyclic, generated by σ, the only subgroup of D of order $|D|/f$ is generated by σ^f.

(b) Suppose F/K is normal and let σ' be the restriction of σ to F. Since $\sigma(\mathfrak{P}) = \mathfrak{P}$, we have $\sigma'(\mathfrak{p} \cap F) = \mathfrak{p} \cap F$, and σ' belongs to the decomposition group of $\mathfrak{p} \cap F$ over K. Moreover,

$$\sigma'(x) \equiv x^q \pmod{\mathfrak{P} \cap F}$$

for all $x \in \mathfrak{O} \cap F$, with $q = |\mathfrak{o}/\mathfrak{p}|$. □

21.10 Kummer's Criterion

We state without proof a theorem, which in principle determines the prime ideals above a given prime ideal in a number field extension L/K, in the case when the prime ideal does not ramify. We omit the proof for reasons of space. Fröhlich and Taylor [46] chapter II.2 theorem 23 refer to it as *Kummer's Criterion.* An analogous result when $K = \mathbb{Q}$ is *Dedekind's Criterion*; see Theorem 12.1. More general results along similar lines deal with some prime ideals that ramify, and there are versions for more general field extensions.

Theorem 21.32. *Let L/K be a number field extension, with $L = K(\alpha)$ for $\alpha \in \mathfrak{O}$. Let $g(t)$ be the minimal polynomial of α over K. Let $\mathfrak{p}$ be a* *prime ideal of* $\mathfrak{o}$. *Suppose that modulo* $\mathfrak{p}$ *the reduction* $\bar{g}$ *of* g *factorises into distinct irreducible polynomials in* $(\mathfrak{o}/\mathfrak{p})[t]$:

$$\bar{g}(t) = \prod_i \bar{g}_i(t)$$

Then:

(a) $\mathfrak{p}$ *does not ramify in* L.

(b) *Let* $\mathfrak{P}_i = \gcd(\mathfrak{p}, g_i(\alpha))$ *in* $\mathfrak{O}$ *where* $g_i(t) \in \mathfrak{o}[t]$ *reduces modulo* $\mathfrak{p}$ *to* $\bar{g}_i(t)$. *Then the* $\mathfrak{P}_i$ *are the distinct prime ideals of* $\mathfrak{O}$ *above* $\mathfrak{p}$.

(c) *We have*

$$f_{\mathfrak{p}_i}(L/K) = \partial \bar{g}_i \qquad \mathfrak{p}\mathfrak{O} = \prod \mathfrak{P}_i$$

□

21.11 Ramification and the Discriminant

In this section we prove a beautiful theorem of Dedekind. A similar result holds for any number field extension L/K, but the proof is more technical; see Fröhlich and Taylor [46] chapter II.2 theorem 22. For simplicity we consider only $K/\mathbb{Q}$.

Theorem 21.33. (Dedekind) *Let* K *be a number field. Then a prime* $p \in \mathbb{Q}$ *ramifies over* K *if and only if* p *divides the discriminant* d_K.

Corollary 21.34. *In particular, only finitely many primes in* $\mathbb{Q}$ *ramify in* K.

The main technical tool is the result that the discriminant commutes with reduction (mod p). We work up to this through several lemmas and then prove Theorem 21.33 and Corollary 21.34.

Lemma 21.35. *For suitable choices of basis of* $\mathfrak{O}$ *and* $\mathfrak{O}/\langle p\rangle$,

$$d_{\mathbb{Z}}(\mathfrak{O}) \pmod{p} = d_{\mathbb{Z}_p}(\mathfrak{O}/\langle p\rangle)$$

Proof: Let $\mathfrak{O}_p = \mathfrak{O}/\langle p\rangle$. Let $\{w_1, \ldots, w_n\}$ be a $\mathbb{Z}$-basis of $\mathfrak{O}$. Let $\bar{w}_i$ be the reduction of w_i (mod p). Then $\{\bar{w}_1, \ldots, \bar{w}_n\}$ is a $\mathbb{Z}_p$-basis of $\mathfrak{O}/\langle p\rangle$.

For $a \in \mathfrak{O}$ let M_a be the matrix of the linear map $l_a : \mathfrak{O} \to \mathfrak{O}$ defined by $l_a(x) = ax$, relative to the basis $\{w_1, \ldots, w_n\}$. Then the matrix of the corresponding linear map $l_{\bar{a}} : \mathbb{Z}_p \to \mathbb{Z}_p$ relative to the basis $\{\bar{w}_1, \ldots, \bar{w}_n\}$ is the reduction $M_{\bar{a}}$ of M_a (mod p).

Clearly

$$\mathrm{T}_{\mathfrak{O}_p/\mathbb{Z}_p}(\bar{x}) = \mathrm{T}(M_{\bar{x}}) = \mathrm{T}(M_x) \pmod{p} = \mathrm{T}_{\mathfrak{O}/\mathbb{Z}}(x) \pmod{p}$$

Thus the (mod p) reduction of $\mathrm{T}_{\mathfrak{O}/\mathbb{Z}}(w_i w_j)$ is $\mathrm{T}_{\mathfrak{O}_p/\mathbb{Z}_p}(\bar{w}_i \bar{w}_j)$. Take determinants and use Lemma 18.44, or recall Proposition 2.37, which relates the discriminant to the trace. (The proof of that proposition generalises to extensions L/K in place of $K/\mathbb{Q}$.) □

Lemma 21.36. *Let R be a ring and let B_1, B_2 be rings containing R as a subring, each being a finitely generated free R-module. Then, for suitably chosen bases over R,*

$$d_R(B_1 \times B_2) = d_R(B_1)d_R(B_2)$$

Proof: Choose bases $\{u_i\}, \{v_j\}$ for B_1, B_2 respectively, so that

$$B_1 = \bigoplus_{i=1}^{m} Ru_i \qquad B_2 = \bigoplus_{j=1}^{n} Rv_j$$

Use the basis $\{u_i\} \cup \{v_j\}$ for $B_1 \times B_2$. Observe that $u_i v_j = 0$ since the sum is direct.

The matrix whose determinant is $d_R(B_1 \times B_2)$ is a block matrix

$$J = \begin{bmatrix} \mathrm{T}_{B_1 \times B_2/R}(u_i u_k) & 0 \\ 0 & \mathrm{T}_{B_1 \times B_2/R}(v_j v_l) \end{bmatrix} \tag{21.10}$$

The direct sum structure implies that

$$\begin{aligned} \mathrm{T}_{B_1 \times B_2/R}(x) &= \mathrm{T}_{B_1/R}(x) \quad \text{if } x \in B_1 \\ \mathrm{T}_{B_1 \times B_2/R}(x) &= \mathrm{T}_{B_2/R}(x) \quad \text{if } x \in B_2 \end{aligned}$$

Therefore

$$J = \begin{bmatrix} \mathrm{T}_{B_1/R}(u_i u_k) & 0 \\ 0 & \mathrm{T}_{B_2/R}(v_j v_l) \end{bmatrix}$$

Now take the determinant. □

Proof of Theorem 21.33

The prime $p \in \mathbb{Z}$ divides $d_{\mathbb{Z}}(\mathfrak{O})$ if and only if $d_{\mathbb{Z}}(\mathfrak{O}) \equiv 0 \pmod{p}$. By Lemma 21.35,

$$d_{\mathbb{Z}}(\mathfrak{O}) \pmod{p} = d_{\mathbb{Z}_p}(\mathfrak{O}_p)$$

Therefore $p|d_{\mathbb{Z}}(\mathfrak{O})$ if and only if $d_{\mathbb{Z}_p}(\mathfrak{O}_p) = \bar{0}$ in $\mathbb{Z}_p$.

Let the prime ideal factorisation of p in $\mathfrak{O}$ be

$$p\mathfrak{O} = \mathfrak{P}_1^{e_1} \cdots \mathfrak{P}_g^{e_g}$$

Combined with Lemma 21.36, this tells us that

$$d_{\mathbb{Z}_p}(\mathfrak{O}_p) = \prod_{i=1}^{g} d_{\mathbb{Z}_p}(\mathfrak{O}/\mathfrak{P}_i^{e_i})$$

Therefore it suffices to prove that if $\mathfrak{P}$ is a prime ideal in $\mathfrak{O}$ and $\mathfrak{P}^e$ divides $p\mathfrak{O}$, then

$$d_{\mathbb{Z}_p}(\mathfrak{O}/\mathfrak{P}^e) = \bar{0} \in \mathbb{Z}_p \quad \Leftrightarrow \quad e > 1$$

The vanishing of a discriminant is independent of the choice of basis. We choose a convenient basis as follows.

Suppose that $e > 1$. Any $x \in \mathfrak{P} \setminus \mathfrak{P}^e$ is nilpotent in $\mathfrak{O}/\mathfrak{P}^e$. Therefore $\bar{x}$ can be part of a $\mathbb{Z}_p$-basis of $\mathfrak{O}/\mathfrak{P}^e$; say $\{\bar{x}_1, \ldots, \bar{x}_n\}$ with $\bar{x}_1 = \bar{x}$.

To simplify notation, write T for the trace map $\mathrm{T}_{(\mathfrak{O}/\mathfrak{P}^e)/\mathbb{Z}_p}$. The first column of the matrix $(\mathrm{T}(\bar{x}_i\bar{x}_j))$ contains the numbers $(\mathrm{T}(\bar{x}_i\bar{x}))$. But x is nilpotent, so $x_i x$ is nilpotent, so the matrix $M_{x_i x}$ is nilpotent, hence has trace zero. Therefore the determinant of the trace matrix is $\bar{0}$, but this is the discriminant $d_{\mathbb{Z}_p}(\mathfrak{O}/\mathfrak{P}^e)$ by Lemma 18.44 or Proposition 2.37 generalised to extensions.

Finally, suppose that $e = 1$. Then $\mathfrak{O}/\mathfrak{P}^e$ is a finite field of characteristic p. Now we must prove that $d_{\mathbb{Z}_p}(\mathfrak{O}/\mathfrak{P}) \neq \bar{0}$. Suppose for a contradiction it is $\bar{0}$. Since $\mathfrak{O}/\mathfrak{P}$ is a field, the trace map $\mathrm{T} : \mathfrak{O}/\mathfrak{P} \to \mathbb{Z}_p$ is identically zero. Consider the polynomial

$$g(t) = t + t^p + t^{p^2} + \cdots + t^{p^{r-1}} \in (\mathfrak{O}/\mathfrak{P})[t]$$

where $p^r = |\mathfrak{O}/\mathfrak{P}|$. Theorem 18.39, together with Definition 18.38, show that $\mathrm{T}(\theta) = g(\theta)$ for all $\theta \in \mathfrak{O}/\mathfrak{P}$. But the degree of the polynomial g is less than the size $|\mathfrak{O}/\mathfrak{P}|$, so it cannot be identically zero (if it were, it would have more zeros than its degree).

Therefore when $e = 1$, the discriminant of a finite extension of $\mathbb{Z}_p$ is nonzero. □

Corollary 21.34 is immediate because the discriminant has finitely many prime factors.

21.12 Exercises

21.1 Verify the claim in Section 22.6 that if $g(t) = t^4 + t^3 + t^2 + t + 1 \in \mathbb{Z}[t]$ then $\bar{g}$ is irreducible over $\mathbb{Z}_3$.

21.2 Show that if $g(t) = t^4 + t^3 + t^2 + t + 1 \in \mathbb{Z}[t]$ then $\bar{g}$ is reducible over $\mathbb{Z}_{11}$. Indeed, it has four distinct linear factors. This is in accordance with Section 22.6 and Theorem 21.32.

21.3 Again let $g(t) = t^4 + t^3 + t^2 + t + 1 \in \mathbb{Z}[t]$. Use a computer algebra package to find those rational primes p in the range (say) $2 \leq p \leq 151$ for which $\bar{g}(t) \in \mathbb{Z}_p[t]$ has at least one linear factor.

What do you notice about these primes? Formulate a conjecture. Use the structure of the multiplicative group $\mathbb{Z}_p^*$ to prove it.

Experiment shows that when there is one linear factor, and $p \neq 5$, there are four distinct linear factors. Relate this to the Galois group of $\mathbb{Q}(\zeta)/\mathbb{Q}$.

21.4 Prove that $\langle p \rangle$ ramifies completely in $\mathbb{Z}[\zeta_p]$; indeed, $\langle p \rangle = \mathfrak{P}^{p-1}$ where $\mathfrak{P}$ is a prime ideal of $\mathbb{Z}[\zeta_p]$.

21.5 Let $q \neq p$ be a rational prime that divides $(a^p - 1)/(a - 1)$ for $2 \leq a \in \mathbb{Z}$. Prove that $\langle q \rangle_{\mathbb{Z}}$ splits in $\mathbb{Z}[\zeta_p]$.

21.6 Let $\zeta = \zeta_p$ for p prime; let $q \neq p$ be prime. Suppose that $f(t) = t^{p-1} + \cdots + t + 1$ has a linear factor (mod q), that is, a zero (mod q). Prove that this zero is simple.

21.7 (a) Show that $f(t) = t^3 + t^2 - 2t + 8$ is irreducible over $\mathbb{Q}$. Let θ be a zero of f, $K = \mathbb{Q}(\theta)$, and $\mathfrak{O}$ the ring of integers of K. Show that $\Delta[1, \theta, \theta^2] = -4.503$ and 503 is prime.

(b) Let $\phi = 4/\theta$. Show that $\phi \in \mathfrak{O}$, that $\phi \notin \mathbb{Z}[\theta]$, that $A = \mathbb{Z} + \mathbb{Z}\theta + \mathbb{Z}\phi$ is a ring, and that $A \supsetneq \mathbb{Z}[\theta]$. Calculate the discriminant of A over $\mathbb{Z}$ and use the result to prove that $\mathfrak{O} = A$.

(c) Show that $\langle 2 \rangle_{\mathfrak{O}} = 2\mathfrak{O}$ is the product of three distinct prime ideals $\mathfrak{p}_1\mathfrak{p}_2\mathfrak{p}_3$ of $\mathfrak{O}$.

(d) Use Kummer's Criterion to deduce that there is no element $\psi \in \mathfrak{O}$ such that $\mathfrak{O} = \mathbb{Z}[\psi]$.

21.8 Let $K = \mathbb{Q}(\sqrt{5}, \mathrm{i})$. Calculate the discriminant of K over $\mathbb{Q}(\mathrm{i})$ and use this to show that the ring of integers is $\mathbb{Z}(\frac{1}{2}(1 + \sqrt{5}), \mathrm{i})$. Prove that the only rational primes that ramify in K are 2 and 5, with ramification indices equal to 2.

21.9 Let θ be the real zero of $t^3 - t + 1$. Let $K = \mathbb{Q}(\theta)$. Use the results of Example 18.17 to show that $\mathfrak{O} = \mathbb{Z}[\theta]$. Deduce that $\mathfrak{O}$ is a principal ideal domain.

21.10 (continued) Show that the only prime that ramifies in K is 23, with ramification index 2. Prove that $23 = \mathfrak{p}^2\mathfrak{q}$ where $\mathfrak{a} \neq \mathfrak{q}$ are prime in $\mathfrak{O}$. Deduce that (for the normal closure) $g \geq 2$ and e is even. Since $efg = 6$, show that $e = 2, f = 1, g = 3$.

21.11 Consider the extension $\mathbb{Q}(\zeta_5)/\mathbb{Q}$. Is the ideal $\langle 2 \rangle$ in $\mathfrak{O}_{\mathbb{Q}}$ inert, split, or ramified in $\mathbb{Q}(\zeta_5)$? Why?

21.12 Consider the extension $\mathbb{Q}(\zeta_7)/\mathbb{Q}$. Is the ideal $\langle 2 \rangle$ in $\mathfrak{O}_{\mathbb{Q}}$ inert, split, or ramified in $\mathbb{Q}(\zeta_7)$? Why?

21.13 As in Exercise 12.6, let θ be a zero of $t^5 - t + 1$. In that exercise we proved that the ring of integers is $\mathbb{Z}[\theta]$, and that every ideal class contains an ideal of norm 1, 2, or 3. Use Corollary 12.6 to show that $\mathbb{Z}(\theta)$ is a principal ideal ring. (*Hint*: Use reduction (mod 2) and (mod 3) to show that $\mathfrak{O}$ contains no ideals of norm 2 or 3.)

21.14 Prove that if $K/\mathbb{Q}$ is a finite extension in which no rational prime ramifies, then $K = \mathbb{Q}$. That is, for every number field except $\mathbb{Q}$ itself, some rational prime ramifies.

(*Hint*: Let $[K : \mathbb{Q}] = n$. Use Corollary 12.6 to show that

$$\frac{n!}{n^n}\left(\frac{4}{\pi}\right)^t \sqrt{|d_K|} \geq 1$$

Define

$$b_n = \frac{n!}{n^n}\left(\frac{\pi}{4}\right)^{n/2}$$

Show that $\sqrt{|d_K|} \geq b_n$, and that $b_{n+1} > b_n$. Deduce that if $n \geq 2$ then $|d_K| > 1$. Now apply Theorem 21.33.)

22

Quadratic Reciprocity

The theory of quadratic residues is one of the great triumphs of the classical period of number theory. An integer k that is prime to a positive integer m is said to be a *quadratic residue modulo* m if there exists $z \in \mathbb{Z}$ such that

$$z^2 \equiv k \pmod{m}$$

and otherwise k is a *quadratic nonresidue.*

The main aim of this chapter is to prove a remarkable theorem about quadratic residues:

Theorem 22.1. (Quadratic Reciprocity Law) *If p, q are distinct odd primes, at least one of which is congruent to 1 modulo 4, then p is a quadratic residue of q if and only if q is a quadratic residue of p. Otherwise, precisely one of p, q is a quadratic residue of the other.*

The reciprocal nature of the relationship between p and q in the first case gives rise to the name of the law. Gauss first proved it in 1796 when he was eighteen years old. The result had been conjectured earlier by Euler and Legendre, though Gauss said that he did not know this at the time. He thought so highly of this theorem that he called it 'the gem of higher arithmetic' and developed eight different proofs. In the 19th century, quadratic reciprocity continued to arouse interest, and more than fifty different methods of proof were found by mathematicians such as Cauchy, Eisenstein, Jacobi, Leopold Kronecker, Kummer, Liouville, and Karl Zeller. In fact it was when Kummer was studying higher reciprocity laws that he devised his partial proof of Fermat's Last Theorem. In 1850 Kummer referred to the higher reciprocity laws as the 'the pinnacle of contemporary number theory', regarding Fermat's Last Theorem as a 'curiosity'; see Edwards [37]. It is only right and proper, therefore, that any text on Fermat's Last Theorem should include a description of the result that so fascinated number theorists, and whose study led Kummer to his proof.

We restate this result using Legendre symbols as Theorem 22.9 below, and following Samuel [115] we prove it in Section 22.3 using cyclotomic fields and ramification theory. We then apply quadratic reciprocity to illustrate ramification theory in an extended example.

22.1 Legendre Symbol

The Legendre symbol was introduced by Legendre, who published two volumes on number theory in 1830. It provides a compact way to express the distinction between quadratic residues and nonresidues.

Definition 22.2. Let p be an odd prime and k an integer not divisible by p. The *Legendre symbol* is

$$\left(\frac{k}{p}\right) = \begin{cases} +1 & \text{if } k \text{ is quadratic residue modulo } p \\ -1 & \text{otherwise} \end{cases}$$

The Legendre symbol is written (k/p) when this is more convenient typographically.

When p is odd, quadratic residues and nonresidues can be characterised in terms of the multiplicative group $G = \mathbb{F}_p^*$ of the finite field $\mathbb{F}_p$, which is isomorphic to $\mathbb{Z}_{p-1}$. In this case $p - 1 = 2k$ is even, so G has a unique subgroup H of index 2. This consists of all elements of the form γ^2 for $\gamma \in G$. We easily deduce:

Proposition 22.3. *Let p be an odd prime.*

(a) *There are $(p-1)/2$ nonzero quadratic residues and $(p-1)/2$ nonresidues modulo p.*

(b) *Let x_1, x_2 be nonzero quadratic residues and y_1, y_2 be quadratic nonresidues modulo p. Then:*

$$\begin{aligned} &x_1x_2 \text{ is a quadratic residue} \\ &x_1y_1 \text{ is a quadratic nonresidue} \\ &y_1y_2 \text{ is a quadratic residue} \end{aligned}$$

Proof: (a) By definition the subgroup H of $G = \mathbb{F}_p^*$ consists of the nonzero quadratic residues, and $|H| = p - 1$. The nonresidues lie in the other coset $G \setminus H$ of H, which contains the other $p - 1$ elements.

(b) There is a homomorphism $G \to \mathbb{Z}_2$ where $\mathbb{Z}_2 = \{\pm 1\}$ under multiplication, with kernel H. The nonzero quadratic residues lie in H, so they map to $+1$. The nonresidues lie in the coset $G \setminus H$ and map to -1. □

Proposition 22.4. *The Legendre symbol is multiplicative:*

$$\left(\frac{ab}{p}\right) = \left(\frac{a}{p}\right)\left(\frac{b}{p}\right)$$

provided a, b are not divisible by p.

Proof: By Proposition 22.3, the product ab is a quadratic residue if and only if both a and b are quadratic residues or both a and b are quadratic nonresidues. □

Proposition 22.5. (Euler's Criterion) *For an odd prime p and an integer k not divisible by p,*

$$\left(\frac{k}{p}\right) \equiv k^{(p-1)/2} \pmod{p}$$

Proof: The group $\mathbb{Z}_p^*$ is cyclic of order $p - 1$, so it has a generator s. Thus $s^{p-1} \equiv 1 \pmod{p}$, and

$$(s^{(p-1)/2} - 1)(s^{(p-1)/2} + 1) = (s^{p-1} - 1) \equiv 0 \pmod{p}$$

Since $s^{(p-1)/2} \not\equiv 1 \bmod p$, we have

$$s^{(p-1)/2} \equiv -1 \pmod{p}$$

Now $k \equiv s^a \pmod{p}$ for some a, so

$$\begin{aligned}\left(\frac{k}{p}\right) &= (-1)^a \\ &\equiv (s^{(p-1)/2})^a \pmod{p} \\ &= (s^a)^{(p-1)/2} \\ &\equiv k^{(p-1)/2} \pmod{p}\end{aligned}$$

□

22.2 Quadratic Subfield of Cyclotomic Field

Let p be an odd prime, and $\zeta = \zeta_p$. The group $G = \text{Gal}(\mathbb{Q}(\zeta))$ is cyclic of order $p-1$, so it has a unique subgroup H of index 2. Then $F = \text{Fix}(H)$ is a quadratic extension of $\mathbb{Q}$, which must be $\mathbb{Q}(\sqrt{d})$ for some squarefree d. By the Galois correspondence, it is the *unique* quadratic subfield of $\mathbb{Q}(\zeta)$.

Example 22.6. (a) Suppose that $p = 5$. Let $\theta = \zeta + \zeta^4$. We saw in Example 19.10 that $\theta = (1+\sqrt{5})/2)$. So $F = \mathbb{Q}(\sqrt{5})$.

(b) Suppose that $p = 7$. Let $\theta = \zeta + \zeta^2 + \zeta^4$. In Exercise 19.11 we show that $\mathbb{Q}(\theta) = \mathbb{Q}(\sqrt{-7})$.

These and similar examples suggest that $F = \mathbb{Q}(\sqrt{\pm p})$ where the sign is $+1$ when $p \equiv 1 \pmod 4$ and -1 when $p \equiv 3 \pmod 4$. Using the Legendre symbol and Euler's Criterion, we can now prove:

Theorem 22.7. *Let F be the unique quadratic subfield of the cyclotomic field $\mathbb{Q}(\zeta_p)$. Then*

(a) *If $p \equiv 1 \pmod 4$ then $F = \mathbb{Q}(\sqrt{p})$.*

(b) *If $p \equiv 3 \pmod 4$ then $F = \mathbb{Q}(\sqrt{-p})$.*

Proof: Define

$$S = \sum_{i=1}^{p-1} \left(\frac{i}{p}\right) \zeta^i$$

Then

$$S^2 = \sum_{i,j} \left(\frac{i}{p}\right) \zeta^i \left(\frac{j}{p}\right) \zeta^j = \sum_{i,j} \left(\frac{ij}{p}\right) \zeta^{i+j}$$

As i ranges over $\mathbb{Z}_p^*$, so does ij for any fixed j, so we can replace i by ij to obtain

$$\begin{aligned} S^2 &= \sum_{i,j} \left(\frac{ij^2}{p}\right) \zeta^{i(j+1)} = \sum_{i,j} \left(\frac{i}{p}\right) \zeta^{i(j+1)} \\ &= \sum_{i} \left(\frac{-1}{p}\right) \zeta^0 + \sum_{j \neq -1} \left(\frac{i}{p}\right) \sum_{j} \zeta^{j(i+1)} \end{aligned}$$

But $1 + \zeta + \cdots + \zeta^{p-1} = 0$, so

$$\sum_{j \neq -1} \left(\frac{i}{p}\right) \sum_{j} \zeta^{j(i+1)} = -1$$

Therefore

$$\begin{aligned} S^2 &= \left(\frac{-1}{p}\right)(p-1) + (-1)\sum_{j\neq -1}\left(\frac{j}{p}\right) \\ &= p\left(\frac{-1}{p}\right) + \sum_{j}\left(\frac{j}{p}\right) \\ &= \left(\frac{-1}{p}\right)p \end{aligned}$$

By Euler's Criterion, Proposition 22.5, this is p if $p \equiv 1 \pmod 4$ and $-p$ if $p \equiv 3 \pmod 4$. □

22.3 Proof of the Quadratic Reciprocity Law

We now use the structure of cyclotomic fields to prove the central result in this area. First, we need:

Lemma 22.8. *Let ζ be a primitive pth root of unity. Let $q \neq p$ be prime. Then q does not ramify in $\mathbb{Q}(\zeta)$.*

Proof: By Theorem 21.33, q ramifies in $K = \mathbb{Q}(\zeta)$ if and only if q divides the discriminant d_K. But $d_K = (-1)^{(p-1)/2}p^{p-2}$ by Theorem 3.10, so $q = p$. □

We now state the Quadratic Reciprocity Law using the Legendre symbol, see Definition 22.2, and we give a proof using cyclotomic fields.

Theorem 22.9. (Quadratic Reciprocity Law) *If p, q are distinct odd primes, then*

$$\left(\frac{p}{q}\right) = (-1)^{(p-1)(q-1)/4}\left(\frac{q}{p}\right) \tag{22.1}$$

Proof: Let $K = \mathbb{Q}(\zeta_q)$. Then $\mathrm{Gal}(K/\mathbb{Q}) \cong \mathbb{F}_q^* \cong (\mathbb{Z}_{q-1}, +)$. This is abelian, so we use the notation of (21.9).

The field K contains a unique quadratic subfield F. No prime $p \neq q$ ramifies in F; if it did, it would ramify in K, contrary to Lemma 22.8. Let

$$q^* = (-1)^{(q-1)/2}q$$

Then Lemma 22.7 implies that $F = \mathbb{Q}(\sqrt{q^*})$. Let $p \neq q$ be prime. Let σ_p be the Frobenius automorphism

$$\sigma_p = \left(\frac{K/\mathbb{Q}}{p}\right)$$

Its restriction to F is $\left(\frac{F/\mathbb{Q}}{p}\right)$ by Proposition 21.31. This restriction is the identity if the residue class of $p \pmod{q}$ is a square in $\mathbb{F}_q^*$; otherwise it is a nontrivial automorphism of F. If we identify $G/H = \mathrm{Gal}(F/\mathbb{Q})$ with $\{\pm 1\}$, we have shown that

$$\left(\frac{F/\mathbb{Q}}{p}\right) = \left(\frac{p}{q}\right)$$

On the other hand, the decomposition of p in F gives further information on $\left(\frac{F/\mathbb{Q}}{p}\right)$. By definition, this is the identity if p splits in F and is nontrivial if p is inert.

By Theorem 21.19, if p is odd then

$$\left(\frac{F/\mathbb{Q}}{p}\right) = \left(\frac{q^*}{p}\right)$$

Thus

$$\left(\frac{p}{q}\right) = \left(\frac{q^*}{p}\right) = \left(\frac{-1}{p}\right)^{(q-1)/2)} \left(\frac{q}{p}\right)$$

But Euler's Criterion, Proposition 22.5, gives

$$\left(\frac{q}{p}\right) = \left(\frac{-1}{p}\right)^{(p-1)/2}$$

which leads to (22.1). □

22.4 Application of Quadratic Reciprocity

As an example of quadratic reciprocity in action, we now prove a result that is used in the proof of Theorem 22.14 below. It is typical of many such results on the representation of primes by quadratic forms, which are proved in a similar manner.

Theorem 22.10. *An odd prime p is of the form $p = a^2 + ab - b^2$ if and only if $p \equiv \pm 1 \pmod{5}$.*

Proof: Let $\tau = (1+\sqrt{5})/2$. By Theorem 12.11 the ring $\mathbb{Z}[\tau]$ is a unique factorisation domain. Also $\mathrm{N}(a+b\tau) = a^2+ab-b^2$.

If p is a rational prime and $p = a^2+ab-b^2$ for $a,b \in \mathbb{Z}$ then $4p = (2a+b)^2 - 5b^2$. If $p = 5k \pm 2$ then $(2a+b)^2 \equiv \pm 3 \pmod 5$, but ± 3 are quadratic nonresidues (mod 5), contradiction.

Otherwise $p = 5k \pm 1$. Then $\left(\frac{5}{p}\right) = \left(\frac{p}{5}\right) = 1$, so 5 is a quadratic residue (mod 5). Thus there exists $x \in \mathbb{Z}$ such that $p|(x^2-5)$. Therefore $p|(x+\sqrt{5})(x-\sqrt{5})$ in $\mathbb{Z}[\tau]$. If p is prime in $\mathbb{Z}[\tau]$ then $p|(x+\sqrt{5})$ or $p|(x-\sqrt{5})$. But $\frac{x}{p} \pm \frac{\sqrt{5}}{p}$ is not in $\mathbb{Z}[\tau]$. Therefore p has a nontrivial factor $a+b\tau$. Now $\mathrm{N}(p) = p^2$, so $\mathrm{N}(a+b\tau) = \pm p$. If the norm is $+p$ then $a^2+ab-b^2 = p$. Otherwise $a^2+ab-b^2 = -p$, but then $p = b^2+(-a)b-(-a)^2$. □

Of course, odd primes $5k \pm 1$ have k even, so they are actually of the form $10l \pm 1$.

22.5 Computation Using the Legendre Symbol

Quadratic reciprocity provides almost enough information to compute the quadratic character of a given number modulo any given prime; that is whether it is a quadratic residue or a nonresidue. Example 22.13 below illustrates the method. When computing (p/q) in this manner, four cases are not covered by odd primes: $p = \pm 1$ (not primes but needed) and $p = \pm 2$. Clearly

$$\left(\frac{1}{q}\right) = 1$$

and

$$\left(\frac{-2}{q}\right) = \left(\frac{2}{q}\right)\left(\frac{-1}{q}\right) \tag{22.2}$$

so only two further items of information are required: $p = -1$ and $p = 2$. The next proposition deals with these cases, which are often called the 'first and second supplements' to quadratic reciprocity.

Proposition 22.11. *If q is an odd prime then*

$$\left(\frac{-1}{q}\right) = (-1)^{(q-1)/2} \tag{22.3}$$

$$\left(\frac{2}{q}\right) = (-1)^{(q^2-1)/8} \tag{22.4}$$

Proof: Equation (22.3) is the case $k = -1$ of Euler's Criterion, Proposition 22.5.

We now prove (22.4). By Theorem 21.19, the prime 2 splits in $\mathbb{Q}(\sqrt{q^*})$ if $q^* \equiv 1 \pmod 8$, and it is inert if $q^* \equiv 5 \pmod 8$. Now

$$(-1)^{(q^2-1)/8} = (-1)^{(q^{*2}-1)/8}$$

which is 1 if $q^* \equiv 1 \pmod 8$ and -1 if $q^* \equiv 5 \pmod 8$. Thus

$$\left(\frac{F/\mathbb{Q}}{2}\right) = (-1)^{(q^2-1)/8}$$

so

$$\left(\frac{2}{q}\right) = (-1)^{(q^2-1)/8}$$

□

Corollary 22.12. *Let p be an odd prime. Then:*

(a) -1 *is a quadratic residue* $\pmod p$ *if and only if* $p \equiv 1 \pmod 4$.
(b) 2 *is a quadratic residue* $\pmod p$ *if and only if* $p \equiv \pm 1 \pmod 8$.
(c) -2 *is a quadratic residue* $\pmod p$ *if and only if* $p \equiv 1, 3 \pmod 8$.

Proof:

(a) We want $(p^2-1)/8$ even, so $p - 1 = 4k$, hence $p = 4k + 1$.
(b) We want $(p-1)/2$ even, so $p \equiv \pm 1 \pmod 8$.
(c) Equation (22.2) states that $\left(\frac{-2}{q}\right) = \left(\frac{2}{q}\right)\left(\frac{-1}{q}\right)$. We want

$$\left(\frac{2}{q}\right) = \left(\frac{-1}{q}\right) = \pm 1$$

where the $\pm$ signs are the same for both expressions. A $+$ sign leads to $p \equiv 1 \pmod 8$, while a $-$ sign leads to $p \equiv 3 \pmod 8$. □

This completes the key properties of the Legendre symbol, leading to a quick method to compute the quadratic character of a number $\pmod p$:

Example 22.13. Is 1984 a quadratic residue modulo 97?

$$\left(\frac{1984}{97}\right) = \left(\frac{44}{97}\right) = \left(\frac{2}{97}\right)^2\left(\frac{11}{97}\right) = (\pm 1)^2\left(\frac{11}{97}\right)$$
$$= \left(\frac{11}{97}\right) = \left(\frac{97}{11}\right) = \left(\frac{9}{11}\right) = \left(\frac{3}{11}\right)^2 = 1$$

(The first step uses $1984 \equiv 44 \pmod{97}$; the second uses multiplicativity of the Legendre symbol; in the third we do not care whether the sign is $+$ or $-$; in the fifth we use quadratic reciprocity; and so on.) Therefore 1984 is a quadratic residue modulo 97. In fact, $1984 \equiv 23^2 \equiv 74^2 \pmod{97}$.

22.6 Extended Example

Having used ramification theory to prove the Law of Quadratic Reciprocity, we can repay the compliment by using quadratic reciprocity as a key step in an extended example that illustrates many of the ideas studied in Chapter 21.

We saw in Example 19.10 that when $\zeta = \zeta_5$ is a primitive 5th root of unity, the subfields of $\mathbb{Q}(\zeta)$ are $\mathbb{Q} \subseteq \mathbb{Q}(\sqrt{5}) \subseteq \mathbb{Q}(\zeta)$, whose degrees over $\mathbb{Q}$ are 1, 2, 4 respectively. The rings of integers are

$$\mathbb{Z} \subseteq \mathbb{Z}[\tau] \subseteq \mathbb{Z}[\zeta]$$

where $\tau = \frac{1}{2}(1+\sqrt{5})$ is the golden number. Also

$$\zeta + \zeta^4 = \frac{-1+\sqrt{5}}{2} = \tau - 1$$

The nontrivial conjugate of τ is

$$\bar{\tau} = \frac{1-\sqrt{5}}{2} = -1/\tau = 1 - \tau$$

The rings of integers are all Euclidean domains, hence principal ideal domains, so every prime ideal is principal, generated by a prime element of the relevant ring. We examine how some prime ideals in $\mathbb{Z}$ extend to $\mathbb{Z}[\tau]$, and how some examples of prime ideals in $\mathbb{Z}$ and $\mathbb{Z}[\tau]$ extend to $\mathbb{Z}[\zeta]$.

First, consider the quadratic field extension $\mathbb{Q}(\sqrt{5})/\mathbb{Q}$, with rings of integers $\mathbb{Z}[\tau]/\mathbb{Z}$.

The prime ideals in $\mathbb{Z}$ are $\langle p \rangle$ for rational primes p.

In $\mathbb{Z}[\tau]$ we observe that if $a + b\tau \in \mathbb{Z}[\tau]$, so $a, b \in \mathbb{Z}$, then

$$\mathrm{N}(a+b\tau) = (a+b\tau)(a+b\bar{\tau}) = a^2 + ab - b^2$$

We claim:

Theorem 22.14. *The units of $\mathbb{Z}[\tau]$ are $\pm\tau^{\pm n}$ for $n \in \mathbb{Z}$.*

The primes of $\mathbb{Z}[\tau]$ are:

(a) *The numbers $\pm\sqrt{5}$.*

(b) *All rational primes of the form* $p = 5k \pm 2$.
(c) *The factors* $a + b\tau$ *for rational primes* $p = 5k \pm 1$.

Proof: Let p be a rational prime and consider the equation

$$p = \mathrm{N}(a + b\tau) = a^2 + ab - b^2 \tag{22.5}$$

which we rewrite as

$$(2a + b)^2 - 5b^2 = 4p$$

If $p = 5k \pm 2$ then $(2a + b)^2 \equiv \pm 3 \pmod 5$ which is impossible. Therefore these primes remain prime in $\mathbb{Z}[\tau]$.

If $p = 5k \pm 1$ then Theorem 22.10 below implies that there is a solution of (22.5) in integers. Now

$$\langle p \rangle = \langle a + b\tau \rangle \langle a + b\bar{\tau} \rangle$$

and each factor on the right has norm p so is prime in $\mathbb{Z}[\tau]$.

Finally,

$$5 = (\sqrt{5})^2$$

and $\mathrm{N}(\pm\sqrt{5}) = 5$ is prime, so $\pm\sqrt{5}$ is prime in $\mathbb{Z}[\tau]$. □

We now move on to prime elements and prime ideals in $\mathbb{Z}[\zeta]$.

An example for Case (c) above is the rational prime 11. The norm of a general element of $\mathbb{Z}[\zeta]$ is complicated, but it is easy to show that

$$\mathrm{N}(a + b\zeta) = b^4 - ab^3 + a^2b^2 - a^3b + a^4$$

In particular, $N(1 + 2\zeta) = 11$. This is prime, so $\langle 1 + 2\zeta \rangle$ is a prime ideal in $\mathbb{Z}[\zeta]$. Working in the normal extension $\mathbb{Q}(\zeta)/\mathbb{Q}$, the prime ideal $\langle 1 + 2\zeta \rangle$ lies above $\langle 11 \rangle$ in $\mathbb{Z}$. Under the Galois group of $\mathbb{Q}(\zeta)/\mathbb{Q}$ it maps to $\langle 1 + 2\zeta^i \rangle$ for $1 \leq i \leq 4$.

By direct calculation or general theory (the Galois group acts transitively on prime ideals above a given one),

$$\langle 11 \rangle = \langle 1 + 2\zeta \rangle \langle 1 + 2\zeta^2 \rangle \langle 1 + 2\zeta^3 \rangle \langle 1 + 2\zeta^4 \rangle \tag{22.6}$$

with each factor on the right a prime ideal of $\mathbb{Z}[\zeta]$. Therefore $\langle 11 \rangle_{\mathbb{Z}}$ splits in $\mathbb{Z}[\zeta]$ into four prime factors.

Now $11 = 10 + 1$, so it comes under case (c) of Theorem 22.14. Thus the ideal $\langle 11 \rangle_{\mathbb{Z}}$ factorises in $\mathbb{Z}[\tau]$, and indeed $11 = 3^2 + 3.1 - 1^2$. So

$$11 = (3 + \tau)(3 + \bar{\tau})$$

and

$$\langle 11\rangle = \langle 3+\tau\rangle\langle 3+\bar{\tau}\rangle$$

in $\mathbb{Z}[\tau]$. Both ideals on the right have norm 11, so they are prime ideals of $\mathbb{Z}[\tau]$. Since these two ideals are different, the prime $11 \in \mathbb{Z}$ splits in $\mathbb{Z}[\tau]$.

From (22.6), each of the prime ideals $\langle 3+\tau\rangle$ and $\langle 3+\bar{\tau}\rangle$ of $\mathbb{Z}[\tau]$ must split in $\mathbb{Z}[\zeta]$. In fact,

$$\begin{aligned}\langle 3+\tau\rangle &= \langle 1+2\zeta^2\rangle\langle 1+2\zeta^3\rangle \\ \langle 3+\bar{\tau}\rangle &= \langle 1+2\zeta\rangle\langle 1+2\zeta^4\rangle\end{aligned}$$

The second equation follows from the equation

$$(1+2\zeta)(1+2\zeta^4) = 4+\sqrt{5} = 3+2\tau = (3+\bar{\tau})\tau^2$$

when we observe that $\tau\bar{\tau} = -1$ so τ is a unit; the first is then obvious.

As regards case (a), we have shown that the prime ideal $\langle 5\rangle$ in $\mathbb{Z}$ factorises as

$$\langle 5\rangle = \langle\sqrt{5}\rangle^2$$

in $\mathbb{Z}[\tau]$. That is, it ramifies (completely).

We now ask what happens to $\langle 5\rangle$ and $\langle\sqrt{5}\rangle$ when extended to $\mathbb{Q}(\zeta)$. This question is answered by Lemma 13.11, which tells us that the ideal $\mathfrak{l} = \langle 1-\zeta\rangle$ satisfies $\mathfrak{l}^4 = \langle 5\rangle$ and has norm $\mathrm{N}(\mathfrak{l}) = 5$. Since this is a rational prime, $\mathfrak{l}$ is a prime ideal, so $\langle 5\rangle_{\mathbb{Z}}$ ramifies completely in $\mathbb{Q}(\zeta)$. The only prime ideal above $\langle 5\rangle_{\mathbb{Z}}$ is $\mathfrak{l}$. Therefore $\langle\sqrt{5}\rangle_{\mathbb{Z}[\zeta]} = \mathfrak{l}^2$ also ramifies completely.

Finally, case (b). As an example we take $p = 3$. This remains prime in $\mathbb{Z}[\tau]$ and does not ramify since 3 does not divide the discriminant. Can it split in $\mathbb{Z}[\zeta]$? We now show that Dedekind's Criterion, Theorem 12.1, implies that it remains prime.

In that theorem, let $K = \mathbb{Q}$, $\alpha = \zeta$, $L = \mathbb{Q}(\zeta)$, $\mathfrak{O} = \mathbb{Z}[\zeta]$, and $\mathfrak{p} = \langle 3\rangle$, considered as an ideal of $\mathfrak{o} = \mathbb{Z}$. The minimal polynomial g of ζ is

$$g(t) = t^4+t^3+t^2+t+1$$

and we claim $\bar{g}$ is irreducible over $\mathbb{Z}_3$.

It has no linear factors since $\bar{g}(0) = 1, \bar{g}(1) = 2, \bar{g}(2) = 1$ in $\mathbb{Z}_3$. Therefore any nontrivial factor $\bar{q}$ must be quadratic; also monic. It must be irreducible over $\mathbb{Z}_3$ since there are nonlinear factors. There are only nine monic quadratics, and eliminating those with linear factors only three of them remain:

$$t^2+1 \qquad t^2+t+2 \qquad t^2+2t+2$$

By long division, none of these divides $\bar{g}(t)$.

Therefore $\bar{g}(t)$ is irreducible. By Theorem 21.32 there is only one prime $\mathfrak{P}$ above $\mathfrak{p}$, and it does not ramify. Therefore it must be $\mathfrak{p}\mathfrak{O}$; that is, $\langle 3\rangle_L$ is a prime ideal in $\mathfrak{O}$.

We have not found the prime ideal factorisations in $\mathfrak{O}$ for all prime ideals in $\mathbb{Z}$, but we have exhibited examples of the three basic types of behaviour:

$\langle 11\rangle_{\mathbb{Z}}$ splits.
$\langle 5\rangle_{\mathbb{Z}}$ ramifies.
$\langle 3\rangle_{\mathbb{Z}}$ is inert.

Theorem 3.10 states that the discriminant of $\mathbb{Q}(\zeta)$ is $(-1)^{(4)/2}\,5^{5-2}$, which is 5^3. By Theorem 21.34 the only rational prime that ramifies in $\mathbb{Q}(\zeta)$ is 5. Therefore all other rational primes either split or are inert.

22.7 Exercises

22.1 Calculate the following Legendre symbols:

$$\left(\frac{60}{17}\right) \quad \left(\frac{60}{19}\right) \quad \left(\frac{223}{59}\right) \quad \left(\frac{2310}{12323}\right)$$

22.2 Is 111111111 a quadratic residue (mod 100003)? (*Hint*: Use the Legendre symbol and quadratic reciprocity. Feel free to use a calculator or computer algebra to do the arithmetic.)

22.3 Let $p^* = p$ when $p \equiv 1 \pmod 4$, and $p^* = -p$ when $p \equiv 3 \pmod 4$. List all odd primes $p < 50$ for which $\mathbb{Q}(\sqrt{p^*})$ has unique prime factorisation but $\mathbb{Q}(\zeta_p)$, which contains $\mathbb{Q}(\sqrt{p^*})$ by Theorem 22.7, does not.

22.4 Gauss found a criterion for quadratic residues of an odd prime p by splitting the nonzero integers modulo p into negative and positive parts:

$$N = \{-(p-1)/2, \ldots, -2, -1\} \qquad P = \{1, 2, \ldots, (p-1)/2\}$$

Let k be an integer, and let $|kP \cap N| = \nu$. Then Gauss's criterion states that $(k/p) = (-1)^\nu$. Use Euler's criterion to prove this.

22.5 Verify Euler's criterion and Gauss's criterion (Exercise 22.4) directly when $p = 11$ and $p = 13$. (Computer assistance recommended.)

22.6 Characterise the primes for which 11 is a quadratic residue.

23

Valuations and p-adic Numbers

The real field $\mathbb{R}$ and the complex field $\mathbb{C}$ have considerable extra structure, in addition to their algebraic properties as fields. In particular, both fields have a notion of 'absolute value': $|x|$ for $x \in \mathbb{R}$ and $|z|$ for $z \in \mathbb{C}$. This brings all the operations and techniques of analysis, such as limits and infinite series, into play; it also endows these fields with a topology. The resulting ideas are of central importance in algebraic number theory, because the humble field $\mathbb{Q}$ of rationals has infinitely many distinct absolute values: the one inherited from $\mathbb{R}$, and a 'p-adic' absolute value for every prime p. These ideas extend to prime ideals in number fields, but this is beyond our scope.

In this chapter we begin with an axiomatic definition of 'valuations', the general term for such functions, and relate them to prime ideals. Given a valuation, we define an associated valuation ring and valuation ideal. We prove that the valuation ring is a discrete valuation ring, and we show that for a given number field K there is a natural bijection between the set of prime ideals and the set of valuations.

Next we define the closely related concept of an absolute value, showing that it induces a metric on the number field. The absolute values related to prime ideals have a very strong property, and induce 'ultrametrics'. Concepts from metric spaces then become relevant, and we focus on two: convergence of sequences, and the notion of a complete metric space. We define the field of p-adic numbers and its ring of p-adic integers in three equivalent ways: as the completion of $\mathbb{Q}$ with respect to the p-adic absolute value, as an inverse limit, and as a concrete representation akin to the standard base-n representation of integers.

We then consider the topology of p-adic numbers and p-adic integers, which is very strange compared to that of $\mathbb{R}$ and $\mathbb{C}$ because of the ultrametric property. For example all open balls are also closed, and all closed balls are also open. Topologically, the p-adic integers form a Cantor set, with fractal structure.

We end by sketching, without proofs, three applications of p-adic numbers. The first is the Hasse Principle about numbers that can be represented by quadratic forms. The second is the Three-Squares Theorem, equivalent to Gauss's proof that every positive integer is the sum of three triangular numbers. The third, given just a brief mention, is the use of p-adic numbers in Galois cohomology, a method central to Wiles's proof of Fermat's Last Theorem.

23.1 Valuations

In Section 20.2 we defined the valuation $v_{\mathfrak{p}}(\mathfrak{a})$ of a fractional ideal $\mathfrak{a}$ relative to a prime ideal $\mathfrak{p}$. In this section we formulate maps like $v_{\mathfrak{p}}$ axiomatically as examples of 'valuations', and develop some basic properties. Valuations open up a different way of thinking about number fields and their rings of integers. For reasons of space we confine attention to the simpler areas of the theory.

As usual let K be a number field with ring of integers $\mathfrak{O}$, and let K^* be the multiplicative group of nonzero elements of K. We concentrate on principal ideals $\langle x \rangle = x\mathfrak{O}$ where $x \in K$, and define

$$v_{\mathfrak{p}}(x) = v_{\mathfrak{p}}(x\mathfrak{O}) \qquad x \neq 0$$

so that $v_{\mathfrak{p}} : K^* \to \mathbb{Z}$.

We have not defined $v_{\mathfrak{p}}(0)$ before since the prime factorisation of 0 has not been defined. It is convenient (though not strictly necessary) to extend the definition of $v_{\mathfrak{p}}$ to include $x = 0$. To do so we extend $\mathbb{R}$ to $\mathbb{R}_\infty = \mathbb{R} \cup \{\infty\}$, where the new element ∞ satisfies the conditions

$$\infty + \infty = \infty \qquad \infty + r = \infty \qquad \infty > r$$

for $r \in \mathbb{R}$. Only the additive and order properties of $\mathbb{R}$ are needed, so we do not consider subtraction or multiplication here. Now $v_{\mathfrak{p}} : K \to \mathbb{R}_\infty$.

Axioms for a Valuation

Definition 23.1. A *valuation* on a number field K is a map $v : K \to \mathbb{R}_\infty$ such that:

(a) $v(x) = \infty$ if and only if $x = 0$.
(b) $v(xy) = v(x) + v(y)$ for all $x, y \in K$.
(c) $v(x + y) \geq \min(v(x), v(y))$ for all $x, y \in K$.

Proposition 23.2. *If $\mathfrak{p}$ is any prime ideal of $\mathfrak{O}$, the map $v_{\mathfrak{p}}$ is a valuation.*

Proof: See Propositions 20.2 and 20.4. $\square$

We observe some simple consequences of the axioms:

Proposition 23.3. *Let v be a valuation. Then*
(d) $v(1) = 0$.
(e) $v(-1) = 0$.
(f) *If $v(x) > v(y)$ then $v(x + y) = v(y)$.*

Proof: (d) By axiom (b), $v(1) = v(1.1) = v(1) + v(1)$, so $v(1) = 0$. (Condition (a) of Definition 23.1 rules out the other possibility $v(1) = \infty$.)

(e) Now $0 = v(1) = v((-1).(-1)) = v(-1) + v(-1) = 2v(-1)$, so $v(-1) = 0$.

(f) By axiom (c), $v(x + y) \geq \min(v(x), v(y))$, which in this case equals $v(y)$.

If $v(x+y) > v(y)$ then $v(y) = v(x+y-x) \geq \min(v(x+y), v(x) > v(y)$, a contradiction. Therefore $v(x + y) = v(y)$. $\square$

Valuations and Prime Ideals

Proposition 23.2 determines a map $\mathfrak{p} \mapsto v_{\mathfrak{p}}$ from prime ideals to valuations. It is also possible to go the other way.

Definition 23.4. If v is a valuation on K then the set

$$\mathfrak{o}_v = \{x \in K : v(x) \geq 0\}$$

is the *valuation ring* of v.

The set

$$\mathcal{P}_v = \{x \in K : v(x) > 0\}$$

is the *valuation ideal* of v.

The set

$$\Gamma_v = \{v(x) : x \in K^*\}$$

is the *valuation group* or *value group* of v.

Before proceeding further, we justify these names:

Proposition 23.5. *The valuation ring is a subring of K. The valuation ideal is an ideal of $\mathfrak{o}_v$. The value group is a group under multiplication.*

Proof: The set $\mathfrak{o}_v$ contains 1 since $v(1) = 0$ by Proposition 23.3(d). Suppose that x, y in $\mathfrak{o}_v$. Proposition 23.3(b) implies that $x + y \in \mathfrak{o}_v$, and Definition 23.1(b) implies that $xy \in \mathfrak{o}_v$.

To prove that $\mathcal{P}_v$ is an ideal, use 23.1(c).

That Γ_v is a group follows from 23.1(b). □

Example 23.6. Let $K = \mathbb{Q}$ and consider the prime $2 \in \mathbb{Z}$, so $\mathfrak{p} = \langle 2 \rangle = 2\mathbb{Z}$. Now $v_{\mathfrak{p}}(x)$ is the power of 2 in the prime factorisation of x (or ∞ when $x = 0$). So, for instance,

$$v_{\mathfrak{p}}(16) = 4 \qquad v_{\mathfrak{p}}(240) = 4 \qquad v_{\mathfrak{p}}\left(\frac{22}{7}\right) = 1 \qquad v_{\mathfrak{p}}\left(\frac{1023}{1024}\right) = -10$$

The valuation ring $\mathfrak{o}_v$ consists of all rational numbers $\frac{p}{q}$ where, in lowest terms, q is odd.

The valuation ideal $\mathcal{P}_v$ is equal to $2\mathfrak{o}_v$, and consists of all rational numbers $\frac{p}{q}$ where, in lowest terms, q is odd and p is even.

The valuation group Γ_v is isomorphic to the integers $\mathbb{Z}$ under addition, since $v(2^k) = k$.

Proposition 23.7. *An element $x \in \mathfrak{o}_v$ invertible (that is, a unit) in $\mathfrak{o}_v$ if and only if $v(x) = 0$, in which case $v(x^{-1}) = 0$ as well.*

Proof: If x is invertible then there exists $y \in \mathfrak{o}_v$ such that $xy = 1$. Therefore $0 = v(1) = v(xy) = v(x) + v(y)$ and $v(x), v(y) \geq 0$. Therefore $v(x) = v(y) = 0$.

Conversely, if $v(x) = 0$ then $x \notin \mathcal{P}_v$. Therefore the ideal $x\mathfrak{o}_v = \mathfrak{o}_v$ since $\mathcal{P}_v$ is maximal, so $1 \in x\mathfrak{o}$. Thus there exists $y \in \mathfrak{o}$ such that $xy = 1$.

□

Valuation rings and their valuation ideals have tightly constrained structures:

Theorem 23.8. *The valuation ring $\mathfrak{o}_v$ is a principal ideal domain with unique maximal ideal $\mathcal{P}_v$, hence a discrete valuation ring. If $\mathbb{F}$ is the group of fractional ideals of $\mathfrak{o}_v$, the map $\psi : (\mathbb{Z}, +) \to \mathbb{F}$ defined by $\psi(m) = \mathcal{P}_v^m$ is a group isomorphism.*

Proof: By Proposition 23.7, $x \in \mathfrak{o}_v$ invertible in $\mathfrak{o}_v$ if and only if $v(x) = 0$. Therefore $\mathfrak{o}_v^* = \mathfrak{o}_v \setminus \mathcal{P}_v$, which implies that $\mathcal{P}_v$ is the unique maximal ideal of $\mathfrak{o}$.

Let $\mathfrak{a}$ be any nonzero ideal of $\mathfrak{o}_v$. Choose $b \in \mathfrak{a}$ such that $v(b)$ is minimal. We claim that $\mathfrak{a} = b\mathfrak{o}_v$. Clearly $a \supseteq b\mathfrak{o}_v$. Conversely, if $c \in \mathfrak{a}$ then $v(b) \leq v(c)$, so $v(b^{-1}c) \geq 0$. Therefore $b^{-1}c \in \mathfrak{o}_v$ and $c = b.b^{-1}c \in b\mathfrak{o}_v$.

Moreover, if $a\mathfrak{o}_v = \mathcal{P}_v$ and $v(b) = n$, then

$$a^n\mathfrak{o}_v = \mathcal{P}_v^n = b\mathfrak{o}_v = \{c \in K : v(c) \geq n\}$$

□

From the proof we read off:

Corollary 23.9. $K^*/\mathfrak{o}_v^* \cong \mathbb{Z}$. □

Theorem 23.10. *Let $\mathfrak{o}$ be a Dedekind domain, with v a valuation such that $\mathfrak{o} \subseteq \mathfrak{o}_v$. Let $\mathfrak{p}_v = \mathcal{P}_v \cap \mathfrak{o}$. Then:*

(a) *$\mathfrak{p}_v$ is a prime ideal of $\mathfrak{o}$.*

(b) *$v_{\mathfrak{p}_v} = v$.*

(c) *$\mathfrak{p}_v\mathfrak{o}_v = \mathcal{P}_v$.*

(d) *$\mathfrak{o}/\mathfrak{p}_v \cong \mathfrak{o}_v/\mathcal{P}_v$.*

Proof: To compare v with $v_{\mathfrak{p}_v}$, we claim that $v_{\mathfrak{p}_v}(z) = 0$ implies $v(z) = 0$. Any such z can be written as $z = a'/b'$ where $a', b' \in \mathfrak{o}$ and $v_{\mathfrak{p}_v}(a') = v_{\mathfrak{p}_v}(b') = l$, say. Let $\pi \in \mathfrak{p}_v^{-l} \setminus \mathfrak{p}_v^{-l+1}$ and define $a = \pi a', b = \pi b'$. Then $z = a/b$ with $a, b \in \mathfrak{o} \setminus \mathfrak{p}_v$; therefore $a, b \in \mathfrak{o} \setminus \mathcal{P}_v$, so $v(z) = 0$.

Consider the ideal $\mathfrak{p}_v\mathfrak{o}_v$. By Theorem 23.8, $\mathfrak{p}_v\mathfrak{o}_v = \mathcal{P}_v^e$ for some $e \geq 1$. But if $0 \neq x \in \mathfrak{o}$ with $v_{\mathfrak{p}_v}(x) = l$, then $x\mathfrak{o} = \mathfrak{p}^l\mathfrak{a}$ where $(\mathfrak{a}, \mathfrak{p}_v) = 1$. Now $\mathfrak{a}\mathfrak{o}_v = \mathfrak{o}_v$, so $x\mathfrak{o}_v \in \mathcal{P}_v^{le}$ so $v(x) = ev_{\mathfrak{p}_v}(x)$. Thus $v = ev_{\mathfrak{p}_v}$. But in fact $e = 1$ since v and $v_{\mathfrak{p}_v}$ have the same value group $\mathbb{Z}$.

Finally, we show that $\mathfrak{o} + \mathcal{P}_v = \mathfrak{o}_v$. Since $\mathfrak{o} \subseteq \mathfrak{o}_v$ we have $\mathfrak{o} + \mathcal{P}_v \subseteq \mathfrak{o}_v$. Conversely, let $x \in \mathfrak{o}_v$: we must show that $z \in \mathfrak{o} + \mathcal{P}_v$. This is clear if $v(z) > 0$, so suppose $v(z) = 0$. Then, as above, $z = a/b$ with $a, b, \in \mathfrak{o} \setminus \mathfrak{p}_v$. Since $\mathfrak{o}/\mathfrak{p}_v$ is a field, there exists $c \in \mathfrak{o}$ such that $bc \in 1 + \mathfrak{p}_v$. Now

$$z - ac = a(b^{-1} - c) = ab^{-1}(1 - bc) \in \mathfrak{o}_v\mathfrak{p}_v = \mathcal{P}_v$$

□

Proposition 23.11. *Let K be a number field, v a valuation. Then $\mathfrak{o} \subseteq \mathfrak{o}_v$.*

Proof: Since $v(-1) = 0 = v(1)$, Equation (20.8) implies that v is non-negative on $\mathbb{Z}$. Since $\mathfrak{o}$ is the integral closure of $\mathbb{Z}$ in K, an element $x \in \mathfrak{o}$ is a zero of some monic polynomial $t^n + a_1 t^{n-1} + \cdots + a_n$ over $\mathbb{Z}$. By (20.8),

$$nv(x) \geq \inf_{0 \leq i \leq n-1} (v(a_{n-1}) + iv(x)) = \inf_{0 \leq i \leq n-1} (iv(x))$$

since each $v(a_{n-i}) \geq 0$. Thus $v(x) \geq 0$ so $x \in \mathfrak{o}_v$. □

Theorem 23.12. *For a number field K the map $\mathfrak{p} \mapsto v_\mathfrak{p}$ is a bijection between the set of prime ideals and the set of valuations.*

Proof: Use Theorem 23.10 and Proposition 23.11. □

Theorem 23.12 tells us that anything that can be stated in the language of prime ideals has an equivalent statement in the language of valuations. Sometimes one version will have a more natural proof than the other.

23.2 Absolute Values and Topology

We now widen the consideration of valuations to include the usual absolute values on $\mathbb{R}$ and $\mathbb{C}$. This is done by placing the entire discussion in a topological context. We assume familiarity with basic metric spaces and topology; in particular with metrics, limits of sequences, Cauchy sequences, the topology of a metric space, and compact topological spaces. See any standard text on the subject, such as Magnus [84] or Sutherland [133].

Definition 23.13. Let $\mathbb{R}^+ = \{x \in \mathbb{R} : x \geq 0\}$. If K is a field, a function $|\cdot| : K \to \mathbb{R}^+$ is an *absolute value* if, for all $x, y \in K$,

(a) $|x| = 0$ if and only if $x = 0$.

(b) $|xy| = |x||y|$.

(c) $|x + y| \leq |x| + |y|$.

If, more strongly than (c), we have

(d) $|x + y| \leq \max(|x|, |y|)$,

the absolute value is *non-Archimedean*. (Another term is *ultrametric*, although strictly this refers to the associated metric, defined by $d(x, y) = |x - y|$. We will abuse terminology and use this for the absolute value as well.) If (d) does not hold, the absolute value is *Archimedean*.

We explicitly exclude the *trivial absolute value* for which $|0| = 0$ and $|x| = 1$ whenever $x \neq 0$.

Proposition 23.14. (a) $|1| = 1$.
(b) $|-1| = 1$.
(c) $|-x| = |x|$.

Proof: (a) $|1| = |1.1| = |1||1|$, so either $|1| = 0$ or $|1| = 1$. But the former implies $1 = 0$, a contradiction.
(b) $1 = |1| = |(-1)(-1)| = |-1|^2$, so $|-1| = 1$ since absolute values are non-negative.
(c) $|-x| = |(-1)x| = |(-1)||x| = 1.|x| = |x|$.

□

An ultrametric has the surprising property that adding something smaller does not increase the norm:

Proposition 23.15. *If* $|\cdot|$ *is an ultrametric, then* $|y| < |x|$ *implies that* $|x+y| = |x|$.

Proof: Certainly $|x+y| \leq |x|$. To show equality, suppose that $|x+y| < |x|$. Then $|x| \leq \max(|x+y|, |-y|) < |x|$, a contradiction. □

Definition 23.16. The absolute value $|\cdot|$ is *discrete* if the set of all absolute values $|K^*| = \{|x| : x \in K^*\}$ is a discrete subgroup of $\mathbb{R}^*$

Examples 23.17.
(a) The usual absolute value $|\cdot|$ on $\mathbb{R}$ is an absolute value. It is Archimedean and not discrete. Properties (a) and (b) are standard, (c) is the triangle inequality, and (d) fails because, for example, $2 = |1+1| > \max(|1|, |1|) = 1$. It is not discrete because the value group is the multiplicative group of positive real numbers.
(b) The usual absolute value $|\cdot|$ on $\mathbb{C}$ is an absolute value. It is Archimedean and not discrete. The reasoning is the same as for $\mathbb{R}$.
(c) Let $K = \mathbb{Q}$, with p a rational prime. Let $0 < \lambda < 1$ be a real number. Then

$$|x|_p = \lambda^{v_p(x)} \tag{23.1}$$

is an absolute value. It is non-Archimedean and discrete. Properties (a), (b), and (d) follow from (20.3) and (20.8) with $\mathfrak{p} = \langle p \rangle$.
(d) More generally, let K be a number field and let $\mathfrak{p}$ be a prime ideal of $\mathfrak{O}$. Then for $0 < \lambda < 1$,

$$|x|_{\mathfrak{p}} = \lambda^{v_{\mathfrak{p}}(x)}$$

is an absolute value. It is non-Archimedean and discrete. Properties (a), (b), and (d) follow from (20.3) and (20.8).

The role of λ here indicates a lack of uniqueness in closely related absolute values. To get round this issue, two absolute values $|\cdot|_1$ and $|\cdot|_2$ are said to be *equivalent* if, for all $x \in K$,

$$|x|_2 = |x|_1^\alpha \qquad \text{for some } \alpha \in \mathbb{R}, \alpha > 0$$

Equivalent absolute values define the same topology. For a given $\mathfrak{p}$, all absolute values of the form $\lambda^{v_\mathfrak{p}(x)}$ are equivalent. It is usual to normalise by taking $\lambda = 1/\mathrm{N}(\mathfrak{p})$:

Definition 23.18. Let $\mathfrak{p}_v$ be the valuation ideal of v. Then $\mathrm{N}(\mathfrak{p}_v) < \infty$, so we can define the *normalised absolute value*

$$|x|_v = (\mathrm{N}(\mathfrak{p}_v))^{-v(x)}$$

In the special case $K = \mathbb{Q}$ with $v = v_p$ this becomes

$$|x|_p = p^{-v_p(x)}$$

We end this section by stating, without proof, a key result that explains why the absolute values $|\cdot|_p$ for prime p are emphasised in algebraic number theory:

Theorem 23.19. (Ostrowski's Theorem) *An absolute value on $\mathbb{Q}$ is equivalent either to $|\cdot|_\mathbb{R}$ or to $|\cdot|_p$ for a prime p.*

Proof: See Fröhlich and Tayor [46] (2.15) p. 67. □

We lack the space to discuss these ideas in more detail; see Fröhlich and Taylor [46] chapter II section 2. As a substitute, Sections 23.3 and 23.4 consider the most important special case, where $K = \mathbb{Q}$.

23.3 Convergence and Completion

Let p be a fixed but arbitrary rational prime. The absolute value $|\cdot|_p$ defines a metric d_p via (23.2). In any metric space there are the usual notions related to convergence of a sequence. We single out three:

Definition 23.20. Let X be a metric space with metric d. Then:

(a) A sequence (a_n) *converges* in X if there exists $a \in X$ such that, for all $\varepsilon > 0$ there exists $N \in \mathbb{N}$ for which $d(a_n, a) < \varepsilon$ for all $n > N$. We say that a is the *limit* of the sequence (a_n), and write

$$\lim_{n \to \infty} a_n = a$$

(b) A sequence (a_n) is a *Cauchy sequence* in X if for all $\varepsilon > 0$ there exists $N \in \mathbb{N}$ such that $d(a_m, a_n) < \varepsilon$ for all $m, n > N$.

(c) The space X is *complete* if and only if every Cauchy sequence in X converges.

A standard theorem states that any metric space embeds isometrically (that is, preserving distances) in a canonically defined complete metric space:

Theorem 23.21. *Let X be any metric space. Then there exists a metric space $\widehat{X}$ and an isometric embedding $\phi : X \to \widehat{X}$ such that $\widehat{X}$ is complete and every point in $\widehat{X}$ is the limit of a sequence in X.*

The space $\widehat{X}$ is unique up to an isometry that is the identity on X.

Proof: We sketch the proof, because of the importance of this result. Let Y be the set of all Cauchy sequences in X. Define two Cauchy sequences (a_n) and (b_n) to be equivalent, written $(a_n) \sim (b_n)$, if $d(a_n, b_n)$ tends to zero. Let $\widehat{X}$ be the set of equivalence classes for $\sim$. There is a natural metric $\hat{d}$ on $\widehat{X}$ defined by $\hat{d}((a_n),(b_n)) = \lim_{n\to\infty} d(a_n, b_n)$. The map ϕ is defined using constant sequences $\phi(x) = (x, x, x, \ldots)$, and the rest of the proof is routine. □

Definition 23.22. The metric space $\widehat{X}$ is the *completion* of X.

This construction should be familiar as one method to construct the real numbers $\mathbb{R}$ from the rational numbers $\mathbb{Q}$, by forming the completion of $\mathbb{Q}$ with respect to the usual metric $d(q, r) = |q - r|$. In this case we must also define the field operations on $\mathbb{R}$, which again is done using limits and verifying that the resulting operations induce corresponding ones on the equivalence classes of Cauchy sequences. Everything here is routine, but the verifications are lengthy. For details see Stewart and Tall [129].

23.4 The p-adic Numbers

We now introduce one of the most important, and surprising, constructions in algebraic number theory. For each prime $p \in \mathbb{Z}$ there is a *complete* field $\mathbf{Q}_p$ for the metric induced by the valuation v_p. In some ways this field is analogous to $\mathbb{R}$ and $\mathbb{C}$; in other ways it is so different that it might have come from an alien planet. This is the field of p-adic numbers, introduced by

Kurt Hensel in 1897. Some 'prehistory' can be detected in earlier number-theoretic research, but the idea was stunningly original.

We describe three distinct approaches to the p-adic numbers: completion of a suitable ultrametric space, an inverse limit, and an explicit representation by infinite sequences of rational integers.

The p-adic Numbers as a Completion

We can gain some intuition about the p-adic world by thinking about how we would distinguish numbers if we knew about only one prime p. In such a situation, two numbers x, y are 'the same' if p occurs to the same power in their prime factorisations. That is, $v_p(x) = v_p(y)$, or equivalently, $v_p(x/y) = 0$.

So, for example, we would be unable to tell the difference between $x = 2$ and $y = \frac{22}{7}$, since y has factorisation $y = 2.11.7^{-1}$ but 7 and 11 are invisible to us. Now $v_2(x) = 1 = v_2(y)$. Equivalently, $x/y = 7/11$ and $v_2(7/11) = 0$.

Now, the p-adic absolute value is not $v_p(x)$, but $|x|_p = p^{-v_p(x)}$. This expression is chosen in order to obtain a metric—indeed an ultrametric. The *p-adic distance* between $x, y \in \mathbb{Q}$ is defined to be

$$d_p(x, y) = |x - y|_p = p^{-v_p(x-y)} \tag{23.2}$$

For example, when $x = 2$ and $y = 22/7$, the distance between x and y is

$$d_2(x, y) = |x - y|_2 = 2^{-v_2(x-y)} = 2^{-3} = 1/8$$

because

$$x - y = 2 - \frac{22}{7} = -\frac{8}{7}$$

with the 2-power factor 2^3.

Intuitively, a number is close to zero in the p-adic metric if it is divisible by a *large* positive power of p. And the closer two numbers become in the p-adic metric, the *larger* is the power of p that divides their difference. The p-adic metric is an 'ultrametric', a concept that may be unfamiliar. It differs in many ways from the usual metric on $\mathbb{Q}$ as a subset of $\mathbb{R}$.

Since $\mathbb{Q}$ is a metric space under the p-adic metric d_p, defined in (23.2), it has a completion. Formally:

Definition 23.23. The set of *p-adic numbers* $\mathbf{Q}_p$ is the completion of $\mathbb{Q}$ in the metric d_p.

It is easy to check that under algebraic operations induced naturally from those of $\mathbb{Q}$, the p-adic numbers form a field. The field operations are easily seen to be continuous, so it is a topological field. By definition it contains $\mathbb{Q}$.

It is also possible to consider completions of $\mathfrak{p}$-adic absolute values from prime ideals $\mathfrak{p}$ of a number field K. These fields turn out to be finite extensions of $\mathbf{Q}_p$. For reasons of space, we omit this topic.

Remark. The standard notation for the p-adic number field is $\mathbb{Q}_p$, with $\mathbb{Z}_p$ for the p-adic integers (defined below). In this book, however, we are using $\mathbb{Z}_p$ for the integers modulo p, and also for the cyclic group of order p. Modern algebraic number theorists change that to $\mathbb{Z}/p\mathbb{Z}$ or just $\mathbb{Z}/p$. Because p-adic numbers are not used elsewhere in the book, and $\mathbb{Z}/p\mathbb{Z}$ is a bit clumsy, we use boldface symbols for the p-adic structures $\mathbf{Q}_p$ and $\mathbf{Z}_p$ to make the distinction clear.

Example 23.24.

In $\mathbf{Q}_p$ define a sequence (s_n) by

$$s_n = 1 - p + p^2 - p^3 + \cdots + (-p)^n$$

We claim that $\lim_{n\to\infty} s_n = \frac{1}{p+1}$.

By the sum of a geometric progression,

$$s_n = \frac{1-(-p)^{n+1}}{1+p} = \frac{1}{1+p} + p^{n+1}\frac{(-1)^n}{1+p}$$

Therefore

$$|s_n - \frac{1}{1+p}|_p = |p^{n+1}\frac{(-1)^n}{1+p}|_p = p^{-(n+1)} \to 0 \text{ as } n \to \infty$$

The p-adic Integers

The field $\mathbf{Q}_p$ has a special subring $\mathbf{Z}_p$, the *p-adic integers*.

Definition 23.25. The set of *p-adic integers* is

$$\mathbf{Z}_p = \{x \in \mathbf{Q}_p : |x|_p \leq 1\}$$

Proposition 23.26. *The p-adic integers form a subring of* $\mathbf{Q}_p$.

Proof: Use (a), (b), (c) of Definition 23.13. For additive inverses, observe that $|-x|_p = |(-1)x|_p = |-1|_p|x|_p = |x|_p$. □

The p-adic Integers as an Inverse Limit

Perhaps the 'cleanest' formulation of the p-adic integers is to view them as an 'inverse limit' of familiar rings.

Definition 23.27. An *inverse limit system* is a sequence of sets X_n and maps f_n for $n \geq 1$, such that

$$\cdots \xrightarrow{f_3} X_3 \xrightarrow{f_2} X_2 \xrightarrow{f_1} X_1$$

Let L be the set of all $(a_n) \in \prod_1^\infty X_n$ such that $f_n(a_{n+1}) = a_n$ for all $n \geq 1$. Then L is the *inverse limit* of the system, denoted by

$$L = \lim_{\longleftarrow} X_n$$

Proposition 23.28. *Suppose that all of the sets X_n are non-empty, and all maps f_n are surjective. Then L is non-empty.*

Proof: Construct an element (a_n) of L inductively as follows. Choose any $a_1 \in X_1$. Choose $a_2 \in f_1^{-1}(a_1)$, then $a_3 \in f_2^{-1}(a_2)$, and so on. Clearly $(a_n) \in L$. □

We now give a construction of a ring L, which turns out to be isomorphic to $\mathbf{Z}_p$, using inverse limits.

Let $X_n = \mathbb{Z}_{p^n}$ for $n \geq 1$, and define $f_n : X_{n+1} \to X_n$ by

$$f_n(x) = x \pmod{p^n}$$

This is well defined since p^n divides p^{n+1}. Clearly each X_n is finite and each f_n is a surjective homomorphism. Thus the inverse limit

$$L = \lim_{\longleftarrow} X_n$$

is non-empty. It is easy to prove that L is a ring, because the f_n are homomorphisms. We now show:

Theorem 23.29. *The inverse limit L is isomorphic to $\mathbf{Z}_p$.*

Proof: Define a map $\phi : L \to \mathbf{Z}_p$ as follows. Let $(a_n) \in L$. For each n let $x_n \in \mathbb{Z}$ be such that $x_n \pmod{p^n}$ is equal to a_n. Then $x_m \equiv x_n \pmod{p^N}$ whenever $m, n \geq N$. Therefore $|x_m - x_n|_p \leq p^{-N}$, so (x_n) is Cauchy. Therefore it converges in the complete metric space $\mathbf{Z}_p$.

Let $\phi((a_n)) = \lim_{n\to\infty} a_n \in \mathbf{Z}_p$. It is easy to see that ϕ is a ring homomorphism. We claim it is a bijection.

Let $a \in \mathbf{Z}_p$. Define a_n to be the image of a under the map

$$\psi : \mathbf{Z}_p \to \mathbf{Z}_p/(p^n\mathbf{Z}_p) \cong \mathbb{Z}_{p^n}$$

Then $\psi : \mathbf{Z}_p \to L$, and it is routine to verify that ψ is the inverse of ϕ. □

Clearly L is a domain, since $\mathbf{Z}_p$ is. Therefore L has a field of fractions, unique up to isomorphism, and this must be isomorphic to $\mathbf{Q}_p$. We can therefore give an alternative definition of $\mathbf{Q}_p$ as the field of fractions of the inverse limit L.

The p-adic Numbers as a Concrete Construction

By examining the structure of L we can derive a concrete representation of $\mathbf{Z}_p$, and then extend it to $\mathbf{Q}_p$. (This is where things become distinctly weird until you get used to the difference between $\mathbf{Q}_p$ and $\mathbb{R}$.) We sketch the ideas involved.

Define a *p-adic series* to be a formal expression

$$a = \sum_{i=m}^{\infty} a_i p^i \tag{23.3}$$

where $m \in \mathbb{Z}$ and $0 \leq a_i \leq p-1$ for all i. In particular, m can be *negative*. Sums and products of such series can be defined in the obvious way; they are 'formal Laurent series' in the formal variable p. This is best done in two steps, starting with $\mathbf{Z}_p$. Here we take $m = 0$ in (23.3), and consider

$$a = \sum_{i=0}^{\infty} a_i p^i$$

(padding out with initial zero terms if $m > 0$).

If we let the coefficients a_i be arbitrary integers, this is the ring of formal power series over $\mathbb{Z}$ in powers of p. Next, we impose an equivalence relation

$$\sum_{i=0}^{\infty} a_i p^i \sim \sum_{i=0}^{\infty} b_i p^i \Leftrightarrow a_i \equiv b_i \pmod{p^i}$$

This preserves the ring operations. We can then take representatives of the equivalence classes, to make the coefficients satisfy $0 \leq a_i \leq p-1$. Effectively, this is like writing integers in base p notation, and adding and multiplying by 'carrying' surplus digits to the next place along.

This construction gives the set R of all p-adic series with $m = 0$ a ring structure. We can identify each such series with an element $(b_n) \in L$. Let S be the set of all such series. Let

$$b_n = a_1 + pa_2 + \cdots p^{n-1}a_{n-1}$$

and define a map $\phi : S \to L$ by $\phi((a_n)) = (b_n)$. It is then routine to verify that ϕ is a ring isomorphism from S to L.

To deal with the p-adic numbers, observe that $\mathbf{Q}_p$ is the field of fractions of $\mathbf{Z}_p$. Indeed, every element of $\mathbf{Q}_p$ has the form $p^m a$ where $a \in \mathbf{Z}_p$. (The important case is when $m < 0$.) Expand a as a p-adic series with all coefficients in the range $0 \leq a_i \leq p-1$. Then multiply by p^m, which shifts a_i to a_{m+i}. Now $\mathbf{Q}_p$ is represented in a similarly concrete fashion by the set R of all formal series (23.3), where now m can be negative.

The p-adic integers $\mathbf{Z}_p$ are the subring of R for which $a_k = 0$ for all $k < 0$. Its units are the elements of $\mathbf{Z}_p$ that satisfy $a_0 \neq 0$. The complement of the units is the unique prime ideal $\mathfrak{p}$ of $\mathbf{Z}_p$, which consists of all p-adic series such that $a_0 = 0$. It is equal to the principal ideal $\langle p \rangle = p\mathbf{Z}_p$. The powers of $\mathfrak{p}$ are all of the proper nonzero ideals of $\mathbf{Z}_p$, and they form a decreasing sequence

$$\mathbf{Z}_p \supseteq \mathfrak{p} \supseteq \mathfrak{p}^2 \supseteq \mathfrak{p}^3 \supseteq \cdots$$

with intersection 0. Thus $\mathbf{Z}_p$ is a complete valuation ring.

Example 23.30. We revisit Example 23.24. What we proved there can be restated in the terms just discussed:

The fraction $\frac{1}{p+1}$ is represented by the p-adic series

$$1 - p + p^2 - p^3 + \cdots$$

23.5 p-adic Topology

We briefly discuss topological features of p-adic numbers and integers. Again, for any undefined topological terms see Magnus [84], Sutherland [133], or any other standard text on point-set topology and metric spaces.

The first two results below would be fallacious in $\mathbb{R}$, where convergence of sequences and series is notoriously subtle. In $\mathbf{Q}_p$ everything is simpler (providing some recompense for other topological peculiarities):

Proposition 23.31. *A sequence (a_n) in $\mathbf{Q}_p$ is Cauchy if and only if for any $\varepsilon > 0$ there exists N such that $|a_{n+1} - a_n|_p < \varepsilon$ for all $n \geq N$.*

Proof: Necessity is clear: the usual Cauchy condition is that $|a_n - a_m|_p < \varepsilon$ for all $m, n \geq N$.

For sufficiency, observe that if $m < n$ then

$$\begin{aligned} |a_n - a_m|_p &= |(a_n - a_{n-1}) + (a_{n-1} - a_{n-2}) + \cdots + (a_{m+1} - a_m)|_p \\ &\leq \max(|a_n - a_{n-1}|_p, |a_{n-1} a_{n-2}|_p, \ldots, |a_{m+1} - a_m|_p) \end{aligned}$$

and the rest is easy. □

Corollary 23.32. *A series $\sum_n a_n$ converges in $\mathbf{Q}_p$ if and only if $|a_n|_p \to 0$ as $n \to \infty$.*

Proof: The corresponding sequence of partial sums is Cauchy, so it converges by completeness. □

Proposition 23.33. *Any open ball in $\mathbf{Q}_p$ is also closed, and any closed ball is also open.*

Proof: Consider an open ball

$$B(a, r) = \{x \in \mathbf{Q}_p : |x - a|_p < r\}$$

Let n be the smallest integer such that $p^{-n} \geq r$. There are no p-adic numbers with absolute value between $p^{-(n+1)}$ and p^{-n} (the value group is discrete and consists of the powers p^k). Thus $B(a, r) = \{x \in \mathbf{Q}_p : |x - a|_p < p^{-n}\}$, which equals $\{x \in \mathbf{Q}_p : |x - a|_p \leq p^{-(n+1)}\}$. Thus $B(a, r)$ is closed. The converse follows similar lines. □

Exercise 22.13 gives a different proof.

Proposition 23.34. *The sets $\mathbf{Q}_p$ and $\mathbf{Z}_p$ are totally disconnected.*

Proof: Let $a \neq b \in \mathbf{Q}_p$, and let $\delta = |a - b|_p$. Let $A = \{x \in \mathbf{Q}_p : |x - a|_p < \delta\}$. Then A is both closed and open, $a \in A$, but $b \notin A$. Therefore $\mathbf{Q}_p$ is totally disconnected. The same argument applies to $\mathbf{Z}_p$. □

Proposition 23.35. *The set $\mathbf{Z}_p$ is compact.*

One way to prove this is to use the inverse limit definition of $\mathbf{Z}_p$. Each space $X_n = \mathbb{Z}_{p^n}$ is finite; in the discrete topology it is compact. By Tychonoff's Theorem, a cartesian product of compact topological spaces is compact in the product topology. But $\mathbf{Z}_p$ is clearly a closed subset of this space, so it too is compact. A number of details need checking; in particular, that the p-adic metric topology is the same as the product topology.

Instead, we give a proof using the metric space structure directly:

Proof: A metric space is compact if and only if it is sequentially compact; that is, every infinite sequence has a convergent subsequence.

Let $X = \{0, \ldots, p-1\}$. Consider an infinite sequence α_n in $\mathbf{Z}_p$. Let

$$\alpha_n = \sum a_{ni} p^i$$

where $a_{ni} \in X$ for all n, i. By the pigeonhole principle there exists $b_0 \in X$ such that $a_{n0} = b_0$ for infinitely many n. Consider the subsequence consisting of all α_n such that $\alpha_{n0} = b_0$. Denote this by (α_{0n}). Inductively, for each k we define a subsequence (α_{kn}) of $(\alpha_{(k-1)n})$ by finding $b_k \in X$ such that $a_{nk} = b_k$ for infinitely many n, and moreover α_n belongs to $(\alpha_{(k-1)n})$.

Let $\beta = \sum b_i p^i \in \mathbf{Z}_p$. Consider the subsequence (α_{nn}). The first k coefficients a_{nn} are the same as those of β. Therefore $a_{nn} \to \beta$ and we have found a convergent subsequence. □

These topological properties may seem strange, but they tell us that topologically, $\mathbf{Z}_p$ is a familiar space. It is homeomorphic to the Cantor set (or 'middle thirds' set). The reason is that *any* perfect compact totally disconnected space is homeomorphic to the Cantor set; see for instance Hocking and Young [64] section 2-15.

Figure 23.1 is a schematic representation of the 5-adic integers. By the discussion of (23.3), any element of $\mathbf{Z}_5$ can be represented as a 5-adic series $a_0 + a_1 5 + a_2 5^2 + \cdots$, with $0 \leq a_i \leq 4$. The five largest discs in the figure correspond to the first digit a_0. The next smallest discs correspond to the second digit a_1, and so on. So each disc has five subdiscs representing the next digit. The resemblance to the Cantor set is apparent, and we also see that in some sense the space is a *fractal*.

Compare Figure 23.1 with Figure 23.2, a schematic representation of the inverse limit structure of the 5-adic integers. Each congruence class modulo 5^n splits into five congruence classes modulo 5^{n+1}. This process continues infinitely often.

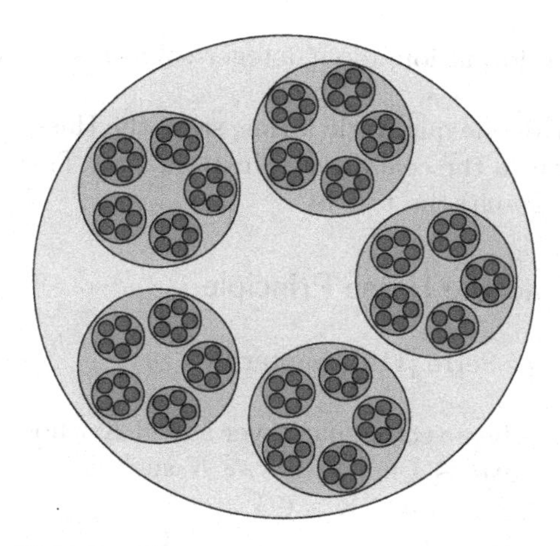

Figure 23.1. Topology of the 5-adic integers (schematic).

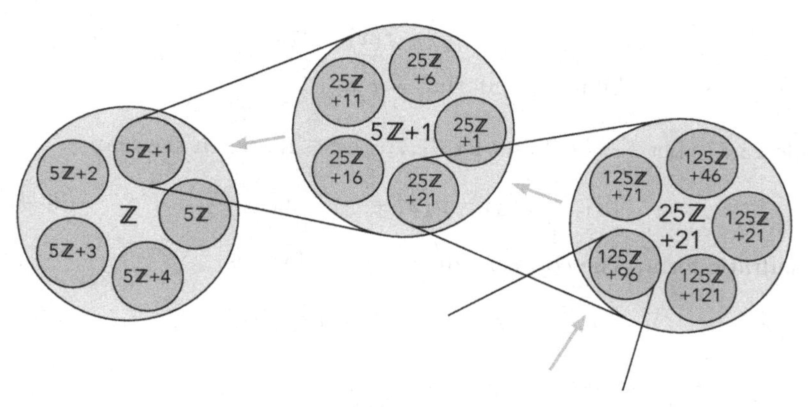

Figure 23.2. Inverse limit structure of the 5-adic integers (schematic).

23.6 Applications of p-adic Numbers

The importance of p-adic numbers in number theory goes back to their introduction by Hensel, and it is stated in the 'local-global principle' of Helmut Hasse: a polynomial equation in one or more variables with rational coefficients has a rational solution if and only if it has a solution in real numbers *and* in p-adic numbers for every prime p. Hasse proved that his principle holds for quadratic equations, but it is known to fail for polynomials in several indeterminates of higher degree. This idea can be seen as a major generalisation of a method we have used several times in earlier chapters: obtaining information about integer solutions of an equation by reduction modulo p.

In this section we define quadratic forms, state the Hasse Principle, and outline its application to the celebrated Three-Squares Theorem of Gauss. We omit (or briefly sketch) the proofs.

Quadratic Forms and the Hasse Principle

For omitted details, see Serre [119] chapter IV.

Definition 23.36. Let V be a vector space over a field K whose characteristic is not 2. A *quadratic form* is a map $f : V \to K$ such that

(a) $f(ax) = a^2 f(x)$ for all $a \in K, x \in V$.

(b) The map $(x, y) \mapsto f(x+y) - f(x) - f(y)$ is bilinear.

Let

$$x.y = \tfrac{1}{2}(f(x+y) - f(x) - f(y))$$

Then $(x, y) \mapsto x.y$ is bilinear, and $f(x) = x.x$.

Example 23.37. Let $V = K^n$ and let $a_1, \ldots, a_n \in K$. Then the map

$$f(x_1, \ldots, x_n) = a_1 x_1^2 + \cdots + a_n x_n^2 \tag{23.4}$$

is a quadratic form. The corresponding bilinear form is

$$x.y = a_1 x_1 y_1 + \cdots + a_n x_n y_n$$

More generally, if A is an $n \times n$ matrix then

$$f(x) = \sum_{i,j} a_{ij} x_i x_j$$

is a quadratic form. Replacing A by $\frac{1}{2}(A + A^{\mathrm{T}})$ we can assume A is a symmetric matrix when the characteristic is not 2.

A quadratic form such as (23.4) is *diagonal*. It can be proved that when the characteristic is not 2, every quadratic form can be transformed into a diagonal one by a linear change of variables. See Serre [119] chapter IV theorem 1.

Definition 23.38. A quadratic form $f : V \to K$ *represents* $a \in K$ if there exists $x \in V$ such that $x \neq 0$ and $f(x) = a$.

For example, the Four-Squares Theorem states that when $V = \mathbb{Q}^4, K = \mathbb{Q}$ the quadratic form $x_1^2 + x_2^2 + x_3^2 + x_4^2$ represents every positive integer. (The condition $x \neq 0$ affects only the case when $a = 0$. It ensures that 'represents zero' requires a nonzero x.)

Lemma 23.39. *A quadratic form f on K^n represents a if and only if the quadratic form $g(x, x_{n+1}) = f(x) - ax_{n+1}^2$ represents 0 with $x_{n+1} \neq 0$.*

Proof: If f represents a then $f(x) = a$, so $g(x, 1) = f(x) - a = 0$, and here $x_{n+1} \neq 0$.

Conversely, if g represents 0 then $f(x) - ax_{n+1}^2 = 0$. If $x_{n+1} \neq 0$ then $f(x/x_{n+1}) = a$. □

Thus we can often reduce the question of which numbers a quadratic form represents to whether a related quadratic form represents 0.

Let f be a quadratic form over $\mathbb{Q}$. Since $\mathbb{Q}$ embeds as a subfield of the p-adic numbers $\mathbf{Q}_p$, we can regard f as a quadratic form over $\mathbf{Q}_p$. Similarly, we can regard f as a quadratic form over $\mathbb{R}$.

Theorem 23.40. (Hasse–Minkowski Theorem) *A quadratic form f on $\mathbb{Q}$ represents 0 if and only if it represents 0 on $\mathbb{R}$ and on $\mathbf{Q}_p$ for all primes p.*

Proof: Trivially, if f represents 0 on $\mathbb{Q}$ then it does so on $\mathbb{R}$, and on $\mathbf{Q}_p$ for all p. The converse is much harder: see Serre [119] chapter IV theorem 8. □

Since $\mathbb{R}$ and $\mathbf{Q}_p$ are complete metric spaces as well as fields, analytic methods can be used to solve polynomial equations over them, or to prove that solutions exist. The Hasse–Minkowski Theorem provides a wealth of new analytic methods for tackling Diophantine equations, motivating the study of analysis over p-adic fields.

The Hasse–Minkowski Theorem does not extend to forms of higher degree. For example, Ernst Selmer [118] proved that the equation $3x^3 + 4y^3 + 5z^3 = 0$ has a solution in real numbers, and in all p-adic fields, but it has no nonzero rational solution.

Three-Squares Theorem

We have proved the Two-Squares and Four-Squares Theorems, Theorem 8.4 and Theorem 8.5, respectively. The number four in the Four-Squares Theorem is best possible, because 7 is not the sum of three squares. The most economical expression of 7 as a sum of squares is

$$7 = 2^2 + 1^2 + 1^2 + 1^2$$

requiring four. More generally, no number of the form $8n - 1$ is a sum of three squares, by considering the expression modulo 8. Gauss kept a small diary to record his most important results. Early in his career we find the cryptic entry:

$$** \text{E Y R H K A. num} = \Delta + \Delta + \Delta$$

Hint: EYRHKA = eureka; Δ means 'triangular number'. Gauss had proved that every (positive) integer is a sum of three triangular numbers. As Cauchy pointed out later on, this result is equivalent to the following theorem, which can be proved using p-adic numbers; see Serre [119] chapter 4, appendix.

Theorem 23.41. (Three-Squares Theorem) *An integer n is a sum of three squares if and only if it is positive and not of the form $4^a(8b-1)$ for $a, b \in \mathbb{N}$.*

Proof: By the Hasse–Minkowski Theorem, n must be represented by f in $\mathbb{R}$ and in all $\mathbf{Q}_p$.

The case of $\mathbb{R}$ leads to $n > 0$.

For $\mathbf{Q}_p$ we must use results about two invariants, one of which depends on 'Hilbert symbols' and is outside our scope. The cases p odd imposes no further condition on n; the case $p = 2$ implies that $-n$ should be a square in $\mathbf{Q}_2$. Other arguments show that this is equivalent to $n \neq 4^a(8b-1)$. □

We can now deduce:

Theorem 23.42. (Gauss's Eureka Theorem) *Every positive integer is a sum of three triangular numbers.*

Proof: Triangular numbers are those of the form $m(m+1)/2$.

Let n be a positive integer. Apply the Three-Squares Theorem to $8n+3$ to obtain $x_1, x_2, x_3 \in \mathbb{N}$ such that $x_1^2 + x_2^2 + x_3^2 = 8n + 3$. Considering this modulo 8, each of the x_i are odd. Write $x_i = 2m_i + 1$. Then

$$\sum_i^3 \frac{m_i(m_i+1)}{2} = \frac{1}{8}\left(\sum (2m_i+1)^2 - 3\right) = \frac{1}{8}(8n+3-3) = n$$

□

Galois Cohomology

Modern algebraic number theory deals with all finite extensions of $\mathbb{Q}$ simultaneously, by considering $\mathbb{A}/\mathbb{Q}$, where $\mathbb{A}$ is the field of all algebraic numbers. This extension has infinite degree. The study of the Galois group of $\mathbb{A}/\mathbb{Q}$ makes heavy use of representations over the p-adic numbers. The machinery relies strongly on p-adic cohomology, which is beyond the scope of this book. The results are central to modern algebraic number theory. For example, they are vital to Wiles's proof of Fermat's Last Theorem, which we summarised in Chapter 17; however, we skipped over the use of cohomology.

23.7 Exercises

23.1 Prove that the series $\sum_{n=0}^{\infty} p^n$ converges in $\mathbf{Q}_p$, and its sum is $\frac{1}{1-p}$.

23.2 Does the sequence $s_n = 1/n!$ converge or diverge in $\mathbf{Q}_p$? Justify your answer. If it converges, what is the limit?

23.3 Prove that $\mathbf{Z}_p$ is both closed and open in $\mathbf{Q}_p$.

23.4 Prove that $\mathbf{Q}_p$ is totally disconnected.

23.5 Prove that every triangle in $\mathbf{Q}_p$ is isosceles. That is, if a, b, c are distinct then at least two of the distances $d_p(a, b)$, $d_p(a, c)$, and $d_p(b, c)$ are equal.

23.6 Prove that three distinct points $\mathbf{Q}_p$ cannot be collinear. That is, if a, b, c are distinct then $d_p(a, c) \neq d_p(a, b) + d_p(b, c)$.

23.7 Can an equilateral triangle exist in $\mathbf{Q}_p$?

23.8 Define a right-angled triangle in $\mathbf{Q}_p$ to be a set of three distinct points whose distances satisfy Pythagoras's Theorem. Do any right-angled triangles exist? If so, state one and prove it is right-angled. If not, prove the impossibility.

23.9 Prove that $\mathbf{Q}_p$ is locally compact.

23.10 How should Figure 23.1 be modified to represent $\mathbf{Q}_p$? (Informal answer only required.)

23.11 Prove that $s = \sum_1^{\infty} n!$ converges in $\mathbf{Q}_p$ for all p. Find the first 10 terms of the 2-adic series for s.

23.12 Prove Proposition 23.33 by observing that open/closed balls are inverse images of open/closed balls in the value group, which is discrete, so any open ball there is closed, and conversely.

23.13 Let p be a rational prime and let r/s be a rational number in lowest terms. Show that if $p \nmid s$ then r/s has a p-adic expansion

$$r/s = \sum_{i=0}^{\infty} a_i p^i$$

that is eventually periodic; that is, there exists $k > 0$ such that $a_{n+k} = a_n$ for all sufficiently large n. Prove that the converse also holds.

If $p|s$ let h be the highest power p^h dividing s. Now show that r/s has a p-adic expansion

$$r/s = \sum_{i=-h}^{\infty} a_i p^i$$

that is eventually periodic, and conversely.

23.14 Prove that $\mathbf{Z}_p$ and $\mathbf{Q}_p$ are uncountably infinite sets, with cardinality equal to that of $\mathbb{R}$.

23.15 Let M be the algebraic closure of $\mathbb{Q}$ in $\mathbf{Q}_p$; that is, the set of all zeros of polynomials $q(t) \in \mathbb{Q}(t)$ in $\mathbf{Q}_p$. Show that $[\mathbf{Q}_p : M]$ is uncountably infinite.

23.16 An analogue of the exponential function can be defined on $\mathbf{Q}_p$ by interpreting the standard series

$$\exp(x) = \sum_{n=0}^{\infty} \frac{x^n}{n!}$$

as a series in $\mathbf{Q}_p$. This series converges for all real numbers, but it does not converge for all p-adic numbers. Prove that the series converges in $\mathbf{Q}_p$ if and only if $x \in p\mathbf{Z}_p$ for p odd, and $x \in 4\mathbf{Z}_2$ for $p = 2$.

23.17 An analogue of the logarithmic function can be defined on $\mathbf{Q}_p$ by interpreting the standard series

$$\log(x) = \sum_{n=1}^{\infty} \frac{(-1)^{n-1}}{n}(x-1)^n$$

as a series in $\mathbf{Q}_p$. Prove that the series converges in $\mathbf{Q}_p$ if and only if $x - 1 \in p\mathbf{Z}_p$.

23.18 When the series in Examples 17 and 18 converge, prove that

$$\begin{aligned}\exp(x+y) &= \exp(x)\exp(y)\\ \log(xy) &= \log(x)+\log(y)\end{aligned}$$

23.19 Let $m \geq 1$ if $p \neq 2$, and $m \geq 2$ if $p = 2$. Prove that exp and log define isomorphisms between the additive group $p^m\mathbf{Z}_p$ and the multiplicative group $1+p^m\mathbf{Z}_p$, and are mutual inverses in this context.

23.20 Show that $\frac{1}{6} \equiv 1 - 5 + 5^2 - \cdots + (-5)^{n-1} \pmod{5^n}$.

Find a similar expression for $\frac{1}{p+1} \pmod{p^n}$ for prime p.

23.21 In ordinary base-10 arithmetic, it is well known that the last k decimal digits of a sum $m+n$ or a product mn depend only on the last k decimal digits of m and n. (For example, $1276 + 4932 = 6208$. Look only at the last three digits: $276 + 932 = 1208$, with last three digits 208. Similarly $1276 \times 4932 = 6293232$, and $276 \times 932 = 257232$ has the same final three digits.)

Prove this statement.

Relate it to the concrete representation of p-adic integers. (Ignore the fact that 10 is not prime, and generalise your proof of the statement.)

Bibliography

[1] H. Anton. *Elementary Linear Algebra* (5th ed.), Wiley, New York 1987.

[2] T.M. Apostol. *Mathematical Analysis*, Addison–Wesley, Reading MA 1957.

[3] A. Baker. Linear forms in the logarithms of algebraic numbers, *Mathematika* **13** (1966) 204–216.

[4] K. Barner. Paul Wolfskehl and the Wolfskehl Prize, *Notices Amer. Math. Soc.* **44** (1997) 1294–1303.

[5] Sister M. Beiter. Magnitude of the coefficients of the cyclotomic polynomial $F_{pqr}(x)$, *Amer. Math. Monthly* **75** (1968) 370–372.

[6] Sister M. Beiter. Magnitude of the coefficients of the cyclotomic polynomial $F_{pqr}(x)$ II, *Duke Math. J.* **38** (1971) 591–594.

[7] Y. Bilu. Catalan's conjecture (after Mihăilescu), *Astérisque* **294** vii (2004) 1–26.

[8] B.J. Birch. Diophantine analysis and modular functions, *Proc. Conf. Algebraic Geometry*, Tata Institute, Bombay 1968, 35–42.

[9] S. Bloch. The proof of the Mordell Conjecture, *Math. Intelligencer* **6.2** (1984) 41–47.

[10] Z.I. Borevič and I.R. Šafarevič. *Number Theory*, Academic Press, New York 1966.

[11] C. Breuil, B. Conrad, F. Diamond, and R. Taylor. On the modularity of elliptic curves over $\mathbb{Q}$: wild 3-adic exercises, *J. Amer. Math. Soc.* **14** (2001) 843–939.

[12] E. Brieskorn and H. Knörrer. *Plane Algebraic Curves*, Birkhäuser, Basel 1986.

[13] J. Brillhart, J. Tonascia, and P. Weinberger. On the Fermat quotient, in *Computers in Number Theory*, Academic Press, New York 1971, 213–222.

[14] J. Buhler, R. Crandall, R. Ernvall, and T. Metsänkylä. Irregular primes and cyclotomic invariants to four million, *Math. Comp.* **60** (1993) 161–153.

[15] R.P. Burn. *Groups: a Path to Geometry*, Cambridge University Press, Cambridge 1985.

[16] A. Cauchy. *Oeuvres 1(X)*, Gauthier–Villars, Paris 1897, 276–285 and 296–308.

[17] A. Cayley. *An Elementary Treatise on Elliptic Functions*, Dover, New York 1961.

[18] H. Chatland and H. Davenport. Euclid's algorithm in real quadratic fields, *Canad. J. Math.* **2** (1950) 289–296.

[19] D.A. Clark. A quadratic field which is Euclidean but not norm-Euclidean, *Manuscripta Mathematica* **83** (1994) 327–330.

[20] H. Cohen. *A Course in Computational Algebraic Number Theory*, Springer, Berlin 1993.

[21] K. Conrad. Pell's Equation, I.
kconrad.math.uconn.edu/blurbs/ugradnumthy/pelleqn1.pdf

[22] K. Conrad. Pell's Equation, II.
kconrad.math.uconn.edu/blurbs/ugradnumthy/pelleqn2.pdf

[23] B. Conrad, F. Diamond, and R. Taylor. Modularity of certain potentially Barsotti–Tate Galois representations, *J. Amer. Math. Soc.* **12** (1999) 521–567.

[24] D. Coppersmith. Fermat's last theorem (case 1) and the Wieferich criterion, *Math. Comp.* **54** (1990) 895–902.

[25] D.A. Cox. Introduction to Fermat's Last Theorem, *Amer. Math. Monthly* **101** (1994) 3–14.

[26] H.S.M. Coxeter. *Regular Polytopes* (2nd ed.), Macmillan, New York 1963.

[27] H.S.M. Coxeter. *Introduction to Geometry* (2nd ed.), Wiley, New York 1969.

[28] J.T. Cross. Primitive Pythagorean triples of Gaussian integers, *Mathematics Magazine* **59** (1986) 106–110.

[29] H. Darmon. A proof of the full Taniyama–Shimura Conjecture is announced, *Notices Amer. Math. Soc.* **46** (1999) 1397–1401.

[30] H. Darmon, F. Diamond, and R. Taylor. Fermat's Last Theorem, in *Current Developments in Mathematics 1995*, International Press, Cambridge MA 1994, 1–154.

[31] R. Dedekind. Über den Zusammenhang zwischen der Theorie der Ideale und der Theorie der höheren Congruenzen, *Abhandlungen der Königlichen Gesellschaft der Wissenschaften in Göttingen* **23** (1878) 3–38. English translation: arxiv.org/abs/2107.08905.

[32] V.A. Demjanenko. L. Euler's Conjecture, *Acta Arithmetica* **25** (1973/4) 127–135.

[33] M. Deuring. Imaginäre quadratischen Zahlkörper mit der Klassenzahl Eins, *Invent. Math.* **5** (1968) 169–179.

[34] F. Diamond. On deformation rings and Hecke rings, *Ann. Math.* **44** (1996) 137–166.

[35] S. Dolan. A very simple proof of the two-squares theorem, *The Mathematical Gazette* **105** (2021) 511.

[36] D.S. Dummit and R.M. Foote. *Abstract Algebra*, Wiley, New York 2004.

[37] H.M. Edwards. The background of Kummer's proof of Fermat's Last Theorem for regular primes, *Arch. Hist. Exact Sci.* **14** (1975) 219–236.

[38] H.M. Edwards. *Fermat's Last Theorem*, Springer, New York 1977.

[39] N. Elkies, On $A^4 + B^4 + C^4 = D^4$, *Math. Comp.* **51** (1988) 825–835.

[40] M.H. Fenrick. *Introduction to the Galois Correspondence*, Birkhäuser, Boston 1992.

[41] J.B. Fraleigh. *A First Course in Abstract Algebra* (7th ed.) Pearson, Harlow 2013.

[42] N. Freitas, B.V. Le Hung, and S. Siksek. Elliptic curves over real quadratic fields are modular, *Inventiones Mathematicae* **201** (2015) 159–206.

[43] E. Frenkel. Recent advances in the Langlands Program, *Bull. Amer. Math. Soc.* **41** 151–184.

[44] G. Frey. Rationale Punkte auf Fermatkurven und gewisteten Modularkurven, *J. Reine Angew. Math.* **331** (1982) 185–191.

[45] G. Frey. Links between stable elliptic curves and certain Diophantine equations, *Ann. Univ. Sarav.* **1** (1986) 1–40.

[46] A. Fröhlich and M.J. Taylor. *Algebraic Number Theory*, Cambridge Studies in Advanced Mathematics **27**, Cambridge University Press, Cambridge 1991.

[47] S. Friedberg. What is the Langlands Program? *Notices Amer. Math. Soc.* **65** (2018) 663–665.

[48] D.J.H. Garling. *A Course in Galois Theory*, Cambridge University Press, Cambridge 1986.

[49] C.F. Gauss. *Disquisitiones Arithmeticae* (translated by A.A. Clarke), Yale University Press, New Haven CT 1966.

[50] H.J. Godwin and P.A. Samet. A table of real cubic fields, *J. London Math. Soc.* **34** (1959) 108–110.

[51] D.M. Goldfeld. Gauss's class number problem for imaginary quadratic fields, *Bull. Amer. Math. Soc* **13** (1985) 23–37.

[52] J.R. Goldman. *The Queen of Mathematics*, A.K. Peters, Wellesley MA 1998.

[53] E.S. Golod and I.R. Šafarevič. On class field towers, *Izv. Akad. Nauk SSSR ser. Mat.* **28** (1964) 261–272; *Amer. Math. Soc. Trans., 2nd ser.* **48** (1965) 91–102.

[54] J. Gray. *A History of Abstract Algebra: from Algebraic Equations to Modern Algebra*, Springer Nature Switzerland 2018.

[55] A. Granville and M. Monagan. The first case of Fermat's last theorem is true for all prime exponents up to 714,591,416,091,839, *Trans. Amer. Math. Soc.* **306** (1988) 329–359.

[56] H. Hancock. *Theory of Elliptic Functions*, Dover, New York 1958.

[57] G.H. Hardy. *A Course of Pure Mathematics*, Cambridge University Press, Cambridge 1960.

[58] G.H. Hardy and E.M. Wright. *An Introduction to the Theory of Numbers*, 4th edition, Cambridge University Press, Cambridge 1960.

[59] M. Harper. $\mathbf{Z}(\sqrt{14})$ is Euclidean, *Canad. J. Math.* **56** (2004) 55–70.

[60] K. Heegner. Diophantische Analysis und Modulfunktionen, *Math. Zeit.* **56** (1952) 227–253.

[61] H. Heilbronn and E.H. Linfoot. On the imaginary quadratic corpora of class-number one, *Quart. J. Math. (Oxford)* **5** (1934) 293–301.

[62] Y. Hellegouarch. Points d'ordre $2p^h$ sur les courbes elliptiques, *Acta Arith.* **26** (1975) 253–263.

[63] C. Hermite. Extrait d'une lettre de M.C. Hermite à M. Borchardt sur le nombre limité d'irrationalités aux laquelles se réduisent les racines des équations à coefficients entiers complexes d'un degré et d'un discriminant donnés, *J. Reine Angew. Math.* **53** (1857) 182–192.

[64] J.G. Hocking and G.S. Young. *Topology*, Addison-Wesley, Reading MA 1961.

[65] J.F. Humphreys. *A Course in Group Theory*, Oxford University Press, Oxford 1996.

[66] T.W. Hungerford. *Algebra*, Graduate Texts in Mathematics **73**, Springer, New York 1974.

[67] K. Inkeri. Über den Euklidischen Algorithmus in quadratischen Zahlkörpern, *Ann. Acad. Scient. Fennicae* **41** (1947) pp.35.

[68] N. Jacobson. *Basic Algebra* vol. 1 (2nd ed.), Dover, Mineola 2009.

[69] N. Jacobson. *Basic Algebra* vol. 2 (2nd ed.), Dover, Mineola 2009.

[70] B. Juran, P. Moree, A. Riekert, D. Schmitz, and J. Völlmecke. A proof of the corrected Sister Beiter cyclotomic coefficient conjecture inspired by Zhao and Xhang, arXiv:2304.09250v1 (2023).

[71] A. Ya. Khinchin. *Continued Fractions*, Dover, Mineola 1997.

[72] R.B. King. *Beyond the Quartic Equation*, Birkhäuser, Boston 1996.

[73] L.J. Lander and T.R. Parkin. Counterexamples to Euler's Conjecture on sums of like powers, *Bull. Amer. Math. Soc.* **72** (1966) 1079.

[74] S. Lang. *Algebraic Numbers*, Addison–Wesley, Reading MA 1964.

[75] S. Lang. *Algebraic Number Theory*, Springer, New York 2000.

[76] R. Langlands. Letter to André Weil, 1967; Institute for Advanced Study, Princeton, publications.ias.edu/rpl/paper/43.

[77] H. Lebesgue. L'oeuvre mathématique de Vandermonde, *Thalès* **4** (1937–1939) 28–42.

[78] G.T. Lee. *Abstract Algebra: an Introductory Course*, Springer Nature Switzerland 2018.

[79] D.H. Lehmer. A Note on trigonometric algebraic numbers, *Amer. Math. Monthly* **40** (1933) 165–166.

[80] H.W. Lenstra Jr. Solving the Pell equation, *Notices Amer. Math. Soc.* **49** (2002) 182–192

[81] S.L. Loney. *The Elements of Coordinate Geometry*, Macmillan, London 1960.

[82] R.L. Long. *Algebraic Number Theory*, Dekker, New York 1977.

[83] I.D. Macdonald. *The Theory of Groups*, Oxford University Press, Oxford 1968.

[84] R. Magnus. *Metric Spaces: a Companion to Analysis*, Springer, New York 2022.

[85] H. Maier. Anatomy of integers and cyclotomic polynomials, in: J.-M. De Koninck, A. Granville, F. Luca (eds.), *Anatomy of Integers*, CRM Proceedings and Lecture Notes **46** Amer. Math. Soc., Providence RI 2008, 89–95.

[86] S. Marques and K. Ward. Cubic fields, a primer, *European J. Math.* **5** (2019) 551–570.

[87] J.M. Masley and H.L. Montgomery. Cyclotomic fields with unique factorization, *J. Reine Angew. Math.* **286/287** (1976) 248–256.

[88] D. Masser. Open problems, *Proc. Symp. Analytic Number Thy.* (ed. W.W.L. Chen), Imperial College, London 1985.

[89] R.D. Mauldin. A generalization of Fermat's Last Theorem: the Beal Conjecture and prize problem, *Notices Amer. Math. Soc.* **44** (1997) 1436–1437.

[90] B. Mazur. Modular curves and the Eisenstein ideal, *Publ. Math. IHES* **47** (1977) 33–186.

[91] B. Mazur. Rational isogenies of prime degree, *Invent. Math.* **44** (1978) 129–162.

[92] B. Mazur. Number theory as gadfly, *Amer. Math. Monthly* **98** (1991) 593–610.

[93] B. Mazur. Questions about powers of numbers, *Notices Amer. Math. Soc.* **47** (2000) 195–202.

[94] H. McKean and V. Moll. *Elliptic Curves.* Cambridge University Press, Cambridge 1999.

[95] P. Mihăilescu. Primary Cyclotomic Units and a Proof of Catalan's Conjecture, *J. Reine Angew. Math.* **572** (2004) 167–195.

[96] L.J. Mordell. *Diophantine Equations*, Academic Press, New York 1969.

[97] W. Narkiewicz. Class number and factorization in quadratic number fields, *Colloq. Math.* **17** (1967) 167–190.

[98] P.M. Neumann. *The Mathematical Writings of Évariste Galois*, European Mathematical Society, Zürich 2011.

[99] P.M. Neumann, G.A. Stoy, and E.C. Thompson. *Groups and Geometry*, Oxford University Press, Oxford 1994.

[100] A. Nitaj. On a conjecture of Erdős on 3-powerful numbers, *Bull. London Math. Soc.* **27** (1995) 317–318.

[101] J. Oesterlé. Nouvelles approches du 'theorème' de Fermat, *Astérisque* **161/162** (1988) 165–186.

[102] C.D. Olds. *Continued Fractions*, New Mathematical Library, Mathematical Association of America, Washington 1963.

[103] S.S. Pillai. On the equation $2^x - 3^y = 2^X - 3^Y$, *Bull. Calcutta Math. Soc.* **37** (1945) 15–20.

[104] C. Reid. *Hilbert*, Springer, Berlin 1970.

[105] P. Ribenboim. *13 Lectures on Fermat's Last Theorem*, Springer, New York 1979.

[106] P. Ribenboim. *Fermat's Last Theorem for Amateurs*, Springer, New York 1991.

[107] K. Ribet. On modular representations of $\mathrm{Gal}(\overline{\mathbb{Q}}/\mathbb{Q})$ arising from modular forms, *Invent. Math.* **100** (1990) 431–476.

[108] K. Ribet. Galois representations and modular forms, *Bull. Amer. Math. Soc.* **32** (1995) 375–402.

[109] B. Riemann. Grundlagen für eine allgemeine Theorie der Funktionen einer veränderlichen complexen Grosse, *Bernhard Riemann's Gesammellte Mathematische Werke* (eds. R. Dedekind and H.M. Weber) 3–45.

archive.org/details/bernardrgesamm00riemrich/page/n7/mode/2up.

[110] D.J.S. Robinson. *A Course in the Theory of Groups*, Springer, New York 1996.

[111] J. Roe. *Elementary Geometry*, Oxford University Press, Oxford 1993.

[112] J.J. Rotman. *An Introduction to the Theory of Groups* (4th ed.), Springer, New York 1995.

[113] K. Rubin and A. Silverberg. A report on Wiles' Cambridge lectures, *Bull. Amer. Math. Soc.* **31** (1994) 15–38.

[114] P. Samuel. About Euclidean rings, *J. Algebra* **19** (1971) 282–301.

[115] P. Samuel. *Algebraic Theory of Numbers*, translated from the French by A.J. Silberger, Hermann, Paris 1972.

[116] R. Schoof. *Catalan's Conjecture*, Springer, New York 2008.

[117] J.L. Selfridge and B.W. Pollock. Fermat's Last Theorem is true for any exponent up to 25,000, *Notices Amer. Math. Soc.* **11** (1964) 97, abstract no. 608–138.

[118] E.S. Selmer. The Diophantine equation $ax^3 + by^3 + cz^3 = 0$, *Acta Mathematica* **85** (1951) 203–362.

[119] J.-P. Serre. A Course in Arithmetic, Graduate Texts in Methematics **7**, Springer, New York 1973.

[120] D. Shanks. The simplest cubic number fields, *Math. Comp.* **28** (1974) 1137–1152.

[121] D. Sharpe. *Rings and Factorization*, Cambridge University Press, Cambridge 1987.

[122] C.L. Siegel. Zum Beweise des Starkschen Satzes, *Invent. Math.* **5** (1968) 169–179.

[123] S. Singh. *Fermat's Last Theorem*, Fourth Estate, London 2002.

[124] G. Smith and O. Tabachnikova. *Topics in Group Theory*, Springer, London 2000.

[125] L. Soicher. Computing Galois groups over the rationals, *J. Number Theory* **20** (1985) 273–281.

[126] H.M. Stark. A complete determination of the complex quadratic fields of class-number one, *Michigan Math. J.* **14** (1967) 1–27.

[127] I. Stewart. *The Problems of Mathematics*, Oxford University Press, Oxford 1987.

[128] I. Stewart. *Galois Theory* (5th ed.), Chapman & Hall/CRC, Boca Raton FL 2023.

[129] I. Stewart and D.O. Tall. *The Foundations of Mathematics* (2nd ed.), Oxford University Press, Oxford 2015.

[130] I. Stewart and D.O. Tall. *Complex Analysis* (2nd ed.), Cambridge University Press, Cambridge 2018.

[131] L. Stickelberger. Über eine neue Eigenschaft der Diskriminanten algebraischer Zahlkörper, in: *Verhandlungen des ersten internationalen Mathematiker-Kongresses, Zürich 1897* (ed. F. Rudio), Teubner, Leipzig 1898, 182–193.

[132] D.J. Struik. *A Concise History of Mathematics*, Bell, London 1962.

[133] W.A. Sutherland, *Introduction to Metric and Topological Spaces* (2nd ed.), Oxford University Press, Oxford 2010.

[134] R.L. Taylor and A. Wiles. Ring theoretic properties of certain Hecke algebras, *Ann. of Math.* **141** (1995) 553–572.

[135] H. te Riele and H. Williams. New computations concerning the Cohen-Lenstra heuristics, *Experimental Math.* **12** (2003) 99–113.

[136] R. Tijdeman. On the equation of Catalan, *Acta Arith.* **29** (1976) 197–209.

[137] E.C. Titchmarsh. *The Theory of Functions*, Oxford University Press, Oxford 1960.

[138] J. Unger. Solving Pell's equation with continued fractions, ir.canterbury.ac.nz/bitstream/handle/10092/10158/unger_2009_report.pdf.

[139] A.-T. Vandermonde. Mémoire sur las résolution des équations, *Histoire de l'Acad. Royale des Sciences* (1771) 365–416.

[140] S. Wagstaff. The irregular primes to 125000, *Math. Comp.* **32** (1978) 583–591.

[141] L.C. Washington. *Introduction to Cyclotomic Fields* (2nd ed.), Springer, New York 1982.

[142] A. Weil. Sur un théorème de Mordell, *Bull. Sci. Math.* **54** (1930) 182–191.

[143] A. Weil. *Number Theory: an Approach Through History from Hammurapi to Legendre*, Birkhäuser, Boston 1984.

[144] R.S. Westfall. *Never at Rest*, Cambridge University Press, Cambridge 1980.

[145] A. Wiles. Modular elliptic curves and Fermat's Last Theorem, *Ann. of Math.* **141** (1995) 443–551.

[146] S.H. Yang. Continued fractions and Pell's equation, math.uchicago.edu/ may/VIGRE/VIGRE2008/REUPapers/Yang.pdf.

Further Reading on Algebraic Number Theory

1. S. Alaca and K.S. Williams. *Introductory Algebraic Number Theory*, Cambridge University Press, Cambridge 2003.

2. E. Artin. *Algebraic Numbers and Algebraic Functions*, Gordon & Breach, New York 1967.

3. Z.I. Borevič and I.R. Šafarevič. *Number Theory*, Academic Press, New York 1966.

4. J.W.S. Cassels and A. Fröhlich. *Algebraic Number Theory*, Academic Press, New York 1967.

5. Ph. Cassou-Noguès and M.J. Taylor. *Elliptic Functions and Rings of Integers*, Birkhäuser, Basel 1987.

6. H. Cohen. *A Course in Computational Algebraic Number Theory*, Springer, Graduate texts in Mathematics **138**, Berlin 1993.

7. H. Cohn. *A Classical Invitation to Algebraic Numbers and Class Fields*, Springer, New York 1978.

8. P.M. Cohn. *Algebraic Numbers and Algebraic Functions*, Chapman & Hall, London 1991.

9. A. Fröhlich and M.J. Taylor. *Algebraic Number Theory*, Cambridge Studies in Advanced Mathematics **27**, Cambridge University Press, Cambridge 1991.

10. G.H. Hardy and E.M. Wright. *An Introduction to the Theory of Numbers*, Oxford University Press, Oxford 1954.

11. E. Hecke. *Lectures on the Theory of Algebraic Numbers*, Springer, New York 1981.

12. F. Jarvis. *Algebraic Number Theory*, Springer, New York 2014.

13. K. Kato, N. Kurokawa, and T. Saito. *Number Theory 1: Fermat's Dream*, Translations of Mathematical Monographs **186**, Amer. Math. Soc., Providence RI 1991.

14. S. Lang. *Algebraic Numbers*, Addison-Wesley, Reading MA 1964.

15. S. Lang. *Algebraic Number Theory*, Springer, New York 2000.

16. R.L. Long. *Algebraic Number Theory*, Dekker, New York 1977.

17. O.T. O'Meara. *Introduction to Quadratic Forms*, Springer, New York 1973.

18. J.S. Milne. *Algebraic Number Theory*, jmilne.org/math/CourseNotes/ANT.pdf.

19. H. Pollard. *The Theory of Algebraic Numbers*, Math. Assoc. America, Buffalo NY 1950.

20. P. Ribenboim. *Algebraic Numbers*, Wiley–Interscience, New York 1972.

21. P. Samuel. *Algebraic Theory of Numbers*, translated from the French by A.J. Silberger, Hermann, Paris 1972. Reprinted Dover, New York 2008.

22. W. Stein. *A Brief Introduction to Classical and Adelic Algebraic Number Theory*, williamstein.org/papers/ant/ 2004.

23. J. Stillwell. *Algebraic Number Theory for Beginners*, Cambridge University Press, Cambridge 2022.

24. H.P.F. Swinnerton-Dyer. *A Brief Guide to Algebraic Number Theory*, Cambridge University Press, Cambridge, 2001.

25. M. Trifković. *Algebraic Theory of Quadratic Numbers*, Springer, New York 2013.

26. L.C. Washington. *Introduction to Cyclotomic Fields* (2nd ed.), Springer, New York 1982.

27. A. Weil. *Basic Number Theory*, Springer, New York 1967.

28. E. Weiss. *Algebraic Number Theory*, Dover, New York 1999.

29. H. Weyl. *Algebraic Theory of Numbers*, Princeton University Press, Princeton NJ 1940.

Internet Resources

There are innumerable useful sources on the web. Since URLs change frequently, new material is often added, and old material can suddenly disappear, we do not list any. Use your favourite search engine.

Index of Notation

Index

Printed in the United States
by Baker & Taylor Publisher Services